TOPICAL ISSUES OF RATIONAL USE OF NATURAL RESOURCES 2019

T0321219

PROCEEDINGS OF THE XVTH FORUM-CONTEST OF STUDENTS AND YOUNG RESEARCHERS UNDER THE AUSPICES OF UNESCO, SAINT-PETERSBURG, RUSSIA, 13 – 17 MAY 2019

United Nations · International Competence Centre
Educational, Scientific and · for Mining-Engineering Education
Cultural Organization · under the auspices of UNESCO

Topical Issues of Rational Use of Natural Resources 2019

Volume 2

Editor

Vladimir Litvinenko

Saint-Petersburg Mining University, Saint-Petersburg, Russia

CRC Press
Taylor & Francis Group
Boca Raton London New York

CRC Press is an imprint of the
Taylor & Francis Group, an **informa** business

A BALKEMA BOOK

CRC Press/Balkema is an imprint of the Taylor & Francis Group, an informa business

© 2020 Taylor & Francis Group, London, UK

Typeset by Integra Software Services Pvt. Ltd., Pondicherry, India

Library of Congress Cataloging-in-Publication Data

Applied for

Published by: CRC Press/Balkema
        Schipholweg 107C, 2316XC Leiden, The Netherlands

First issued in paperback 2023

ISBN: 978-0-367-85713-4 (Set hbk)
ISBN: 978-1-003-01452-2 (ebk)

DOI: https://doi.org/10.1201/9781003014522

Volume 1:
ISBN: 978-0-367-85719-6 (hbk)
ISBN: 978-1-003-01457-7 (ebk)

DOI: https://doi.org/10.1201/9781003014577

Volume 2:
ISBN: 978-1-03-257098-3 (pbk)
ISBN: 978-0-367-85720-2 (hbk)
ISBN: 978-1-003-01463-8 (ebk)

DOI: https://doi.org/10.1201/9781003014638

Publisher's Note
The publisher has gone to great lengths to ensure the quality of this reprint but points out that some imperfections in the original copies may be apparent.

Table of Contents

The latest management and financing solutions for the development of mineral resources sector

Environment protection and sustainable nature management

New approaches to resolving hydrocarbon sector-specific issues

Topical Issues of Rational Use of Natural Resources 2019 – Litvinenko (Ed)
© *2020 Taylor & Francis Group, London, ISBN 978-0-367-85720-2*

Preface

Dear friends!

The mineral and raw materials complex has strategic and key value for the economy of any state. It is correct both for countries where mining constitutes the basis of gross domestic product and for the countries importing raw materials. In either case, the stability and predictability of these countries development directly depends on raw materials.

At present, the global public opinion strives to reinforce adherence to environmental standards, to reduce the harmful impact on the environment, sometimes completely rejecting raw materials extraction in favor of switching to alternative energy sources, utterly denying exploration of the Arctic and oceans.

These trends affect the prestige and authority of our professions, causing a serious damage to the image of the entire industry, which inevitably leads to shortage of young qualified personnel employed in the raw materials sector.

However, it must be recognized that main reasons for most of the major recent environmental disasters are the so-called human factor, design mistakes, errors in technological solutions, and low qualification of the management and technical staff.

That's why it is so important now to stand up to the challenge of assessing professional competencies of those specialists in the mineral resource complex who are responsible for the fundamental decisions that sometimes are crucial for the environment and lives of workers involved in the exploitation of mineral deposits and the construction of industrial facilities.

In our work, we absolutely must use the latest achievements of the technological progress, such as new sources of raw materials and energy, digitalization opportunities, revolutionary approaches to all stages of the technologies of raw materials treatment processes.

These kinds of problems are solved by our young scientists – participants of the international Forum-Contest "Topical issues of rational use of natural resources 2019". The main idea of this event was to provide future specialists with the opportunity to talk about their research, discuss their results with colleagues, get an appraisal of their studies from the competent jury; such events give the industry a chance to replenish its human resources with the first-rate specialists.

This major scientific assembly is designed to develop a training system for qualified specialists in mineral resources and fuel and energy complexes, and also to promote international scientific interaction among young scientists from various technical universities on the matters of natural resources use, and to assess the quality of training of young specialists at the world's leading mining universities and identify the most talented ones.

The format of the forum-contest is quite unique. On the one hand, it allows the young researchers to listen to lectures of distinguished scientists and, on the other hand, to show their own knowledge and achievements.

Each year, the number of participants and participating countries and universities is going up. If in the 2017 Forum (the largest forum in the preceding university's history), 226 reports were presented, including 127 foreign and 99 national ones, in 2019, 410 participants from 85 universities and companies got registered. This is an absolute record.

This year, the format of the forum has fundamentally evolved. The forum agenda included:

- 7 thematic sessions;
- Case championship;
- Participants' presentations in TED format (a public speaking contest)

- A separate section for high school students
- Master classes by leading experts.

At the thematic sessions, the winners were determined in two categories: "Best Student" and "Best Graduate Student and Young Scientist". A total of 88 prizes were awarded.

Within the framework of this Competition Forum, for the first time ever, a scientific meeting of young scientists on the "Applied Value of the Mendeleev Periodic Table" was held.

I would like to note that in the walls of Mining University we held this competition for the 15th time. We have come a long way; now the Forum is taking place in the new Multifunctional Complex, which has a congress hall with the capacity for more than 2000 people; all its halls and auditoriums are equipped with the state-of-the-art media and technical apparatus.

We updated our approaches to the organization of the competition, tightened the requirements for competitive works. This allowed us not only to publish collections of articles indexed in the SCOPUS international database, but also to hold a forum under the auspices of UNESCO.

The Competence Center of this world-famous organization with an impeccable reputation was opened at the University in 2018. It was designed to bring together the world's leading universities specializing in mining education to work together on, first of all, unifying educational and professional standards, developing an integrated system of international assessment of professional competencies with assigning qualifications to engineers and managers employed at enterprises of the mineral and energy sectors, and forming a unified system of professional certification for the qualification of a "Professional Engineer".

I am sure that our two-volume article digest made by one of the most prestigious publishers and containing 130 articles will be useful to a wide circle of readers and specialists interested in the sustainable development of the industry, be they researchers, industrialists or fellow scientists from various universities around the world.

In our articles we tried to highlight the most promising areas of the research and scientific and technological development of the mineral resources complex.

We plan to continue such work in the future.

I want to wish everyone good luck, daring in scientific work and confidence in its necessity.

Vladimir Litvinenko
Doctor of Engineering, Professor;
Rector of Saint Petersburg Mining University

Topical Issues of Rational Use of Natural Resources 2019 – Litvinenko (Ed)
© 2020 Taylor & Francis Group, London, ISBN 978-0-367-85720-2

Organizers

International Competence Centre for Mining Engineering Education under the auspices of UNESCO

United Nations Educational, Scientific and Cultural Organization

International Competence Centre for Mining-Engineering Education under the auspices of UNESCO

Ministry of Science and Higher Education of the Russian Federation

Saint Petersburg Mining University

Breakthrough technologies of integrated processing of mineral hydrocarbon and technogenic raw materials with further production of new generation materials

Topical Issues of Rational Use of Natural Resources 2019 – Litvinenko (Ed)
© 2020 Taylor & Francis Group, London, ISBN 978-0-367-85720-2

Investigation of the influence of foaming reagents of different group chemical composition on the efficiency of coal flotation

V.N. Petukhov & Ya.S. Batyaev
Nosov Magnitogorsk State Technical University, Magnitogorsk, Russia

ABSTRACT: Effective use of coal of various technological brands is possible only after their enrichment, since increased ash and sulfur content reduce the usefulness of the coal, and if they are used in coking with ash more than 11%, the metallurgical coke obtained from these coals will not meet the requirements of blast-furnace production. Indicators of coal fine coal enrichment products are largely determined by the reagent mode used. In this work, studies were conducted to study the effect of the group chemical composition of the blowing agent reagents on the flotation rates. As the initial flotation feed, coal fines were investigated at the central processing plant Belovskaya with an ash content of 18.2%. It has been established that to obtain high indicators of coal flotation products, it is necessary to apply the technical product of petrochemistry Oxal T-80 as a blowing agent reagent. Its application allows to increase the extraction of combustible mass in concentrate by 1.9-2.5% and reduce the loss of organic matter of coal with flotation waste.

1 INTRODUCTION

The flotation indicators are largely determined by the flotation activity and the selectivity of the reagents used. Taking into account the ever increasing demands on the quality of coal concentrates supplied to coking, it is necessary to conduct research on new effective reagent modes. A significant amount of research has been done to assess the flotation properties of surfactants, when used as blowing agent reagents (Petukhov, Kubak, 2016; Deberdeev, Picat-Ordynsky, Rudanovskaya, 1986; Chizhevsky, Vlasova, Savinchuk, 1981). An important direction in the search for reagents for coal flotation is the study of the flotation activity of pure chemical compounds, which is of theoretical and practical interest. Abramov A.A., McFadzean B., Sahbaz O., Solo-Zhenkin P.M., Ryaboy V.I., Karimian A., Pan Lei., Et al. Were searching for the dependence of flotation activity and selectivity of reagents on the elemental composition and chemical structure of the compound (Abramov, 2016; McFadzean, Mhlanga, O,Connor, 2013; Sahbaz, Cinar, Kelebek, 2016; Ejtemaei, Nguyen, 2017; Solozhenkin, Solozhenkin, Sanda, 2011; Ryaboy, Shepeta, Kretov, Golirov, 2014; Karimian, Rezaei, Mazoumi, 2013; Pan Lei, Yoon Rol-Hoan, 2014). In a number of studies, it was found that compounds that are simultaneously highly hydrophobic and capable of forming foams that are optimal in terms of stability and degree of dispersion of the air phase are more suitable for flotation (Petukhov, Kubak, Semenov, 2014; Limarev, 2011; Popova, Petukhov, 1978; Popova, 1980). We have conducted studies of the flotation activity of technical products of petrochemistry, differing in group chemical composition, used as blowing agent reagents. In the study, Ekofol-440 reagent was used, which is used as a blowing agent reagent under the conditions of the central processing plant Belovskaya, MMK-COAL, as well as the technical product of petrochemistry Oxal T-80, which is used as a blowing agent reagent flotation of polymetallic ores. The technological product of petrochemistry was selected for research as a blowing agent reagent due to its group chemical composition. In a number of works (Popova, Petukhov, 1978; Popova, 1980) high foaming and water-repellent properties of alkyl substituted 1,3 - dioxacyclanes were established. Due to the fact that the reagent of the Oxal

T-80 foaming agent is contained in the predominant amount of a mixture of dioxane and pyran alcohols which have a high surface activity, providing an increased foaming ability of Oxal T-80, and the presence in polar molecules 1,3 dioxacyclanes of their oxygen atoms have a positive effect on their adsorption on the positive centers of the coal surface. The Ecofol-440 reagent contains in the group chemical composition a mixture of polar chemical compounds: -2-ethyl-1-hexanol, alcohols of normal structure, isostrate olefins, esters and other chemical compounds (Table 1). The boiling point of Ecofol-440 is 190-420 °C, the density is 0.85-0.95 g / cm3, the freezing temperature is minus 30 °C, the flash point in a closed crucible is more than 45 °C, the kinematic viscosity at 20 °C is 3-7 mm2 / s . The physicochemical parameters of Ecofol-440 indicate the possibility of its use throughout the year.

The technical petrochemical product Oxal T-80 in the chemical group composition mostly contains oxygen-containing compounds of a cyclic structure (Table 2).

The technical product Flotek-2342, manufactured according to TU 2452-006-62494573-2012, LLC Chemical Company Vektor, was used as a collector reagent. G.Kemerovo, Teresh-kova st. house 41. The reagent is a mixture of hydrocarbons and minor amounts of polar chemical compounds.

As the source of power, coal fines of the "Zh" technological grade were used, which are supplied to the flotation in the conditions of the central processing plant "Belovskaya" and "MMK-COAL". The sieve characteristic of the source power of the central processing plant "Belovskaya" is given in Table 3. According to the results of the study, it can be concluded

Table 1. Chemical group composition and physicochemical properties of Ecofol-440.

Chemical group composition	
Blend composition	Content, % of the mass
2-ethyl-1-hexanol	18,3-18,6
Normal alcohols C_{10}-C_{16}	5,6-6,0
Alcohols of isomeric structure C_{11}-C_{20}	14,0-14,5
2- (decyloxy) -ethanol to 2- (decyloxy) -ethanol	28,0-28,3
Esters of fatty acids (C10-C_{12}) -	2,2-2,5
Olefins of isomeric structure (C_{11}-C_{16})	28,0-33,0
Unidentified compounds	0,3-0,7
Fractional composition	
Boiling point, °C	110-130
Flash point, °C	77
Ignition temperature, °C	290
Pour Point, °C	— 60
Physical and chemical indicators	
Density at 20 °C, g/cm^3	0,87
Dynamic viscosity at 20 °C, mPa*s	5
pH value at 20 °C (at 5.3 g/l H_2O)	5

Table 2. Chemical group composition of the blowing agent Oxal T-80.

Components of the blowing agent Oxal T-80	Mass % of compound
Light chemical compounds	0,05-0,8
4,4- dimethyldioxane	0,2-0,3
The amount of pre-pyran alcohol compounds	3,5-9,5
Pyran alcohol	2,0-4,0
Methylbutanediol	0,03-1,5
The sum of "x" dioxane alcohols	35,0-65,0
Distillation residue	15,7- 59,1

Table 3. Characteristics of the particle size distribution of the coal under investigation, grade ZH.

Size grade, mm	Yield (γ), %	Ash content	Total yield,%	Total ash content, %
+0,5	29,1	20,4	29,1	20,4
0,5-0,25	24,7	17,8	53,8	19,2
0,25-0,16	17.2	16,4	71,0	18,5
-0,16	29,4	16,8	100,0	18,2
Total	100	18,2		

that the mineral particles are evenly divided into coal size classes. The greatest amount of mineral impurities contains the class (+0.5) mm, the ash content of which is 20.4%. The lowest ash class is the class of grain size (0.25-0.16) mm, the ash content of which is 16.4%. Judging by the granulometric composition, in the initial flotation feed in the smallest amount there is a class of 0.25-0.16 mm (Table 3).

2 METHODS

The study of the flotation activity of selected technical products used by us as flotation agents was carried out in a mechanical type flotation machine "Mechanobr" with a chamber volume of 500 cm. According to flow chart 1. The rotation speed of the impeller is 1500 rpm and the slurry temperature is 20 ° C permanent. Air consumption was also kept constant and controlled by a float-flow rotameter.

In the experiments, we used a sample of coal weighing 50 g with a particle size of — 0.5 mm. The charge was placed in the chamber of the flotation machine and filled with the necessary amount of water. Reagent-collector filed drops of a known mass fractionally, three times. If the number of drops required for the experiment was more than three, then most of them were served first, reducing their number with each successive time. The foamy product forming the concentrate was also removed fractionally. The reagent blowing agent was supplied in the form of an aqueous emulsion by measuring pipette. Emulsion with a concentration of 1 mg / ml, was prepared in an electric emulsifier. According to the chosen flotation pattern (Figure 1), the time of contacting the coal with water was two minutes. According to their incident, the first part of the collector reagent was fed into the pulp, the contact time of which with coal was equal to seconds. After that, a blowing agent was introduced into the process, which was in contact with coal for twenty seconds. The reagent blowing agent was administered once in full.

After three minutes after the introduction of all reagents, air was supplied to the chamber of the flotation machine and the process of coal flotation began, which lasted one minute. Accordingly, the time of removal of the concentrate was also a minute. After that, the air supply was stopped, the concentrate was removed, and the second part of the collector reagent was supplied. Further, the time and sequence of the operations performed were the same as when the first supply of the collector reagent.

The collected concentrate was dried on an hot plate with an isolated heating element to constant weight. The dried concentrate was weighed on a technical scale with an accuracy of 0.01 g, carefully ground, mixed and tested for ash. Ash content was determined by taking a sample of each concentrate and calcining it in a muffle furnace (GOST 11022-95). Investigations were conducted in parallels and if there are discrepancies in the ash content of products by more than 0.2%, the sample was re-selected. In order for the results to be adequate, we conducted a series of three identical experiments. The effect of various reagents on the flotation process of coal particles was compared according to the results of a series of experiments carried out on the same day.

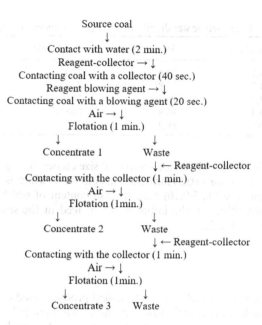

Figure 1. Diagram of coal flotation using a blowing agent.

In flotation studies, the conduct of a large number of experiments is justified by the goal: a better study of the flotation properties of reagents for these coals. These experiments reveal the nature and degree of activity of the interaction of reagents with a coal surface, and also establish the nature of changes in the properties of this surface arising as a result of this interaction.

The following were determined as qualitative indicators of flotation: -concentration and ash content; waste disposal and ash content; extraction of combustible mass in concentrate; extraction of the mineral part to waste; the efficiency ratio of the flotation process according to the formula:

$$\eta = 0,01 * \sqrt{\varepsilon e.c. * \varepsilon e.m.}$$

where $\varepsilon e.c.$ - extraction of combustible mass in concentrate, %; $\varepsilon e.m.$ - extraction of mineral mass to waste, %.

3 EXPERIMENTAL PART

The study of the effectiveness of the reagents' vaprivateley during coal flotation found that the group chemical composition of the blowing agent reagent has the most significant effect on the performance of flotation. During the flotation of coal fines, the central processing plant Belovskaya, using the ecofol-440 foaming agent in the amount of 0.10 kg/t and the flotek 2432 collector, obtained very low flotation rates. The concentrate yield was 60.5% with an ash content of 7.4%, and the ash content of the waste was 34.8%. The use of the Oxal T-80 foaming agent with an equal consumption of reagents made it possible to increase the concentrate yield by 7.5%. The ash content of flotation waste increased from 34.8% to 41%. At the same time, the use of the Oxal T-80 foaming agent made it possible to increase the extraction of the combustible mass into concentrate from 68.5% to 76.9% while simultaneously increasing the efficiency ratio of the coal flotation process from 0.717 to 0.746 (Table 4). Increasing the consumption of blowing agent reagents to 0.20 kg / ton of coal leads to a significant

Table 4. Effect of consumption of reagents of blowing agents of various group chemical composition on indicators of flotation of coal breeze at the central processing plant Belovskaya.

Reagent Mode		Reagent consumption, kg/t			Flotation rates					
Collecting agent	Blowing agent	Collecting agent	Blowing agent	Total	Flotation products	Yield %	Ash Content, A^d,%	Combustible recovery into concentrate, ε e.c.%	Mineral matter recover to waste, $\varepsilon\varepsilon$.m.%	Efficiency ratio
	Oxal	2,16	0,10	2,26	Concentrate	68,0	7,4	76,9	72,3	0,746
	T-80				Waste	32,0	41,0			
	Ecofol-440	2,16	0,10	2,26	Concentrate	60,5	7,4	68,5	75,1	0,717
					Waste	39,5	34,8			
	Oxal	2,16	0,20	2,36	Concentrate	80,7	8,0	90,7	64,4	0,764
	T-80				Waste	19,3	60,7			
Flotek-2432	Ecofol-440	2,16	0,20	2,36	Concentrate	67,5	7,6	76,2	71,8	0,739
					Waste	32,5	40,2			
	Oxal	2,16	0,30	2,46	Concentrate	85,5	9,2	94,9	56,8	0,734
	T-80				Waste	14,5	71,3			
					Concentrate	76,2	8,0	85,7	66,4	0,720
	Ecofol-440	2,16	0,30	2,46	Waste	23,8	50,7			
					Baseline	100,0	18,2			

Table 5. The results of the properties investigation of the two-phase and three-phase foams formed by the reagents under study.

Chemical compound or technical product	Reagent concentration in water		Parameters of the formed 2-phase foam		Fracture velocity 2-phase foam, $m*10^{-3}$ sec.	Three-phase foams stability	
	mg/l	mol/l $*10^{2}$	Height, $m*10^{-8}$	Foam formation time, sec.	Reagent	The initial height of the foam column, $m* 10^{-3}$ — Reagent	The final height of the foam column after vacuuming, $m*10^{-3}$% foam destruction
Oil acid 20/87,2	217	246	37	30	4,9	4,4-dimethyl 1,3-dioxane	156
Butyl aldehyde 46/77	215	293	41	50	5,3	T-80	198
Methyl-ethyl ketone 119/44	208	289	89	59	7.7	Butyl alcohols distillation residues (KOBS)	210
Butyl alcohol 128/55	207	277	156	73	7,5	4,4-dimethyl 1,3-dioxane *	285
Hexyl alcohol 160/48	218	217	284	96	8,3	Oxal T-80*	305
4,4-dimethyl 1,3-dioxane 180/42	255	220	45	29	5.6	Butyl alcohols distillation residues (KOBS)*	310

* During fl°tati°n, a mixture °f a bl°wing reagent with a c°llect°r °f tract°r ker°sene was used (rati° 5:95)

increase in coal flotation rates. However, the patterns in the effectiveness of the blowing agents remain. The use of the Oxal T-80 foaming agent as a reagent can improve the floatability of coal fines. The use of the Oxal T-80 foaming agent in the amount of 0.20 kg/ton of coal as a reagent made it possible to increase the concentrate yield from 67.5% to 80.7% at an ash content of 8.0%. The ash content of the flotation waste increased from 40.2% to 60.7% compared to the use of the Ekofol-440 blowing agent as a reagent (Table 4). The highest flotation rates were obtained with the consumption of blowing agent reagents in the amount of 0.30 kg/ton of coal. It also follows the pattern of change in the floatability of coal fines depending on the blowing agent reagent used. The use of the Oxal T-80 foamer allows to increase the flotation rates as compared with the use of the Ekofol 440 foamer used at the central processing plant Belovskaya (Table 4). The use of the Oxal T-80 foaming agent instead of Ekofol-440 allowed to increase the concentrate yield to 85.5% with its ash content of 9.2%, which meets the requirements of the coke-chemical industry for the concentrates supplied for coking. Removing the combustible mass in the concentrate was 94.9%, which is 9.2% more compared to the use of the Ekofol 440 blowing agent. The ash content of the flotation waste when using the Oxal T-80 blowing agent increased to 71.3%, which indicates a significant decrease in the loss of organic matter from the flotation waste.

One of the main reasons for the high flotation activity of the Oxal T-80 expander is the presence of dioxane and pyran alcohols in the group chemical composition of the technical product. the activity of chemical compounds positively influencing their adsorption on the coal surface due to the formation of intermolecular reagent complexes with polar E centers coal surface, but also have a positive effect on the formation of two-phase and three-phase foams (Table 5).

The study of the properties of two-phase and three-phase foams formed by investigated reagents showed that the stability of foams formed using alkyl substituted 1,3-dioxanals and the technical product Oxal T-80 is slightly lower compared to the blowing agents in the group chemical composition of which aliphatic oxygen-containing chemical compounds are present. For example, a two-phase foam formed by 4,4 - dimethyl - 1,3 - dioxane collapses in 5.6 seconds, and at the same concentration in water, the two-phase foam formed by hexyl alcohol collapses in 8.3 seconds (Table 5). In the case of using the cobber foaming agent containing aliphatic oxygen-containing chemical compounds in the group chemical composition, the percentage of three-phase foam destruction was 42%, using the Oxal T-80 foaming agent is destroyed by evacuating 48% of the formed three-phase foam (Table 5). The decrease in the stability of the three-phase foam causes an increase in the selectivity of the separation of the organic mass of coal from mineral particles. The mineral components of coal carried to the froth product, at high flotation rates, with coal floccules due to the low retention strength in the foam layer fall into the pulp, which reduces the ash content of the flotation concentrate. Research confirms the above position. The use of the Oxal T-80 foaming agent as a reagent not only leads to an increase in the extraction of the combustible mass into the concentrate, but also improves the efficiency of the flotation process. The indicator of the efficiency of the flotation process in the case of the use of the reagent of the Oxal T-80 blowing agent increases from 0.717-0.739 to 0.734-0.764 (Table 4).

4 CONCLUSION

The study of the influence of the group chemical composition of the blowing agent reagents on the flotation efficiency has established an increased efficiency and selectivity of the technical product of petrochemistry Oxal T-80 compared with the use of the Ekofol-440 blowing agent reagent used at the central processing plant Belovskaya MMK-COAL. It was established that with equal consumption of reagents, the use of the Oxal T-80 foaming agent as a reagent makes it possible to increase the extraction of the combustible mass of coal into a concentrate from 68.5 - 85.7% to 76.9 - 94.9% compared to the use of Ecofol-440, With obtaining the ash content of the concentrate in the range of 7.4 - 9.2%, which satisfies the requirements for concentrates supplied to coking. At the same time, the ash content of the flotation waste when using Oksal T-80 rises from 40.2 50.7% to 60.7 - 71.3%, which indicates a decrease in the organic mass loss of coal containing flotation waste. The study made it possible to establish that when using the Flotek 2432 collector as a reagent, the consumption of blowing agent reagents to reduce losses of organic matter of coal with flotation waste should be more than 0.100 g/t, and Oxal T-80 should be used as a reagent of the sprayer. The high efficiency and selectivity of the Oxal T-80 reagent reagents are due to the presence of dioxane alcohols in the reagent, which provide not only increased surface activity of the reagent but also positively affect their adsorption on the coal surface due to the formation of intermolecular coal surface centers. The results of the study allow us to recommend industrial testing using the Oxal T-80 foaming agent in the central processing plant Belovskaya as a reagent.

REFERENCES

Abramov, A.A. 2016. Flotation enrichment methods - Mountain Book.

Chizhevsky, V.B., Vlasova, N.S., Savinchuk, L.G. 1981. Flotation properties of T-66, T-80 and T-81. Non-ferrous metallurgy 18: 12–14.

Deberdeev, I.Kh., Picat-Ordynsky, G.A., Rudanovskaya, L.A. 1986. New flotation reagent KETGOL. Coke and Chemistry 11: 13–15.

Ejtemaei, M., Nguyen, A.V. 2017. Characterisation of sphalerite and pyrite surfaces activated by copper sulphate. Minerals Engineering 100: 223–232.

Karimian, A., Rezaei, B., Mazoumi, A. 2013. The effect of mixed collectors in the rougher flotation of sungun copper. Life Science Journal 10: 268–272.

Limarev M.S. 2011. Research of new organosilicon compounds as reagents of blowing agents during coal flotation, *8th round of dressers from CIS countries. Moscow, Feb. 28-March 2, 2011*: Collection of materials 2. MISiS: 171–176.

McFadzean, B., Mhlanga, S.S., O,Connor, C.T. 2013. The effect of thiol collector mixtures on the flotation of pyrite and galena. Minerals Engineering Vol.50-51: 121–129.

Pan Lei, Yoon Rol-Hoan, 2014. Direct measurement of hydrodynamic and surface forces in bubble-particle. Proc .of the XXV11 IMPC, Santiago, Chile 2: 88.

Petukhov, V.N., Kubak, D.A. 2016. The search for effective blowing agent reagents for coal flotation based on the calculation of the basic quantum chemical characteristics of the complexes "OMU-water" and "OMU-reagents". Coke and Chemistry 2: 26–33.

Petukhov, V.N., Smirnov, A.N., Kubak, D.A. 2017. Use of structural and quantum-chemical parameters of reagents to substantiate their flotation activity. *Materials International. scientific-practical the conference "50 years of the Russian scientific school of integrated development of the earth's depths", November 13- 16, 2017.*

Petukhov, V.N., Kubak, D.A., Semenov, D.G. 2014. Study of the influence of group chemical composition of complex reagents on the efficiency of coal flotation. Vestnik Nosov Magnitogorsk State Technical University 1: 34–39.

Petukhov, V.N., Smirnov, A.N., Voloshchuk, T.G., Kubak, D.A. 2018. Analysis of Intermolecular Complexes of Coals Organic Mass in the Search for Composite Flotation Reagents. Coke and Chemistry, 2018. 61(2): 64–71.

Popova, L.A., Petukhov, V.N. 1978. The mechanism of adsorption of organic compounds of different composition and structure on the coal surface. News of higher educational institutions. Mining Journal 5: 158–163.

Popova, L.A. 1980. Investigation of the influence of the physicochemical properties of hydrocarbons and 1,3-dioxanes on the floatability of coal at different stages of metamorphism. Lyubertsy: 23.

Ryaboy, V., Shepeta, E., Kretov, V., Golirov, V. 2014. New dialkyldithiophosphates for the flotation of copper, gold and silver containing ores. Proc. of the XXV11 IPMPC, Santiago, Chile, 3: 1–8.

Sahbaz, O., Cinar, M., Kelebek, S. 2016. Analysis of flotation of unburned carbon from botton ashes. Acta Montanistica Slovaca 21(2): 93–101.

Solozhenkin, P.M. & Solozhenkin, O.I. & Sanda Krausz, 2011. Study of sulphydric flotation reagents isomery. Annals of the University of Petrosanu, Mining Engineering 12(XXX1V). Universitas Publishing Hause ISSN 1454-9174: 65–73.

A selection of emulsifiers for preparation of invert emulsion drilling fluids

S.R. Islamov, A.V. Bondarenko & D.V. Mardashov
Saint Petersburg Mining University, St. Petersburg, Russia

ABSTRACT: In the present context, a decline in oil production and a steady increase in its costs require a different approach to the selection of type and technological parameters of well killing fluids that provide safe operating conditions for underground repairs and have no negative impact on the filtration properties of the bottomhole formation zone. Practice shows that the most promising solution to this problem is the application of invert emulsion drilling fluids that have become widespread in various branches of the oil industry. In this paper, authors provide a comprehensive methodology of emulsifier selection for the preparation of invert emulsion drilling fluids for designated production well parameters. As compared to previously used emulsifiers, the agents specified in this paper allow not only to improve the characteristics of well killing fluids, but also, other parameters being equal, to reduce the concentration of emulsifier in the final solution and thereby to lower the cost of production well killing.

1 INTRODUCTION

In the course of well exploitation, especially at the late stages of field development, the issue of deteriorating reservoir properties in the bottomhole formation zone is becoming more and more pressing.

Layered and zonal heterogeneity of oil beds, complex structure of filtration channels, presence of immiscible liquids, low values of open porosity and absolute permeability all play their part in deterioration of production well performance.

Efficiency of production wells is determined, among other things, by the conditions of reservoir completion and the composition of well killing fluids used in repair operations.

Currently manufacturers of chemical agents offer a large variety of different blocking substances for well killing, among which a special place belongs to invert emulsion drilling fluids (IEDF).

Practice shows that successful use of such systems requires external dispersion medium to be a hydrocarbon fluid (oil, diesel, gas condensate etc.). A dispersed phase can be represented by water, various solutions of acids, alkalis, polymers and cements, which makes it possible to use different technologies of bottomhole zone treatment, ensuring preservation, restoration and improvement of the filtration properties of reservoir rocks (Orlov et al. 1991; Petrov et al. 2008; Zhelonin et al. 2015; Dandekar 2013).

The preparation of aggregate stable emulsions is achieved by adding to the initial composition a sufficient amount of emulsifying agent, most effective for the required combination of components. Rising costs of well killing technology dictate the need for correct selection of an emulsifier (Nikolaev & Leusheva 2016; Ryabokon 2009; Tokunov & Saushin 2004).

Thus the emulsifier should be cheap, and its concentration in the final solution should be minimal. It is also crucially important that the prepared emulsion must have properties that fulfill all the requirements of well killing operations.

Fulfillment of this task is especially relevant for the preparation of invert-emulsion drilling fluids for well killing at an oil-gas condensate field in Russia characterized by the following conditions: porous-fissured carbonate reservoir; open and extended horizontal section of the wellbore in the pay formation; acid fracturing and massive hydrochloric acid treatment. Production requires large-scale injection of well killing fluids, which raises the cost of well killing operations in case of underground well repair.

The aim of this research was to cut the costs of well killing in case of underground repair operations by reducing the concentration of emulsifying agent in the final solution. This has been achieved by selecting the most efficient emulsifier for the designated combination of components in the IEFD solution.

The authors present a comprehensive methodology of physical and chemical studies carried out for various emulsifying agents, which allows to choose the best options and to draw recommendations for their subsequent application.

The selection of emulsifiers has been carried out according to the following method:

1. Selection of a group of emulsifiers basing on the values of interfacial tension at the boundary "emulsified oil – water solution of calcium chloride".

2. Estimation of thermal and electrical stability of emulsions prepared with selected emulsifiers.

3. Estimation of rheological properties of emulsions prepared with selected emulsifiers.

2 MATERIALS AND METHODS

2.1 *Estimation of interfacial tension at the boundary "emulsified oil – water solution of calcium chloride"*

The device has been controlled with a software package, which by means of automatic video recording allows to define the shape and size of the drop and to estimate surface tension.

Water solution of calcium chloride with a density of 1 220 kg/m^3 (density varies depending on the required values of bottomhole pressure and IEDF viscosity) was poured into a standard measuring cuvette with a total volume of 25 cm^3. The oil from the oil-gas condensate field was mixed with an emulsifier at the following volumetric concentrations: 0.01; 0.025; 0.05; 0.1; 0.2; 0.5%. Prepared mixture was collected with a syringe fixed in a tensiometer so that the tip of the needle (with a rounded end) was visible on the computer monitor. Then the needle was placed in a cuvette with the water solution of calcium chloride, and test liquid was uniformly displaced from the syringe into the cuvette with the formation of floating drops. The entire process of extrusion was video recorded and later analyzed in a software package. As a result of this analysis, six values of interfacial tension were calculated for each liquid sample (for each emulsifier concentration), then the maximum and minimum values were removed from the range. An arithmetic mean of the remaining four values was used to plot the relationship between interfacial tension at the boundary "emulsified oil – water solution of calcium chloride" and the concentration and type of emulsifier.

2.2 *Estimation of thermal and electrical stability of emulsions*

IEDF solutions were prepared using commercial oil from the oil-gas condensate field (20% vol.), water solution of calcium chloride with a density of 1.220 g/cm^3 (78% vol.) and emulsifiers (2% vol.), recommended in section 2.1. Selected concentrations of the components are explained as follows.

Practice shows that the 80/20% ratio of water and hydrocarbon phases is optimal from the viewpoint of preparing an emulsion of greater stability. Emulsifier concentration (2% vol.) was selected experimentally for the sake of comparative analysis within the framework of the studies described below.

Prepared solutions were placed in a thermostat at a set temperature of 37 °C (reservoir temperature) and held for 7 days. Every 24 hours visual assessment of the emulsion state and the amount of the separated water phase was performed. If the layer of precipitated water phase is absent, emulsion solution is considered stable. At high temperatures, a small release of hydrocarbon layer (the upper dark layer) is tolerated, which is easily removed by mere shaking or stirring of the composition.

Further, the emulsion electrical stability was determined. Electrical stability is an important indicator characterizing resistance of inverted emulsions to degradation. Estimation of IEDF electrical stability was carried out using electrical stability tester. This device is designed to assess direct current breakdown voltage of inverse emulsions, which characterizes electrical stability of the emulsion.

Operation principle of the device is based on measuring electrical resistance of the fluid as the current of certain magnitude passes through it.

2.3 *Estimation of thermal and electrical stability of emulsions*

In the scope of these experiments, IEDF solutions have been prepared with emulsifiers, selected according to the results of interfacial tension tests at the boundary "emulsified oil – water solution of calcium chloride" and thermal stability assessment. For the sake of comparability, the concentration of emulsifiers in the solution remained constant (2% by volume).

A rheological test was performed to determine the effective viscosity and static shear stress of the IEDF solutions.

The procedure of conducting rheological tests involved the following stages:

1. Construction of the initial rheological composition curve (at 37 °C).
2. Estimation of static shear stress (at 37 °C).

The rheological curve was constructed in the controlled-rate mode (CR test). The main idea of the method was to establish dependence between the shear stress and the shear rate with gradual increase of the latter parameter from 0 to 300 s^{-1} (consistency curve).

For a more substantial analysis of consistency curves for IEDF solutions prepared with selected emulsifiers, the dependence of effective viscosity on the number of cylinder rotations has been constructed.

The procedure of estimating static shear stress of invert emulsions is the following:

- viscometer is set to "constant shear rate" mode;
- the test liquid is poured into the cylinder (45 ml);
- cylinder rotation speed is set to 200 rpm and the composition is stirred at the set speed for 10 s, then the device is switched off;
- after 1 and 10 minutes of quiescent state, the maximum value is taken before the destruction of sample structure at the rotation speed of 3 rpm.

3 RESULTS AND DISCUSSION

In total, 18 emulsifiers of different brands and manufacturers (EM-01 – EM-18) have been used in the study of interfacial tension. Also, for the sake of comparison, the values of interfacial tension have been estimated for pure oil and EM-N emulsifier currently used at the oil-gas condensate field for IEFD preparation (Figure 1).

According to the data shown in Figure 1, it is possible to highlight emulsifiers (Table 1) that provide a more effective decrease in interfacial tension at the phase boundary at lower concentrations, as compared to others. This finding contributes to the creation of an emulsion with greater thermal and aggregate stability, since low interfacial tension at the phase boundary is one of the crucial conditions for obtaining a stable emulsion. In addition, surface activity of emulsifiers directly affects the preservation of the bottomhole filtration properties and shortens the period of well development and their withdrawal to the operating mode after underground repair operations (Babalyan et al. 1962; Jouenne et al. 2016).

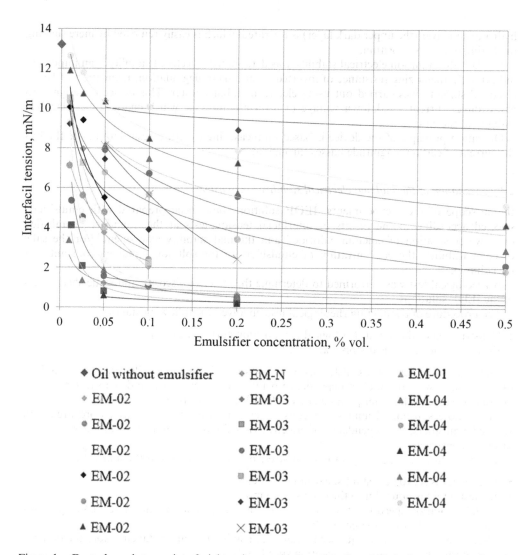

Figure 1. Dependence between interfacial tension on the boundary "emulsified oil – water solution of calcium chloride" and the concentration and type of emulsifier.

Table 1. Final results of interfacial tension for the most efficient six emulsifiers.

No.	Emulsifier code name	Interfacial tension (mN/m) at emulsifier concentrations	
		C=0.05% vol.	C=0.2% vol.
1	EM-N	3.77	1.45
2	EM-02	1.22	0.67
3	EM-04	1.89	0.41
4	EM-06	0.82	0.31
5	EM-07	1.58	0.52
6	EM-10	0.61	0.22
7	EM-13	1.25	0.57

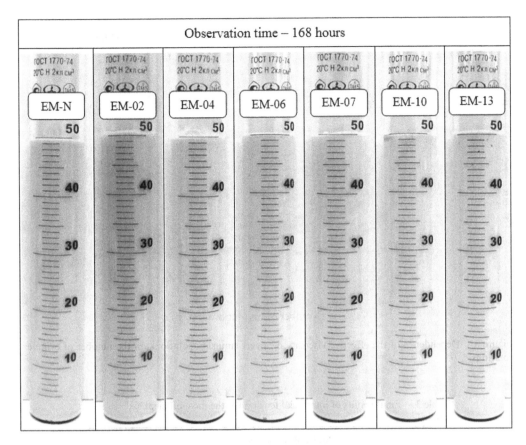

Figure 2.   Estimation results of IEDF thermal stability at 37 °C.

Table 2.   Final results of interfacial tension for the most efficient six emulsifiers.

No.	Emulsifier code name	Breakdown voltage, V
1	EM-N	40
2	EM-02	50
3	EM-04	70
4	EM-06	40
5	EM-07	70
6	EM-10	70
7	EM-13	80

Estimation results of IEDF thermal stability are presented in Figure 2. All the emulsion solutions are stable at the temperature of 37 °C for 7 days.

The estimation results of electrical stability values for invert emulsion fluids are presented in Table 2.

Research has demonstrated that emulsion fluids prepared with emulsifying agents with numbers in their code name tend to have greater electrical stability than those stabilized with EM-N agent. Maximal values of breakdown voltage are observed in IEDF solutions prepared with EM-04, EM-07, EM-10 and EN-13 emulsifiers.

In the scope of rheological tests, IEDF solutions have been prepared with emulsifiers (EM-02, EM-04, EM-06, EM-07, EM-10, EM-13 and EM-N), selected according to the results of

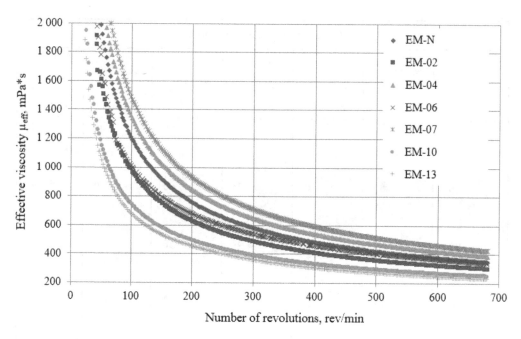

Figure 3. Dependence of the effective viscosity of invert emulsion solutions on the number of cylinder rotations.

Table 3. Final results of interfacial tension for the most efficient six emulsifiers.

No.	Emulsifier code name	Static shear stress (D=3 rpm)		Change in the effective viscosity (compared to viscosity of the solution with EM-N), %
		in 1 min	in 10 min	
1	EM-N	22.7	30.7	0
2	EM-02	18.4	27.3	-16
3	EM-04	26.3	44.7	11
4	EM-06	29.8	40.4	-7
5	EM-07	29.3	29.6	22
6	EM-10	14.1	13.5	-33
7	EM-13	11.3	24.4	-38

interfacial tension tests at the boundary "emulsified oil – water solution of calcium chloride" and thermal stability assessment. Research results are shown in Figure 3.

Results of static shear stress tests at the reservoir temperature are presented in Table 3.

Higher values of effective viscosity indicate a better gas-retaining ability of the composition and its sedimentation stability in case of using a solid filler, as well as reduced risk that the composition might penetrate the bottomhole formation zone. As the effective viscosity increases, so does the structural strength of the emulsion (Demakhin et al. 2015; Elkatatny 2016; Lirio Quintero et al. 2017).

Results of rheological tests show that the greatest effective viscosity is demonstrated by the IEDF solution prepared with EM-07 emulsifier. At 300 rpm, effective viscosity of this solution exceeds μ_{eff} of the emulsion prepared with EM-N by 22% (Figure 3, Table 3). The solution

Table 4. Final results of interfacial tension for the most efficient six emulsifiers.

No.	Emulsifier code name	Emulsifier real name	Manufacturer
1	EM-04	Yalan E-2 Brand A	LLC «Sintez-TNP»
2	EM -06	Emitrit	JSC «Polyeks»
3	EM -07	CSE-1013	LLC MNC «HimServiceEngineering»

stabilized by EM-04 agent is also characterized by a higher viscosity (11%) than the reference one. Solutions prepared with EM-02, EM-06, EM-10 and EM-13 emulsifiers have lower values of effective viscosity (μ_{eff} below the reference value by 16, 7, 33 and 38%, respectively).

Static shear stress is an indicator of the strength that internal structure of the well killing fluid has formed during the quiescent period. Its value is defined by the voltage that must be exceeded in the quiescent fluid to destroy its internal structure and set the fluid in motion (Sharath et al. 2017). The higher the value of static shear stress of the well killing fluid, the lower the probability of its absorption in the bottomhole formation zone. This is especially true for the wells, whose pressure gradients are significantly lower than the potential of well killing fluids to exert required repression on the oil bed. In case of normal or increased reservoir pressure, a large amount of static shear stress reduces the likelihood of fluid breakthrough into the well (Vikrant Wagle et al. 2018).

As a result of ultimate static shear stress estimation, it has been established that solutions stabilized with EM-04 and EM-06 emulsifiers are characterized by the greatest strength of their internal structure. Emulsions prepared with EM-02 and EM-07 agents have static shear stress values comparable to those of IEDF solutions stabilized with EM-N emulsifier.

Thus, a comprehensive analysis of the results of physical and chemical studies of emulsion compounds makes it possible to highlight the following emulsifiers and to recommend them for application at the oil-gas condensate field: EM-04, EM-06 and EM-07 (Table 4).

4 CONCLUSION

Based on the laboratory tests of physical and chemical properties of emulsifiers aimed at the selection of the most effective agent to stabilize IEDF solutions for well killing operation at the oil-gas condensate field, the following findings have been established:

1. Emulsifiers EM-02, EM-04, EM-06, EM-07, EM-10 and EM-13 reduce the interfacial tension on the border "emulsified oil – water solution of calcium chloride". The values of the interfacial tension for these emulsifiers are 2-3 times lower as compared to those of EM-N agent currently used at the oil-gas condensate field.

2. Invert emulsion solutions stabilized by all emulsifiers listed above are heat-stable at 37 °C during 7 days. The maximum value of the electrical stability is observed in invert emulsion solutions prepared with EM-04, EM-07, EM-10 and EM-13 agents.

3. The higher the static shear stress, the greater is the pressure gradient that the emulsion is able to withstand without collapsing. The same conclusions are valid for the increase of effective viscosity of the invert emulsion, since it increases its structural strength. Thus, as a result of rheological studies, it has been established that the use of EM-07 and EM-04 increases effective viscosity of IEDF by 22 and 11%, respectively. Relatively small reduction of the effective viscosity (by 7%) as compared to the reference value of EM-N emulsifier, is observed in solutions stabilized with EM-06 agent.

4. The greatest value of static shear stress is observed in emulsions stabilized with EM-04 and EM-06. The values of static shear stress for solutions prepared with EM-02 and EM-07 are comparable to those stabilized with EM-N emulsifier.

5. As an alternative to EM-N, it is recommended to use emulsifiers Yalan E-2 Brand A, Emitrite and CSE-1013. IEDF solutions prepared with the latter two agents have comparable

rheological properties, which is the reason why the authors recommend these emulsifiers to be used as components in emulsion solutions.

Thus, the use of specified emulsifying agents in the composition of invert emulsion fluids, as compared to EM-N, will allow not only to improve the characteristics of well killing fluids, but also, other parameters being equal, to reduce the concentration of emulsifier in the final solution and thereby to lower the cost of production well killing.

REFERENCES

Babalyan, G.A., Kravchenko, I.I., Marhasin, I.L. & Rudakov, G.V. 1962. *Physical and Chemical Principles of the Application of Surfactants in the Development of Oil Reservoirs. Moscow: Gostoptekhizdat.*

Dandekar, A.Y. 2013. *Petroleum Reservoir Rock and Fluid Properties.* Boca Raton: CRC press.

Demakhin, S.A., Merkulov, A.P., Kasyanov, D.N., Malayko, S.V., Anfinogentov, D.A. & Chumakov, E.M. 2015. Well bluffing with block packs is an effective method of preserving filtration properties of the reservoir. *Neft-Gaz-Novacii* 192(1): 66-69.

Elkatatny, S.M. 2016. Determination the Rheological Properties of Invert Emulsion Based Mud on Real Time Using Artificial Neural Network. *SPE Kingdom of Saudi Arabia Annual Technical Symposium and Exhibition, doi: 10.2118/182801-MS, 25-28 Aprili.* Saudi Arabia: Dammam.

Jouenne, S., Klimenko, A. & Levitt, D. 2016. Tradeoffs Between Emulsion and Powder Polymers for EOR. *SPE Improved Oil Recovery Conferencei, doi: 10.2118/179631-MS, 11-13 April.* USA: Tulsa.

Lirio Quintero, Ramakrishna Ponnapati & Mary Jane Felipe. 2017. Cleanup of Organic and Inorganic Wellbore Deposits Using Microemulsion Formulations: Laboratory Development and Field Applications. *Offshore Technology Conference, doi: 10.4043/27653-MS, 1-4 May.* USA: Houston.

Nikolaev, N.I. & Leusheva, E.L. 2016. Development of drilling fluids composition for efficiency increase of hard rocks drilling. *Journal of Mining Institute* 219(3): 412-420.

Orlov, G.A., Kendis, M.Sh. & Glushchenko, V.N. 1991. *Application of Invert Emulsions in Oil Production.* Moscow: Nedra.

Petrov, N.A., Soloviev, A.Ya., Sultanov, V.G., Krotov, S.A. & Davydova, I.N. 2008. *Emulsion Solutions in Oil and Gas Processes.* Moscow: Chemistry.

Ryabokon, S.A. 2009. *Technological Liquids for Completion and Servicing of Wells.* Krasnodar: Prosveshcheniye-Yug.

Sharath, S., Donald, W. & Jonathan, W. 2017. Acid-Soluble Lost Circulation Material for Use in Large, Naturally Fractured Formations and Reservoirs. *SPE Middle East Oil & Gas Show and Conference, doi: 10.2118/183808-MS, 6-9 March.* Kingdom of Bahrain: Manama.

Tokunov, V.I. & Saushin, A.Z. 2004. *Technological Liquids and Compositions for Increasing the Productivity of Oil and Gas Wells.* Moscow: Nedra.

Vikrant Wagle, Abdullah S. Al-Yami & Ali AlSafran. 2018. Designing Invert Emulsion Drilling Fluids for HTHP Conditions. *SPE Kingdom of Saudi Arabia Annual Technical Symposium and Exhibition, doi: 10.2118/192192-MS, 23-26 April.* Saudi Arabia: Dammam.

Zhelonin, P.V., Mukhametshin, D.M., Archikov, A.B., Zvonarev, A.N., Krayevsky, N.N. & Gusakov, V.N. 2015. Justification of the algorithm for selection of well killing technologies. *Scientific and Technical Bulletin of OJSC "Rosneft"* 39(2): 76-81.

Topical Issues of Rational Use of Natural Resources 2019 – Litvinenko (Ed)
© 2020 Taylor & Francis Group, London, ISBN 978-0-367-85720-2

PGO Processing with azeotropic rectification to extract naphthalene

Y.A. Bulauka & S.F. Yakubouski
Polotsk State University, Novopolotsk, Belarus

ABSTRACT: In this study, azeotropic rectification method was attempted to extract industrial naphthalene from fractions of liquid pyrolysis' products of the hydrocarbon feed and obtain refined naphthalene. Extraction of naphthalene from fractions of liquid pyrolysis products (LPP) from hydrocarbons including the atmospheric and then the vacuum distillation of heavier cut with the extraction of naphthalene concentrate that is having azeotropic rectification and then it is coming for the stages of crystallization and pressing. In order to obtain condition for industry naphthalene purification using azeotropic rectification method, the Belarussian heavier cut's chromatography were studied. The optimization of the process of azeotropic rectification to obtain refined naphthalene is mainly discussed to provide a theoretical for the actual production and technology improvement

1 INTRODUCTION

The general power of pyrolysis processes in the world exceeds 130 million tons per year (Nakamura, 2007). The process of getting light olefins is accompanied by forming about 20% coproducts. The usage of these coproducts is a serious technical and economic problem that is associated with increased profitability of production. In order to remain competitive in the ethylene business for steam crackers, more effort must be made to upgrade all of the by products that are formed by liquid crackers. Producers who do not upgrade these by-products will face growing tension on the plant margins owing to competition from the world's low-cost regions (Paliashkevich, 2017).

One of these coproducts is the heavier cut of pyrolysis gas oil (PGO). The heavy distillates steam-cracked naphtha contains aromatic hydrocarbons that boils above 180 Celsius degrees (Apicella, 2003.). Only in Russia the production of PGO exceeds 325000 tons per year. Belarussian petrochemical Plant "Polymir" which is part of JSC "Naftan" is able to produce from 12000 up to 16000 tons of PGO yearly (Bulauka, 2018).

The issue of rational use of PGO is relevant for Belarus due to the future plans to increase the capacity of the enterprise, which will lead to an increase in the amount of by-products and degradation of problems associated with their marketing.

Nowadays, PGO uses as a source of boiler heater. It is possible to obtain from the heavy pyrolysis tar not only boiler fuel but also carbon black, inactive carbon black, coke, dark petroleum resins, concrete superplasticizing agents, plasticizers, bitumen materials and to extract individual aromatic hydrocarbons (Naphthalene, 1-methylnaphthalene, 2-methylnaphthalene, etc.).

With the increasing global market demand for naphthalene, naphthalene defining technology is extensively studied internationally. Industrial naphthalene from coal tar accounts for 85% of total naphthalene production in the world. Naphthalene is produced from a high heteroatoms content coal-tar resin, in order to remove these heteroatoms expensive cleaning operations are used (Kershaw, 1993; Gargiulo, 2015; Granda, 2003; Gargiulo, 2016; George, 2010).

Up to date, there is still a problem to release of pure naphthalene from coal tar which related with its separation from components close to boiling point Naphthalene (217,97°C), such as Thionaphthene (219,90°C) and 2,3-Xylenol (216,87°C). It should be noted that the PGO of a wide fraction of light hydrocarbons, in contrast to coal tar, does not contain hetero atomic compounds, including Thionaphthene and Xylenols, and therefore is the preferred raw material for producing high purity Naphthalene.

There is a method for producing naphthalene from liquid pyrolysis products in two vacuum columns (Gentry, 2009; Apicella, 2017; Petlyuk. 1965). Naphthalene separation process is shown in Figure 1.

The first column is designed to remove the lighter fraction than naphthalene with the column overhead. In the second column, the heavy fraction of liquid pyrolysis products are removed by the bottom product, and the naphthalene fraction is separated from the top of the column, which is sent to crystallization. The disadvantage of this process is not a high degree of purity of the product.

There is a method for separating naphthalene from the C_9-C_{11} fraction of liquid pyrolysis products. The fraction is exposed to catalytic hydrostabilization and then the C_{10}-C_{11} aromatic fraction is separated by means of rectification. The obtained product is subjected to thermal hydrodealkylation together with a hydrostabilized and hydrotreated C_6-C_8 aromatic fraction. Hydrodealkylate is processed to rectification with the release of naphthalene as one of the products. The disadvantage of the process is the complicated technological scheme of processing liquid pyrolysis products and the use of expensive catalysts.

There is a method of producing naphthalene from naphthalene-containing fractions of liquid pyrolysis products by hydro-purification of unsaturated hydrocarbons in the presence of an aluminum palladium sulphide catalyst. The disadvantages of this method of producing naphthalene are high pressure, the use of hydrogen for the hydrogenation of unsaturated compounds and expensive catalysts.

There is a method of producing naphthalene from naphthalene-containing fractions of liquid pyrolysis products by purification of unsaturated hydrocarbons by the method of their polymerization in the presence of an aluminum-cobalt-molybdenum catalyst. The process of

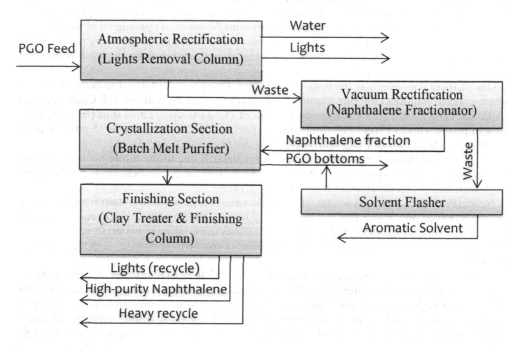

Figure 1. Naphthalene separation process.

polymerization of unsaturated hydrocarbons is carried out at a temperature of 150-210°C, pressure up to 0.1 MPa, for 0.5-2 hours. The main disadvantage of this method is the need to use hydrogen and expensive catalyst, low degree of purity of naphthalene.

There is a method of separating naphthalene from a fraction of 190-250°C of liquid pyrolysis products, which is previously subjected to catalytic polymerization in the presence of an aluminum-cobalt-molybdenum catalyst. The disadvantage of this method is the need to use a catalyst, which loses activity during the process, which leads to the need for regeneration and, as a consequence, the use of complex technological schemes, including the regeneration of the catalyst.

There is a method of separating naphthalene from the fraction that maintained at a temperature of 200-300°C, a pressure of 0.1-1.0 MPa for 2-10 hours, the treated fraction sent to atmospheric & vacuum simple distillation. The distillate is sent to the separation of naphthalene by crystallization in a known manner. The disadvantage of this method is the need for preliminary heat treatment in the reactor for liquid pyrolysis products at high temperatures up to 300°C and pressures up to 1.0 MPa, and involving expensive initiators or unsaturated individual aromatic hydrocarbons in the case of polymerization of reactive non-saturated compounds and high-boiling resins at low temperatures 200-280°C.

Current methods for the naphthalene extraction are not used in manufacture because of the high energy inputs, expensive catalysts, or low naphthalene purity.

2   PURPOSE AND OBJECTIVES OF THE STUDY

The purpose of the research is to develop an efficient method for obtaining pure naphthalene by azeotropic distillation of naphthalene-containing fraction of liquid products of pyrolysis of hydrocarbons. The proposed method of separation of naphthalene from liquid products of the pyrolysis of hydrocarbons is made in laboratory conditions using PGO produced at the factory "Polymir" with a naphthalene content of more than 18% wt.

3   RESEARCH METHODS

The composition of the fractions of PGO production was investigated by gas chromatography. Gas chromatography identifies resin ingredients by peak area, evaluate their quantitative content with high accuracy. The possibility of extracting naphthalene by the method of azeotropic distillation with ethylene glycol has been studied.

4   RESULTS AND DISCUSSION

As a result of fractional distillation of PGO according to Engler, the yield of fractions is the following: b.b.-180°C was 1.89% wt., fraction 180-210°C was 18.76% wt., fraction 210-230°C was 14.45% wt. and semi-solid non-distillable residue of polymeric nature (pitch) was 64.90% wt. We have analyzed the Belarussian heavy pyrolysis resin (tar) and identified individual substances. PGO liquid concentrate is a mixture of various groups of hydrocarbons, primarily aromatic, both monocyclic and polycyclic. Also, all fractions contain isoparaffin, unsaturated, naphthenic and paraffinic hydrocarbons.

Table 1 presents data on the group hydrocarbon composition of individual fractions of heavy pyrolysis resin, produced at the factory "Polymir".

While the containing of aromatic hydrocarbons in PGO reaches to 68%wt., in particular, naphthalene up to 18 % wt.. More than 75 individual aromatic hydrocarbons were found in the PGO liquid concentrate and their content increases with the weighting of the fractional composition. Table 2 shows the main individual aromatic components that make up the liquid concentrate of PGO boiling up to 230°C.

Table 1. The group hydrocarbon composition of individual fractions of PGO.

Groups of hydrocarbons	Fractions of PGO, % wt.			
	b.b.-180°C	180-210°C	210-230°C	Total fraction
Paraffins	2,04	0,94	0,43	0,79
Isoparaffins	10,96	13,29	14,04	13,47
Aromatics	62,82	66,30	70,47	67,82
Naphthenes	7,30	5,26	1,94	4,00
Olefins	13,09	5,27	3,43	6,70
Unknown	3,79	5,64	9,69	7,22

Table 2. Individual composition of aromatic hydrocarbons of separate fractions of PGO.

Individual aromatic hydrocarbon	Fractions of PGO, % wt.			
	b.b.-180°C	180-210°C	210-230°C	Total fraction
naphthalene	2,98	14,97	23,91	18,00
1-methyl-2-isopropylbenzene	4,16	13,69	9,28	11,36
2-methyl naphthalene	0,44	2,62	7,53	4,52
4-methylindane	1,23	4,03	4,91	4,24
1-methyl naphthalene	0,30	1,69	5,24	3,08
1-methyl-3-isopropylbenzene	1,19	3,02	1,35	2,23
n-pentylbenzene	0,00	1,48	1,98	1,61
2,3-dihydroindene	8,12	1,25	0,74	1,41
tert-butylbenzene	1,03	1,88	0,37	1,21
1,2,3-trimethylbenzene	0,62	1,56	0,51	1,08
1-methyl-3-n-propylbenzene	0,54	1,40	0,59	1,02
1,2,4-trimethylbenzene	0,69	1,49	0,45	1,02
biphenyl	0,08	0,50	1,63	0,94

The main component of the liquid product of PGO with boiling point up to 230°C is naphthalene and its alkyl derivatives.

Naphthalene plays an irreplaceable role in Fine Chemical Industry. Naphthalene is used for the synthesis of sulfonic acids, phthalic anhydride, azo dyes, plasticizers, decalin, tetralin, naphthol and others. Sulfonic acids from naphthalene are good surface-active substances (surfactants). The way of the derivation of superplatifikators for concrete from naphthalene is actively developing now.

The output of naphthalene at the factory "Polymir" from the PGO fraction with boiling point up to 230°C can be about 1000 tons per year. Naphthalene recovery is economically feasible. The results of chromatographic analysis show that in the process of obtaining high-purity naphthalene from PGO, separation of homogeneous azeotropes (naphthalene & 1-methyl naphthalene; naphthalene & 2-methyl naphthalene; naphthalene & biphenyl and others) might be a problem.

Various product derivatives can further increase the profitability of naphthalene recovery (Xu, 2012). Methylnaphthalenes are used as insecticides, solvents and starting materials in the synthesis of dyes, to produce sulfonic acids of mono- and dimethylnaphthalenes, used as surface active substances. Besides:

- 2-Methylnaphthalene is a valuable raw material for the production of synthetic vitamin K3 (2-methyl-1,4-naphthoquinone, menadione), which is widely used in medicine as a drug to increase blood clotting.

- 1-Methylnaphthalene is a reference when determining the cetane number of diesel fuel (for 1-methyliaphthalin, it is assumed to be zero).
- 1,4-dimethylnaphthalene is used to suppress the germination of potatoes and vegetables.
- 2,6-dimethylnaphthalene is oxidized to 2,6-naphthalene dicarboxylic acid used in the production of polyesters and polyamides.

The theoretical output of 2-methylnaphthalene at the factory "Polymir" from a liquid product PGO can reach 250 tons per year, for 1-methylnaphthalene is about 170 tons per year, for 1,4-dimethylnaphthalene is about 18 tons per year, for 2,6-dimethylnaphthalene is about 15 tons per year.

Cymols of PGO liquid product can be widely used for the synthesis of cresols, highly effective antioxidants, phthalic acids (mainly isophthalic and terephthalic acids), flavors, etc. The use of cymols in petrochemical synthesis allows sprawling of the raw material base for the production of alkylaromatic hydrocarbons. The theoretical output of cymols at the factory "Polymir" from PGO with boiling point up to 230°C can be: for 1-methyl-2-isopropyl benzene is about 630 tons per year, for 1-methyl-3-isopropyl benzene is about 125 tons per year and for 1-methyl-4-isopropyl benzene is about 2 tons per year.

Indane (2,3-dihydroinden) is the starting material for the synthesis of 2-, 4- and 5-indanols, which are used in the preparation of medicines. The theoretical output of indane from the liquid fraction of PGO can be up to 80 tons per year.

Tert-butylbenzene is the starting compound in the preparation of valuable fragrances, and is also used as a solvent and raw material for alkyl polystyrenes. It is potentially possible to organize the production of tert-butylbenzene from PGO liquid product up to 68 tons per year.

Pseudocumene (1,2,4-trimethylbenzene) is used in the production of trimellitic acid and its anhydride, pseudokumidin, vitamin E. The theoretical output of 1,2,4-trimethylbenzene from the liquid fraction of PGO can be more than 55 tons per year.

Biphenyl is used as a precursor in the synthesis of polychlorinated biphenyls, as well as other compounds used as emulsifiers, insecticides and dyes. It is possible to extract about 50 tons of biphenyl per year from PGO liquid product.

It is expedient to use the residue of the distillation of PGO as a raw material for the production of pitches and carbon fibers.

The objective of this investigation is to develop a method for the separation of naphthalene of a high degree of purity from the liquid products of PGO while reducing energy consumption and material resources.

The problem is solved by the fact that method of obtaining naphthalene includes atmospheric and then vacuum distillation of PGO with separation of naphthalene concentrate, which is subjected to azeotropic distillation, and then sent to the crystallization and pressing steps.

Scheme of PGO Processing with azeotropic rectification to extract high purity naphthalene is shown in Figure 2.

The inventive method of naphthalene extraction from liquid pyrolysis products of hydrocarbon raw materials in industrial conditions can be realized as follows: naphthalene concentrate is released in a known way by atmospheric and vacuum-intelligent distillation, naphthalene concentrate is mixed with ethylene glycol in a mass ratio of 1:1 and sent to the column of azeotropic distillation. In the azeotropic distillation column ethylene glycol-naphthalene azeotrope is separated as a distillate, and the balance of the upper product (azeotrope) is mixed in a mixing tee with wash water to dissolve ethylene glycol. Subsequently, naphthalene is subjected to crystallization and pressing on a vacuum filter press in a known method.

A distinctive feature of the proposed method of extracting naphthalene from the fraction of PGO from the existing ones is the use of an additional stage of azeotropic distillation of naphthalene concentrate, the absence of a stage of polymerization of reactive unsaturated compounds.

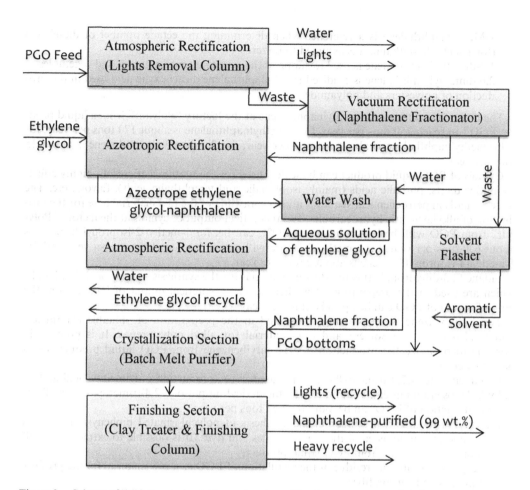

Figure 2. Scheme of PGO Processing with azeotropic rectification to extract naphthalene.

As a result of this method, naphthalene is obtained with a purity of 99% wt. (solidifying point more 79,6°C). The purity degree of naphthalene meets the requirements of GOST 16106 for "Naphthalene-purified", and the naphthalene can be used as a feed for petrochemical synthesis.

Table 3 presents the assessment of the competitiveness of naphthalene production from different manufacturers from CIS countries according to the price factor.

Table 3. Assessment of the competitiveness of naphthalene producers.

Manufacturer name	Qualification (purity, wt.%)	Price per ton, USD
Phenolic plant LLC Inkor and Co.	99	4500
West-Siberian Metallurgical Plant	99	4200
Avdeevsky Coke Chemical Plant	99	3800
Gubakhinsky Coke Chemical Plant	100	4100
Magnitogorsk Metallurgical Plant	100	3700
Nizhny Tagil Metallurgical Plant	99	4500
Novolipetsk Metallurgical Plant	99	4000
Enakievsky Coke Chemical Plant	99	4050

As it can be seen from Table 3, the price for 1 ton ranges from $ 3,800 to $ 4,500.

The State Organization SO «BELISA» has developed a business plan for the project of this processing station with a planning horizon for 5 years, investment costs are about $ 3.1 million net present value is $ 6.9 million, internal rate of return is 74%, dynamic payback period is 2.67 years, product profitability is 28%. These facts prove the rationality of investing in this project.

5 CONCLUSION

In order to increase the profitability of pyrolysis units, it is recommended to organize complex technological schemes for the processing of PGO, including the processes of primary fractionation into narrow fractions and pitches, azeotropic rectification, water washing and crystallization of the resulting distillate in order to obtain pure naphthalene.

REFERENCES

Apicella B., Barbella R., Ciajolo A., Tregrossi, A. 2003. Comparative analysis of the structure of carbon materials relevant in combustion. *Chemosphere.* 51(10): 1063–1069.
Apicella B., Tregrossi A., Popa C., Mennella V., Russo C. 2017. Study on the separation and thin film deposition of tarry aromatics mixtures (soot extract and naphthalene pitch) by high-vacuum heating *Fuel.* 209: 795–801.
Bulauka Y.A., Yakubouski S.F. 2018. Rational refining of heavier cut of pyrolysis gas oil. *Oil and Gas Horizons:* abstract book of *10th International Youth Scientific and Practical Congress, Moscow, November 19–22, 2018,* Gubkin Russian State University of Oil and Gas (National Research University): Moscow: 61.
Gargiulo V., Apicella B., Alfe M., Russo C., Stanzione F., Tregrossi A., et al.2015. Structural characterization of large polycyclic aromatic hydrocarbons. Part 1: the case of coal tar pitch and naphthalene-derived pitch. *Energy Fuels.* 29: 5714–5722.
Gargiulo V., Apicella B., Stanzione F., Tregrossi A., Millan M., Ciajolo A., et al.2016. Structural characterization of large polycyclic aromatic hydrocarbons. Part 2: solvent-separated fractions of coal tar pitch and naphthalene-derived pitch. *Energy Fuels.* 30 (4): 2574–2583.
Gentry J.C. & Zeng, M. 2009. Pygas upgrading for European steam crackers. *PTQ. Q1:* 103–108.
George A., Morgan T.J., Alvarez P., Millan M., Herod A.A., Kandiyoti R. 2010. Fractionation of a coal tar pitch by ultra-filtration, and characterization by size exclusion chromatography, UV-fluorescence and laser desorption-mass spectroscopy. *Fuel.* 89: 2953–2970.
Granda M., Santamaria R., Menendez R. 2003 Coal-tar pitch: composition and pyrolysis behavior. *Chemistry and physics of carbon.* 28: 263–330.
Kershaw J.R., Black K.J.T. 1993. Structural characterization of coal-tar and petroleum pitches. *Energy Fuels.* 7 (3): 420–425.
Nakamura D.N. 2007. Global ethylene capacity increases slightly in 06. *Oil and Gas Journal.* 105 (27): 45–48.
Paliashkevich P.M., Yakubouski S.F., Bulauka, Y.A. 2017. The research on the blend composition of heavy pyrolysis gas oil. *European and national dimension in research. technology: materials of ix junior researchers' conference, Novopolotsk, April 26–27, 2017.* Polotsk state university: Novopolotsk: 212–213.
Petlyuk F.B., Plantonov V.M. 1965. Thermodynamically optimal method for separating multicomponent mixtures. *International Chemical Engineering.* 5: 555–561.
Xu G.Y., Gao Z.X., Zhang J., Bai, J.F. 2012. Preliminary Study of Refined Naphthalene Production with Fractional Crystallization. *Advanced Materials Research.* AMR.554–556.51: 51–55.

Topical Issues of Rational Use of Natural Resources 2019 – Litvinenko (Ed)
© 2020 Taylor & Francis Group, London, ISBN 978-0-367-85720-2

Factors affecting on the extraction of alumina from kaolin ore using lime-sinter process

A.B. ElDeeb
Metallurgy Department, Saint-Petersburg Mining University, Saint-Petersburg, Russia
Mining and petroleum department, Faculty of Engineering, Al-Azhar University in Cairo, Egypt

V.N. Brichkin, R.V. Kurtenkov & I.S. Bormotov
Metallurgy Department, Saint-Petersburg Mining University, Saint-Petersburg, Russia

ABSTRACT: The extraction of alumina from the kaolin ore excavated from the Irkutsk region, Russia by a combination of Pyro- and Hydro-metallurgical Processes was investigated. The kaolin ore was processed using lime-sinter process by sintering kaolin-limestone charge in the temperature range 800 -1400°C for 1hr. Factors affecting the percent recovery of alumina from kaolin ore such as the briquetting pressure and the sintering temperature were investigated. The sintering process leads to the dehydroxylation of the kaolinite and activates its transformation to $12CaO \cdot 7Al_2O_3$ - $CaO \cdot Al_2O_3$ phases from which alumina can by extracted easily using aqueous solutions of sodium carbonate. The highest recovery of alumina of about 80% was obtained at 1360°C sintering temperature and 5 MPa briquetting pressure. The obtained sludge composed mainly from $2CaO \cdot SiO_2$ phase with a suitable chemical and mineralogical composition that makes it suitable to be used as cementation material in the production of the Portland cement.

1 INTRODUCTION

Currently most of the world's commercial alumina is produced by the Bayer process using bauxite as raw material. However, bauxite production is concentrated in a limited number of countries rich in bauxite ores and the bauxite is not readily available to meet the increasing industrial demand for the production of alumina. Some countries which do not possess this mineral have identified alternate local resources for the extraction of alumina. It is therefore necessary to look for the production of this important raw material through other available resources of raw materials that contain high alumina and low iron oxide (Arsentyev et al., 2017, ElDeeb & Brichkin, 2018, ElDeeb et al., 2019).

Huge reserves of low-grade and non-traditional aluminum-containing materials, such as high-silicon bauxites (Sizyakov 2016), kaolin (Suss et al., 2014, Rakhimov et al., 2016), anorthosite (Knudsen et al., 2012) and coal fly ash (Shemi et al., 2015), can serve as an efficient source of raw materials for alumina production.

Clay is one of the numerous aluminous raw materials that are distributed on a large scale in the world. Many of the clays contain as much as 25 - 40% alumina which can serve as a suitable substitute for bauxite ore from which alumina of high purity can be extracted (Al-Ajeel et al., 2014, Gorbachev & Krasnikova, 2015).

One of the promising technologies suitable for the processing of non-bauxitic and low grade aluminum ores that contain large percentage of silicon oxide is the lime sinter method. This method has many advantages that make it more economical and applicable (ElDeeb et al., 2019).

In this work, the extraction of alumina from kaolin ore using the lime sinter process was investigated. Factors affecting the percent recovery of alumina from kaolin ore were studied.

2 MATERIALS AND METHODS

2.1 *Materials*

The Kaolin ore sample was obtained from Irkutsk mining site, while Limestone was obtained from Pikalevo region, Russia. Chemically pure grade Sodium carbonate was used for leaching process.

The as-received kaolin sample was thoroughly mixed then coning and quartering were done to obtain a representative sample for performing chemical and mineralogical characterization. The limestone was grinded on a vibration shaker for 5 minutes.

The representative sample was characterized chemically using X-Ray Fluorescence Spectrometry XRF, and physically using a Bruker D8 Focus X-Ray Diffractometer XRD with Cu Ka radiation and λ = 1.5418 Å (40 kV and 30 mA). The particle size analysis for kaolin ore was performed by laser microanalysis using the domestic Microsizer 201C analyzer, which allows determining the particle size distribution in the range from 0.2 to 600 μm. Differential Thermal Analysis DTA and Thermal Gravimetric Analysis TGA for kaolin ore and kaolin-limestone charge were investigated using a SDT Q600 V20.9 Build 20 instrument in order to clarify the thermal decomposition behavior of kaolin ore and kaolin-limestone charge.

2.2 *Methods*

2.2.1 *The charge preparation*
The kaolin-limestone charge was prepared according to the stoichiometric calculation on the basis of the molar ratios of oxides [$CaO/SiO_2 = 2$, $CaO/Al_2O_3 = 1.8$ and $CaO/Fe_2O_3 = 1$]. This ratio provides the required phases suitable for the leaching process (ElDeeb et al., 2019).

The equations below describe the primary reactions between limestone and kaolin (Al Ajeel et al., 2014). The Kaolin and limestone charge was effectively mixed in a drum mixer for 4hr.

$$2CaCO_3(s) + SiO_2(s) \rightarrow 2CaO \cdot SiO_2(s) + CO_2 \uparrow \qquad (1)$$

$$12CaCO_3(s) + 7Al_2O_3(s) \rightarrow 12CaO \cdot 7Al_2O_3(s) + CO_2 \uparrow \qquad (2)$$

The charge obtained from the mixing process was formed on the form of cylindrical briquettes. Briquetting was carried out on the hydraulic press "Laptuls" using the press mould of 30 mm diameter and 30 mm height. It was noted that the maximum possible pressure, at which the briquette would not break down and can bear the pressure was in the range from 2.5 - 5MPa.

2.2.2 *The sintering process*
The sintering process of the charge was performed in a laboratory high-temperature chamber furnace in closed air conditions in the temperature range from 800°C to 1400°C and for 1hr. In The sintering process, the briquettes was put in crucible then put in the furnace that start to heat at the predetermined temperature with a rate of 10°C/minute. After reaching the desired temperature, the duration of the isothermal holding at the given temperature was 1hr. After finishing the sintering time, the furnace was allowed to cool down to the room temperature in order to allow the sintered material to be annealed and achieve self-disintegration process. The obtained sinter material was characterized chemically using XRF.

2.2.3 *The leaching process*
The obtained sinter was leached using a freshly prepared Na_2CO_3 solution with a concentration of 120 g/l. The leaching process was carried out in the HEL Auto-Mate II reactor system equipped with mechanical stirring sufficiently high to prevent the settling of the pulp. This reactor system is used to carry out the leaching process in a parallel mode.

All leaching experiments were carried out using the following conditions, the at 70°C leaching temperature, 1/5 solid to liquid ratio, 30 min. leaching time using 120 g/l Sodium carbonate solution and 600 rpm stirring speed. During the leaching process the solid alumina in the sintered kaolin in the form of $12CaO \cdot 7Al_2O_3 - CaO \cdot Al_2O_3$ transformed into sodium aluminate ($NaAlO_2$) in the liquid form and passed into the solution according to equations (3, 4).

$$12CaO \cdot 7Al_2O_3 + 12Na_2CO_3 + 5H_2O \rightarrow 14NaAlO_2 + 12CaCO_3 + 10NaOH \qquad (3)$$

$$CaO \cdot Al_2O_3 + Na_2CO_3 \rightarrow 2NaAlO_2 + CaCO_3 \qquad (4)$$

When the leaching time finished the produced pulp was filtered out using vacuum. The pregnant solution was separated and its volume was measured and the sludge was then washed with hot distilled water. The washing solution was separated and its volume was measured. The sludge was separated and weighted then dried in the draying furnace. The obtained sludge's were analyzed chemically by XRF. The alumina extraction degree was determined as the ratio of the Al_2O_3 passed into the solution after leaching compared to its content in the sintered kaolin.

3 RESULTS AND DISCUSSION

3.1 The mineralogical and chemical analysis of the Kaolin sample

The mineralogical analysis of the Kaolin ore using XRD analysis is shown in Figure 1, revealed that it is composed mainly of the kaolinite mineral barring aluminium oxide and quartz mineral.

The chemical analysis of the Kaolin and limestone ores using XRF analysis shown in Table 1. It indicates that it contains 31.9% Al_2O_3, which consider as an economic percentage for the extraction of alumina in comparison with bauxite and Nepheline ores.

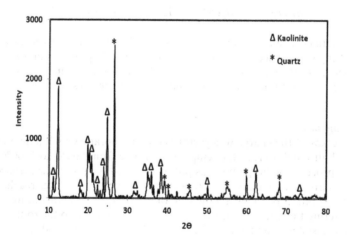

Figure 1. XRD analysis of the kaolin ore.

Table 1. The chemical analysis of the kaolin and limestone ores.

Materials	SiO$_2$	Al$_2$O$_3$	Fe$_2$O$_3$	TiO$_2$	CaO	Na$_2$O	K$_2$O	P$_2$O$_5$	MgO	others	L.O.I
Kaolin ore	52.2	31.9	1.4	0.58	0.59	0.15	0.15	0.06	0.53	0.065	13
Limestone	2.01	0.41	0.56	0.58	53.3	-	-	-	-	-	43.72

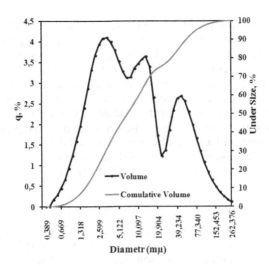

Figure 2. The particle size analysis of the kaolin ore.

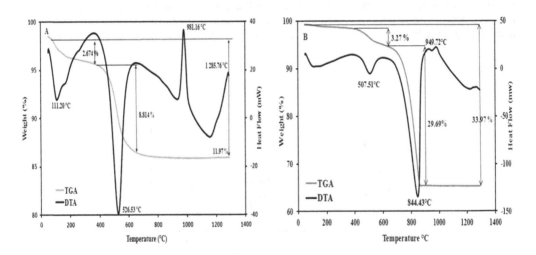

Figure 3. DTA and TGA analysis for (A) the kaolin ore (B) the kaolin-Limestone charge.

The particle size analysis of the kaolin ore is shown in Figure 2. It is shows that about 93.1% of kaolin (-67.523 μm). The last studied indicates that when high percentage of the particle size of the kaolin ore is (-75 μm), it is better for the effective extraction of the alumina from the kaolin ore so it doesn't need further grinding (Yan et al., 2017).

The DTA and TGA analysis of the kaolin ore is shown in Figure 3a and for the kaolin-limestone charge is shown in Figure 3b. In Figure 3a, it is clear that there is a significant weight loss stage within the temperature range (350–600°C) and the corresponding endothermic peak and mass signal peak at 526.53°C were observed, which can be attributed to the removal of structural water in the kaolin ore. The DTA curve of the kaolin sample show two well-defined weight loss regions, the first due to the loss of physisorbed water below 200°C, and the seconded due to the dehydroxylation of coordinated and structural water above 450°C. The sharp exothermic peak at 981.16°C could be assigned to the phase transformation from meta-kaolinite to Al-Si spinel or the mixture of γ-alumina, amorphous silica and mullite (Zhang et al., 2015).

The DTA and TGA analysis for the kaolin-limestone charge indicate that, two obvious weight loss stages in the TGA curve were displayed within the temperature range of 300-850°C as shown in Figure 3b. In the first weight loss stage (300-650°C), one endothermic peaks in the DTA curve were showed at 507.51°C. Therefore, the weight loss between 300 and 650°C can be attributed to the dehydroxylation of kaolinite. In the second weight loss stage (650-850 °C), one endothermic peak was visible in the DTA curve corresponding with one mass signal peak at 844.43°C in the TGA curve that can be attributed to the further release of CO_2 from $CaCO_3$ and the complete dissociation of the limestone. Compared with the decomposition temperature of pure $CaCO_3$ (above 1000°C), the existence of kaolinite obviously decreased the decomposition temperature of $CaCO_3$ in the mixture, which suggested that the solid-state reaction between kaolinite and $CaCO_3$ occurred during the heating process (Zhang et al., 2015, Evtushenko & Sysa, (2006).

3.2 The effect of the briquetting pressure on the extraction of alumina

Briquetting pressure has an effective influence on the extraction of alumina from the sinter prepared at briquetting pressure in the range from 2.5 to 5 MPa. It is cleared from Figure 4 that the percent recovery of alumina in the solution increases with increasing the briquetting pressure. The total recovery of alumina in the solution was 78.72% at 2.5 MPa then increase up to 80.49% at 5 MPa. The briquetting pressure of 5 MPa was chosen as the best briquetting pressure because it has the maximum recovery of the alumina. The increasing of the alumina recovery with increasing the briquetting pressure can be attributed to increasing the cohesion and inter-reaction between kaolin and limestone with increasing the briquetting pressure.

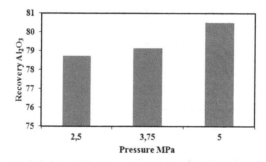

Figure 4.   The effect of the briquetting pressure on the percent recovery of alumina in the solutions.

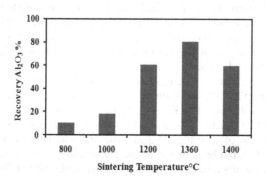

Figure 5.   The effect of the sintering temperature on the percent recovery of alumina in the solutions.

3.3 *The effect of sintering temperature on the extraction of alumina*

The most important factor affecting on the extraction of alumina from kaolin-limestone charge is the sintering temperature. The sintering temperature has a significant effect on the efficiency of the self-disintegration process and the completeness of the phase transformations, hence the extraction of alumina from kaolin-limestone. The effect of the sintering temperatures on the extraction of alumina was investigated in the range from 800 to 1400°C as shown in Figure 5. At temperatures 800, 1000 and 1200°C, it was noted that there weren't any self-disintegration of the briquettes occurred. At 800°C there were no changes in the briquettes, but at 1000 and 1200°C there were only changes in the cooler of the briquettes and small cracks occurred on their surfaces. It was noted that the self-disintegration process of the briquettes started from 1300-1400°C and there was a clear transformation from briquettes into a fine powder.

The sintering temperature is the most important factor affecting on the Physico-chemical processing of kaolin ore for the extraction of alumina because it is responsible for converting the kaolin-limestone charge mainly into C_2S and $C_{12}A_7$ compounds. Alumina can be extracted from $C_{12}A_7$ by leaching with sodium carbonate solution. The dicalcium silicate C_2S plays extremely important roles. First, its development is complete enough to tie up nearly all of the silica present in a non soluble form and separate them in the sludge. Second, it undergoes a crystallographic transformation on cooling with an increase in the molar volume due to the transformation of $2CaO \cdot SiO_2$ from β to γ form as shown in equation 5, which results in the transformation of the sintered briquettes into powder, that increases the solubility of the alumina containing compounds without any need for further grinding (Stange et al., 2017).

$$\beta - Ca_2SiO_4 \overset{675°C}{\rightarrow} \gamma - Ca_2SiO_4 \tag{5}$$

As shown in Figure 5, the recovery of alumina increases with increasing the sintering temperature. The percentage recovery of alumina increases from 10.51% at 800°C to 80.49 at 1360°C then decrease to 59.70% at 1400°C. This can be attributed to the formation of C_2S and $C_{12}A_7$ compounds in the sinter at 1360°C. C_2S tie up nearly all the silica present in the charge in the form of nonsoluble compounds and alumina can be extracted easily from $C_{12}A_7$ compound by leaching solution.

4 CONCLUSION

The obtained results indicate the affectivity of the lime sinter process for the economic extraction of alumina from kaolin ore and the kaolin ore can be used as a good alternative for Bauxite in the production of alumina.

The optimum conditions for sintering the kaolin-limestone charge to produce efficient self-disintegration and at the same time obtaining the highest recovery of alumina are; 1360 °C Sintering temperature and 5 MPa briquetting pressure. The highest recovery of alumina about 80% was obtained using 120 g/l Na_2CO_3 solution at 70□C for 30 minutes.

One of the most important characteristics of lime-sinter process for processing the kaolin ore is the self-disintegration process making it more economical compared with those used in the processing of Bauxite and Nepheline ores.

ACKNOWLEDGMENT

The study was carried out with the financial support of the Ministry of Education and Science of the Russian Federation (registration number of the project 11.4098.2017/ПЧ of 01.01.2017)

REFERENCES

Al-Ajeel, A.A. & Abdullah, S.Z. & Muslim, W.A. & Abdulkhader, M.Q. & Al-Halbosy, M.K. & Al-Jumely, F.A. 2014. Extraction of Alumina from Iraqi colored kaolin by lime-sinter process. Iraqi Bull. Geol. Min. 10 (3): 109–117.

Arsentyev V.A. & Gerasimov A.M. & Mezenin A.O. 2017. Kaolines beneficiation technology study with application of hydrothermal modification. Obogashchenie Rud 2: 3-9.

ElDeeb, A.B.S. & Brichkin, V.N. 2018. Egyptian aluminum containing ores and prospects for their use in the production of Aluminum. Int. J. Sci. Eng. Res. 9 (5): 721–731.

ElDeeb, A.B. & Brichkin, V.N. & Kurtenkov, R.V. & Bormotov, I.S. 2019. Extraction of alumina from kaolin by a combination of Pyro- and hydrometallurgical Processes. Applied Clay Science 172: 146-154.

Evtushenko E.I. & Sysa O.K. 2006. Structural modification of the raw clay under hydrothermal conditions. Izvestiya Vuzov. Severo-Kavkazskiy Region. Tekhnicheskiye Nauki 2: 82–86.

Knudsen, C. & Wanvik, J. & Svahnberg, H. 2012. Anorthosites in Greenland: A possible raw material for aluminium? Geol. Surv. Denmark Greenland Bull 26: 53–56.

Rahier, H. & Wullaert, B. & Van, M.B. 2000. Influence of the degree of dehydroxylation of kaolinite on the properties of aluminosilicate glasses. J. Therm. Anal. Calorim. 62: 417-427.

Rakhimov R. Kh. & Rashidov Kh. K. & Yermakov V. P. & Rashidov J. Kh. & Allabergenov R. J. 2016. Resource saving and energy efficient technology of the production alumina from secondary kaolin with Angren deposit. Computational Nanotechnology 1: 45–51.

Shemi, A. & Ndlovu, S. & Sibanda, V. & Van Dyk, L.D. 2015. Extraction of alumina from coal fly ash using an acid leach-sinter-acid leaches technique. Hydrometallurgy 157: 348-355.

Sizyakov V.M., 2016. Chemical and technological mechanisms of a alkaline aluminium silicates sintering and a hydrochemical sinter processing. Proceedings of the Mining institute. Scientific Journal 217: 102-112.

Stange, K. & Lenting, C. & Geisler, T. 2017. Insights into the evolution of carbonate-bearing kaolin during sintering revealed by in situ hyperspectral Raman imaging. Journal of the American Ceramic Society:1-14.

Suss, A.G. & Damaskin, A.A. & Senyuta, A.S. & Panov, A.V. & Smirnov, A.A. 2014. The influence of the mineral composition of low-grade aluminum ores on aluminium extraction by acid leaching. In: Grandfield, J. (Ed.), Light Metals. TMS: 105–109.

Yan, K. & Guo, Y. & Fang, Li. & Cui, Li. & Cheng, F. & Li, T. 2017. Decomposition and phase transformation mechanism of kaolinite calcined with sodium carbonate. Applied Clay Science 147: 90-96.

Zhang, S. & Ou, X. & Qiang, Y. & Niu, J. & Komarneni, S. 2015. Thermal decomposition behavior and de-intercalation mechanism of acetamide intercalated into kaolinite by thermoanalytical techniques. Appl. Clay Science 114 (10): 309-314.

Gorbachev B.F. & Krasnikova E.V. 2015. State and possible ways of development of raw material base of kaolins, refractory and high-melting clays in the Russian Federation. Stroitel'nye Materialy 4: 6–17.

Topical Issues of Rational Use of Natural Resources 2019 – Litvinenko (Ed)
© 2020 Taylor & Francis Group, London, ISBN 978-0-367-85720-2

Improving efficiency of gravity separation of fine iron ore materials using computer modelling

A.V. Fomin & M.S. Khokhulya
Mining Institute of Kola Science Center of Russian Academy of Sciences, Apatity, Russia

ABSTRACT: The paper presents the results of the research aimed at improvement of efficiency of gravity separation of fine fractions of iron ore materials produced at the mineral processing plant of Olkon JSC. Using computer modelling it was shown that spiral separators provide more effective processing of fine fractions of tailings of primary magnetic separation compared to jigging. The modelling results were confirmed with industrial spiral separation tests. A separation technology for processing the tailings of primary magnetic separation was developed.

1 INTRODUCTION

At present, the most part of the richest ores, which can be easily processed, has been depleted. So, a problem of separation of fine-disseminated ores with complex material composition has arisen. The problem requires the research and development of technologies and devices including gravity separation for processing the fine fractions of materials (Bogdanovich & Fedotov, 2007).

The mineral processing plant of Olkon JSC, located at the Murmansk region, produces magnetite-hematite concentrate. The processing technology includes several magnetic separation stages to produce magnetite concentrate and two jigging stages to produce hematite concentrate from a non-magnetic fraction (t ailings) of primary magnetic separation. At present time the jigging separation does not provide for acceptable content and recovery of hematite to a concentrate in processing hematite ores (Khokhulya et al., 2013). Table 1 presents values of yield, content, and recovery of hematite iron in products of jigging which were achieved during sampling of this operation at the processing plant.

As can be seen from Table 1, the recovery of hematite iron in tailings is about 50% and this fact means that about half of hematite is lost in this operation. Most part of hematite losses with tailings is particles with diameter less than 0.2 mm, as far as jigging is effective in separating relatively coarse particles (Mukherjee et al., 2006). This problem requires development of a new technology for effective separation of fine fractions.

Today one of the new research directions of separation processes is computer modelling based on Computational Fluid Dynamics (CFD) methods (Skorokhodov et al., 2013). The computer modelling allows a researcher to investigate hydrodynamics of the water flow and predict separation characteristics such as yield, content and recovery of valuable mineral (Khokhulya & Fomin 2017). In this study CFD-modelling was applied to determine the efficiency of using the spiral separators instead of jigging.

2 RESEARCH OBJECTS AND METHODS

A sample of non-magnetic fraction of primary magnetic separation, which is the feed of jigging, was collected at the mineral processing plant. The grain sizes, mineral and chemical compositions were analyzed to set correctly parameters of computer modelling. According to the results of mineral analysis, the sample has a following mineral composition: hematite - 10%, magnetite - 1%, quartz and feldspar - 73-75%, amphiboles, mica - 10-12%, others (garnet,

Table 1. Separation indicators of jigging at the processing plant.

Products	Yield, %	Content of hematite iron, %	Recovery of hematite iron, %
Concentrate	4.7	42.5	37.2
Middlings	2.9	24.0	12.9
Tailings	92.4	2.9	49.9
Total:	100.0	5.4	100.0

Table 2. The results of grain size analysis and content of different forms of iron in a sample.

Grain size classes, mm	Yield, %	Content of form of iron, %			Distribution of form of iron, %		
		Total	Magnetic	Hematite	Total	Magnetic	Hematite
+2.5	0.5	5.1	1.6	1.7	0.3	1.0	0.1
-2.5+1.6	0.5	4.1	0.7	1.5	0.2	0.4	0.1
-1.6+1.0	1.1	4.3	0.7	1.8	0.5	1.0	0.3
-1.0+0.6	3.7	5.7	1.0	3.6	2.2	4.7	1.8
-0.6+0.4	9.8	8.3	1.0	6.4	8.5	12.3	8.3
-0.4+0.3	15.8	9.2	1.0	7.2	15.2	19.9	15.0
-0.3+0.2	23.2	8.8	0.7	7.0	21.4	20.4	21.4
-0.2+0.1	29.6	10.2	0.6	8.4	31.6	22.4	32.7
-0.1+0.071	7.9	10.7	0.6	7.9	8.8	6.0	8.2
-0.071	7.9	13.7	1.2	11.6	11.3	11.9	12.1
Total:	100.0	9.6	0.8	7.6	100.0	100.0	100.0

epidote, calcite, single grains of sulphides) - 3-4 %. The results of grain size analysis and content of different forms of iron are presented in Table 2.

According to data above, 53% of hematite iron in the sample is distributed in size class -0.2 mm which is not processed effectively using the existing jigging machines.

CFD-modelling was conducted using ANSYS software to estimate the efficiency of spiral separation for processing the primary magnetic separation tailings. Two spiral concentrators (VSR-500 and SHV-500) were selected as the modelling objects, both with diameter 500 mm. The geometry of separators is presented in Figure 1.

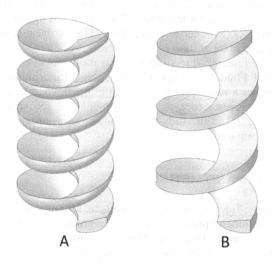

A   B

Figure 1. 3D geometry of spiral separators (A - VSR-500, B - SHV-500).

The difference between separators is quantity of turns, distance between turns and surface profile geometry. VSR-500 has a profile in form of a highly inclined curve and SHV-500 has a profile in form of a lowly inclined curve.

The model development process consists of following stages: creation of 3D geometry of an object, generation of a mesh, setting simulation parameters and boundary conditions, calculation, analysis of results and adequacy of a model.

The computational meshes consisting of approximately 1,000,000 tetrahedral cells were generated. The cell size is 8 mm, but in the suspension movement zone, where greater accuracy of calculations is required, the cell size is less than 1 mm.

The model parameters and boundary conditions were set up by ANSYS Fluent software. To model the continuous phase movement, a Volume Of Fluid (VOF) model (Hirt & Nichols, 1981) was used; the turbulence of the flow was simulated using a k-ε model (Menter, 1994), and the trajectories of mineral particles were calculated with a Discrete Phase Model (DPM) (Cundall & Strack, 1979). The mineral particles flowrate (250 kg/h) and mass content of solid in the feed (30%) were set as a boundary condition for inlet. Granular and mineral characteristics of the feed were set according to the analyses results presented above.

3 COMPUTER MODELLING RESULTS

Calculation of the models designed in ANSYS Fluent provided for the following results of the simulations: velocity of suspension, volume concentration of phases, trajectories of movement and other.

Figure 2 presents distribution of water velocity magnitude at spiral surface of the separators. The maximum velocity is observed at the outer side of the spiral surface and the

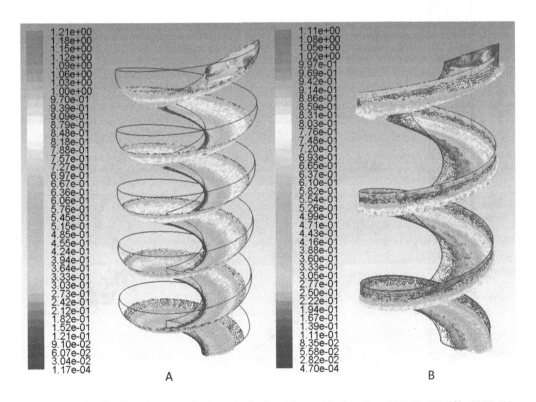

Figure 2. Distribution of water velocity magnitude, m/s, at spiral surface of VSR-500 (A), SHV-500 (B).

511

minimum velocity is observed at its inner side. Such velocity distribution is consistent with the studies of water flow characteristics in the spiral separators conducted by other authors (Ivanov and Prokop'ev, 2000).

Figure 3 presents mineral particle distribution in a computational domain colored by their density; hematite particles are colored with red. Figure 3 demonstrates the distribution of particles which is typical for spiral separation (Sadeghi et al., 2016). The hematite particles have a tendency to be concentrated at the inner side of the spiral surface and be recovered in the concentrate. Most part of the particles with intermediate density (colored by light blue) is recovered to the middlings. Quartz and feldspar particles colored by blue pass mostly to the tailings and less to the middlings.

When developing a mineral processing technology, it is important to predict the separation indicators such as the yield, the content and the recovery. The combination of a DPM model and macros developed in C++ allow getting these characteristics (Table 3).

The simulation results showed the advantage of spiral separators against jigging (Table 1) due to more efficient separation of fine fractions. SHV-500 provided for a higher yield, content and recovery of hematite in a concentrate and VSR-500 provided a richer middling product. The recovery of the valuable mineral in the tailings was 9-16%.

In order to verify the adequacy of the developed models, laboratory experiments were conducted using these two spiral separators. The conditions of the experiments coincided with the conditions of modelling. The adequacy of the modelling was estimated on the basis of the calculation of an absolute simulation error considered as a module of the difference in the yields

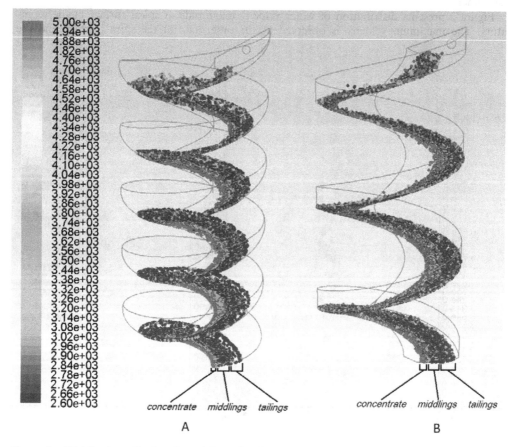

Figure 3. Distribution of mineral particles colored by density, kg/m3, in models of VSR-500 (A), SHV-500 (B).

Table 3. Predicted indicators of spiral separation of primary magnetic separation tailings.

Products	Yield, %	Hematite content, %	Hematite recovery, %	Yield, %	Hematite content, %	Hematite recovery, %
	VSR-500			SHV-500		
Concentrate	6.5	78.5	51.2	8.7	81.9	71.5
Middlings	42.0	7.8	32.8	42.4	4.5	19.2
Tailings	51.5	3.1	16.0	48.9	1.9	9.3
Total:	100.0	10.0	100.0	100.0	10.0	100.0

of products and the contents of hematite obtained during modelling and laboratory experiments, respectively. The absolute simulation error for the concentrate yield did not exceed 1%; for the middlings and tailings yields it was less than 10%. The error of the hematite iron content in the concentrate was less than 8%, in the middlings - less than 2% and in the tailings - less than 1%. These facts verify the adequacy of modeling under given operating conditions.

Thus, on the basis of modelling results, the spiral separators can be recommended for processing the tailings of primary magnetic separation.

4 INDUSTRIAL TESTS RESULTS

In order to confirm the obtained results, the industrial tests of spiral separation were carried out at the processing plant of Olkon JSC. The spiral separator SHV-500 was used in tests as far as the results of modelling and laboratory experiments showed its higher efficiency. Table 4 presents the results of industrial tests.

During the tests the plant processed the ore from the Olenegorsky open pit. According to the technology, the hematite is recovered from these ores only. The results of tests 1-3 were received as a result of processing of the ore having abnormally high content of hematite. These results are not representative for the plant operation during long time. Nevertheless, the results of these three tests showed high efficiency of spiral separation.

The usual content of hematite iron in the tailings of primary magnetic separation is less than 10% (tests 4-7). Comparison of results of these tests with results of jigging (Table 1) showed that spiral separation provides more effective processing of primary magnetic separation tailings and decreases losses of fine hematite fractions with tailings.

Most part of hematite in spiral separation tailings is hematite particles with diameter less than 0.045 mm which are not processed effectively by gravity separation. Also this fact explains relatively high recovery of hematite in tailings during test 7 (29.5%). During test 7 content of particle size fraction -0.045 mm in the feed increased to 20%, while during other tests its content was less than 10%.

Thus, results of industrial tests proved the results of computer modelling and confirmed the feasibility of replacing the jigging machines by the spiral separators.

5 DEVELOPMENT OF A SEPARATION TECHNOLOGY

As presented in Table 4, spiral separation provides for the production of the rough hematite concentrate with different content of hematite iron (from 27.7% to 58.2%) depending on content of the valuable mineral in the feed. Further separation of produced products (rough concentrate and middlings) was conducted with the use of two operations of spiral separation and shaking table concentration. The middlings of the second stage of the spiral separation were classified by grain size of 0.2 mm. The oversize product was milled and the undersize product was sent to the third stage of spiral separation.

Table 4. The results of industrial tests of spiral separation.

Products	Yield, %	Content of hematite iron, %	Recovery of hematite iron, %
Test №1			
Concentrate	25.6	58.2	80.6
Middlings	33.3	6.7	12.1
Tailings	41.1	3.3	7.3
Total:	100.0	18.5	100.0
Test №2			
Concentrate	34.0	56.7	91.0
Middlings	37.3	3.2	5.6
Tailings	28.7	2.5	3.4
Total:	100.0	21.2	100.0
Test №3			
Concentrate	26.5	58.2	78.9
Middlings	36.2	7.5	13.9
Tailings	37.3	3.8	7.2
Total:	100.0	19.6	100.0
Test №4			
Concentrate	13.3	52.0	85.3
Middlings	40.0	2.4	11.8
Tailings	46.7	0.5	2.9
Total:	100.0	8.1	100.0
Test №5			
Concentrate	8.4	31.3	78.9
Middlings	51.3	0.9	13.8
Tailings	40.3	0.6	7.3
Total:	100.0	3.3	100.0
Test №6			
Concentrate	6.1	36.1	65.5
Middlings	29.6	2.4	21.1
Tailings	64.3	0.7	13.4
Total:	100.0	3.4	100.0
Test №7			
Concentrate	3.9	27.7	48.1
Middlings	35.9	1.4	22.4
Tailings	60.2	1.1	29.5
Total:	100.0	2.2	100.0

The flowsheet of the developed technology to produce the hematite concentrate from tailings of primary magnetic separation is presented in Figure 4.

While developing the technology, such problems raised as high variation of the hematite iron content in the feed and necessity of separation of the material with low content of hematite iron (down to 2%). At the first stage of the spiral separation the quality of the concentrate and middlings is widely varying (Table 4) and requires different amount of further recleaning operations. To solve this problem, the preliminary concentration operation is used at the first stage of the spiral separation. At this stage only two products are produced: tailings and combined concentrate and middlings. This decision allows separation of both the rich and poor feed through stabilization of the product quality sent to the second stage of the spiral separation where three products are produced (concentrate, middlings and tailings). The rough concentrate is processed using the shaking table. The main advantage of this equipment is that shaking tables are less sensitive to the feed quality and can produce hematite concentrate with required content of valuable mineral even from the feed containing 30-40% of hematite iron which is important when processing ores with low content of hematite. As an alternative to shaking tables, the spiral separation and high intensity magnetic separation were considered.

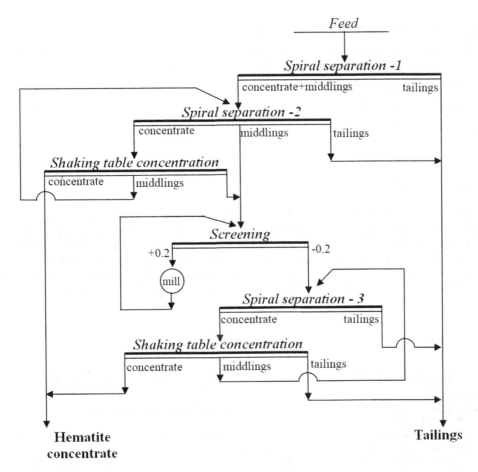

Figure 4.   Flowsheet of the separation technology for primary magnetic separation tailings.

Both of them showed good results, but the use of the shaking table gave the highest values of hematite content and recovery.

The middlings produced at the second stage of the spiral separation could not be effectively separated without milling. More than a half of hematite particles in this product were aggregates with waste rock minerals. The results of mineralogical analyses of middling products showed that more that 90% of hematite was liberated in size class -0.2 mm, so a screening operation was used to separate the liberated hematite particles. The particles with diameter more than 0.2 mm were sent to milling. The undersize screen product was processed through the spiral separators and shaking tables designed for fine material separation.

The characteristics of separation of primary magnetic separation tailings with the use of the developed technology are presented in Table 5.

The primary magnetic separation tailings containing 5.2% of hematite iron and 8.2% of total iron were selected as the feed to make sure that the developed technology will provide for the necessary content and recovery of hematite to a concentrate when processing ores with low content of hematite. The final concentrate containing more than 62% of total iron and more than 54% of hematite iron was produced. The final tailings contained 4% of total iron and 1.4% of hematite iron. The parameters of the total iron content in these products were set by requirements of Olkon JSC. The through recovery of hematite to the final concentrate of about 75% was achieved.

515

Table 5. Results of separation of tailings of primary magnetic separation according to developed technology.

| Products | Yield, % | Content of form of iron | | Recovery of form of iron | |
		Total,%	Hematite,%	Total,%	Hematite,%
Concentrate	7.2	62.2	54.5	54.6	75.4
Tailings	92.8	4.0	1.4	45.4	24.6
Feed	100.0	8.2	5.2	100.0	100.0

6 CONCLUSION

Results of computer modelling of spiral separation of primary magnetic separation tailings showed that spiral separators provide more effective separation of fine fractions of the material in comparison with jigging machines used at the mineral processing plant. The modelling results were confirmed by the industrial tests. A separation technology of primary magnetic separation tailings was developed. The technology provides for the production of the hematite concentrate with the content of total and hematite iron more than 62% and 54% respectively and recovers more than 75% of hematite.

REFERENCES

Bogdanovich A.V. & Fedotov K.V. 2007. The main trends in the development of equipment and technology of gravity separation of sand and finely disseminated ores. *Mining Journal*. №2. P. 51-57. In Russian.

Cundall P.A. & Strack D.L. 1979. A Discrete Numerical Model for Granular Assemblies. *Geotechnique*. V.29. № 1.P. 47-65.

Hirt C.W. & Nichols B.D. 1981. Volume of fluid (VOF) method for the dynamics of free boundaries. *Journal of Computational Physics*. V.39. № 1. P.201-225.

Ivanov V.D. & Prokop'ev S.A. 2000. *Spiral separators for ores and sands processing in Russia*. Moscow: "Daksi". In Russian.

Khokhulya M.S., et al. 2013. Intensification of the process of segregation separation of fine fractions of ore minerals by hydraulic separation. *Innovative processes of complex and deep processing of mineral materials: proceedings of International meeting "Plaksinsky readings-2013"*. P. 246-248. In Russian.

Khokhulya M.S. & Fomin A.V. 2017. The use of computational fluid dynamics in the processes of gravity separation of various types of mineral materials. *Mining informational and analytical bulletin (scientific and technical journal)*. № S23. P. 474–482.In Russian.

Menter F.R. 1994. Two-equation eddy-viscosity turbulence models for engineering applications. *AIAA Journal*. V. 32. № 8. P.1598-1605.

Mukherjee A.K., et al. Role of water velocity for efficient jigging of iron ore. *Minerals engineering*. V.19 № 9 P. 952-959.

Sadeghi M. et al. Radial distribution of iron oxide and silica particles in the reject flow of a spiral concentrator. *International journal of mineral processing*. V.153. P. 51-59.

Skorokhodov V.F. et al. Computational fluid dynamics methods in research and analysis of mineral separation. *Journal of Mining Science*. V. 49. № 3. P. 507-513. In Russian.

Topical Issues of Rational Use of Natural Resources 2019 – Litvinenko (Ed)
© 2020 Taylor & Francis Group, London, ISBN 978-0-367-85720-2

Revisiting the initialization of the kinetic models for flotation of carbonaceous raw materials

T.D. Kalmykova & A.O. Romashev
Mining university, Saint-Petersburg, Russia

ABSTRACT: Investigation of flotation kinetics is one of the important tasks in the theory and practice of modern beneficiation. At the moment, there is no universal model that satisfactorily describes the process, due to the presence of many influencing factors.

The experiments of flotation of carbonaceous raw materials were carried out to initialize the model type, the data on the extraction of the ash product were subjected to statistical processing. The coefficient of determination was chosen as a criterion for selecting the optimal function.

According to the results of the studies, the use of an integral method of data processing has increased the accuracy of forecasting. The best results were achieved using specialized packages that perform regression processing. The modified Kelsall model showed the highest accuracy of the considered models. From a practical point of view, the most convenient method is the integral one.

1 INTRODUCTION

The study of flotation kinetics is one of important tasks in the theory and practice of modern mineral processing. At the moment, the universal model that satisfactorily describes the process is not suggested because of many influencing factors.

To present, there is no unified theory on how the flotation constant is calculated and which parameters have a significant impact on the process. In terms of Metrology and standardization no direct feature of the flotation of particles with the standardized measurement technology has been proposed.

The most commonly used model for predicting kinetic data is the approximation of experimental points by various functions. In practice, the exponential equation of the first order (the model of K. F. Beloglazov) is widely used.

The advantages of this model are the ease of finding the kinetic constant by linearization of the obtained dependencies, but, there are almost always significant differences in practical data, which makes this model only suitable for rough estimation. (Beloglazov, 1938)

The dependence of distribution of rate constants from distribution of particle sizes may cause the problem of the accurate estimation of the distribution of rate constants. Many researchers have proposed more accurate models to describe the kinetics of flotation of various minerals.

The most widespread and practical use found models: Klimpel, Kelsall, Gamma and Fully mixed model, as well as their modifications.

The representation of the rate constant distribution is the difference between the classical model and the Klimpel model. In contrast to the classical model, Klimpel model constant has a fixed or constant value over a limited property range. For low and high values of the property, the rate constant is zero. (Klimpel et al., 1980)

Kelsall (Kelsall, 1961) supposed that the use of two rate constants, describing a fast floating component and a slow floating component, gives a better approximation to the distribution of particle floatabilities than single rate constant.

The modified Kelsall model was proposed by Jowett in 1983. This model describes non linear rate data by the sum of two straight lines.

The modified gamma function proposed by Loveday (Loveday, 1966) and Inoue and Imaizumi (Imaizumi, 1963) can be simplistically described as being made up of the sum of P exponential distributions.

Fully mixed model describes the first order time-recovery of a component from a monodisperse feed with an exponential distribution of floatabilities.

Coefficients included in these models, were explained by the authors of these equations as some physical parameters, depending on the speed or other parameters of the process of flotation separation. Meanwhile practical determination of these parameters can cause some difficulties. In the case of a single coefficient this problem can be reduced to solving a system of equations with one unknown parameter. However, in more complicated models, such a solution may not exist and the use of more complex mathematical apparatus requires to find the most suitable values of the coefficients. Generally, such problems are solved by finding the extremum (minimum) of the objective function:

$$\Delta = y_t - y_e \rightarrow \min$$

where y_t – extraction at time t by researched model at the current value of its coefficients; y_e – experimental values obtained at flotation time t.

In mathematical analysis, this optimization task is called curve-fitting. Steepest descent method (SDM) and Newton's method (NM) are developed and well known for it for long. Unhappily, these methods have a number of significant disadvantages, especially — steepest descent method can converge very long at the end of optimization, and Newton's method requires calculation of the second derivatives, which requires a lot of calculations.

Also often you often have to deal with the least square method (LSM) when finding unknown coefficients. This method minimizes the sum of the error square, l.e. the objective function is represented as:

Table 1. The mostly used models.

Model	Equation	Equation parameters
Classical	$R = R_\infty(1 - e^{-kt})$	k - flotation rate constant [1/min]
Klimpell	$R = R_\infty\left(1 - \frac{1}{kt}(1 - e^{-kt})\right)$	k - flotation rate constant [1/min]
Kelsall	$R = (1 - \varphi)(1 - e^{-k_f t}) + \varphi(1 - e^{-k_s t})$	φ - fraction of flotation components with the slow rate constant k_f - fast flotation rate constant [1/min] k_s - slow flotation rate constant [1/min]
Modified Kelsall	$R = R_\infty((1 - \varphi)(1 - e^{-k_f t}) + \varphi(1 - e^{-k_s t}))$	φ - fraction of flotation components with the slow rate constant k_f - fast flotation rate constant [1/min] k_s - slow flotation rate constant [1/min]
Gamma	$R = R_\infty\left[1 - \left(\frac{\lambda}{\lambda + t}\right)^P\right]$	λ - kinetic constant [min] P - exponent
Fully mixed	$R = R_\infty\left[1 - \left(\frac{1}{1 + \frac{t}{k}}\right)\right]$	k - flotation rate constant [1/min]

Note: in all equations R - recovery of mineral (dependent variable); R_∞ - infinite (equilibrium) recovery - maximum possible recovery of mineral; t - flotation time [min]

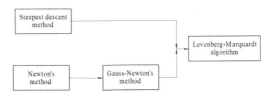

Figure 1. Methods and algorithms.

$$\frac{1}{2}\sum_{i=1}^{N}[y_T - y_э]^2 = \frac{1}{2}\sum_{i=1}^{N}\Delta^2 \rightarrow min$$

The Levenberg-Marquardt algorithm is a nonlinear least-squares method and it based on the methods given in the flowchart below (Figure 1). This method is more stable than the Gauss-Newton's method and many times more effective.

This algorithm can be implemented in most modern programming languages. In this paper, the calculation was carried out in MatLab, as one of the most convenient tool for developing the mathematical model. (Kovalchuk & Baburin, 2018, Kovalchuk & Poddubniy, 2018)

These models formed the basis of specialized computer packages: HCS Chemistry, JKSim-Float MODSIM™, USIM PAC™, etc. Among alternative approaches, the integral method of experimental data processing shows good convergence results, as well as the selection of the type of regression dependence using specialized programs (CurveExpert, DataFit, Minitab, etc), which contain a database of models and allow for fast automatic processing of experimental data and make statistical check of the received models. (Amelunxen & Runge, 2014)

2 MATERIALS AND METHODS

Two representative samples of carbonaceous raw materials were selected as the object of study. The choice of them as an object of research is due to the importance of them as a source of strategic metals such as noble, rare, scattered and non-ferrous metals.

Figure 2. Laboratory flotation machine.

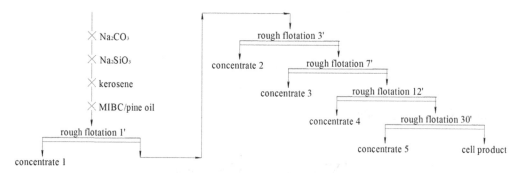

Figure 3. Flotation circuit.

According to the researchers, this type of material is one of the most promising from an economic point of view and in the near future will play a key role. (Cherepovitsyn et al. 2017, Smirnova et al. 2017)

The experiments were carried out on a laboratory flotation machine Laarmann.

As flotation reagents were used: Na_2CO_3, Na_2SiO_3, kerosene, MIBC. Time intervals of control selections were 1, 3, 7, 12 and 30 minutes from the beginning of the process.

3 RESULTS AND DISCUSSION

According to the form of the obtained kinetic dependences, it is difficult to draw a conclusion about the form of the kinetic dependence. To initialize the model view, the obtained data were sequentially determined constants for the K.F. Beloglazov equation, constants of models at an integral method of processing, and also specialized software for the selection of regression dependence was used. As a criterion for the selection of the optimal features have been selected coefficient of determination – R^2.

Using this parameter as a selection criterion is allowed due to the presence of only one predictor – flotation time t in the considered functions.

Experimental and simulated data are presented in Table 2 and Figures 4, 5, 6.

It should be noted that R2=0.809 was achieved by minimizing the squared deviations using the simplex-method of finding solutions. Determination of the coefficient k as the slope tangent of the linearized dependence led to a decrease in the accuracy and deterioration of statistical indicators, this indicates the deviation of the experimental dependence on the shape of the curve by the proposed model.

The use of the integral data processing method allowed to increase the prediction accuracy in comparison with the "classical" model. The increase in the order of the kinetic equation leads to an increase in the coefficient of determination by 0.112 while reducing the standard error to 0.01, but as you can see from Figure 1 as the flotation time increases, the difference between the experimental data and the regression curve increases, and the error exceeds the level of 5% (time t=1800 seconds), which can lead to prediction errors and underestimation of the valuable component extraction.

The best results were achieved by using specialized packages that perform regression processing. In the analysis of data from enumeration polynomials of high degrees (more than 2) were excluded, since despite the high coefficient of determination of such models, the curve has a "sinusoidal" appearance and cannot be used for forecasting. More than 67 models were analyzed to select the most suitable equation. The models were selected by the value of the determination coefficient and the standard error. At close values of indicators, statistical models were compared according to the Akaike information criterion. The Hoerl model showed the highest accuracy of the considered models:

Table 2. Results of experiments and calculations on kinetic models.

| № | t, s | ε, % | Kinetic models | | Integral method: order | | | |
			Classical model	Hoerl curve	0,5	1	1,5	2
1	60	13,58	4,32	14,07	19,48	15,26	13,74	13,48
2	180	28,08	12,40	27,25	26,30	25,88	27,39	29,39
3	420	42,47	26,58	43,10	36,24	38,88	42,20	44,35
4	720	55,07	41,13	54,85	46,49	50,38	53,04	52,74
5	1800	67,90	73,40	67,91	77,13	78,25	72,37	62,71
R^2			0,809	0,998	0,886	0,965	0,997	0,998
Standard error			10,842	0,08	0,7	0,14	0,03	0,01

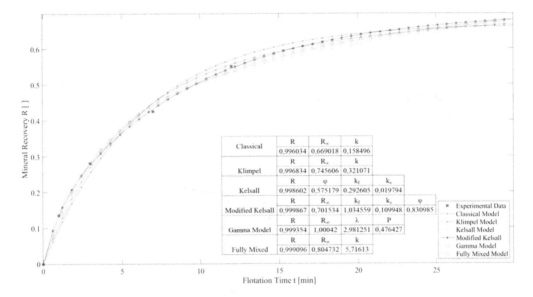

Figure 4. Results obtained using the Levenberg-Marquardt algorithm.

$$\varepsilon = a \cdot b^t \cdot t^c$$

The coefficients of the model found: a=1,05; b=0,9996; c=0,6389. The coefficient of determination is 0.99; the standard error is 0.833. The shape of the curve has no pronounced kinks and kinks between the points and corresponds to the physical nature of the real process.

The curves approached the real experimental data due to using the Levenberg-Marquardt algorithm.

The best results were achieved with the use of modified Kelsall model. The coefficient for this model was chosen by using the Levenberg-Marquardt algorithm. This algorithm has significant advantages before the "classical" optimization methods: the method of the steepest descent and Newton's method. In this research finding the coefficients of the model with using this algorithm was carried out by using the developed algorithm in MathLab program.

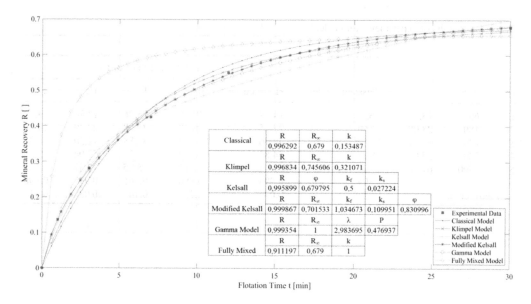

Figure 5. Results obtained using the least square method.

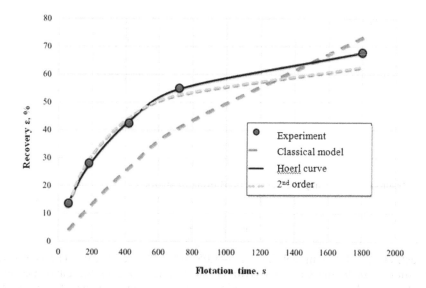

Figure 6. The results of the experiments and simulated dependences.

4 CONCLUSIONS

1. The flotation experiments were conducted and kinetic dependences were plotted. It is difficult to draw a conclusion about the form of the kinetic dependence according to the form of the obtained kinetic dependences

2. R2 is about 0.8 was achieved by minimizing the squared deviations using the simplex-method of finding solutions.

3. A decrease in the accuracy and deterioration of statistical indicators because of determination of the coefficient k as the slope tangent of the linearized dependence in the classical model.

4. The use of an integral method of data processing allows to achieve fairly accurate results ($R2$ is about 0,9). This method is most easily applicable because it can be used on free software.

5. The increase in the order of the kinetic equation leads to an increase in the coefficient of determination, but as the flotation time increases, the difference between the experimental data and the regression curve increases, which can lead to prediction errors and underestimation of the valuable component extraction.

6. The highest accuracy using automated regression processing showed the Hoerl model. The shape of the curve has no pronounced kinks and kinks between the points and corresponds to the physical nature of the real process.

7. The best results were achieved with the use of modified Kelsall model. The coefficient for this model was chosen by using the Levenberg-Marquardt algorithm.

8. As a result of the study was developed the algorithm of calculation in MathLab which allows to find the most common kinetic model. Among the advantages of this approach can be noted its high accuracy in the definition of the constants and the ability to add new models.

ACKNOWLEDGEMENT

The work is executed at financial support of Russian Scientific Foundation (project No. 19-17-00096).

We want to express our gratitude to Dr. Prof. Aleksandrova Tatyana Nikolaevna, head of mineral processing department of Mining university.

REFERENCES

Afanasova A.V., Aleksandrova T.N., Nikolaeva N.V. 2018, Processing of carbonaceous ores containing ultra-dispersed metals with using high-frequency current. *International Multidisciplinary Scientific Geo-Conference Surveying Geology and Mining Ecology Management, SGEM 2018, pp. 119–124.*

Aleksandrova T.N., Nikolaeva N.V., Aleksandrov A.V., Pavlova U.M. 2017. Beneficiation of carbonaceous rocks. *International Multidisciplinary Scientific GeoConference Surveying Geology and Mining Ecology Management, SGEM, 17 (11), pp. 781–788.*

Aleksandrova, T.N., Nikolaeva, N.V., Potemkin, V.A. 2018. Beneficiation of carbonaceous rocks: New methods and materials. *Innovation-Based Development of the Mineral Resources Sector: Challenges and Prospects - 11th conference of the Russian-German Raw Materials, pp. 391–398.*

Amelunxen, P. & Runge, K. 2014. Innovations in froth flotation modeling. *Mineral Processing and Extractive Metallurgy: 100 Years of Innovation.*

Beloglazov K.F. 1938. The kinetics of flotation process. *Tezisy dokladov sessii po fizikohimicheskim problemam obogasheniya,* Moscow: Metallurgizdat [In Russ].

Cherepovitsyn A.E., Ilinova A.A., Smirnova N.V. Key stakeholders in the development of transboundary hydrocarbon deposits: the interaction potential and the degree of influence. *Academy of Strategic Management Journal. T. 16.*

Imaizumi T. 1963. Kinetic consideration of froth flotation. *Proceedings of the Sixth International Mineral Processing Congress, 1963. Pergamon Press, 1963. pp. 581–593.*

Kelsall D.F. 1961. Application of probability in the assessment of flotation systems.

Klimpel R.R. 1980. Selection of chemical reagents for flotation. *Mineral processing plant design. T. 2. pp. 907–934.*

Kovalchuk M.S., Poddubniy D.A., Diagnosis of Electric Submersible Centrifugal Pump/*IOP Conference Series: Earth and Environmental Science,* № 115, 2018. C 1–6.

Kovalchuk M.S., Baburin S.V., Modelling and control system of multi motor conveyor/*IOP Conference Series: Materials Science and Engineering,* № 327, 2018. pp. 1–6.

Lavrik N.A., Litvinova N.M., Aleksandrova T.N.N, Stepanova V., Lavrik A. 2018. Platinum mineralization comparative characteristics of the some Far East deposits. *E3S Web of Conferences. – EDP Sciences, 2018. – V. 56. – p. 04017.*

Loveday B.K. 1966. Transactions of the IMM, 75.

Nikolaeva N.V., Aleksandrova T.N., Romashev A.O. 2018. Effect of grinding on the fractional composition of polymineral laminated bituminous shales. *Mineral Processing and Extractive Metallurgy Review. – 2018. – V. 39. – No 4. – pp. 231–234.*

Smirnova N.V., Rudenko G.V. Tendencies, problems and prospects of innovative technologies implementation by Russian oil companies. *Journal of Industrial Pollution Control. T. 33. №. 1. pp. 937–943.*

Topical Issues of Rational Use of Natural Resources 2019 – Litvinenko (Ed)
© 2020 Taylor & Francis Group, London, ISBN 978-0-367-85720-2

Methods and technologies for the processing of water-hydrocarbon emulsions and technogenic raw materials of metallurgical and petrochemical enterprises: A review

S.I. Khusnutdinov
Montanuniversität, Leoben, Austria
Mining University, Saint-Petersburg, Russia

J. Schenk
Montanuniversität, Leoben, Austria

I.Sh. Khusnutdinov & A.F. Safiulina
Kazan National Research Technological University, Kazan, Russia

V.Y. Bazhin & O.A. Dubovikov
Mining University, Saint-Petersburg, Russia

ABSTRACT: Currently, there are several processes during which high-stable water - hydrocarbon emulsions are formed. These include heavy pyrolysis resin emulsions formed during the extraction and preparation of highly viscous oils and natural bitumen; sludge, emulsion cleaning and decomposition of cutting fluids. At the petrochemical, oil-producing, steel-rolling and metallurgical enterprises of Russia annually produce about 40 million tons of liquid and solid waste oil, in general, in the world, the volume of their formation reaches hundreds of millions of tons. This article reviewed the classic and non-classical methods of processing such raw materials.

1 INTRODUCTION

The need for recycling sludge is due to several reasons:

1) they lead to pollution of the lithosphere, air and water basins and pose a threat to public health;
2) sludge collectors are also dangerous in terms of fire (Zhuravlev 2010)
3) barns occupy significant areas, and due to their lack of oil waste is often burned without purification of waste gases (Peganov 2001);
4) waste contains valuable hydrocarbon raw materials.

The problem of processing waste lubricating oils is no less acute worldwide, since, along with other hydrocarbons, used lubricating oils significantly pollute the biosphere. Unlike petroleum and other petroleum products, used motor oils, when released into the environment, are even less naturally neutralized (oxidation, photochemical reactions, biodegradation). During operation, the quality of the oil changes due to thermal decomposition and oxidation. As a result of these processes, asphalt-resinous compounds, soot particles, various salts, acids, surfactants, particles of metals and oxides accumulate in oils. To this it is necessary to add that the additives contained in the oils retain contaminants that enter or form in the oils during operation. That is why the combustion of waste oils pollutes the atmosphere with heavy metals, soot, sulfur dioxide, and stable chemical compounds. In this regard, in many countries, waste oils are burned only after the removal of environmentally harmful substances from them.

2 METHODOLOGY

To denote the totality of production and consumption wastes, which at this stage in the development of science and technology can be used in the national economy, the most general concept of "secondary material resources" has been introduced (Mazlova 2001).

Petroleum wastes which belong to toxic products of organic origin with possible mineral impurities are singled out in a special group. They can be combustible (liquid combustible waste), non-combustible or limited combustible (oil sludge, sludge from sewage treatment plants, black oil earth, etc.).

Table 1. The main sources of a waste.

Name	Waste composition
Municipal solid waste	Organic matter - 60-70% (carbon - 35%), ash-content - 30-40%, humidity of the total mass - 40-50%
Oil sludge from sedimentation tanks of oil refineries, railway enterprises, tank farms and repair plants	Oil products - 20-30%, water - 20-30%, mechanical impurities - 40-50%
Soil contaminated with petroleum products of the territories of state-owned enterprises, tank farms and repair plants	Oil products - 0.1-5 g/kg, humidity - 40-50% of the total mass
Coal slurry	Carbon - 10-30%, ash content - 70-90%
Waste oils and lubricants, paper filters of machines and mechanisms	Oil products - 90%, moisture - 8%, metal and mineral inclusions - 2%

2.1 *Classification of waste treatment and disposal methods*

There are many approaches for classifying petroleum waste treatment methods. According to the first approach, sludge processing methods can be divided into non-destructive and destructive.

Non-destructive methods: controlled open unloading; burial; the use of oily sludge in agriculture; the introduction of sludge as an organic fertilizer.

Destructive methods include incineration on site or with household waste with preliminary dehydration; inclusion in the cement in its production by wet; aerobic treatment (Krasnogorskaya 2004)

According to another classification (Zhuravlev 1990), at present, there are mainly three ways of using heavy watered oil residues:

1) preliminary dehydration, thermal or press-drying of water-cut sludge and further processing of the oil products obtained according to known schemes;
2) processing of sludge into gas and steam and gas;
3) burning of oil sludge in the form of aqueous emulsions and the use of heat generated.

2.2 *Gravitational upholding*

In the territory of each oil refinery that has been in operation for decades, there are sludge ponds - natural sedimentation tanks, in which oil waste accumulates in large quantities.

In industry, the method is implemented in vessels of a horizontal or vertical type of large capacity, where oil is settled at the expense of consists of a reflector, partitions, a droplet generator, coils for heating and unloading devices.

For example, a sedimentation tank construction with intermediate partitions is known, which are made with a decrease in the height of the lower edges in the direction of fluid flow, which allows operating at a low height of the oil-water interface, reduces the probability of the separated water entering the oil compartment again by a transfusion through the partition. To reduce the sludge time, it is also proposed to pretreat the emulsion with a swirler.

Gravitational sedimentation is implemented in the dynamic mode in the apparatus of the type SDS (gravitational-dynamic separator). The separation of two fluids occurs due to a specially organized movement, which achieves accelerated coalescence of small droplets of oils and petroleum products, and the environment, including under the action of natural gravity forces. (Burlaka, 2008)

Advantages of the method of gravitational upholding - does not require large capital and operating costs; may be part of the combined method. Disadvantages - low separation efficiency and duration of the process; the scope is limited; large number of residues.

2.3 Extraction. rinsing with water

Extraction is used to extract the oil component, based on the selective solubility of petroleum products in organic solvents. Solvents must be fully and fairly simply regenerated, with low energy consumption. It is known to use freons, alcohols, and aqueous solutions of surfactants as solvents (Krasnogorskaya 2004)

A process was proposed for the treatment of reservoir sludges by the method of LANSCO by an American firm to bring them to a mobile state with subsequent separation by centrifuging (A.N. Spider).

It was also found that by mixing an oil sludge emulsion with fuel oil, it is possible to provide partial dehydration of the raw material (Kovalsky 2011).

In recent years, Baroid has developed special installations that allow the purification of drilling sludge by three-step washing of sludge with various solvents in the fully closed Unitired Solids Control system.

In the course of research (Palgunov 1990.), the process of destruction of various emulsions was carried out using a demulsifier and an aqueous solution of salts. It is shown that the combination of the thermochemical method with the influence of a saline solution intensifies the dehydration of stable emulsions.

The disadvantages of the extraction process are the use of expensive solvent, the need for its further regeneration and incomplete extraction of the hydrocarbon component.

2.4 Electromagnetic and wave effects

Recently, waste disposal technology has been actively developed using microwave heating, which has several advantages over other methods: non-contact heat supply, rapid heating throughout the volume, full process automation, no secondary waste, simplicity and reliability of operation.

The method of microwave exposure improves the quality of separation of water-oil emulsion by increasing the number and intensity of contact between each other of water droplets in the water-oil emulsion stream, which is under the influence of microwave energy in the annular zone. The separation efficiency is ensured by installing the structure of the elements, which divides the flow of water-oil emulsion in the annular zone into separate channels. It can be made in the form of a spiral, mesh or plates from materials of a round or flat streamlined profile (Petrov 1984)

The disadvantages of this group of methods include high energy costs and the lack of commercial samples introduced into the industry.

2.5 Mixing with additives, adsorbents to obtain marketable products

In this direction, the following trends have emerged.

Liquid sludge is subjected to the processes of homogenization and emulsification using a vibro-cavitation grinder to prepare a stable water-oil emulsion and burn it in boilers (Krel 1980).

This process is simple, affordable, highly efficient and economical. A similar plant works stably at the Atfrau refinery, at OJSC Ufimsky Refinery and OJSC Khabarovsk Refinery, at OJSC Shymkentnefteorgsintez. The disadvantage of this method is that fuel oil should be used immediately for its intended purpose, i.e. burn until the emulsion is stable and reverse processes occur.

Another trend involves the use of waste oil, for example, waste oils, as components of boiler fuels or their co-processing with oil. However, the presence of contaminants and additives in waste oils adversely affects the operation of electrical desalting plants, worsens the separation of oil, increases the content of oil products in wastewater.

The third direction involves the use of oil waste in construction, as components of road materials, or the disposal of sludge. The most detailed questions about the disposal or further use of oil sludge in construction are disclosed by the author.

However, the scope of application of this method is limited, since preliminary preparation of this oil sludge is required, their use as components may adversely affect the properties of the final product, the environmental problem remains unsolved, or huge areas for landfills are required.

2.6 Chemical methods

The essence of the chemical method is the neutralization of oily waste by alkaline earth metal oxides. This forms calcium hydroxide in the form of a powder, with a high specific surface, which is able to absorb oil hydrocarbons. Each granule is covered with a layer of calcium carbonate, which serves as a hydrophobic shell.

In, oil sludge is mixed with burnt milled lime in a ratio of 1: 1-2, respectively, then frozen in layers and thawed in natural conditions. In the dehydrated after freezing and thawing the mixture is added partially dehydrated clay or drill cuttings in a ratio of 1: 0.5.

Fest Alpine and Leo Consult (Germany) have jointly developed an installation for the chemical curing of oily waste, varnishes, paints, acidic resins, etc. The installation works on the principle of mixing waste with special hydrophobic additives based on lime (the so-called "DSR - process"). The company "Meisner Grundbau" considered the technology of chemical processing and disposal of oily waste (oil-containing sludge, acid tars, oil-polluted soil, emulsion sludge, etc.

Along with the main component of the reagent (quicklime), it may contain synthetic surfactants based on octyl sulfates, sulfanol, etc.. It is also proposed to add water-soluble surfactant and film-forming component. To achieve a stable suspension state of the mixture include a stirrer turbulizer. Turbulization provides the maximum use of air oxygen as an oxidizing agent, which contributes to the formation on the surface of the calcium-containing component of a strong three-dimensional film.

Recently, systems using so-called magnetic fluids have been tested to collect petroleum products and petroleum wastes from the surface of reservoirs, as well as to extract petroleum products from wastewater. They are produced on the basis of such components as water, hydrocarbons, fluorinated hydrocarbons, mineral oils, vacuum oils, silicon organic liquids, surfactants, and also on the basis of various magnets, such as iron, magnetite (Fe_3O_4), cobalt.

However, the application This method requires special equipment, a significant amount of reagent, the environmental problem is not fully solved, irreversibly lost valuable hydrocarbons contained in oil sludge.

2.7 Biological methods

The essence of the biological method of utilization of oil waste is that microorganisms convert petroleum hydrocarbons into simpler compounds, accumulate these organic products and involve them in the carbon cycle.

A key role in biodegradation is assigned to microorganisms that contribute to the intracellular oxidation of petroleum hydrocarbons.

The main methods of environmental biotechnology are bioremediation, bioprocessing and biodegradation. The following methods have been developed: in-situ biodegradation, processing in the liquid and solid phases. In the first case, microorganisms are introduced directly into contaminated soil and water.

In, the method includes the steps of introducing organic components into the oil sludge, forming the porosity of the material being processed, completing the multicomponent interrelation of oil-oxidizing microorganisms with a compostable composition changing in structure.

The reactor for disposal is equipped with a heating system, and there is a special nozzle in it to secure the anaerobic association of microorganisms. As a result of the activity of microorganisms, gas and surfactants are formed, which create favorable conditions for the sedimentation of mechanical impurities from the product. The hydrocarbon phase accumulates in the upper part of the reactor, and water accumulates in the lower part. The residual sludge is sent after the anaerobic block to the biological stabilizer. In this reactor, with the constant presence of a culture of aerobic bacteria, residual petroleum product is utilized to form lipids and other products.

2.8 *Combined methods*

Often for the disposal of oil sludge use complex technologies. For example, prior to feeding waste onto centrifuges or filter presses, water-soluble polymer electrolytes are introduced into them, i.e. flocculants. As a result of their action, water particles are desorbed from the surface of solid contaminants and, due to a decrease in surface tension, the latter coalesce.

Giprotyumenneftegaz OJSC has developed a trap emulsion processing technology. Raw materials are treated with a reagent, heated in a preheater and sent back to the tank or retreated with a reagent. The emulsion thus treated after several cycles of circulation settles in the tank. Processing is then possible in an electrocoalescentor or in a centrifuge.

The company Bird proposes to heat the sludge to 80 ° C, mix with a demulsifier and polyelectrolyte in the system of flocculation of the solid phase of the drilling fluid. Next, in a three-phase centrifuge, oil and water are separated from the solid phase.

The most effective, though not always cost-effective, consider the thermal method of disposal of sludge. Of the thermal methods of processing oil waste, combustion, gasification, and pyrolysis are most often used. Combustion is carried out in an oxidizing atmosphere, gasification - in a partially oxidizing, pyrolysis - without access of air. Also in this group can be attributed processes based on the evaporation of the aqueous and light hydrocarbon phases of oil sludge.

Thermal methods differ in the organization of instrumentation, technological regime and the nature of the raw materials used.

2.9 *Burning*

During incineration, furnaces of various designs were used For waste containing no more than 20% of solid impurities, fluidized bed furnaces are widely used. When burning oil sludge containing up to 70% of impurities, rotary kiln-type rotary kilns became widespread.

An example of a large plant (New Jersey, USA) for the incineration of liquid waste is an incinerator with a capacity of 4 m3 waste/h, incineration is carried out at 1000-1200° C, the residence time in the combustion zone is at least 2.5 seconds. The unit is equipped with a Venturi type scrubber unit, a cooling scrubber and an aerosol trap. (Belousov 1987).

2.9.1 *Pyrolysis. Cracking. Coking.*

A distinctive feature of thermal cracking processes is the production of gaseous, liquid (resin) and solid (coke) from raw materials, in contrast to combustion processes in which valuable hydrocarbon components of the raw materials are irretrievably lost (Kondrasheva 2019). The processes of this group also differ in instrumentation, mode and nature of the raw materials used.

Published several works devoted to the processing of solid and semi-liquid waste by the method of pyrolysis to produce low-calorie fuel gas and tar. The process of low-temperature pyrolysis is usually carried out at a temperature of 500-550° C to obtain combustible gases and solid residue. Pyrolysis is advisable to use when disposing of solid oil sludge with a low water content (less than 3%).

There is a pyrolysis unit (Khusnutdinov 2018) consisting of a rotating cylindrical cavity and a combustion chamber located around this cavity. A special feature is the tangential entry of fuel and oxidizing agent into the combustion chamber, which leads to the formation of flame vortices.

It is also proposed to carry out joint cracking of oil sludge with tires (Khusnutdinov 2018), coking mixtures of liquid and solid products of organic origin in various ratios (Khusnutdinov 2018), which ultimately will provide additional sources of oil products. For the processing of acid tars, the technology of thin-film cracking is patented, with simultaneous production of liquid fuel oil and coke from organic components.

2.9.2 *Drying*

Thermal methods also include the drying of solid and pasty oil waste. The indisputable advantage of drying is the preservation of valuable organic components of raw materials, accompanied by a decrease in the volume of finished products by 2-3 times, the environmental friendliness of the process. However, the process is also characterized by high fuel consumption.

The most typical scheme for drying oil waste consists of the following stages: mixing of waste, evaporation of water and light hydrocarbon fractions; mixing sludge with a mixture; preparation of the slip, pressing the mixture; drying briquettes; roasting products. For example, the method of processing sludge is that drilling waste is mixed with sludge, the mixture is granulated by a press, and the granules are fired in a rotary kiln at a temperature of 1150° C to produce building ceramic granules

2.10 *Dehydration*

The evaporation in a thin film is implemented in the following inventions. There is a known method and device for separating water and impurities from oil, where dehydration of water-hydrocarbon emulsions occurs due to evaporation, which occurs by spraying a heated liquid either in the form of small droplets or a thin layer above the surface (Khusnutdinov 2018). Dry (superheated) liquid is used as an evaporation surface and coolant, and also maintains a uniform temperature of the evaporation surface, rather than volume.

To reduce the viscosity of the raw material in the original water-oil emulsion add solvent. Then, the raw material is sprayed through a nozzle over a heated surface, and the aqueous phase and light hydrocarbons are evaporated.

The disadvantages of this solution are the need to use specialized equipment and a solvent, which leads to an increase in operating costs; the absence of mechanical impact and turbulization of the volume of boiling liquid, which allows to stabilize the boiling process in volume and eliminates the possibility of boiling up and transfer of fluid. Also, the use of spray and nozzles does not allow to dehydrate emulsions with a high content of mechanical impurities.

In some inventions direct evaporation is realized in tanks or tanks. However, within the framework of such installations, there is a risk of boiling and emulsion transfer. For example, there is an installation for the processing of sludge, where the dewatering of sludge is performed by direct evaporation of water during its laminar flow from tank to tank, arranged in a cascade version.

The oil sludge dewatering device consists of a tank and a refrigerator interconnected by a pipeline, characterized in that heaters are installed in the bottom of the tank in a plane parallel to the fluid filled into it (Khusnutdinov 2018).

For utilization of oil sludge by evaporation, the method of preliminary mixing of the latter with the prepared oil and further evaporation is also used. There is a known method of preparing oil for processing, where the dewatering of the emulsion is performed by direct evaporation of the aqueous phase of the emulsions in the presence of a demulsifier and partially dehydrated field oil.

The main advantages of the thermal method of disposing of liquid and solid petroleum wastes are: reduction of the amount of petroleum wastes for disposal, smaller volumes of ash; economic benefits; the possibility of obtaining a porous granular material, which can later be used in building materials or road surfaces; high neutralization efficiency; heat recovery.

The common and significant drawbacks of these methods are a large amount of harmful emissions, as well as the need to dispose of the resulting ash, relatively high energy and material costs.

3 CONCLUSION

Analysis of the literature data allows us to conclude that the use of thermal methods for the disposal of sludge can be complicated by the high water content of the oil sludge; high content of mechanical impurities in the sludge (up to 65%; complexity of sludge extraction from sludge collectors and transportation to the sludge-burning installation; complexity of high-quality spray in the sludge-burning furnace due to the variability of its mechanical and physicochemical composition, high viscosity, similar values of phase density.

The above problems that arise when disposing of liquid oil sludge are eliminated when applying a mechanical effect to boiling liquid, which leads to turbulization of the entire volume of liquid and prevent foaming and transfer of the emulsion.

REFERENCES

A. A. Alekseeva, S. I. Khusnutdinov, S. M. Petrov, I. Sh. Khusnutdinov, xA. G. Safiulina, and N. Yu. Bashkirtsev properties and applications of distillate fractions from igily Stable dispersions of liquid pyrolysis products Chemistry and Technology of Fuels and Oils, Vol. 54, No. 3, July 2018 (Russian Original No.3, May - June, 2018).

A.A. Petrov Hydrocarbons oil. M .: Chemistry, 1984.

A.E. Yurchenko Secondary material resources of the refining and petrochemical industries (education and use). M .: Economy. 1984. 143 p.

A. G. Safiulina, R. R. Zabbarov, S. I. Khusnutdinov, A. A. Alekseeva, I. Sh. Khusnutdinov and S. M. Petrov thermomechanical dehydration of highly-stable dispersions Of liquid pyrolysis products Chemistry and Technology of Fuels and Oils, Vol. 54, No., July, 2018 (Russian Original No.3, May - June, 2018).

A.I. Belousov Evaluation of intermolecular interactions in hydrocarbons of oil // Chemistry and technology of fuels and oils. 1987.-№1.-C.26.

A.N. Spider Development of technology for processing oil sludge, industrial and household waste into oil products. Author. dis. on the competition uch. step. Cand. tech. sciences. Tyumen: Tyum. state oil and gas. un-t 2003. 20c.

B.M. Rybak Analysis of oil and petroleum products. M .: Gostoptekhizdat, 1962.

E.A Mazlova, Problems of disposal of sludge and methods for their processing. M .: Noosphere. 2001. 56 p.

Krasnogorskaya N.N., Magid A.B., Trifonova N.A.//Oil and gas business. 2004. V. 2. P.217-222.

Khaidarov F. R., Khisaev R.N., Shaydakov V.V., Kashtanova L.E. Oil sludge. Methods of recycling and disposal. Ufa: Monograph. 2003. 74 p.

E.Krel, Manual laboratory distillation. M .: Chemistry, 1980.P.P Palgunov. Utilization of industrial waste. M .: stroiizdat. 1990. 352c.

Kovalsky, B.I. Modern methods of cleaning and regeneration of used lubricating oils: enterprise / B.I. Kovalsky and [other]. - Krasnoyarsk: Sib. Department. University, 2011-104 s.

G.A. Morozov // Physics of wave processes and radio engineering systems. T.10. Number 3. 2007. P.125-129.

Natalia K. Kondrasheva, Viacheslav A. Rudko, Dmitriy O. Kondrashev, Renat R. Gabdulkhakov, Ivan O. Derkunskii*, and Rostislav R. Konoplin Effect of Delayed Coking Pressure on the Yield and Quality of Middle and Heavy Distillates Used as Components of Environmentally Friendly Marine Fuels Energy Fuels, 2019, 33 (1), pp 636–644.

Palgunov, P.P. Utilization of industrial waste / P.P. Palgunov. - M. Stroyizdat, 1990. -352 pp., Ill.

S.I. Khusnutdinov Thermomechanical Dehydration of Highly-Stable Dispersions of Liquid Pyrolysis Products / S.I. Khusnutdinov, A.G. Safiulina, R.R. Zabbarov, I. Sh. Khusnutdinov, A.N. Gaffarov // Chemistry and Technology of Fuels and Oils No. 3, pp. 15–118, May – June, 2018.

Smolyanov V.M. // Oil refinery. and petrochemistry. 2002. №8. P.29-32.

V.A. Burlaka, Patent of the Russian Federation No. 237608. 2008.

V.M. Zhuravlev, Chernyavskaya E.N., Kulagina T.A., Klymenkov S.I., Pisareva E.N. Modern opportunities for ensuring environmental safety in the handling of industrial waste // Proc. International Scientific and Technical Congress "Energy in the global world." Krasnoyarsk. 2010. S.279-282.

V.N. Peganov, A new approach to the study of the composition of oil sludge and the development of technology for their processing // Proc. report 2 International WasteTech Congress on Waste Management - 2001. M .: SIBIKO Int. 2001. P.264-265.

Topical Issues of Rational Use of Natural Resources 2019 – Litvinenko (Ed)
© 2020 Taylor & Francis Group, London, ISBN 978-0-367-85720-2

Studying the process of non-sulfide flotation of ferruginous quartzites

D.I. Krizhanovskaia
Saint-Petersburg Mining University, Saint Petersburg, Russia

ABSTRACT: The article describes laboratory experiments comprising ore dressing and concentration of mineral raw materials by flotation. Samples of ferruginous quartzite and flotation separation products were studied by using precision methods such as X-ray fluorescence analysis, laser diffraction to determine the particle size distribution of the samples.

It has been proven that in the grinding process aminoacetic acid acts as a surfactant reducing the durability properties of the material. The experiments show that at the subsequently stage physical and chemical activation allows a greater amount of silicon dioxide turn into the form of silicic acid at the aqueous phase that leads to chemisorption improvement of the reagent - the collector of minerals in flotation.

1 INTRODUCTION

The demand for processing refractory and poor ores is increasing in the modern world. This trend is a consequence of the high frequency mining rate of mineral resources, which results in ore depletion, and size reduction of mineral impregnation. This is the reason for the increase in the number of processing plants and mills built to extract valuable components effectively.

About 28% of the earth's crust consists of silicon, which is contained, in such ores as quartz ($SiO2$). This mineral makes up most of the host rock deposits. It becomes obvious that the research on technological solutions aimed at separating valued components from the waste rock and quartz in particular. In addition, the intensification of the extraction of silicon dioxide will allow processing mills to market this product for further use in industry.

In mineral processing, the main separation factor is the physical property of a particular component. For quartz, the property, which allows successful separation processes, is the floatability k depending on the wetting angle (Tikhonov, 2008). Thus, the most suitable method for extracting quartz is flotation.

The versatility and the efficiency of flotation are proven, however, due to low environmental friendliness and large volumes of water required it is not suitable for all ores and climatic conditions. Nevertheless, flotation method is used to obtain valuable concentrates with optimal content and extraction parameters for example during reverse cationic flotation at the Mikhailovskiy MPP, or cleaning quartz from impurities at the Ramenskiy MPP.

The purpose of this work is to research the effect of mechanochemical activation on the process of non-sulfide flotation of quartz by using aminoacetic acid (AA).

2 MATERIALS AND METHODS

The research object is the ore of the Mikhailovskiy deposit represented in magnetic enrichment concentrate of ferruginous quartzite (hereinafter referred to as material), as well as by samples of silica.

The first stage included grinding of monomineralic silica samples in a laboratory ball mill for 5, 10 and 20 minutes in a dry medium without reagents and with the addition of aminoacetic acid of 750 g/t. The measurement of the specific surface area is carried out with a laser diffraction analyzer Mastersizer 2000. The flotation experiments are carried out with the following reagent mode: the medium controller is caustic soda up to pH ≈ 9.5; cationic collector – Lilaflot (LF) and Tomamin; the volume of air supplied is 40 m3/h.

Based on the results, obtained in the first stage, the following experiments are conducted. The grinding of the material in the aminoacetic acid medium is carried out for 5 minutes with different consumption of the amino reagent (750, 1000, 1250 g/t). The experiments on flotation separation are conducted be using the Kano model with one constant and two variable factors at three levels. The reagent mode is represented by a medium regulator - caustic soda to pH ≈ 9.5; cationic collector - Lilaflot; depressor - alkaline starch - 100 g/t; the volume of air supplied is 40 m3/h.

Flotation products contents are measured on scales with precision of 0.01 g.

3   ORE PREPARATION RESULTS

The addition of aminoacetic acid in the process of ore preparation influences the particle size distribution of the grinding product. This effect is reached due to the adsorption of the surfactant and causes "wedging". Aminoacetic acid got into the microcracks of the mineral causes decrease in the interfacial energy of the body, i.e. reduces strength - Rehbinder effect.

Figure 1 shows the particle size characteristics of grinding products. The distribution is more uniform and reduces the maximum particle size during the processing of aminoacetic acid material.

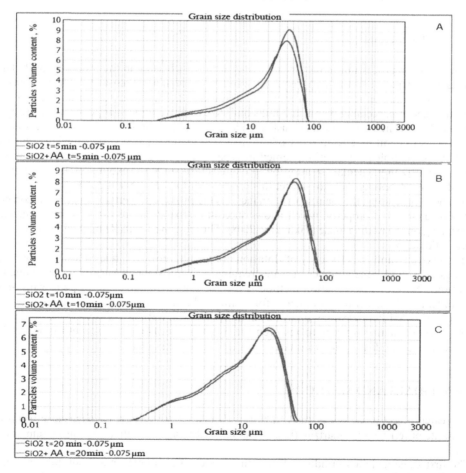

Figure 1.   Grinding products particle size distribution. A) t = 5 min, B) t = 10 min, C) t = 20 min.

Table 1. Silicon dioxide grinding products parameters.

Grinding time, min	Specific surface area without aminoacetic acid, sm²/g	Specific surface area with aminoacetic acid, sm²/g	k_{ac}
5	0,737	0,872	0,237
10	0,862	0,939	0,109
20	1,340	1,400	0,052

The grinding product is analyzed for an increase in specific surface area. The main indicator of growth is presented in the table as the activation coefficient (Pavlova, 2018).

The results of the experiment suggest that grinding silicon dioxide in the aminoacetic acid medium gives the greatest result at grinding time t = 5 min.

On the basis of the data obtained at the previous stage, the grinding time t = 5 min is taken for the material when changing the aminoacetic acid consumption: 750, 1000, 1250 g/t. The final graphs of the particle size distribution are presented in Figure 3.

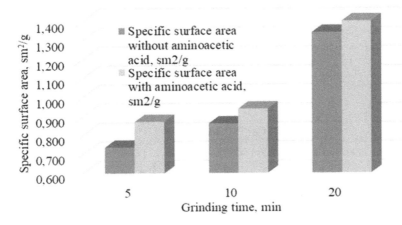

Figure 2. Specific surface area alteration depending on grinding time.

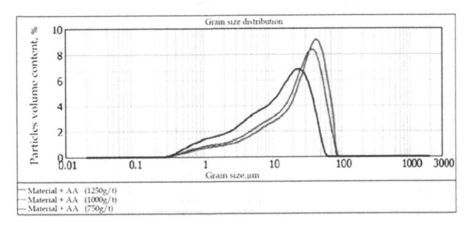

Figure 3. Granulometric composition of the grinding products.

535

4 FLOTATION RESULTS

One of the most common methods of flotation enrichment of ferruginous quartzite is the method of reverse cationic flotation, its advantages are described in the researches of many scientists (Varichev, 2017; Pelevin, 2019; Gubin, 2018). In this paper, the technological scheme was developed, which includes the main flotation stage using cationic collectors.

The experiments of flotation separation are performed according to the canonical scheme (Samygin, 2013) (Figure 4).

Optimal collector reagent and the flow rate are found in an experimental analysis of the above scheme. For the calculated size grade of -75 + 0 μm, the yields of γк and γхв concentrates and tails, respectively, are found.

During the flotation of monomineralic raw materials, the content of the calculated component is 1 share of units or 100%.

The selection of the collector reagent is carried out on the basis of the results of flotation experiments, ceteris paribus (pH ≈ 9.5; collector consumption, 100 g/t; air volume, 40 m3/h).

Based on previous experience, the cationic collector Lilaflot is chosen.

The next step is to determine the optimal collector consumption; the results are shown in Table 3.

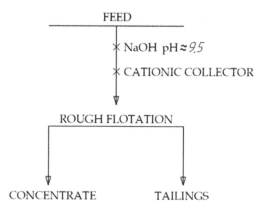

Figure 4.   Technological scheme of flotation separation.

Table 2.   The results of the experiment for different collectors.

Collector	Tomamin	Lilaflot
γ_{conc}, %	93,5	94,8
γ_{tails}, %	6,5	5,2

Table 3.   The results of the experiment for different consumption of LF.

Collector consumption, g/t	50	100	150	200
γ_{conc}, %	91,9	94,8	97,2	97,6
γ_{tails}, %	8,1	5,2	2,8	2,4
ε_{conc}, %	91,9	94,8	97,2	97,6

The graph (Figure 5) shows the visual dependence of the extraction rate on the consumption of the collector.

The graphic image makes it clear that when the collector consumption changes, the extraction of silicon dioxide increases by only 0.4%, which is economically unprofitable due to the cost of the reagent.The flotation products of the treated material in the aminoacetic acid environment have the following parameters:

The mechanism for fixing the cationic collector on the surface of quartz is carried out by chemical adsorption or chemisorption (Kondratiev, 2014). The increased recovery rate is a consequence of the grinding of the material in the aminoacetic acid medium causes the increase in the number of chemically active areas. When immersed in the pulp (alkaline solution), "silicic acid" is formed on the surface of crushed quartz (Figure 6), which interacts with

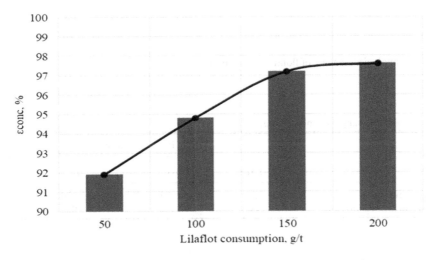

Figure 5.   Dependence of extraction in concentrate on collector reagent consumption.

Table 4.   Technological parameters of flotation products in the aminoacetic acid medium.

Technological parameter	Material with aminoacetic acid	Material without aminoacetic acid
$\gamma_{\text{к}}$, %	97,2	99,1
$\gamma_{\text{хв}}$, %	2,8	0,9
$\varepsilon_{\text{к}}$, %	97,2	99,1

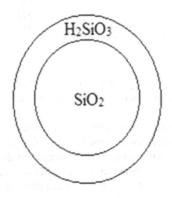

Figure 6.   "silicic" acid formation on the surface of crushed quartz.

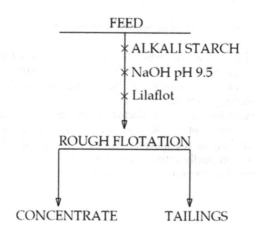

Figure 7. Technological scheme of flotation separation.

Table 5. Factors and ranges of their change.

Parameter	Label	Parameter levels		
		-1	0	1
Lilaflot consumption	X_1	50	100	150
Aminoacetic acid consumption	X_2	750	1000	1250

Table 6. Planning matrix.

№	Parameter levels combination		Output optimization parameters		
	X_1	X_2	Y_{CP}	ε_{XB}	β_{XB}
1	1	1	20,397	63,406	20,397
2	-1	1	24,952	29,956	24,952
3	1	-1	19,143	58,714	19,143
4	-1	-1	24,563	27,441	24,563
5	1	0	21,670	67,177	21,670
6	-1	0	27,952	34,208	27,952
7	0	1	29,952	70,717	29,952
8	0	-1	28,395	65,261	28,395
9	0	0	31,712	75,625	31,712

the collector reagent (Crainiy, 2013; Zhuravleva, 2017). Thus, physicochemical activation at the grinding stage allows higher number of quartz to pass into the aqueous phase in the form of an acid.

To establish the most optimal flotation separation regime the experiments are conducted by using the Kano model with two factors in three levels (Table 5).

The planning matrix and the results of the experiments are shown in Table 6.

After computer processing of experimental data in Microsoft Excel spreadsheets, mathematical models are obtained that adequately describe the dependence of the content and extraction of silicon dioxide in the flotation tails on the costs of the collector and aminoacetic acid.

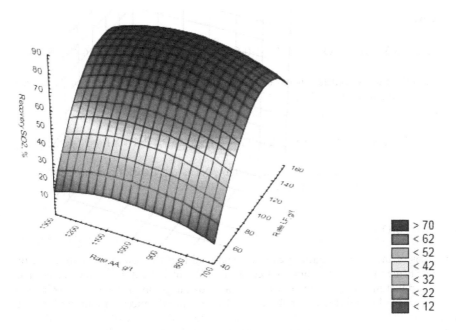

Figure 8.   Graph of the surface dependence of the recovery on the consumption of reagents.

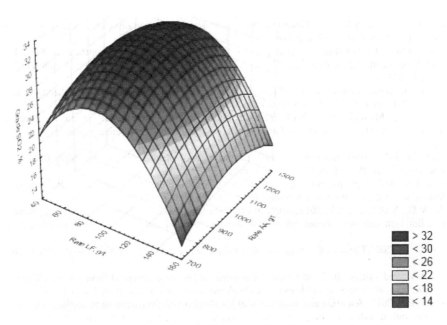

Figure 9.   A graph of the surface dependence of the content β on the consumption of reagents.

Based on the obtained regression equations, the response surfaces and level line maps are presented in Figures 8 and 9.

To recovery:

$$Y\varepsilon = 75.63 + 16.28\ X1 + 2.11\ X2 + 0.54\ X1X2 + 23.72\ X1 * X1 + 6.42\ X2 * X2, \quad (1)$$

where $Y\varepsilon$ is the output extraction parameter, X1 is the Lilaflot consumption level, X2 is the aminoacetic acid consumption level.

For grade:

$$Y\beta = 31.72 + 2.71\ X1 + 0.53\ X2 + 0.2\ 2\ X1X2 + 6.9\ 1\ X1 * X1 + 2.54\ X2 * X2, \quad (2)$$

where $Y\beta$ is the output parameter of the content, X1 is the Lilaflot consumption level, X2 is the aminoacetic acid consumption level.

5 CONCLUSION

On the basis of the research data, it can be concluded that the processed material in the aminoacetic acid medium due to improved chemisorption of the cationic collector is taken into the foam product in a larger volume with constant consumption of the water-repellent agent.

As a result of the additional enrichment of the magnetic concentrate by means of reverse cationic flotation, a low-silica product suitable for the further stage of direct metallization can be obtained. It seems obvious that the introduction of this reagent will improve the quality of iron ore concentrates.

REFERENCES

Avdokhin V.M., Gubin S.L. 2006. Reverse cationic flotation of fine iron ore concentrates. *Gorny Analytical Bulletin (scientific and technical journal) Vol. 5.*

Gubin GV, et al. 2018. Ways of further improving the quality of iron ore concentrates at PJSC Poltava Mining Plant in modern conditions. *Vestnik Zhytomyr State University. Series: Engineering Vol. 1 (81): p.232-239.*

Crainiy, A. A. 2013. Flotation of tailings of wet magnetic separation of unoxidized ferruginous quartzites. *Bulletin of BSTU. V.G. Shukhov Vol. 5.*

Kondratiev, S. A., Moshkin, N. P. 2014. The hydrophobicity of the mineral surface and the selectivity of the flotation separation of minerals. *Fundamental and applied issues of mining sciences, Vol. 2(1), p. 217-223.*

Pavlova, U.M. 2018. Intensification of flotation separation of black shale raw materials using physico-chemical effects. St. Petersburg Mining University.

Pelevin A. Ye. 2019. Ways to improve the efficiency of iron ore dressing technology. *Black metallurgy. Bulletin of scientific, technical and economic information Vol. 2: p. 137-146.*

Samygin, V.D., & Belyaev, A.V. 2013. Application of the topological method for the calculation and analysis of flotation schemes. *Mining information and analytical bulletin (scientific and technical journal), Vol. 10.*

Tikhonov O. N. 2008.Theory of separation of minerals. St. Petersburg State. Mountain Institute. GV Plekhanov.

Varichev A.V. and others. 2017. Innovative solutions in the production of iron ore at Mikhailovsky GOK. *Physical and technical problems of the development of mineral resources Vol. 5:* p. 141-153.

Zhuravleva, E.S. 2017. Scientific and experimental justification of electrochemical methods of improving the technological indicators of processing of magnetic concentrates. National research Technology University "MISIS".

Elevation of gold-bearing sulphide flotation efficiency

V.V. Kuznetsov

Saint-Petersburg Mining university, Saint-Petersburg, Russian Federation

ABSTRACT: The article presents the results of a study on the elevation of gold-bearing sulphide flotation efficiency by using oxidants. The study is due to the need for a clearer understanding of the processes occurring when the oxidizer acts on the surface of the minerals and the various inclusions within, and what range of factors play a key role in this process. Evaluation of process efficiency was conducted by analysis of eH-potential and air-flow rate influence on the sulphide minerals recovery in the concentrate.

1 INTRODUCTION

Industry of most countries in the world faces the problem of widespread depletion of rich metal deposits, so that their extraction is becoming more and more time consuming and, which is important, a costly task. Thus, the increasing relevance of the extraction of metals from ores, where they are represented as inclusions and cannot be extracted using the extraction scheme of the target component. The processing of poor ores can be carried out by changing the established methods of enrichment by adding new methods of working or revising the optimal conditions for the application of these operations.

This work is dedicated to problem of gold-bearing sulphide ores processing. The specificity of such ores is that gold is not in a free form, but is closely associated with sulfides. It is presented in the form of fine particles or embedded in the crystal lattice of sulphides (Aleksandrova, Heide, Afanasova, 2019).

During collective sulfide flotation of such ores, various sulfide minerals, gold-bearing and non-gold-bearing, get to the concentrate. That causes serious problems in further extraction of gold from that kind of material, because each mineral has an individual chemical composition, e.g. arsenic content in arsenopyrite, which is undesirable in the resulting product.

One of the solutions to this problem can be the selective flotation of pyrite-arsenopyrite by the separation of collective concentrate into products of homogeneous mineral composition. For the separation of pyrite-arsenopyrite, it is possible to use oxidatns that change the flotation properties of minerals by oxidizing the surface of the mineral, which changes its floatability properties or cause oxidation of the collector.

2 METHODS AND MATERIALS

As part of this study, a method was considered to increase the efficiency of the flotation enrichment method by adding oxidizing agents. The mechanism of this process goes in two main directions. The first is that the oxidizer, when added to the reactant mixture, interacts with the surface with the solid phase of the pulp. The oxidized film alters the floatability of the mineral, which in turn affects its removal through flotation.

The second is that during the oxidation process, the crystal lattice is disturbed, which allows the release of associated useful components that are enclosed in the interstitial space of the crystal lattice of the mineral.

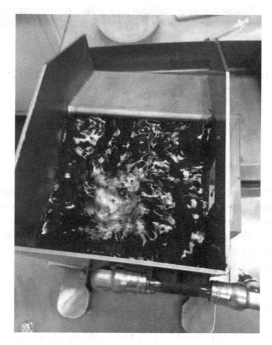

Figure 1. Pyrite monofraction pulp.

Selective recovery by flotation is possible with oxidants appliance due to surface properties adjustment by oxidation. Assuming that on the mineral surface chemical reaction takes place, we can evaluate the process through the formation of various products of that reaction. By analyzing the Pourbaix diagram for that range of pH and chemical elements engaged in the process we can find that range of eH-potential in which the probability of oxidation is as high as possible, corresponded to the zones of oxidations products existence. (Aleksandrova, Romanenko, Arustamian, 2019).

To study the effect of aeration intensity on the process, the amount of air flow rate varied. To assess the effect of the specific intensity of pulp aeration on the efficiency of monofraction extraction, the theoretical value of extracting "true" flotation was established to compare with experimental results. By true flotation in this connotation, we mean the extraction of useful without taking into account the mechanical removal.

The experiment was conducted by following plan. Pyrite monofraction was milled for 15 minutes with P_{80} = 0.80 µm. Flotation of the obtained product lasted 4 minutes. Potassium permanganate was used as an oxidizing agent. Choosing of oxidizer due to its low price. To study the effect of eH potential on the process, the flow rate of the medium regulator was varied, and the pH values did not go beyond the 10.0 - 10.6 interval.

3 RESULTS AND DISCUSSION

In case of pyrite flotation with oxidants presence the eH must be in range of sulphate ions existence as the main products of pyrite surface oxidation and so the main criteria of high possibility of oxidation. The figure of merit for comparing the results is the area of existence of these ions. The probability of oxidation of the surface of the mineral. PH, reagent consumption. It was found that the area of sulphate zones increases by 7.16%, which makes it possible to obtain soluble potassium sulphates (Figure 2, Figure 3).

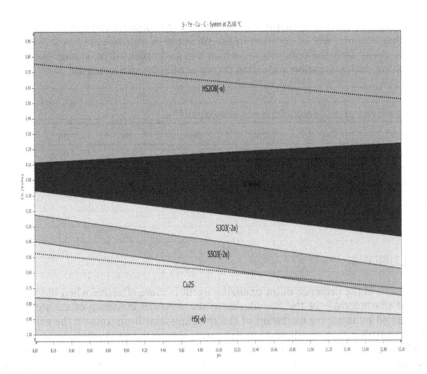

Figure 2. Pourbaix diagram for pyrite.

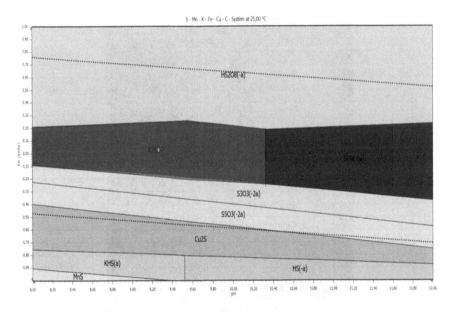

Figure 3. Pourbaix diagram for pyrite-potassium permanganate.

When analyzing the effect of the redox potential, the equations for the dependence of the monofraction extraction on the redox potential of the medium in the pH range 10.0-10.6 were established. Despite the theoretical results, the highest values of monofraction extraction were obtained for the values of redox potentials characteristic of the boundary zone of the existence of sulfate ions.

During studying the effect of aeration intensity on the monofraction extraction process, it was found that the intensity of aeration in the presence of an oxidants has less effect on the extraction of pyrite into the concentrate, which can be explained by similar mechanisms of sulfide oxidation during aeration and in the presence of an oxidizing agent.

Table 1.   Recovery of pyrite monofration in concentrate, % .

Air flow rate*	10	20	30	40	50	60
Theoretical	94.5	95.8	96.6	97.1	97.6	97.9
Without ox.	91.8	94.8	95.9	97.6	97.6	97.9
Using oxidants	97.6	97.8	97.9	98.2	98.1	98.5

*m^3/h

4   CONCLUSIONS

Sulfide flotation in the presence of an oxidizing agent is more effective when the values of the eH potential characteristic of the boundaries of the zones of existence of sulfate ions, which can be explained by intensive oxidation of the resulting transition ions on the surface of minerals, which contributes to the alignment of flotation properties of the surface.

The lack of correlation between the values of aeration intensity and extraction into the froth product and their high relative values during flotation in the presence of an oxidizing

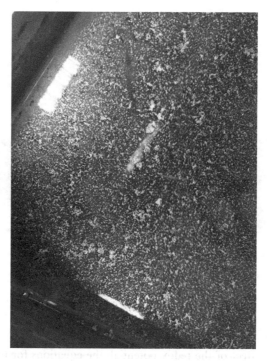

Figure 4.   Pyrite monofraction concentrate.

agent indicate that the oxidizing agent and oxygen in the supplied air perform the same functions on the effect on the floatability of sulphides. This also allows using this process when it is necessary to implement sulphide flotation at low values of aeration intensity. This work was supported by the Russian Science Foundation (project № 19-17-00096).

REFERENCES

Aleksandrova T.N., Romanenko S.A.A, Arustamian K.M. Research of slurry preparation before selective flotation for sulphide-polymetallic ores //IMPC 2018 – 29th International Mineral Processing Congress, pp. 2071-2078.

Aleksandrova T.N., Heide G., Afanasova A.V. Assessment of Refractory Gold-bearing Ores Based on Interpretation of Thermal Analysis Data. Journal of Mining Institute. 2019. Vol. 235, p. 30-37. DOI: 10.31897.PMI.2019.1.30.

Aleksandrova T.N., Afanasova A.V. Fine-dispersed particles of noble metals in sulphide carbonaceous ores and its beneficiation prospects //IMPC 2018 – 29th International Mineral Processing Congress, pp. 2368-2376.

Forrest K., Yan D., Dunne R. Optimisation of gold recovery by selective gold flotation for copper-gold-pyrite ores //Minerals Engineering. – 2001.

Nguyen A. V., Ralston J., Schulze H. J. On modelling of bubble–particle attachment probability in flotation //International Journal of Mineral Processing. – 1998. – Т. 53. – №. 4. – С. 225-249.

O'Connor C. T., Dunne R. C. The flotation of gold bearing ores—a review //Minerals Engineering. – 1994. – Т. 7. – №. 7. – С. 839-849.

Monte M. B. M., Lins F. F., Oliveira J. F. Selective flotation of gold from pyrite under oxidizing conditions //International journal of mineral processing. – 1997. – Т. 51. – №. 1-4. – С. 255-267.

Yalcin E., Kelebek S. Flotation kinetics of a pyritic gold ore //International Journal of Mineral Processing. – 2011. – Т. 98. – №. 1-2. – С. 48-54.

Dunne R. Flotation of gold and gold-bearing ores //Developments in mineral processing. – 2005. – Т. 15. – С. 309-344.

Gül A. et al. Beneficiation of the gold bearing ore by gravity and flotation //International Journal of Minerals, Metallurgy, and Materials. – 2012. – Т. 19. – №. 2. – С. 106-110.

Polymer-modifier for the road bitumen manufacture of the fourth generation

T.I. Lebedzeva & Y.A. Bulauka
Polotsk State University, Novopolotsk, Belarus

ABSTRACT: This study proposes a method for obtaining fourth-generation road binders, that is, Polymer-modified bitumen. The method is different from the existing ones due to the usage of petrochemical production wastes. Modifying bitumen with petrochemical waste will expand the range of polymer modified bitumens, reduce the involvement of imported additives (such as styrene-butadiene-styrene type polymer), reduce the environmental load andprovide a positive economic effect by using cheaper and more available components compared to industrially used analogs. In addition, obtained polymer modified bitumens, in terms of their basic performance indicators, comply the requirements of modern standards for modified road bitumens, ensuring their reliable operation in the composition of asphalt mixes.

1 INTRODUCTION

Petroleum bitumens have been and still rema`in the main type of binding materials used in road construction. Increasing technical level of modern vehicles, the annual growth of the car park and the increase in load on the roadway, the growth of road networks in areas with drastic fluctuations in temperature, require an increase in the production of road bitumen and improvement of their performance characteristics. All of the above have become the result of the adoption of a new standard GOST 33133, that establishes significantly increased requirements for the main indicators of bitumen. However, the introduction of the processes aimed at deeper oil refining at oil refinery plants leads to a sharp deterioration in the hydrocarbon group analysis of oil residues used as raw material for road binders, due to the increased gas-oil extraction in the vacuum distillation columns of fuel oil. Moreover, heavy and paraffinic oils are being processed. A number of oil refineries are already forced to process tars from VU_5^{80} around 1500-1800 s (at a "historical" rate of up to 60s). All of this requires refineries to retool and modernize the existing manufactures to produce better-quality road bitumen. At the same time, bitumen, in contrast to light petroleum products, make almost no profit for oil refineries and such upgrades will be unprofitable.

From a technological and economic point of view (Quintero, 2012), the most appropriate way to solve the problem of improving the performance properties of bitumen materials for road construction is to add modifiers to bring the main quality indicators of road bitumen to the requirements of modern standards.

The fourth-generation road binders produced today, which is bitumens modified with polymers, provide a high level of performance indicators, such as: crack resistance, heat resistance, shear resistance, long-term strength and resistance to low-temperature cracking. The implementation of polymer additives into bitumen enables regulating the structural and mechanical properties of the material directionally. Polymer-bitumen binders combine the qualities inherent in polymers (high elasticity, heat resistance, strength, frost resistance) and oil road bitumens. More than 25% of asphalt concrete surfaces used in Europe contain polymer additives, in the northern states of the USA this figure is 50% (Thomas, 2008), and in China since 2000 usage of polymer-bitumen binders in the construction of new highways has become mandatory (Zhang, 2013).

Various kinds of thermoplastic elastomers (Styrene-butadiene-styrene (SBS) copolymers, Styrene-Isoprene-Styrene (SIS) copolymers, Styrene-Butadiene Rubber (SBR) copolymers and Styrene-Ethylene/Butylene-Styrene (SEBS) copolymers), plastomers (Ethylene Vinyl Acetate (EVA), Ethylene Butyl Acrylate (EBA), Low-Density Polyethylene (LDPE), High-Density Polyethylene (HDPE), Ethylene-Propylene-Diene Monomer (EPDM), Polypropylene (PP), Ethylene Methacrylate (EMA), Polyvinyl Chloride (PVC) and Polystyrene (PS)), natural rubber, crumb rubber, etc., have been investigated as modifiers of road bitumen (Behnood A., 2019). SBS copolymers, because of their excellent engineering properties, are the most frequently used to modify the properties of road bitumen (Schaur, 2017; Laukkanen, 2018; Kumar, 2016; Nian, 2018; Hao, 2017; Brasileiro, 2019). Polymer-bitumen binders on the territory of Russia are produced on the basis of bitumens and block copolymers of the styrene-butadiene-styrene type according to GOST R 52056-2003; in Belarus – according to STB 1220-2009, in the European Union – according to EN 14023: 2010.

Among the main drawbacks of polymer-bitumen binders is the use of expensive plasticizers (for e.g., commercial industrial oil) and polymers (for e.g., thermoplastic elastomers such as styrene-butadiene-styrene rubbers). Adding a small amount of about 1% wt. styrene-butadiene-styrene polymer into bitumen increases its cost by more than 2 times (Coplantz, 1993).

Spent petroleum oils can be used instead of expensive industrial oil to reduce the cost of the final product (Fernandes, 2018); however, this waste is characterized by instability of the composition and may contain worn out additives, wear products and similar substances that may affect performance characteristics and durability of asphalt binders.

In view of the foregoing, the main objective of the study was formulated. The objective is to develop a polymer-bitumen composition based on road bitumen and a polymer modifier from petrochemical waste, which would differ from industrially used analogues by using cheaper and more available components, and approach the requirements for modified road bitumen, ensuring their reliable operation in the composition of asphalt concrete mixtures.

2   OBJECTS, TASKS AND STAGES OF THE STUDY

The research task is to develop a method for producing a polymer modifier from petrochemical waste (low molecular weight polyethylene (LMWPE) and the residue of pyrolysis heavy resin) and determining its optimal adding dosage into road bitumen to obtain polymer bitumen compositions (PBC) with improved performance properties (wide plastic range and lower brittleness temperature, improved adhesion properties) compared to original road bitumen. The average molar-mass of LMWPE is in the range of 1000...5000 g/mol, the double-hump molecular mass distribution ranges from 50 to 5250 and LMWPE is an amorphocrystalline substance (crystallinity from 3 to 16%). The total content of CH_3- groups is approximately 3 times higher than their content in high pressure polyethylene (low density poly ethylene) and is 4...8 units per 100 carbon atoms. A significant part of them falls on the side branches, among which a large proportion are ethyl and butyl radicals. The molecules of LMWPE, in addition to the skeletal CH_2-groups, also contain carbonyl, vinylene and other unsaturated bonds and terminal CH_3-groups. LMWPE is a white to grayish yellow ointment or wax-like product with no foreign inclusions and a structured polymer. It is quite resistant to various factors, including atmospheric: water, salts, some acids and alkalis, soluble in aliphatic and aromatic hydrocarbons at a temperature of more than 80 ° C; is non-polar and hydrophobic, has high adhesion to various materials, i.e. properties inherent in lubricating and bituminous oil materials.

The introduction of a plasticizer into bitumen can improve the plasticity of the polymer-bitumen compositions and the compatibility of the polymer with bitumen. As a plasticizer, a distillation residue under atmospheric pressure of a heavy hydrocarbon pyrolysis resin, which is a mixture of high-boiling multi-core aromatic hydrocarbons (the fraction boiling above 230 °C), was taken. It should be noted that for the study it is proposed to use only that part of the pyrolysis heavy resin of hydrocarbon feed from which the liquid fractions boiling to 230°C are separated by atmospheric distillation, since this fraction can be effectively used to separate individual aromatic hydrocarbons, in particular, naphthalene (Bulauka, 2018).

Scientific research included four successive stages: the selection and study of the properties of the raw components of road bitumen, as well as production wastes to obtain a modifier: low molecular weight polyethylene and heavy pyrolysis resin. At the second stage, the modifier itself was prepared from the waste. At the third stage, the bitumen was compounded with a polymer modifier. At the final stage, we tested the main indicators of the quality of bituminous mixtures.

3   RESEARCH METHODS

A polymer modifier was obtained by mixing the residue of a heavy pyrolysis resin of a hydrocarbon feedstock (plasticizer) and low-molecular-weight polyethylene (polymer) with the help of an anchor-type mechanical mixer in a ratio of 1: 1 ... 2 and subsequent heat treatment (105 ± 5)° C for 120 min).

The resulting polymer additive in concentrations of 1, 3 and 5% wt. was mixed with bitumen in a cylindrical mixer with a mechanical stirrer with adjustable heating of the entire outer surface at a temperature higher than the softening temperature of bitumen, but not more than 150 °C for 120 minutes with constant stirring at an rotation speed of 60 r/min until a homogeneous state of the mixture was obtained.

Low molecular weight polyethylene with a dropping point of 90 °C, manufactured according to TU RB 300041455.031-2004 at the Polimir plant of Naftan OJSC (Belarus) was used.

As a plasticizer, a distillation residue of pyrolysis heavy resin (PHR), non-distilled under atmospheric pressure, manufactured according to TU RB 300041455.002-2003 at the "Polimir" plant of OJSC "Naftan" (Belarus) was used.

4   RESULTS AND DISCUSSION

The properties of the obtained polyethylene bitumen compositions in comparison with bitumen BND 70/100 produced by Naftan and quality standards are presented in Table 1.

The softening temperature of the binder (according to GOST 11506 (EN 1427)) determines the resistance to rutting, which characterizes the degree of mobility and suitability of bitumen for use in various temperature conditions, i.e. plastic and thermal qualities of bitumen. The results showed that the involvement of only low molecular weight polyethylene in bitumen leads to a decrease in temperature and, as a consequence, a decrease in heat resistance of bitumen, this effect is probably due to the fact that the LMWPE macromolecules are in the form of rolled tangles and globules, polymer dissolution and homogenization. Involving only the PHR residue also leads to a decrease in the softening temperature, which is caused by an increase in the content of liquid oil components of bitumen.

The use of a thermally prepared combined additive consisting of 1 part of a plasticizer and 1 ... 2 parts of polymer leads to an increase in the softening temperature of bitumen (up to 5 °C) and, as a result, its heat resistance. This fact can be explained by good solubility of LMWPE in the components of the PHR pitch. It is well known that the better the polymer is dissolved in a solvent, the higher viscosity of the solution itself is. The polymers adsorb the hydrocarbons of the PHR residue and form a separate dispersed phase, which leads to a decrease in the oil/asphaltene ratio in the bitumen, as a result of which an increase in viscosity is observed.

Penetration or penetration depth of a bitumen needle (according to GOST 11501 (EN 1426)) is a conditional indicator, a characteristic of the inverse viscosity value, as well as a bitumen flow behavior index. The results of the analyses show that the involvement of the combined additive in all samples leads to an increase in the penetration, as a result, to a decrease in the hardness of the bitumen. The greater softness of the modified bitumen is due to the fact that the LMWPE macromolecules are stretched in the residue of the PHR, and the maltenous part of the bitumen, which leads to the formation of an elastic structural network in the bitumen.

Table 1. Properties of the obtained polymer-bitumen compositions in comparison with commercial bitumen and quality standards.

Main factors	Penetration according, 0,1 mm at 25°C	Softening Temperatures on a Ring and a Sphere, °C	Tensile Properties cm at 25 °C	Temperature of Fragility, °C	Penetration Index
Norm acc. to STB EN 12591*	70-100	43-51	Not defined	<-10	-1.5...0.7
Norm acc. to GOST 33133*	71-100	>47	>62	<-18	-1...1
Norm acc. to STB 1220**	71-100	>52	Not defined	<-20	Not defined
Norm acc. to GOST R52056 ***	A minimum of 60	>54	A minimum of 25	<-20	-
Norm acc. to EN 14023 ****	45-80	>65	-	<-20	-
Actual value	69	44.9	61.5	-15	-1.84
Bitumen * modified by combined additive consisting of PHR residue and LMWPE in mass ratio 1:2					
1% wt.	86.3	46.0	64.5	-25	-0.92
3% wt.	73.5	47.5	57.5	-19	-0.92
5% wt.	72.7	49.3	52.2	-15	-0.47
Bitumen * modified by combined additive consisting of PHR residue and LMWPE in mass ratio 1:1					
1% wt.	90.0	43.3	86.5	-26	-1.67
3% wt.	80.0	44.0	61.9	-20	-1.74
5% wt.	74.0	44.3	48.4	-16	-1.86

* for viscous road oil bitumen of brand BND 70/100
** for modified road bitumen of brand MRB 70/100
*** for polymer-bitumen binder of brand PBB 60
****for Polymer Modified Bitumen of brand PMB 45/80-65

The penetration index (according to GOST 22245) characterizes the degree of bitumen colloidal nature or the deviation of its state from purely viscous. According to the penetration index (PI), bitumens are divided into three groups:

- PI is less than -2, sol-type bitumens, as a rule, bitumens not having a dispersed phase or containing strongly peptized asphaltenes.
- PI from -2 to +2, sol-gel bitumens (as a rule, residual and slightly oxidized bitumens).
- PI is greater than +2 have pronounced colloidal properties of "gels" (usually, oxidized bitumens).

The penetration index according to GOST 22245 is determined by the formula:

$$PI = \frac{30}{1 + 50A} - 10 \quad (1)$$

where

$$A = \frac{2,9031 - \log P}{T_p - 25} \quad (2)$$

where P - needle penetration depth at 25°C, ×0,1 мм; and T_p - softening temperature, °C.

The requirements of modern GOST 33133-2014 for viscous road bitumen provide for a change in the peneration index from -1 to +1. The calculations of the index characterizing the thermal sensitivity of bituminous binders, that is, the generation index, have been performed. According to the obtained results, 1, 3, and 5% combined additive consisting of 1 part of the PHR and 2 parts of LMWPE got into this range. The dispersed structure of modified bitumen with such an additive is closest to the "sol-gel" type, optimal from the point of view of the quality of road bitumen.

Stretching property (ductility) (according to GOST 11505) is the property of bitumens to stretch into thin threads under the influence of an applied tensile force. It is characterized by absolute elongation to break the thread (in cm) of a sample of bitumen at a given temperature. The greater the extensibility, the more elastic the bitumen is, the better the adhesion to dry surfaces when molten material is applied to them and the better the bonding properties (cohesion) are. Analysis of the research results has shown that the use of a 1% combined additive consisting of 1 part plasticizer and 1 ... 2 parts of polymer leads to an increase in bitumen tensile properties, which makes it possible to predict an increase in elastic and adhesive properties (cohesion).

The results of the analysis of the indicators of penetration and ductility confirm the presence of a synergistic interaction between the polymer and the plasticizer in the combined additive. The joint effect of the components of the combined additive on the structure of bitumen can significantly improve its strength and heat resistance at elevated temperatures, as well as plasticity and elasticity.

At negative temperatures bitumens become brittle. The temperature of fragility (according to GOST 11507 (EN 12593)) characterizes the crack resistance and low-temperature properties of bitumen: the lower it is, the better the quality of bitumen is. The results of the research show that the use of combined additives based on the residue of PHR and LMWPE can improve low-temperature properties, the depression fragility temperature reaches 10°C, as a result, the road surface will work better in harsh climate and cold weather.

The plasticity interval is an important characteristic of road bitumen, it is estimated approximately from the difference between softening temperatures and fragility temperatures. The wider this range is, the better the road surface behaves when the ambient temperature fluctuates. Bitumens with a wide plasticity interval also have a higher deformation capacity, resistance to cracking at low temperatures and shear stability at elevated temperatures. With an increase in the plasticity interval, the adhesion properties of bitumens also increase. Depending on the climatic conditions, binders with a plasticity interval of 60...90°C should be used for the construction of the upper layers of pavements, and binders with a plasticity interval of 50 ... 70°C should be suitable for the lower layers of the pavement. The results of research indicate that the use of combined additives increases the plasticity range from 60°C (for the original commercial bitumen) to 70 ... 82 °C, which will lead to an increase in the deformation capacity, resistance to cracking at low temperatures and resistance to shear at high summer temperatures.

Adhesion (according to GOST 11508) to mineral materials determines the most important quality of the bituminous binder and it is the parameter that determines the durability of road pavements. Adhesion is caused by the formation of a double electric field at the interface of the footage of bitumen and rock material, depends on the polarity of the components and is determined by the electrical conductivity of the solutions of these substances in non-polar solvents. The study found that the involvement of the combined additive leads to an improvement in the adhesion properties of the original bitumen (the surface of the mineral material is covered with bitumen more than ¾, which is shown in Figure 1), probably due to an increase in the intermolecular interaction forces between the bitumen components.

It has been experimentally proven that a synergistic effect, leading to an improvement in the properties of bitumen compositions, occurs only after preliminary mixing of the components of the polymer modifier and plasticizer and their subsequent heat treatment and at a temperature of (105 ± 5)°C for 120 minutes with constant mixing.

a   b

Figure 1.   Degree of adhesion of modified bitumen to mineral material before (a) and after (b) adhesion testing.

Pyrolysis heavy resin in the preliminary mixing with low molecular weight polyethylene and their subsequent heat treatment improves the compatibility of the oligomer with bitumen and increases the plasticity of polyethylene bitumen compositions.

The resistance of bitumen to aging, to a greater degree thermal and chemical, is characterized by thermal oxidative stability. The analysis of the resistance of the modified bitumen to solidification at 163°C (according to EN 12607-1) by the percentage of weight loss after heating and changing the softening temperature was performed. It was found that the values obtained are within the acceptable limits normalized by standards (GOST 33133, STB EN 12591 and EN 14023). At the same time, with an increase in the concentration of the combined additive, the thermo-oxidative stability to aging processes improves.

Similar studies, in addition to the BND 70/100 bitumen, were carried out for the bitumen of viscous oil road BND 40/60, BND 60/90 and BND 90/130, the above regularities were confirmed.

For industrial implementation it is proposed to involve 1 . . . 3% wt. the combined additive obtained from 1 part of the residue PHR and 2 parts of LMWPE. The cost of the raw material components of the polymer modifier and plasticizer is about $ 215 per ton, which is commensurate with the cost of the bitumen itself at about $ 220 per ton.

5   CONCLUSIONS

During the study, the combined effect of the components of the combined additive of the polymer modifier and plasticizer on the structure of road bitumen was studied. It was revealed that the proposed waste of petrochemical plants can increase the softening temperature and at the same time needle penetration depth, increase ductility, reduce the temperature of fragility, ensure plasticity interval and penetration index required by the standards, improve adhesion to the surface of mineral materials, with satisfactory aging resistance of polyethylene bitumen composition. All this together will increase the strength and heat resistance of polymer-bitumen compositions, resistance to rutting at elevated temperatures, as well as ductility, elasticity, crack resistance, which allows us to predict high quality road surface.

REFERENCES

Behnood A., Gharehveran M.M. 2019. Morphology, rheology, and physical properties of polymer-modified asphalt binders. *European Polymer Journal*. 112: 766–791.

Brasileiro L.L., Moreno-Navarro F., Martínez R.T., Sol-Sanchez M. 2019. Study of the feasability of producing modified asphalt bitumens using flakes made from recycled polymers. *Construction and Building Materials*. 208: 269–282.

Bulauka Y. A. & Yakubouski S.F. 2018. Rational refining of heavier cut of pyrolysis gas oil, *Oil and Gas Horizons: abstract book of 10th International Youth Scientific and Practical Congress, Moscow, November 19–22, 2018*, Gubkin Russian State University of Oil and Gas (National Research University): Moscow: 61.

Coplantz J.S., Yapp M.T., Finn F.N., 1993. Review of Relationships between Modified Asphalt Properties and Pavement Performance, *SHRP-A-631*: 243.

Fernandes S.R. M., Silva H.M.R.D., Oliveira J.R.M. 2018. Developing enhanced modified bitumens with waste engine oil products combined with polymers. *Construction and Building Materials*. 160: 714–724.

Hao J., Cao P., Liu Z., Wang Z., Xia S. 2017 Developing of a SBS polymer modified bitumen to avoid low temperature cracks in the asphalt facing of a reservoir in a harsh climate region. *Construction and Building Materials*. 150: 105–113.

Kumar K., Singh A., Maity S., Srivastava M., Garg M. 2016. Rheological studies of performance grade *bitumens* prepared by blending elastomeric *SBS* (styrene butadiene styrene) co-polymer in base bitumens. *Journal of Industrial and Engineering Chemistry*. 44: 112–117.

Laukkanen O., Soenen H., Henning Winter H., Seppala J. 2018. Low-temperature rheological and morphological characterization of SBS modified bitumen. *Construction and Building Materials*. 179: 348–359.

Nian T., Li P., Wei X., Wang P., Guo R. 2018. The effect of freeze-thaw cycles on durability properties of SBS-modified bitumen. *Construction and Building Materials*. 187: 77–88.

Quintero L.S. & Sanabria, L.E. 2012. Analysis of Colombian Bitumen Modified With a Nanocomposite. *Journal of Testing and Evaluation (JTE)*. 40 (7): 93–97.

Schaur A., Unterberger S., Lackner R. 2017. Impact of molecular structure of SBS on thermomechanical properties of polymer modified bitumen. *European Polymer Journal*. 96: 256–265.

Thomas K.P. & Turner, T.F. 2008. Polyphosphoric-acid modification of asphalt binders: Impact on rheological and thermal properties. *Road Materials and Pavement Design*. 9 (2): 181–205.

Zhang C. & Wang, Y. 2013. Thermal, mechanical and rheological properties of polylactide tough-ened by epoxidized natural rubber . *Materials & Design*. 45: 198–205.

Topical Issues of Rational Use of Natural Resources 2019 – Litvinenko (Ed)
© 2020 Taylor & Francis Group, London, ISBN 978-0-367-85720-2

Manufacturing process and optical properties of zinc oxide thin films doped with zinc oxide nanoparticles

W. Matysiak, M. Zaborowska & P. Jarka
Institute of Engineering Materials and Biomaterials, Silesian University of Technology, Gliwice, Poland

ABSTRACT: Due to much better physical properties of ZnO compared to the ones of titanium dioxide (TiO_2), which is currently the most used material in dye sensitized solar cells, efforts are being made to fabricate DSSCs with thin films and/or nanostructures, including nanowires, nanofibres and nanoparticles of zinc oxide. In this paper, zinc oxide thin films were prepared using sol-gel and spin coating methods from $Zn(CH_3COO)_2$ x $2H_2O$ dissolved in ethanol and acetic acid with ZnO monocrystalline nanoparticles of 0 and 10% (wt.) relative to the final concentration of produced solutions. The effect of calcination process on ZnO thin films at 600°C were examined using atomic force microscope to investigate the morphology of semiconductor coatings. Besides, optical properties were analysed on the basis of absorbance in the function of wavelength spectra and the values of energy band gaps were studied.

1 INTRODUCTION

In recent years, renewable energy sources, such as solar, geothermal or wind energy have attracted much attention from scientists around the world, due to the possibility of replacing fossil fuels which cause rising air pollution. The negligible impact on living organisms, the increasing ease of production of devices using natural energy with the development of technology around the world, the profitability of conversion natural energy into electricity, as well as the lack of damaging effect on environment are the main factors influencing the increase of interest among scientists (Ellabban et al. 2014). The discovery in 1991 of dye-sensitized solar cells made by O'Reagen and Gratzel introduced the possibility of producing clean energy by converting solar energy into electricity (O'Regen et al. 1991). Dye-sensitized solar cells are made of few closely cooperating elements: photoanode, which is made of thin metal oxide, semiconductor film; dye, responsible for absorbing electromagnetic radiation photons; electrolyte, the REDOX substance and platinum electrode. The DSSC principle of operation is based on the absorption of solar energy photons by a dye adsorbed on the surface of metal oxide layer, which in the process of current flow is reduced by electrolyte ions, being p-n junction with semiconductor thin film (Grätzel et al. 2003). It has been recently observed, that zinc oxide thin films are gaining much popularity, particularly in applications such as toxic gas sensors, photocatalytic materials and photovoltaic cells. Due to much better physical properties of ZnO compared to the ones of titanium dioxide (TiO2) (Mali et al. 2013, Tański et al. 2016, Matysiak et al. 2017), which is currently the most used material in dye sensitized solar cells, which include energy band gap of 3.22 eV, excitons excitation energy of 60 meV and electron mobility of 155 cm2 V-1 s-1 compared to the value of 10-5 cm2 V-1 s-1 for TiO2, efforts are being made to fabricate DSSCs with thin films (Govindarajan et al. 2015) and/or nanostructures, including nanowires (Baxter et al. 2006, Matysiak et al. 2019), nanofibers (Kim et al. 2015) and nanoparticles of zinc oxide (Lai et al. 2010). The growing interest in replacing titanium dioxide forces researches to try to produce DSSCs based on zinc oxide having an increased solar energy into electricity conversion efficiency (Zhan et al. 2009, Kim et al. 2007, Hongsith et al. 2015). Only in 2000 – 2016 years 860 publications about usefulness of zinc oxide in production of dye-sensitized solar cells were recorded. Moreover, another important aspect deciding on the advantage of using ZnO over TiO_2 is the lower energy losses resulting from the physical properties of nanostructures and thin zinc oxide layers. The main parameter determining the efficiency of dye sensitized solar cell is the efficiency of converting solar energy into electricity, which is closely related to the topography of the surface of the film acting as an electrode with the adsorbed dye. The multitude of thin-layer manufacturing methods optimizes semiconductor coating manufacturing processes with a specific surface roughness, as well as a sufficiently large specific surface area for the adsorption of a dye responsible for absorbing light energy at a certain wavelength.

This paper describes a method for the production of thin semiconductor zinc oxide films using sol-gel and spin coating technique, as well as a study of the surface morphology and optical properties. The analysis was carried out for possible use as electrodes in dye sensitized solar cells.

2 MATERIALS AND METHODOLOGY

Thin semiconductor zinc oxide films were produced using the sol-gel technique and the spin coating method, from solutions for which: ethyl alcohol (EtOH, purity of 99.8%, Sigma Aldrich), zinc acetate dihydrate ($Zn(CH_3COO)_2$ x $2H_2O$, ZAD, Sigma Aldrich), acetic acid (CH_3COOH), Sigma Aldrich) and distilled water (H_2O) was used successively. In addition, mono-crystalline zinc oxide (ZnO-NPs) nanoparticles were used to prepare the solution, which created thin ZnO films decorated with zinc oxide particles.

The first solution was prepared by adding zinc oxide precursor in form of $Zn(CH_3COO)_2$ x $2H_2O$ (0.22 g) to 5 mL of ethanol. Distilled water and acetic acid were then added to the ZAD/EtOH mixture, in molar ratios: $CH_3COOH/ZAD = 5$ and $H_2O/ZAD = 67$. The resulting solution was stirred on magnetic stirrer for 30 minutes. The measured amount of ZnO-NPs powder was added to 5 mL of ethanol to create 10% (wt.) suspension with zinc oxide nanoparticles. Such obtained mixture was sonicated for 15 minutes to break down the ZnO agglomerates. The next step was to add the remaining reagents in the amounts used to prepare the first solution and set them to stirring on a magnetic stirrer for half an hour. Then, the solutions were subjected to a spin coating process using parameters of 3000 rpm and 30 seconds. Nest, the thin films of zinc oxide were annealed at 600°C for 1 hour under atmospheric air and the allowed to cool at room temperature.

3 RESULTS AND DISCUSSION

3.1 *The morphology analysis of ZnO thin films*

The surface morphology images of thin spin coated, semiconductor zinc oxide films, which were obtained before and after annealing process are presented in Figure 1 and Figure 2. The topography analysis of the ceramic surfaces obtained using atomic force microscope (AFM) showed an increase in coating surface roughness in the case of thin ZnO films and ZnO containing monocrystalline ZnO-NPs thin films after calcination process. Moreover, all ZnO and ZnO/ZnO-NPs coatings are characterized by nanometric thicknesses.

The increase in surface roughness of thin films is caused by the increase of single zinc oxide grains during the annealing process at high temperature. All thin films are characterized by a columnar structure of crystallites with growth orientation along the c axis during crystallization, with extended ZnO grains placed perpendicular to the surface of the substrate (Chaitra et al. 2017). The significant increase in roughness of ZnO/ZnO-NPs coatings after the calcination process as compared to the roughness of ZnO coatings after the annealing process is caused by the addition of monocrystalline nanoparticles acting as a reinforcing phase, which also increases the amount and degree of crystallite dispersion on the surface of zinc oxide films decorated with ZnO nanoparticles. The analysis of AFM images of semiconductor ZnO/ZnO-NPs thin films with a surface area of 100 μm^2 allows to conclude that 10% (wt.) of zinc oxide in relation to final concentration will contribute to increased mobility of electrons in the area of the obtained

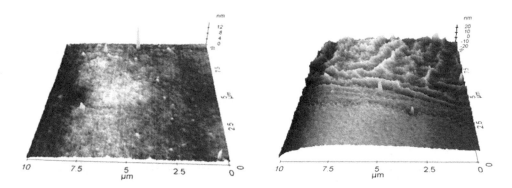

Figure 1. AFM surface image of ZnO thin films; before calcination (at the top) and after calcination process (at the bottom).

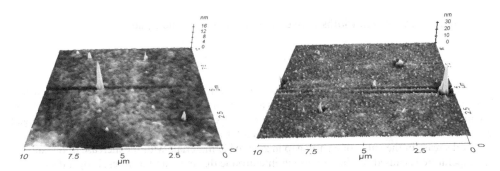

Figure 2. AFM surface image of ZnO/ZnO-NPs thin films; before calcination (at the top) and after calcination process (at the bottom).

monolayers. This significant property will allow the use of thin polycrystalline ZnO/ZnO-NPs films as a photoanode in dye sensitized solar cells.

3.2 The optical properties of ZnO and ZnO/ZnO-NPs thin films

The optical properties of the obtained thin ZnO coatings were examined using UV-Vis spectroscopy by plotting the absorbance graphs as a function of wavelength and determining the energy gap values of the zinc oxide films (Figure 3). The analysis of absorbance spectra of thin ZnO and ZnO/ZnO-NPs coatings showed that the sharp edge of absorption of all semiconductor films was at 340 nm, which probably corresponds to the charge transfer process from the valance band to the conduction band in semiconductor metal oxides (Hernandez et al. 2007). Moreover, the annealing process at 600°C resulted in shifting the sharp absorption edge towards longer waves, corresponding to the infrared range (Zhao et al. 2005). In contrast, the absorption maximum of the obtained zinc oxide films is for a 300 nm wavelength, which corresponds to the near-ultraviolet waves range. For the electromagnetic radiation wavelengths in the visible range, thin layers of ZnO show absorption at the level of 0.2 – 0.5 degree, thus they remain transparent in the spectrum region from circa 400 to 700 nm (Zhao et al. 2006). Moreover, in the case of ZnO/0% ZnO-NPs coatings, a decrease in the absorption level after annealing process was observed at 600°C. This decrease may be caused by an increase in coating roughness, resulting in increased scattering of radiation reaching the zinc oxide coating (Nagayasamy et al. 2013). Due to the fact, that about 25% of solar radiation reaching the Earth's surface are magnetic waves corresponding to ultraviolet wavelengths, a high level of absorption characterizing the ZnO and ZnO/ZnO-NPs thin films obtained gives the possibility to use semiconductor coatings in the construction of electrodes in dye sensitized solar cells.

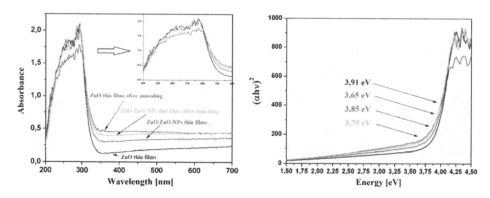

Figure 3. (on the left) UV-Vis spectra obtained for ZnO and ZnO/ZnO-NPs thin films and (on the right) the energy gap widths of ZnO and ZnO/ZnO-NPs thin films.

The values of energy gap widths were determined based on the equation:

$$(\alpha h \nu) = A(h\nu - E_g)^{\frac{1}{2}},$$

where A is a constant, E_g is the value of the energy gap, ν is the radiation frequency and h is Planck's constant. Figure 3 shows a $(\alpha h \nu)^{1/2}$ as a function of the energy of photons of electromagnetic radiation. For the obtained semiconductor coatings, the lowest value of the energy band gap was estimated for ZnO and ZnO/ZnO-NPs films after annealing – 3.65 eV and 3.75 eV, respectively. A decrease in the value of energy gap in zinc oxide materials with the increase in the annealing temperature is caused by the increase in interatomic distances and as a result by a decrease in the average potential seen by electrons (Rai et al. 2012). Higher value of energy gap obtained for ZnO films with 10% (wt.) concentration of zinc oxide nanoparticles than this one for ZnO thin films is caused by a slight increase in absorption in the visible light area of the ZnO/ZnO-NPs coating after annealing (Raghu et al.). The immediate effect of the decrease in band gap width in semiconductor materials is the increase in crystalline particle size of ZnO in thin films. The atoms, which form particles increase in size, which as a result cause moving of the valence and conduction band electrons to be more to the ions core of this particles and thus decreasing the energy band gap of ZnO and ZnO/ZnO-NPs thin films (Ibrahim et. al.).

4 CONCLUSIONS

This paper presents the process of producing ZnO and ZnO films decorated with zinc oxide nanoparticles using sol-gel technique and spin coating method from solutions containing ZnO precursor in the form of zinc acetate dihydrate with a ZnO monocrystalline nanoparticles with concentration of 0 and 10% (wt.), respectively. The surface topography analysis made using atomic force microscopy showed an increase in roughness as the content of metal oxide nanoparticles increases in the coating material. Moreover, the increase in the roughness of films with nanometric thickness was influenced by the annealing process, whereby ZnO coating decorated with ZnO nanoparticles showed increased roughness due to the increase in the amount and dispersion of crystallites with a column structure on the surface of the coatings. The analysis of optical properties carried out on the basis of absorbance spectra as a function of wavelength of electromagnetic radiation showed a decrease in the absorption level in the near-ultraviolet waves region after annealing. The applied calcination temperature of ZnO coatings also resulted in a decrease in the value of the energy gap width of the metal oxide material, which has an impact on the increase in the conductivity of the thin films and more efficient electron mobility within the films. The decrease in energy band gap width caused by calcination process is due to the crystallization of zinc oxide thin films. Interatomic interactions in calcined films caused shift in valence and conduction bands electrons closer to the ions core and thus decrease in activation energy in semi-conductor films.

REFERENCES

Baxter, J.B., et al., *Synthesis and characterization of ZnO nanowires and their integration onto dye-sensitized solar cells*, Nanotechnology 17 (2006), S304–S312.

Matysiak, W. et al., *Novel bimodal ZnO (amorphous)/ZnO NPs (crystalline) electrospun 1D nanostructure and their optical characteristic*, Applied Surface Science, 474 (2019) 232–242.

Chaitra, U., et al. *Effect of annealing temperature on the evolution of structural, microstructural and optical properties of spin coated ZnO thin films*, Ceramics International,43 (2017) 7115–7122.

Ellabban, O., et al., *Renewable energy resources: Current status, future prospects and their enabling technology*, Renewable and Sustainable Energy Reviews. 39 (2014), 748–764.

Elilarassi, R. & Chandrasekaran, G., *Preparation and optical characterization of ZnO thin film for opto-electronic applications*, 2009 International Conference on Emerging Trends in Electronic and Photonic Devices & Systems, ELECTRO'09.

Govindarajan, R., et al., *Sol gel dip coated ZnO films for DSSC applications*, International Journal of Applied Engineering Research, 10 (2015), 79–81.

Grätzel, M., *Dye-sensitized solar cells*, Journal of Photochemistry and Photobiology C: Photochemistry Reviews, 4 (2003), 145–153.

Hernandez, A., et al., *Sol-gel synthesis, characterization and photocatalytic activity of mixed oxide ZnO-Fe2O3*, Journal of Sol-Gel Science and Technology, 42 (2007) 71–78.

Hongsith, K, et al., *Efficiency Enhancement of ZnO Dye-sensitized Solar Cells by Modifying Photoelectrode and Counterelectrode*, Energy Procedia, 2015;79;360-365, DOI: doi:10.1016/j.egypro.2015.11.503.

Khan, Z., et al., *Optical and Structural Properties of ZnO Thin Films Fabricated by Sol-Gel Method*, Materials Sciences and Applications, 2 (2011) 340–345.

Kim I.D., et al., *Dye-sensitized solar cells using network structure of electrospun ZnO nanofiber mats*, Appl. Phys. Lett. 2007;91, DOI: 10.1063/1.2799581.

Kim J.H., et al., *Electrospun ZnO Nanofibers as a Photoelectrode in Dye-Sensitized Solar Cells*, J Nanosci Nanotechnol. 15 (2015), 2346–2350.

Lai M.H., et al., *ZnO-Nanorod Dye-Sensitized Solar Cells: New Structure without a Transparent Conducting Oxide Layer*, International Journal of Photoenergy, 2010 (2010), 5 pages.

Mali S.S., et al, *Novel Synthesis and Characterization of Mesoporous ZnO Nanofibers by Electrospinning Technique*, ADS Sustainable Chem. Eng., 1 (2013), 1207–1213.

Matysiak, W., et al., *Analysis of the Optical Properties of PVP/ZnO Composite Nanofibers*, Properties and characterization of modern materials. Eds.: Andreas Ochsner, Holm Altenbach. Singapore: Springer, 2017, s. 43–49.

Nagayasamy, N., et al., *The Effect of ZnO Thin Film and Its Structural and Optical Properties Prepared by Sol-Gel Spin Coating Method*, Open Journal of Metal, 3 (2013) 8–11.

O'Regan. B & Grätzel, M., *A low-cost, high-efficiency solar cell based on dye-sensitized colloidal TiO2 films*, Nature, 353 (1991), 737–740.

Raghu, P., et al, *Enhanced Optical Band-gap of ZnO Thin Films by Sol-gel Technique*, AIP Conf. Proc. 1728, 020469-1-020469-4.

Rai, R. C., et al., *Elevated temperature dependece of energy band gap of ZnO thin films grown by e-beam deposition*, Journal of Applied Physics 111 (2012).

Tański, T., et al., *Manufacturing and investigation of physical properties of polyacrylonitrile nanofibre composites with SiO2, TiO2 and Bi2O3 nanoparticles*, Beilstein J. Nanotechnol. 2016, 7, 1141–1155.

Tański, T., et al., *Analysis of optical properties of TiO2 nanoparticles and PAN/TiO2 composite nanofibers*, Materials and Manufacturing Processes, 10.11.2016, DOI: 10.1080/10426914.2016.1257129.

Zhang, W, et al., *Facile construction of nanofibrous ZnO photoelectrode for dye-sensitized solar cell applications*, Appl. Phys. Lett. 2009;95, DOI: 10.1063/1.3193661.

Zhao, Z. W. & Tay, B. K., *Optical properties of nanocluster-assembled ZnO thin films by nanocluster-beam deposition*, Appl. Phys. Lett. 87 (2005).

Zhao, L., et al, *Structural and optical properties of ZnO thin films deposited on quartz glass by pulsed laser deposition*, Applied Surface Science, 252 (2006) 8451–8455.

Ibrahim, A, et al, *The Influence of Calcination Temperature on Structural and Optical Properties of ZnO Nanoparticles via Simple Polymer Synthesis Route*, Science of Sintering, 49 (2017) 263–275.

Topical Issues of Rational Use of Natural Resources 2019 – Litvinenko (Ed)
© 2020 Taylor & Francis Group, London, ISBN 978-0-367-85720-2

Influence of capillary pressure on the restoration of the bottomhole zone permeability at the filtrate-oil interfacial phase

V.I. Nikitin, V.V. Zhivaeva, O.A. Nechaeva & E.A. Kamaeva
Samara State Technical University (SSTU), Samara, Russia

ABSTRACT: The paper presents a study on the restoration of permeability of the bottomhole formation zone after intervasion of the filtrate of drilling flushing fluid. The author conducted a study on samples of natural core using clay, potassium chloride, polymer clay and polymer flushing fluids. The theoretical part of the study is devoted to the choice of flushing fluid based on the saturation and radius of its invasion into the reservoir. These parameters are determined by calculation based on the theory of two-phase filtration. The formula for the distribution of phase fluxes with regard to capillary pressure is derived and its effect on the saturation and the radius of the filtrate invasion are established. A link has also been established between the restoration of permeability and interfacial tension at the interface of oil and water filtrate.

1 INTRODUCTION

At the drilling of productive strata on repression using water-based drilling flushing fluids (Morenov & Leusheva 2016, Blinov & Dvoynikov 2018), the filtrate penetrates into the pore space of the reservoir. As a result, a zone with a lower permeability, compared to the natural, appears (Nikitin & Zhivaeva 2016, 2017). In order to improve the quality of the opening of productive layers, using mathematical modeling methods (Blinov & Podoliak 2016), a criterion for the choice of flushing fluid was introduced. The proposed criterion, which includes the saturation of the pore space with the filtrate and the radius of its invasion (Zhivaeva & Nikitin 2016), allows to select the flushing fluid, taking into account the minimum decrease in permeability of the bottomhole formation zone during the initial drilling-in (Nikitin 2017, Morenov & Leusheva 2019). The dimensionless measure of the reduction in filtration properties of the bottomhole formation zone caused by the invasion of the drilling fluid filtrate is determined as follows:

$$\xi = \overline{S}_f r_f / r_w \qquad (1)$$

where $\overline{S}_f$ - average saturation of the bottomhole zone filtrate, r_f - filtrate penetration radius, r_w - well radius. The criterion for choosing the optimal flushing fluid is the condition of the minimum indicator ξ:

$$\xi \to \min \qquad (2)$$

2 RESEARCH QUESTIONS

Modeling the process of the filtrate invasion into the reservoir is based on the theory of two-phase filtration (Churakov & Nikitin 2017). The function of the distribution of phase flows is the ratio of the flow rate of the filtrate penetrating into the formation to the total flow rate of fluids (Nikitin et al. 2017). Based on the form of the distribution function of the flows of the contacting liquids, the parameters and are determined (Nikitin 2016). The classic form of this function is called the Buckley-Leverett equation and does not take into account capillary pressure:

$$f_f = \frac{1}{1 + \frac{k_{ro}\mu_f}{k_{rf}\mu_o}} \qquad (3)$$

where k_{ro}, k_{rf}, - relative phase permeability for oil and filtrate respectively, μ_o, μ_f,- dynamic viscosities for oil and filtrate appropriately. The phase flux distribution function is derived taking into account capillary pressure:

$$f_f = \frac{1 - \frac{2\pi h k k_{ro}}{q_t \mu_o} P_c}{1 + \frac{k_{ro}\mu_f}{k_{rf}\mu_o}} \qquad (4)$$

where q_t- total consumption, k - natural permeability of the formation, P_c - capillary pressure on the section of two liquids. If we neglect the capillary pressure, then the function (4) coincides with the Buckley-Leverett equation (3) (Kegang Ling et al. 2015). In subsequent calculations, capillary pressure is taken into account.

3 METHODS

In mathematical modeling, data from filtration experiments are used to saturate natural core samples from the Novofedorovskoye field with filtrates of drilling flushing fluids of various types, including clay, calcium chloride, polymer clay and polymer systems (Zhivaeva et al. 2018). The filtration experiment was carried out on the PMC-RP-1-40-AP/PP installation of the "Geologika" company in a package of SSTU (Figure 1.).

The experiment (Urshulyak et al. 2016) was carried out taking into account reservoir conditions, in connection with this, the initial water saturation was created in the core, the reservoir temperature and pressure were taken into account. The permeability recovery coefficient $K_{P.R.}$ was determined by injecting the filtrate into the oil-saturated core and restoring the inverse filtration of oil with access to the stationary mode. To determine the interfacial tension, a tensiometer of a rotating drop SVT manufactured by DataPhysics (Figure 2.) is used.

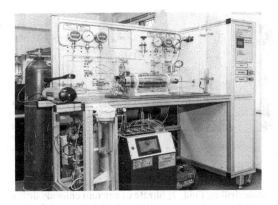

Figure 1. Appearance of PMC-RP-1-40-AP/PP unit, of the JSC "Geologika" in the complete set SSTU.

Figure 2. Measuring cell and optical zoom of SVT tensiometer manufactured by DataPhysics.

4 FINDINGS

It was established that with increasing capillary pressure at the border of the filtrate and oil, the radius of the filtrate invasion decreases, but at the same time the average saturation of the filtrate increases (Figure 3-4).

It has been established that for flushing fluids whose filtrates (Nikolaev & Leusheva 2016) have the highest interfacial tension at the boundary with oil, the permeability recovery

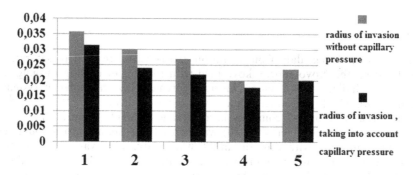

Figure 3. The radius of the filtrate invasion of drilling flushing systems with and without capillary pressure 1) filtrate of clay shale drilling fluid, 2) filtrate of calcium chloride drilling fluid 3) filtrate of polymer clay shale drilling fluid 4) filtrate of polymer drilling fluid 5) filtrate of polymer drilling fluid.

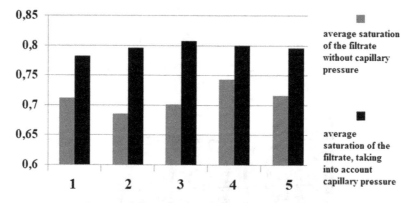

Figure 4. The average saturation of the filtrate of drilling flushing systems with and without capillary pressure 1) filtrate of clay shale drilling fluid, 2) filtrate of calcium chloride drilling fluid 3) filtrate of polymer clay shale drilling fluid 4) filtrate of polymer drilling fluid 5) filtrate of polymer drilling fluid.

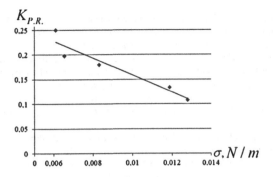

Figure 5. Correlation dependence between surface tension and coefficient of permeability recovery.

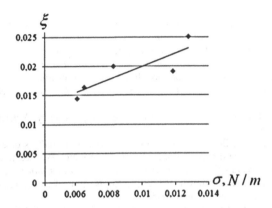

Figure 6. Correlation dependence between surface tension and by parameter ξ.

coefficient has the lowest values (Figure 5). The proposed criterion for the choice of flushing fluid on the basis of the invasion radius and saturation of the filtrate is also associated with capillary pressure (Figure 6).

5   CONCLUSION

From the results of this work, it follows that in addition to the rheological parameters of the drilling fluid itself, the interfacial tension for the filtrate should be monitored, since at high rates there is minimal recovery of permeability. Thus, fluids with a high interfacial tension rate are the least preferred for opening productive formations. The phase distribution function presented in the work can also be applied to simulate two-phase filtration.

REFERENCES

Blinov, P.A., Dvoynikov, M.V. Rheological and filtration parameters of the polymer salt drilling fluids based on xanthan gum, *Journal of Engineering and Applied Sciences* 2018: 13(14), pp. 5661-5664.
Blinov, P.A., Podoliak, A.V. The method of determining the effects of drilling fluid on the stability of loose rocks, *International Journal of Applied Engineering Research* 2016: 11(9), pp. 6627-6629.
Churakov, N.N. & Nikitin, V.I. 2017. Selecting the optimum parameters for reservoir penetrating fluid at Pokrovsko-Sorochinskoe filed. *Oil. Gas. Novation:* 10: 56-57.

Churakov, N.N. & Nikitin, V.I. 2017 Study of reservoir properties for the core material of the well 78 of the Pokrovsko-Sorochinskoye field *Oil and gas complex: problems and innovations abstracts of the II scientific-practical conference with international participation*: 97.

Kegang Ling &He Zhang & Zheng Shen 2015. A new approach to estimate unvasion radius of water-based drilling fluid filtrate to evaluate formation damage caused by overbalanced drilling. *SPE Drilling & Completion Publisher: Society of Petroleum Engineers*: 27-36.

Morenov, V., Leusheva, E. Influence of the solid phase's fractional composition on the filtration characteristics of the drilling mud (2019) *International Journal of Engineering, Transactions B: Applications* 32 (5), p. 794-798.

Morenov, V., Leusheva, E. Development of drilling mud solution for drilling in hard rocks, *International Journal of Engineering, Transactions A: Basics* 2017: 30(4), pp. 620-626.

Nikitin, V.I. 2017 Calculation of the volume of filtrate penetrated into the reservoir during the initial dissection. *Bulat readings: materials of the I International scientific-practical conference*: 195-197.

Nikitin, V.I. & Milkova, S.Y. & Ossiala V. B. A. 2017 Study of filtration rate of drilling fluids in laboratory studies Ashirovskie reading: *Ashirovskie readings: Collection of labor of a modern scientific-practical conference*:134-138.

Nikitin, V.I. 2016. Simulation of two-phase filtration of drilling flushing systems at the opening of the productive formation. *Oil and Gas Complex: Problems and Innovations: Abstracts of a Scientific-Practical Conference with International Participation:* 34.

Nikitin, V.I. & Zhivaeva, V.V. 2017. Dynamics of filtrate invasion of water-based drilling flushing systems into the formation. *Construction of oil and gas wells onshore and offshore* 11: 40-42.

Nikitin, V.I. & Zhivaeva, V.V. 2016. Filtration modeling of drilling flushing systems at the opening of the formation. *Drilling in complicated conditions: Materials of the International Scientific and Practical Conference*: 79-81.

Nikolaev, N.I., Leusheva, E.L. Increasing of hard rocks drilling efficiency, *Neftyanoe Khozyaystvo - Oil Industry* 2016: (3), pp. 68-71.

Urshulyak, R.V., Buslaev, G.V., Bortey, P.D. Laboratory tests and analysis of core samples from ruptured carbonate deposits in the timan-pechora oil and gas province exposed to acid fracturing liquids, *Society of Petroleum Engineers - SPE Russian Petroleum Technology Conference and Exhibition* 2016.

Zhivaeva, V.V. & Nechaeva O.A. & Nikitin, V.I. 2018. Application of calculated criteria to select reservoir penetrating fluid. *Oil. Gas. Novation* 6: 48-51.

Zhivaeva, V.V. & Nikitin, V.I. 2016. Model for calculating the radius of drilling fluid invasion at the opening of the reservoir. *Modern science-based technologies* 6: 250-254.

Topical Issues of Rational Use of Natural Resources 2019 – Litvinenko (Ed)
© 2020 Taylor & Francis Group, London, ISBN 978-0-367-85720-2

Activity of the catalyst obtained by processing of high-magnesia siderites

A.N. Smirnov, S.A. Krylova, V.I. Sysoev, V.M. Nikiforova & Zh.S. Zhusupova
Nosov Magnitogorsk State Technical University, Russia

ABSTRACT: The principles of the integrated processing of high-magnesia siderites have been developed and tested in enlarged laboratory experiments. These principles allow one to increase the level of using the ores in metallurgy and to obtain additional products. It is established that a porous oxide material forms at the stage of roasting high-magnesia siderites accordingly to the patented scheme. The catalytic activity of the material obtained was investigated in the reactions of steam conversion of ethanol and conversion of carbon dioxide by hydrogen. The main products of the catalytic conversion of ethanol by steam in the temperature range of 470 - 550 °C are hydrogen and acetone. The investigated catalyst demonstrate high activity in the reverse water gas shift reaction at a temperature of 820 °C. When activity of the catalyst is reduced it can be agglomerated and used as a burden component in the blast furnace process.

1   INTRODUCTION

Catalysts and catalytic processes form the basis of the chemical industry. Therefore, the search for new catalysts and improvement of their efficiency are urgent problems of modern science. Gas-phase reactions are often catalyzed by solid catalysts based on rare and noble metals, such as platinum, thorium, rhodium, rhenium etc. For more efficient use of these expensive components they should be evenly dispersed on the surface of a solid support. In addition, the structure of a catalyst often become more complicated when additional components are introduced with the aim of reducing the formation of side products and increasing the catalyst's selectivity. Thus the formation of a catalyst using conventional methods (soaking, co-precipitation from solutions, crystallization, agglomeration) is a rather complicated and costly process.

The cost of the catalyst can be reduced by using natural materials for its production. In this case the minerals should be processed in an energy-efficient and simple way as the energy efficiency and simplicity of the processing scheme are of high importance for large-scale production. The most famous, simple and cheap catalysts are iron and iron oxides based ones. So the catalysts made from iron ores attract considerable attention of researchers (Roiter 1968, Ioffe 1960, Tretyakov et al. 1998, Krylov 2014, Antsiferov 1999, Sharypova 2011). The activity of the catalyst obtained by the integrated processing of high-magnesia siderites (iron-magnesium oxide catalyst, IMOC) was tested in this research in the reactions of steam conversion of ethanol and conversion of carbon dioxide by hydrogen. (Klochkovskiy et al. 2013, 2014, Smirnov et al. 2017, 2019).

2   OBJECT AND METHODS

The basis of siderite ores is iron carbonate - $FeCO_3$, which scarcely can be found naturally in pure form and often has an isomorphic admixture of magnesium, as well as manganese, calcium and other elements (Dziubinska et al. 2004). Iron carbonate forms a continuous isomorphic series with magnesium carbonate - from pure $FeCO_3$ (siderite) to pure $MgCO_3$ (magnesite). In the isomorphic series, stable members of the series are distinguished depending on the $MgCO_3$ content: sideroplesite - up to 30%; pistomesite - from 30 to 50%; mesitite - from 50 to 70% and breynerite - from 70% and higher.

The object of our study was the sideritic ore of the Bakal group of deposits (the Chelyabinsk region). At present, these high-magnesian siderites, whose reserves are more than one billion tonnes, are used in metallurgy just to a limited extent because of their relatively low iron content (28-30%) and high magnesium oxide content (12% and higher) (Zhunyov et al. 1982). The main iron-bearing mineral, sideroplesite, contains up to 25% of the total magnesium oxide content in these ores. Iron in the Bakal ore is also present in small amounts in the form of hydroxides, oxides and pyrite. The content of enclosing rocks (dolomite, calcite, aluminosilicates, schists and others) ranges from 20 to 25% (Krylova et al. 2017).

The specificity of the ores of the Bakal group of deposits is determined not only by a high content of magnesium in the ore, but also by the fact that both the ore and its roasting products are isomorphic mixtures of iron and magnesium carbonates with some manganese carbonates, or their corresponding oxides in the roasting products. Using traditional methods of beneficiation, it is impossible to separate components having a common crystal lattice, and hence, to significantly reduce the content of magnesium oxide in the concentrate. The high content of magnesium oxide in the concentrate adversely affects the course of blast-furnace process, therefore the quality of the product obtained by using traditional beneficiation methods is not high enough to carry out blast furnace smelting using a single component burden (Bai et al. 2012, Kurkov et al. 2009, Leontyev et al. 1997).

The scheme of integrated processing of siderite ore developed by our team (Klochkovskiy et al. 2012, Kolokoltsev et al. 2014, Savchenko et al. 2013) allows one to efficiently separate iron and magnesium oxides obtaining a high-quality iron ore concentrate for metallurgy (Fe_{total} = 58-60%, MgO = 6-8%). The separation is achieved by means of so called "soft" roasting (roasting in the temperature range from 550 to 650 °C without an access of atmospheric oxygen) with subsequent selective leaching of magnesium oxide from the roasted material by the solution of carbon dioxide in water (Fernandez 1999, Medvedev 2005). The thermal treatment of the resulting solution gives an additional end product – a magnesia (MgO = 98% and higher) suitable for high-quality refractories and other applications (Demidov et al. 2013).

In the process of the conducted studies at the stage of ore "soft" roasting a porous oxide material was formed with a defective crystal structure and a non-stoichiometric composition of the phases (XRD pattern - Figure 1). The specific surface of the calcined material, depending on the

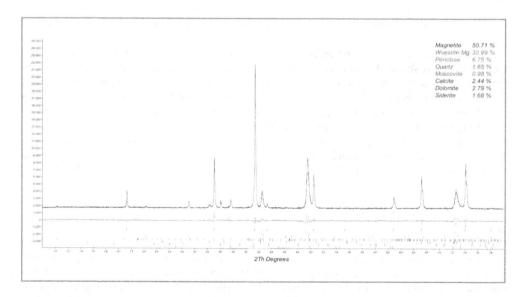

Magnetite	50.71 %
Wuestite-Mg	32.99 %
Periclase	6.75 %
Quartz	1.65 %
Muscovite	0.98 %
Calcite	2.44 %
Dolomite	2.79 %
Siderite	1.68 %

2Th Degrees

Figure 1. An XRD pattern of the ore roasted in "mild" conditions.

564

specific processing conditions, ranges from 5 to 18 m^2/g according to the data obtained by the low-temperature nitrogen adsorption method (BET method). The chemical compositions of the ore and the product of its processing (Table 1) were determined using an ARL QUANT'X EDXRF spectrometer (Thermo Scientific; WinTrace software package, standard tube anode – Rh, additional anode – Ag). Chemical compositions of the phases (Figure 2) were determined using a CAMEBAX– MICROBEAM microanalyzer at an accelerating voltage of 20 kV and beam current of 10^{-8} A. The spot size was 1–5 μm^3. The non-conductive samples were first coated with a conductive material (gold, silver, or carbon). Element contents were measured using the ZAF correction method with an accuracy of ±0.5÷1.0 rel. wt%. The above structure features of the obtained material, its chemical composition and the uniformity of distribution of iron and magnesium oxides which is illustrated by the Figure 2 can be considered as the basis for its application not only in metallurgy, but also as a ready-to-use catalyst.

The catalyst was tested in the temperature range of 430-490 °C at an atmospheric pressure and a fluid flow rate of 700 h^{-1} in a laboratory flow reactor (Figure 3). The feed of the reactor used in the experiments was 1 : 3 (mol.) ethanol-water mixture.

The liquid phase of the reaction products was analyzed using gas-liquid chromatography ("Chromatec Crystal 5000" gas chromatography system). For the analysis a packed column containing the PEG-1000 stationary liquid phase applied onto the Spherochrome porous adsorbent was used. The chromatographic analysis was performed at a temperature of 60 °C; the carrier gas used was nitrogen (30 ml/min). The gas phase of the reaction products was analyzed using gas-adsorption chromatography. The set of 3 packed columns was used: the Carboxen precolumn, the NaX (argon, 20 ml/min) and HayesepQ (helium, 15 ml/min) columns. The analysis was performed at a temperature of 100 °C.

The qualitative determination of components of the reaction products mixture was carried out using retention times with the confirmation of identification by using additives of pure substances. For the quantitative determination of the components' content the absolute calibration method was used.

Table 1. Main components of the Bakal sideritic ore and the product of its integrated processing.

Material	Content of a component, %						
	Fe$_{total}$	FeO	Fe$_2$O$_3$	SiO$_2$	CaO	MgO	LOI*
Initial sideritic ore	27.4	34.0	1.4	3.2	7.3	12.9	35.9
Catalyst (IMOC)	51.6	21.9	49.4	2.5	1.7	17.8	3.5

*) – loss on ignition.

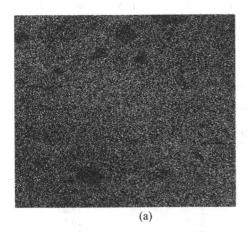
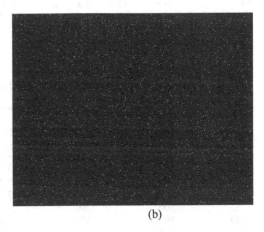

(a)   (b)

Figure 2. Distribution of iron (a) and magnesium (b) in the ore roasted in "mild" conditions.

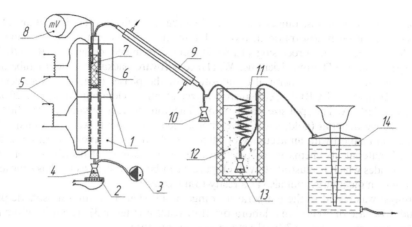

Figure 3. The scheme of the laboratory setup
1- electric tube furnace; 2-electric heating plate; 3-dosing micropump; 4-vaporizer flask; 5-laboratory autotransformer; 6-layer of catalyst; 7-thermocouple; 8 – millivoltmeter; 9-downward condenser; 10-receiver flask; 11-cooling coil; 12-cooling mixture; 13-cooled receiver flask; 14-gasometer.

3 RESULTS AND DISCUSSION

The results of conversion of ethanol at various temperatures are in the Table 2. The results of steam conversion of ethanol at a temperature of 470 °C using IMOC, in comparison with CoO-, MgO- and Cu/CeO$_2$ based catalysts, are in the Table 3. IMOC allowed acetone in the yield which is generally comparable to the yields obtained by using other catalysts while the conversion of ethanol was higher (up to 97.4%).

Table 2. The composition of the products of catalytic conversion of ethanol at various temperatures.

Temperature, °C	Content of a substance, % (vol.)									
	Gas phase						Liquid phase			
	H$_2$	CO$_2$	CO	C$_2$H$_4$	CH$_4$	C$_2$H$_6$	Ethanol	Acetone	AA*	EA**
430	80.5	16.5	0.3	1	0.1	1.6	26.2	12.6	0.2	5.9
450	81	15.9	1.5	1.1	0.3	1.3	3.7	17.1	0.2	2.9
470	78.9	17.2	0.7	1.1	0.1	1.7	3	18.2	0.03	2.1
490	80.5	17.5	0.3	1	0.06	1.6	3	12.6	0.2	2.0

*) – acetaldehyde;
**) – ethylacetate.

Table 3. The results of steam conversion of ethanol on various catalysts (Abuzerli 2016, Lapin et al. 2011, Yakimova 2010).

Catalyst	Temperature, °C	Volumetric flow rate, h^{-1}	Hydrogen yield, %	Acetone yield (balance by carbon), %	Ethanol conversion, %
IMOC	470	700	33.3	19.0	97.4
Cu(5%)/CeO$_2$	500	3000	30.5	< 15.4	84.6
CoO	400	1200	13.2	20.8	41.5
	500	1200	45.9	-	77.6
MgO	500	1200	24.3	17.2	76.5
	550	1200	30.8	22.9	84.7

Table 4. Comparison of catalytic and non-catalytic reaction ratesat a temperature of 820 °C and a CO_2 flow rate of 125 ml/min.

	Content of a component, % (vol.)					
	Non-catalytic reaction			Catalytic reaction		
	H_2	CO_2	CO	H_2	CO_2	CO
Before the reaction	51	49	0	49.7	50.3	0
After the reaction	49.8	45.3	4.9	33.8	33.1	33.1
Equilibrium composition	33.3	33.3	33.4	33.3	33.3	33.4

Table 5. The results of determining the possible flow rate of reagents for IMOC.

	Before the reaction			After the reaction		
V, h^{-1}	H_2	CO_2	CO	H_2	CO_2	CO
240	51.8	48.2	-	31.2	38.7	30.0
360	48.8	51.2	-	32.6	35.2	32.2
480	50.7	49.3	-	34.5	33.4	32.1
960	50.9	49.1	-	36.5	30.0	33.6
1920	48.3	51.7	-	32.1	35.8	32.1
3840	53.1	46.9	-	37.5	30.2	32.3

The experiments on the conversion of carbon dioxide by hydrogen using IMOC were carried out in a laboratory electric tube furnace. The sample of material investigated was placed into the isothermal zone of the tube furnace.

The experiments were carried out for the reverse direction of water gas shift reaction so the catalytic reaction investigated was as follows:

$$CO_2 + H_2 = CO + H_2O \qquad (1)$$

This reaction is used for obtaining a reducing gas with specified reducing activity or synthetic gas. The content of gas mixture both at the input and output of the reactor was determined using gas chromatography. Every determination was fulfilled thrice; the results were processed using statistical methods for gross errors elimination and averaged.

The obtained experimental results are in the Table 4. In this series of experiments the volume of the catalyst loaded into the reactor was 25 cm^3 and the reagent mixture flow rate was 125 ml/h which corresponds to 300 h^{-1}.

The table demonstrates that without a catalyst the reaction proceeds at a relatively low rate and at the experimental conditions the system is far from equilibrium while the reaction which was conducted using the ferromagnesian catalyst is achieved its equilibrium state. The activity determined thus in these experiments was analyzed in the next experimental series, the results of which are in the Table 5. The residence time of reagents in the reactor was varied in this series of experiments by gradual reducing the thickness of layer of the material (+0,5-1 mm) and respective increasing of the CO_2 flow rate. In these experiments the CO_2 flow rate was progressively raised from 240 h^{-1} to 3840 h^{-1} but it didn't result in any significant shift from the equilibrium. It witnesses that the flow rate of reagents can be as high as about 4000 h^{-1} when using the ferromagnesian catalyst. When activity of the catalyst is reduced it can be agglomerated and used as a burden component in the blast furnace process.

Further investigation of activity of the catalyst obtained by the integrated processing of high-magnesia siderites looks highly promising from our point of view. There are large deposits of natural and technogenic raw materials in the Chelyabinsk region which have just a limited use nowadays. These are high-magnesia sideritic ores, lignites and subbituminous coals, as well as significant volumes of coke plant wastes. The energy and economically efficient processing schemes for these materials would allow one to extend their industrial use. With the aim of creating such schemes we plan to investigate the processes of catalytic conversion of the Kuznetsk basin coal concentrate as well as the coal tar produced by the coal plant of JSC "MMK" (Magnitogorsk Iron and Steel Works). The experimental results obtained at the preliminary stage of our research indicate its soundness.

4 CONCLUSIONS

The product of "mild roasting" of natural high-magnesia siderites demonstrates catalytic activity in the reaction of steam conversion of ethanol. The main products of the catalytic process are acetone and hydrogen in 33.3% (from stoichiometric amount) and 19% (balance by carbon) yields respectively at a temperature of 470 °C and a flow rate of reagents of 700 h^{-1}.

In the reaction of reverse water-gas shift (conversion of carbon dioxide by hydrogen) the tested catalyst demonstrates high activity bringing the system to equilibrium at the reactant load of up to 4000 h^{-1}.

Within the framework of the developed technology, it is possible to vary the structure, phase, and chemical composition of the catalyst in a wide range which affect the catalytic activity of the material. The main advantages of using the investigated catalyst on industrial scale are its low cost and relatively simple method of preparation (by processing of natural material), which is especially important for large-scale production.

REFERENCES

Abuzerli, F. Z. 2016. *Kimya Problemleri*. 1: 80–85.

Antsiferov, A.V. 1999. *GIAB*, 8: 107–109.

Bai, S.J, Wen, S.M., Liu, D.W., Zhang, W.B. 2012. Carbothermic reduction of siderite ore with high phosphorus content reinforced by sodium carbonate. *The Can. J. of Metall. and Mater*. 51(4): 376-382.

Demidov, K.N., Borisova, T.V., Vozchikov, A.P., et al. 2013. High-magnesia fluxes for steel-smelting production. *Ural worker*, Satka, Russia: 280.

Dziubinska, B., Narebski, W. 2004. Siderite concretions in paleocene series of polish part of the eastern flysch Carpathians. *Mineral. Pol.* 35(2): 79-91.

Ferna´ndez, A.I., Chimenos, J.M., Segarra, M., et al. 1999. Kinetic study of carbonation of MgO slurries. *Hydrometall.* 53: 155–167.

Ioffe, V. B. 1960. Osnovy proizvodstva vodoroda [Basics of hydrogen production]. Leningrad: Gosudarstvennoe nauchno-tekhnicheskoe izdatel'stvo neftyanoi i gorno-toplivnoi literatury, Leningr. otdelenie.

Klochkovsky, S.P., Smirnov, A.N., Kolokol'cev, V.M. 2014. Method of processing siderite ores (Variants). Pat. RF No. RU2536618.

Klochkovsky, S.P., Smirnov, A.N., Savchenko, I.A. 2012. Development of physicochemical foundations of the use of high-magnesia siderites. *Vestn. of Nosov Magnitogorsk State Tech. Univ.* 49 (1): 26-31.

Kolokoltsev, V., Klochkovsky, S., Smirnov, A. 2014. Physical Chemistry of Integrated High-Magnesia Siderites Processing. *Defect and Diffusion Forum* 353: 171-176.

Krylov, I.O. 2014. *Gornyi informatsionno-analiticheskii byulleten'*, 11: 115–126.

Krylova, S.A., Sysoev, V.I., Alekseev, D.I., Sergeev, D.S., Dudchuk, I.A. 2017. Physical and chemical characteristics of high-magnesia siderites. *Bull. of the South. Ural State Univ.. Ser.: Metall.* 17(2): 13-21.

Kurkov, A.V., et al. 2009. Application of pyro-and hydrometallurgical technologies for purification of Bakalsky sideroplesite ores of magnesium. *Proc. of the Int. Meet. "Plaksin read.":* 198-199.

Lapin, N. V., Bezhok, V. S. 2011. *Zhurn. prikladnoi khimii.* 84(6): 983–987.

Leontyev, L.I., Vatolin, N.A., Shavrin, S.V., Shumakov, N.S. 1997. Pyrometallurgical processing of complex ores. Metallurgy, Moscow: 432.

Medvedev, A.S. 2005. Leaching and ways of its intensification. MISIS, Moscow: 205.

Roiter V. A. (red.) 1968. Kataliticheskie svoistva veshchestv: spravochnik (Catalytic properties of substances: reference book). Kiev: Naukova dumka.

Savchenko, I.A., Klochkovsky, S.P., Smirnov, A.N., Abdrakhmanov, R.N. 2013. Chromatographic analysis of the gas phase formed at "soft roasting" of high-magnesia siderite ore. *The Theory and Process Eng. of Metall. Prod.* 13 (1): 13-15.

Sharypova, V.I. 2011. *Journal of Siberian Federal University Chemistry*, 4: 319–328.

Smirnov A.N., Klochkovsky, S.P., Krylova, S.A., Sysoev, V.I. 2017. Catalytic Activity of Highly Magnesian Siderite Roasting Products. *Bull. of Bashkir Univ.*, 22(3): 657-665.

Smirnov, A.N., Klochkovskii, S.P., Bigeev, V.A., Kolokol'tsev, V.M., Bessmertnykh, A.S. 2013. Sposob pererabotki sideritovykh rud. Pat. RF No. 2471564.

Smirnov, A.N., Klochkovsky, S.P., Krylova, S.A., Sysoev, V.I. 2019. Gasification of the Kuznetsk basin coal concentrate using oxide iron-magnesium catalysts. *Journal of Chemical Technology and Metallurgy*, 54(2): 286-291.

Smirnov, A.N., Klochkovsky, S.P., Krylova, S.A., Sysoev, V.I., Strogonov, D.A.2017. On a possibility of application of oxide iron-magnesia catalysts in the water gas shift reaction. *Curr. Probl. of Mod. Sci., Equip. and Educ.* 2: 66-70.

Smirnov, A.N., Sysoev, V.I., Krylova, S.A. 2017. Investigation of the Catalytic Activity of the Middlings of High-Magnesian Siderites Processing. *Recent Pat. on Mater. Sci.* 10 (2): 136-141.

Tret'yakov, A.S., Chelpanov, I.P., Svetkina, E.Yu., Antsiferov, A.V. 1998. Katalizator dlya okisleniya dioksida sery. Pat. RF No. 2111790.

Yakimova, M. S. 2010. *Vestnik MITKhT*. 5(4): 93–97.

Zhunyov, A.G., Avdoshin, G.G. 1982. Preparation of siderite ores of the Bakalsky deposit for blast-furnace smelting. *Gorn. Z.* 11: 20-22.

Topical Issues of Rational Use of Natural Resources 2019 – Litvinenko (Ed)
© 2020 Taylor & Francis Group, London, ISBN 978-0-367-85720-2

DNA Linkers: The weakest link in the artificial nanomachines

S.A. Toshev
University of Mining and Geology "St. Ivan Rilski", Sofia, Bulgaria

A.R. Loukanov & S. Nakabayashi
Graduate School of Science and Engineering, Saitama University, Saitama, Japan

ABSTRACT: DNA linkers in artificially designed nanomachines are short oligonucleotides, which connect two or more individual nanoparticles and often resulted in fabrication of nano-devices with a wide variety of unique properties and functions. For engineering of any oligo-nucleotide chain with specific sequence as a linker it must be taken under consideration its functional stability in bio-environment, where deoxyribonucleases (DNases) are present too. DNases are enzymes that degrade DNA in cells and biological fluids in sake of protecting them from "intruder" DNA. Depending on the role of a particular DNA linker, its degrad-ation may decrease or completely abolish the functionality of nanomachine. Therefore, the protection of DNA linkers by chemical modifications, packaging or using DNase inhibitors is necessary in order to keep the nanomachine functionality. The present comprehensive study reports the common designs of DNA linkers in nanocomposites, the mechanisms for their enzymatic degradation and methods of protection against DNases.

1 INTRODUCTION

The development of nanomaterial science, nanobiotechnology engineering and DNA syn-thesis within last two decades enabled invention of specifically-designed DNA linkers (spacers), which are able to connect nanoparticles in a flexible manner, and in some cases, provide specific biorecognition capacity for nanoparticle targeting (Park et al. 2008). DNA linkers vary in length from a few nucleotides to several hundred base pairs. The nucleotide sequence of the linker vary depending on its role and function. It may include areas for recognition by enzymes, cell surface molecules (for example membrane proteins) or homology to other DNA. As shown in Figure 1, the linker can be single-stranded or double-stranded, a single linker per nanostructure or a poly/multi-linker. A linker can be covalently bound to nanoparticle surface, or by hold by hydrogen bond (for example, if it is assembled by hybridization of the linker components). Linkers may contain hairpins or branches to allow connecting more nanoparticles or targeting of the linkers to other oligonucleotides or protein clusters on the cell membrane. DNA can easily be conjugated to metal nanoparticles (gold, silver, etc.) and self-assembled nanode-vices with unique nanomechanical and nanoelectronic properties can be created (Alivisa-tos et al. 1996). They possess various applications in the sensor and biomedical technology (Yan et al. 2003). Many of these applications involve the utilization of oligo-nucleotide chains tethered to gold or various biocompatible nanoparticles (Mirkin et al. 1996). They may be additionally hybridized with one another and thus more complicated nanocomposites or nanomachines are formed. However, their use in in vivo cellular environment is a great challenge because of the degradation by deoxyribonucleases (DNases) enzymes. To protect the fullness of these nanostructures the DNA linkers must be modified or packaged, which is the general aim of the current report. The alternative

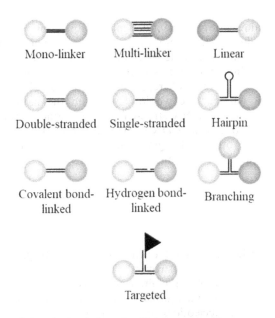

Figure 1. Types of DNA linkers in the design of artificial nanomachines.

approach we demonstrated is to keep the nanomachine functionality in cells or biological fluids by simple inhibition of DNases enzymes.

2 METHODOLOGY

Because of its simplicity, the used method to measure DNA linker degradation was electrophoresis in agarose or polyacrylamide gel. Transmission electron microscope (JEOL, 120 kV) was used to demonstrate the untagled individual nanoparticles. TEM experiment were performed in similar manner as Chou et al (Chou et al. 2016), where the plasmonic core-satellite coupling "off" after DNase exposure. TEM approach was used also to characterize DNA-linked gold nanoparticle assemblies. The combination of gel electrophoresis and TEM enable to the strongest result in confirmation of DNA linker degradation (Akiyama et al. 2015). The meltingannealing analysis was also used to study DNA linker hybrids. If the linked nanoparticles are labeled, more sophisti-cated methods based on the label detection can be applied to assess dissociation of the particles due to DNA linker degradation.

3 EXPERIMENTAL

Physical and chemical agents were used to destroy the DNA linkers as a control experiment. If the linker was not completely covalently-bound and DNA fragments are kept together only by hydrogen bonds, a simple heat would dissociate these bonds and disassemble the linker. In this case, the melting curve of DNA linker was different from the one of naked free DNA. The physical agents used were able to destroy covalent bonds in DNA linkers included sheer (mechanical) stress, and UV or gamma irradiation, while chemical agents were hydrogen peroxide and some drugs producing it, for example, bleomycin. Very few of these agents are available in natural biological systems. Practically, all of them are either reactive oxygen species (ROS) or agents acting through the generation of ROS. Overall, these agents rarely induced DNA fragmentation (via double-strand DNA breaks) because they resulted mostly in single-strand breaks in forms of nicks (broken DNA chain without nucleotides missing) or gaps (with several nucleotides missing). These breaks did not interfere with the linker's function (of binding of the nanoparticles) and were easily repairable by cellular DNA polymerases and DNA ligases. In some cases, initial modification of DNA by bulky adducts or UV light may further lead to transient DNA breaks by DNA repair endonucleases. Cleavage of the bond that is attaching DNA to a nanoparticle would depend on the nature of the bond. One example would be desorption of DNA strand from the surface through thiol-gold bond disruption (Zagorovsky et al. 2016). If DNA linker was bound to a nanoparticle through only one strand, while the other is hanging, the latter can be potentially destroyed by exonucleases that act from DNA terminus.

4 RESULTS

4.1 DNA degradation by DNases

Any intrusion of a foreign DNA in the cell or the body is faced by groups of enzymes, which sometimes are correctly called "DNA destroyers". This mechanism is universal regardless of the nature or configuration of DNA. In addition to DNA linkers it, certainly, includes free-floating DNA in blood and body fluids (urine, lymph, and saliva), linear and circular DNA, long DNA fragments and oligonucleotides, single-stranded and double-stranded DNAs. The natural role of DNA destroyers is the protection of the cell and body against infectious viral, bacterial, or naked DNA arriving from dead cells. The main idea was that all foreign DNA, packed or free floating, needs to be destroyed. This is because at the cellular level, foreign DNA can be cytotoxic mutagenic and carcinogenic. From the whole-body perspective, circulating DNA is may be immunogenic viscous and promote thrombosis of blood vessels. The mammalian cell has an arsenal of DNases able to destroy DNA linker of any size and structure. Most efficient DNases in degradation of linkers were endonucleases, which degraded DNA at any bond in the molecule. However, the importance of exonucleases shall also be considered. For example, Zagorovsky (Zagrosky et al. 2016) described exonuclease activity in blood plasma that was destroying DNA linkers containing open DNA termini. Exonucleases are having no substrate if DNA ends were not exposed – due to the stricture of the linker, or because of previous action of ROS (leaving sugar residue) or endonucleases (leaving phosphate at the end the strand). In further description of DNases, we are considering of endonucleases only. Nine DNA endonucleases are known in mammalian cells (Table 1). All of them except endonuclease G (EndoG) are strict DNases, while EndoG is a nuclease (DNase/RNase).

DNases usually have sequences of preferential cleavage at the beginning of the reaction. For example, DNase I started cutting DNA in AT-rich sequences, while EndoG initiated DNA degradation through cleaving G-rich sentences first (because of which the enzyme got

Table 1. Human and mammalian DNA endonucleases.

Name	Human gene	MW (kDa)	Substrate	Product termini
Deoxyribonuclease 1 (DNase I)	16p13.3	34	DNA	3'OH/5'P
Deoxyribonuclease X (DNase X)	Xq28	35	DNA	3'OH/5'P
Deoxyribonuclease 1-like 2 (DNase1L2)	16p13.3	36	DNA	3'OH/5'P
Deoxyribonuclease gamma (DNase γ)	3p14.3-p21.1	32-36	DNA	3'OH/5'P
Deoxyribonuclease 2 alpha (DNase 2α)	19p13.2	35-45	DNA	5'OH/3'P
Deoxyribonuclease 2 beta (DNase 2β)	1p22.3	35	DNA	5'OH/3'P
L- Deoxyribonuclease 2 (L-DNase 2)	6p25.2	35-42	DNA	5'OH/3'P
Endonuclease G (EndoG)	9q34.1	27-32	DNA, RNA	3'OH/5'P
Caspase-activated DNase (CAD)	1p36.3	40	DNA	3'OH/5'P

its name). However, further cleavage continued seemingly without any sequence specificity. For the purpose of inactivating a DNA linker in a nanomachine, a single double-strand DNA break will serve the function. This was achieved by coinciding single-strand DNA breaks on the opposite strands, like the most of the endonucleases did, or a one-time double-strand DNA break characteristic to DNase II alpha (Yu et al. 2015). Because of DNase activity, DNA linker stability observed in pure suspensions of linked nanomaterials was quickly converted to low stability when injected or exposed with cell cultures or injected *in vivo*. Degradation of nanoparticle-bound DNA depended on several factors. One of them was the lengths of the DNA linker; the sorted was the oligo, the more resistant it was the DNase-mediated degradation (Zagorovsky et al. 2016). For a given DNA length, the interparticle separation within a nanoparticle dimer was controlled primarily by the DNA density and the number of linking DNAs (Zagorovsky et al. 2016).

4.2 *Protection of DNA linkers by DNA modification*

Since there were two components of the reaction, DNA and DNases, the linker DNA stability was improved by targeting any of them. Third potential target could be DNase co-factors, such as bivalent cations. However, this approach was not practical when the nanomachines were used in cells or *in vivo*. Phosphorothioation previously applied for modification of oligonucleotides was not practical either because it protects only termini nucleotides from exonucleases. Protection of DNA linkers from enzymatic degradation was not much different from protection of any DNA for the purposes of transfection in cultured cells or *in vivo* delivery. For this, traditional use of lipophilic substances, PNAs and DNA-PNA Hybrids (Eidelshtein et al. 2012) was quite reasonable. Boylan et al (Boylan et al. 2012) showed that coating of DNA with low MW polyethylenglycol (PEG) coating provided partial protection against DNase I digestion and exhibited the highest gene transfer to lung airways following inhalation in mice. The protection of DNA bound to nanomaterials in our experiment dependent on the thickness of the PEG layer (Zagorovsky et al. 2016). Mao et al (Mao et al. 2001) also showed that the clearance of the PEGylated DNA-chitosan nanoparticles in mice following intravenous administration was slower than unmodified nanoparticles. Ponnuswamy et al (Ponnuswamy et al. 2017) recently reported a new DNA-protecting approach, oligolysine-based coating, which is applicable to DNA nanostructures, provides stability against denaturation at physiological Mg^{2+} concentrations without noticeable distortion. However, the authors admit coating with oligolysine only modestly protects DNA against nucleases.

4.3 *Protection of DNA linkers by using DNase inhibitors*

So far, the use of DNase inhibitors for linker protection has been very limited. This was because the useful, specific and highly active DNase inhibitors are yet to be developed. Some DNases have very specific cellular protein inhibitors, which bind and inactivate them. In our experiment, G-actin was a well-known inhibitor of DNase I (Lacks et al. 1981). The two proteins irreversibly bind and sequester each other. Apoptotic caspase-activated DNase (CAD) has its inhibitor ICAD that modulates CAD's activity. However, not all DNases have such inhibitors, and, even more importantly, pharmaceutical use of the protein inhibitors would be impractical because it will be hard to deliver these molecules intracellularly. Much more realistic are approaches based on RNAi or chemical inhibitors. There are numerous reports indicating strong DNase silencing activity in works utilizing cultured cells (Tsuruta et al. 2007). However, cultured cells were an easy model because they usually have low DNase activity. Although, some reports show appreciable silencing of enzymes by using siRNAs *in vivo* (Davidson et al. 2005) (Spankuch et al. 2004), more practical would be using chemical inhibitors. A few recently developed DNase inhibitors to be named are: inhibitor of DNase gamma, DR396 (4-(4,6-dichloro-[1,3,5]-triazin -2-ylamino)-2-(6-hydroxy-3-oxo-3H-xanthen-9-yl)-benzoic acid) (Yamada et al. 2011) DNase I/ DNase II/EndoG inhibitors JR-132 (1,4-phenylene-bis-aminoguanidine hydrochloride) and IG-17 (1,3-phenylene-bis-aminoguanidine hydrochloride), and highly potent EndoG inhibitors

PNR-3-80 (5-((1-(2-naphthoyl)-5-chloro-1H-indol-3-yl)methylene)-2-thioxodihydropyrimidine-4,6(1H,5H)-dione) and PNR-3-82 (5-((1-(2-naphthoyl)-5-methoxy-1H-indol-3-yl)methylene)-2-thioxodihydropyrimidine-4,6 (1H,5H)-dione). These inhibitors were able to act in cells in micromolar concentrations, and, because DNases are enzymes promoting cell death, the inhibitors were protective against toxic (spontaneous) cell death. This was an additional advantage for the nanomaterial delivery to the cells. In the absence of specific chemical inhibitors of other DNase, some non-specific molecules were used too. Perhaps, the best candidates of them are Zn^{2+} salts and Zn-chelates. Zn-chelate Zn-DIPS was proved to protect against radiation injury *in vivo*. Another Zn-chelate, Zn-NAC was active in the improvement of cold storage of kidneys.

5 CONCLUSION

In summary, DNA linkers are the important part of nanomachines. They provide binding between the elements of nanomachines, and ensure their functionality. In the present report we demonstrated that the oligonucleotide linkers were the "weakest link" in nanomachines because are open to attacks by DNases, which protect cell and body from the intrusion of "foreign" DNA. However, we proposed various methods for protection of DNA linkers by either modifying the linker itself, or by inhibiting the DNases.

REFERENCES

Akiyama, Y. & Shikagawa, H. & Kanayama, N. & Takarada, T. & Maeda, M. 2015. Modulation of Interparticle Distance in Discrete Gold Nanoparticle Dimers and Trimers by DNA Single-Base Pairing. Small 11(26), 3153–3161.

Alivisatos, A.P. & Johnsson, K.P. & Peng, X. & Wilson, T.E. & Loweth, C.J. & Bruchez, M.P.Jr. & Schultz, P.G. 1996. Organization of "nanocrystal molecules using" DNA. Nature 382 (6592), 609–611.

Boylan, N.J. & Suk J.S. & Lai S.K. & Jelinek, R. & Boyle, M.P. & Cooper, M.J. & Hanes, J. 2012. Highly compacted DNA nanoparticles with low MW PEG coatings: in vitro, ex vivo and in vivo evaluation. J Control Release 157(1), 72–79.

Chen, X.J. & Sanchez-Gaytan, B.L & Qian, Z & Park, S.J. 2012. Noble metal nanoparticles in DNA detection and delivery. Wiley Interdiscip Rev Nanomed Nanobiotechnol 4(3), 273–290.

Chou, L.Y. & Song, F. & Chan, W.C. 2016. Engineering the Structure and Properties of DNA-Nanoparticle Superstructures Using Polyvalent Counterions. J Am Chem Soc 138(13), 4565–4572.

Davidson, B.L. & Harper, S.Q. 2005. Viral delivery of recombinant short hairpin RNAs. Methods Enzymol 392, 145–173.

Eidelshtein, G. & Halamish, S. & Lubitz, I. & Anzola, M. & Giannini, C. & Kotlyar, A. 2012. Synthesis and properties of conjugates between silver nanoparticles and DNA-PNA hybrids. J Self-Assembly Mol Electronics 1, 69–84.

Lacks, S.A. 1981. Deoxyribonuclease I in mammalian tissues. Specificity of inhibition by actin. J Biol Chem 256(6), 2644–2648.

Mao, H.Q. & Roy, K. & Troung-Le, V.L. & Janes, K.A. & Lin, K.Y. & Wang, Y. & August, J.T. & Leong K.W. Chitosan. 2001. DNA nanoparticles as gene carriers: synthesis, characterization and transfection efficiency. J Control Release 70(3), 399–421.

Mirkin, C.A. & Letsinger, R.L. & Storhoff, J.J. 1996. A DNA-based method for rationally assembling nanoparticles into macroscopic materials. Natura 382(6592), 607–609.

Park, S.Y. & Lytton-Jean, A.K.R. & Lee, B. & Weigand, S. & Schatz, G.C. & Mirkin, C.A. 2008. DNA-programmable nanoparticle crystallization. Nature 451, 553–556.

Ponnuswamy, N. & Bastings, M.M.C. & Nathwani, B. & Ryu, J.H. & Chou, L.Y.T. & Vinther, M. & Li, W.A. & Anastassacos F.M. & Mooney, D.J. & Shih, W.M. 2017.Oligolysine-based coating protects DNA nanostructures from low-salt denaturation and nuclease degradation. Nat Commun 8, 15654.

Spankuch, B. & Matthess, Y. & Knecht, R. & Zimmer, B. & Kaufmann, M. & Strebhardt, K. 2004. Cancer inhibition in nude mice after systemic application of U6 promoter-driven short hairpin RNAs against PLK1. J Natl Cancer Inst 96(11), 862–872.

Tsuruta, T. & Oh-Hashi, K. & Ueno, Y. & Kitade, Y. & Kiuchi, K. & Hirata, Y. 2007. RNAi knock-down of caspase-activated DNase inhibits rotenone-induced DNA fragmentation in HeLa cells. Neurochem Int 50(4), 601-606.

Yamada, Y. & Fujii, T. & Ishijima, R. & Tachibana, H. & Yokoue, N. & Takasawa, R. & Tanuma, S. 2011. DR396, an apoptotic DNase gamma inhibitor, attenuates high mobility group box 1 release from apoptotic cells. Bioorg Med Chem 19(1), 168-171.

Yan, H. & Patk S.H. & Finkelstein, G. & Reif, J.H. & LaBean, T.H. 2003. DNA-templated self-assembly of protein arrays and highly conductive nanowires. Science 301(5641), 1882-1884.

Yu, H. & Lai, H.J. & Lin, T.W. & Lo, S.J. 2015. Autonomous and non-autonomous roles of DNase II during cell death in C. elegans embryos. Biosci Rep 35(3),e00203.

Zagorovsky, K. & Chou, L.Y. & Chan, W.C. 2016. Controlling DNA-nanoparticle serum interactions. Proc Natl Acad Sci U S A 113(48), 13600-13605.

The latest management and financing solutions for

the development of mineral resources sector

Topical Issues of Rational Use of Natural Resources 2019 – Litvinenko (Ed)
© 2020 Taylor & Francis Group, London, ISBN 978-0-367-85720-2

The estimation of investment appeal of mining enterprises

Z.G. Arakelova & A.A. Kravchenko
Donetsk National Technical University, Donetsk People Republic

ABSTRACT: The article considers the problem of assessing the investment appeal of industrial enterprises. The choice of an appropriate method of assessment, a set of factors, their
grouping and importance evaluation for Donbass enterprises, which have a difficult financial
position for a long period – are the issues that form the problems of study. The article proposes a method of integrated assessment of investment appeal of coal mining enterprises,
taking into account the peculiarities of the DPR state mines functioning. Among the 5 considered state enterprises, the most and the least investment-attractive ones were identified.

1   INTRODUCTION

Every industrial enterprise is looking for opportunities how to increase the competitiveness
and to achieve economic growth. Enterprises seek outside sources of financing in a context of
limited resources. This demonstrates that question of investment opportunity is becoming
more relevance.

Object of study: plot the investment appeal of mining enterprises.

Purpose: analysis of main approaches and assesses the investment appeal of coal mining
enterprises.

The main tasks: the determination and calculation of integrated assessment of investment
appeal of Donetsk mining enterprises.

2   LITERATURE REVIEW

Although, many authors pay attention to this subject (Blank I.A. et al. 2011), scientists still
have not found common determination of investment appeal. More objective and comprehensive definition of investment appeal is belong to Kuvshinov & Kalacheva 2015), they said:
«the assessment of the investment appeal should take into account all the differentiation of
requirements for the effect of investment and risk level, as well as reflect all the positive
aspects of the known research in this area and eliminate the outstanding issues». So that, the
investment appeal is commonly understood to mean an integrated assessments of enterprise as
investment object. It includes productive capacity, financial status, management of enterprise,
industry sector and concerns of investment players.

3   MATERIALS AND METHODS

There are many approaches and methods to analyze the investment appeal, which are applicable for any economic system. Karanina & Vershinina (2015) in their work noticed that despite its universality, most of these methods only assess the financial side of object. As for the
coal enterprises, the financial state plays an important but not the main role in the investment
appeal, because difficult financial position is due to a lack of funding development of the

mine, deterioration of mountain-geological conditions, mismanagement and other factors. Rostislavov R. A. (2009) highlights following basic approaches:

- an approach based on an analysis of the financial indicators of enterprises;
- an integrated approach, which takes into account many aspects of the business entity;
- market-based approach, which is based on the stock market analysis;
- cost-based approach, in which the criterion of company's investment appeal is the growth of its value.

With regard to the coal enterprises it is reasonable to use integrated assessment of investment appeal.

The first thing that needs to be done is to form a matrix A of dimension n × m:

$$|A| = \begin{vmatrix} a_{11} & \dots & a_{1j} & \dots & a_{1m} \\ \dots & \dots & \dots & \dots & \dots \\ a_{i1} & \dots & a_{ij} & \dots & a_{im} \\ \dots & \dots & \dots & \dots & \dots \\ a_{n1} & \dots & a_{nj} & \dots & a_{nm} \end{vmatrix} \qquad (1)$$

where a_{ij} – is j-th index value for i-th object; n - the number of mines; m - the number of indicators taken into account.

Each row $\{a_i\}$ corresponding to the same set of indicators for the various coal enterprises. Initial data for the calculation of the integral index is the value matrix of indices of all the mines.

Secondly, we need to determine a reference organization. For this purpose, in each column of the matrix A optimal value is selected (minimum or maximum index):

- if improvement value of the index shows a positive trend, a reference index will be considered with a maximum value;
- if improvement value of the index indicates the negative trends, the reference index will be considered with a minimum value. Further, for the reference mine, reference values are taken for all indicators. Thus, the maximum achievable values for all parameters are determined according to current operating conditions.

Due to the fact that most of the indicators have different dimensions and are not compatible, there is a necessity to bring them to the relatively dimensionless form (formula 2):

$$\delta_{ij} = \left| \frac{a_j^{ref} - a_{ij}}{a_j^{max} - a_j^{min}} \right| \qquad (2)$$

where $a_j^{ref}, a_{ij}, a_j^{max}, a_j^{min}$– the reference, the actual, maximum and minimum values of mines indicators correspondingly.

Initial matrix is replaced by a matrix of relative deviations of the same dimension, and further calculations are carried out on this matrix:

$$|\delta| = \begin{vmatrix} \delta_{11} & \dots & \delta_{1j} & \dots & \delta_{1m} \\ \dots & \dots & \dots & \dots & \dots \\ \delta_{i1} & \dots & \delta_{ij} & \dots & \delta_{im} \\ \dots & \dots & \dots & \dots & \dots \\ \delta_{n1} & \dots & \delta_{nj} & \dots & \delta_{nm} \end{vmatrix} \qquad (3)$$

Thirdly, it is necessary to reduce relative deviations from a reference of all factors for every mine to get integral index K_i.

Table 1. Data in relative form for the calculation of K_i^{int} of DPR coals for 2018.

			Mine's indicators					
Group	The import-ance of group	Indicators	The importance of the indicator	SE DUEK	SE Makee-vugol	SE Torezan-tratsit	SE Zasyadko	SE Komsomo-lets Donbass
Mining engineering	0.7	Annual coalproduction, ths.t.	18	0.82	0.80	0	1	0.55
		The number of active stopes, pcs.	12	0.57	0.86	0	1	0.43
		Number of KMZ, units.	14	0.50	0.83	0	1	0.33
		Per face output, t.	16	1	0.85	0.44	0	0
		Total length of faces, m.	12	0.59	0.87	0	1	0.43
		KMZ line, m.	14	0.56	0.86	0	1	0.39
		Average sweep of length of faces, m.	11	1	0.26	0	0.49	0.65
		The ash content of produced coal,%	13	0.35	0.72	0.43	1	0
		The ash content of coal, shipped to customers,%	15	0.58	0.03	0.23	0	1
		Opening and preliminary development, m.	13	0.75	0.64	0	1	0.78
		Doing preliminary development, m.	10	0.79	0.74	0	1	0.83
		Horsepower-to-weight ratio of coal loading, requiring mechanized operation, %	8	0.55	0.35	1	0	0.21
		The volume of commodity output, ths. t.	15	0.79	0.84	0	1	0.36
		The cost of 1t of marketable coal products, rub.	18	0.74	1	0.50	0.76	0
		Price for 1 ton, rub	18	0.22	0.37	0.96	0	1
		Products at wholesale prices, ths.rub.	16	0.75	0.81	0	1	0.42
		Average people number, p.	10	0.36	0.64	1	0	0.14
Economic	0.3	The share of GROZ in the FP number,%	11	0.67	1	0.65	0.93	0
		The share of heading man in the FP number,%	11	0.02	1	0.39	0.16	0
		Productivity of coal miner, t/person.	16	0.88	1	0.38	0.72	0
		Productivity of heading man, m/person.	15	0.78	0.22	0	1	0.56
		Average monthly earnings, rub	11	0.47	1	0.22	0.46	0.00
		Coal stocks, ths.t.	12	0.11	1	0.68	0	0.05

$$K_i = \sqrt{\sum_{j=1}^{m} \delta_{ij}^2} \to \min \tag{4}$$

The principle of integral index is: the less the value of the integral index is, the better enterprise matches version of the reference. The value of integral index will reflect level of its declining and estrangement from the reference. It is important to emphasize that the assessment objectivity is associated with the inclusion of different importance of individual indicators within the group. Without taking into account the importance of particular indicators, it is impossible to solve the problem of allocation of the most investment-appeal enterprise.

Indicators of different importance rate are defined as follows:

$$\varphi_j^{un} = \frac{\varphi_j}{\varphi_{av}} = \frac{m\varphi_j}{\sum_{j=1}^{m} \varphi_j} \tag{5}$$

Integral index of investment appeal is calculated as follows:

$$K_i = \sqrt{\sum_{j=1}^{m} \left(\delta_{ij} \frac{\varphi_j}{\varphi_{av}} \right)} = \frac{1}{\varphi_{cp}} \sqrt{\sum_{j=1}^{m} \left(\delta_{ij} \varphi_j \right)^2} \tag{6}$$

Table 2. The values of group and general integral indexes of DPR enterprise's investment appeal.

Name SE	Significance group indicators		
	K_{mi}	K_{eci}	K_i^{int}
SE DUEK	2.65	2.01	1.95
SE Makeevugol	2.72	2.66	2.06
SE Torezantratsit	0.95	1.87	0.87
SE Zasyadko	3.13	2.21	2.29
SE Komsomolets Donbass	1.98	1.52	1.46

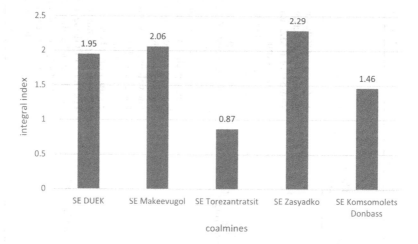

Figure 1. The level of investment attractiveness of the coal-mining enterprises.

where φ_j - the importance of the index in the group; φ_{av} - the average value of importance of all parameters; m - the number of indicators; $\frac{\varphi_j}{\varphi_{av}}$ - the indicators importance level of j-th figure in the group.

Thus, the integral index is calculated separately for each group of mining engineering and economic indicators:.K_{mi}, K_{eci}.

At the next stage the total integral indicator is calculated individually for each mine:

$$K_i^{int} = \sqrt{(K_{mi}\lambda^m)^2 + (K_{eci}\lambda^{ec})^2} \tag{7}$$

where K_{mi}, K_{eci}- integral assessment for each mining engineering and economic indicators correspondently; λ^m, λ^{ec}- the relative importance of the corresponding groups of indicators.

There are a lot of indicators characterizing the activity of coal mines. The first step of an integrated assessment is a determination of indicators that impact on investment attractiveness of coal mines. All indicators can be divided into two groups: mining engineering and economic. The mining engineering indicators have the greatest impact on the integrated assessment. A set of economic indicators characterizes the economic activity of the coal enterprises from the market point of view. These indicators are characterized as economic activity of the enterprise. These groups of indicators provide a comprehensive analysis of coalmines, because they characterize both: economic activity and quality of coal production.

In our opinion, financial indicators, which are used in the approaches above, are not acceptable to assess the investment appeal of coal mines. Because they are a consequence of the enterprise's technical and economic conditions. The financial condition of Donbass coal mines has been traditionally difficult for many years, and accounting for financial indicators can lead to the equalizing of the investment potential of mines and inaccurate assessment of investment appeal.

Five state coal enterprises of the Donetsk people republic were analyzed according to this method. According to the mines, collected information were analyzed and financial indicators were calculated for 2017 (Galenko & Polovyan, 2018). Initial data for determining the integral index of investment appeal are given in relative form in the Table 1 (matrix A) In the Malkin's work (2010) the importance level of the mining engineering and economic indicators of the 20-point scale is justified and numerically expressed. The indicators level and group importance have been determined by the experts. The experts involved production workers – employees of technical and economic services of coal mining enterprises, as well as scientists, who investigate problems of the industrial development of the Donetsk People Republic.

4 RESEARCH RESULTS AND DISCUSSION

As a result of calculations the vector of indicator values of investment appeal assessment in coal mining enterprises is presented in Table 2:

Calculation results can be presented graphically as (Figure 1):

We have come to accept that the most investment attractive enterprise is SE "Torezantratsit". SE "mine Komsomolets Donbass" and SE DUEK are less investment attractive. And the least investment attractive are SE "Makeevugol" and SE A. F. Zasyadko. Enterprises, which integral index is the worst can be characterized by low production efficiency, exhaustion of technical resource of equipment and areas, staff reduction, etc.

5 CONCLUSIONS

Finally, it can be observed, that the developed technique is an assessment tool of investment appeal of coalmines according to the sector-specific. This classification of mines by investment appeal allows:

– to identify the mine with more useful impact for investing;
– to carry out a comprehensive assessment of investment appeal, taking into account a variety of factors, which helps to reduce risks in making investment decisions;
– to take into account the preferences of a particular user (investor) in choosing the importance of individual factors in determining weighting group factors;
– to perform modification of the model when changing the requirements and preferences of the investor, as it is not static.

The resulting model can be used:

– to assess and compare the investment appeal of industrial enterprises';
– in the process of investment management.
– during the step-by-step selection of investment-attractive enterprises,

On the basis estimates of individual groups of factors may be identified those that are unsatisfactory and should be direct the control action in order to increase the investment appeal of industrial enterprise.

The next step in the research is development possible variants of investment projects on increasing production capacity of a mine to assess economic efficiency in the mines of different investment appeal degrees.

REFERENCES

Blank, I. A. 2011. Financial management. Moscow: Omega –L.
Endovitsky, D. A. 2010. Analysis of investment attractiveness of the organization: scientific publication. Moscow: KNORUS.
Galenko, E. I. 2017. Report of the Minister of coal and energy of the DPR Golenko E. And "Results of the coal industry".URL: https://dnrsovet.su/ministr-uglya-i-energetiki-eduard-golenko-podvel-itogi-raboty-za-2017-god/
Jordan, B. D. 2009. Fundamentals of investments: valuation and management. New York: McGrow-Hill.
Kuvshinov, M.S., Kalacheva, A. G. 2015. Development of the analysis of investment appeal of industrial enterprises. Chelyabinsk: Vestnik: 74–80.
Karanina, E. V. 2015. Improvement of the methodology of assessment and ensuring the investment attractiveness of the enterprise. "Economics and management" (12):175–181.
Malkin, A. S. 2010. Design of mines. Moscow: Min. Sciences Akademy.
Polovyan, A.V. 2018. The economy of the Donetsk People Republic: state, problems, solutions: scientific report. Donetsk: Institute of economic researches.
Rostislavov, R. A. 2009. Methods of assessment of investment attractiveness of the enterprise. Tula: Publishing house of Tula state University.– 153–157.
Sharpe, W. F. 2007. Investors and markets: portfolio choices and investment advice. Princeton: Princeton University Press.

Formation and selection of strategic alternatives in the management of production facilities in the mining and smelting company

M.A. Artiaeva
Saint-Petersburg Mining University, Russia

ABSTRACT: In the context of high competition in the steel Russian market in recent years, an effective tool to increase the efficiency and effectiveness of the company is the successful implementation of investment projects. The right generation and formation of a list of strategic alternatives in the current internal and external environment of the company is the choice of investment activities within the framework of the strategic development of the company. This article discusses the method of formation and selection of strategic alternatives in Russia, as well as quantitative assessment of the effectiveness of the business unit, based on the valuation of the business unit in the structure of Russian vertically integrated company.

1 INTRODUCTION

In an environment with a high degree of competition in both the world and the Russian market of ferrous metals, the traditional form of organization of activities in the metallurgical industry are vertically integrated companies (VIC). The prerequisites for the creation of a VIC in the mining business served as a fairly concentrated market, as well as certain opportunities possessed by companies of this form of organization. In these companies, combined mining and metallurgical business of the company become the mining and metallurgical. Iron ore is the raw material in the steel industry, so the functioning of companies depends on sufficient volumes of supply and production of raw materials.

In Russia, some VIC have the problem of shortage of their own raw materials, in a concentrated market competitors do not provide supplies in sufficient volumes, this leads to incomplete loading of processing capacity, the inability of strategic development and to reduce the efficiency of the company. Therefore, to ensure the supply of VIC, it is necessary to expand its own mineral resource base and implement effective investment projects to increase the production capacity of mining enterprises (Ponomarenko et al. 2016).

When choosing a project, in addition to traditional resource factors, additional factors should be taken into account, which include: determination of the total demand of the VIC for mineral raw materials and processing products, projected production and processing volumes, taking into account their balance. Interconnection of technological redistribution, efficient logistics, optimal investment strategy within the VIC, project financing conditions, transfer pricing, exclusion of competing projects. These factors influence the efficiency of the project and the value of the company as a whole.

Methods of economic justification of investment projects in the mining industry are well developed. Significant attention in the evaluation of projects is paid to external factors, including price factors, market demand, etc.

Strategic decisions in the VIC depend on the results of the evaluation of investment projects, taking into account additional factors. The formation of the system of strategic alternatives should take into account the influence of additional factors on the formed effects. Each strategic alternative provides different opportunities and is characterized by different costs and results (effects). Choosing an alternative should maximize the effects.

The main attention in the evaluation of investment projects was paid to the methods of determining the effects of a separate project, while the methods of project evaluation in the implementation of the project complex (program, portfolio) are not sufficiently developed. In the project approach, the methodology for the evaluation of investment projects largely determines the management decisions for the implementation of projects in strategic management.

Hence, the question arises about the need to create a system of strategic alternatives that will allow the company to maximize the effects. Each strategic alternative provides the organization with different opportunities and is characterized by different costs and results (effects), which ultimately determines the need for strategic choice.

The modern concept of strategic management considers the strategy not only as a process of achieving strategic goals, but also as a process of finding alternative ways to achieve these goals. The decision-making process in the mining industry is not unstructured and formalized, as there is considerable uncertainty about the further development of market conditions. The peculiarity of such semi-structured problems is the impossibility of their future comprehensive evaluation using one criterion, which determines the multi-criteria nature of the choice. The implementation of the chosen strategic alternatives has an impact on the company's performance indicators (Ponomarenko et al. 2016). Therefore, the choice of strategic alternatives should be carried out both taking into account the performance indicators of investment projects, and taking into account the criteria and performance indicators of the company's operating activities (Tarasova et al. 2014).

2 MATERIALS AND METODS

The theoretical basis of the work was the approaches to the formation of strategic management systems, methods of assessing the effects of the investment project in the group of companies. Methods of comparative, strategic, technical and economic analysis, methodology of system approach, methods of management decision-making are used.

3 STRATEGIC ALTERNATIVE

Alternative - a fundamental approach in the formation of the strategy, since the external environment in the market economy is multifaceted and extremely dynamic, which causes a multivariate choice. Therefore, the process of forming a specific strategy is reduced to the choice of one or more strategic alternatives from a certain set.

Strategic alternatives are a set of strategies that are consistent with the stated mission and goal of the company, taking into account the basic strategy of the company. Each alternative is characterized by possibilities and limitations and has a certain result. Strategic alternatives allow the company to achieve its goals within the chosen direction and resource and institutional constraints. Based on the formulated mission and goal (s), as well as the results of the analysis of the external and internal environment, various options for the development of the company are formulated (developed), that is, possible ways to achieve the goals in the implementation of the identified opportunities are determined.

The generation of strategic alternatives means the formation of a variety of alternatives that will have a positive effect on their subsequent implementation in certain market conditions for this company. A positive effect is an expected economic result determined based on the interests of stakeholders. When generating strategic alternatives, it is advisable to consider several ways to implement the strategy, since specific features, different costs and results characterize each alternative. Generation of alternatives is carried out by a group of top management of the company, or with the help of specially trained specialists (Cartright et al. 2002). In the process of generating alternatives, formed the maximum number of alternatives. In the development of alternatives, the most common scenario approach.

In the development of alternatives, the most common is scenario approach. At the same time, a developed model of the future state of the company taking into account the forecast state of the external and internal environment. When modeling in the horizon of 5-10 years, alternative variants are formed of the future with each of the strategic alternatives. (Efremov et al. 2008)

Strategic choice is made exclusively from alternative strategies of the company. In the absence of alternatives, the meaning of the choice ceases to exist. When choosing strategic alternatives, it is necessary to justify the factors that most significantly affect the achievement of the company's goals.

Further, from the selected strategic alternatives that meet the General goal of the company, an analysis is made for compliance with objective restrictions, which include:

1) sufficiency of financial resources. If you need to use borrowed funds, the company should take into account the relevant risks. When justifying the capital structure, it is necessary to determine the maximum possible level of financing for the company in the chosen strategic alternative.

2) the level of acceptable risk, including insufficient cash flows, financial dependence, etc. This factor can significantly narrow the range of strategies;

3) the potential of the company. When choosing a strategic alternative, it is necessary to assess the opportunities and threats for a specific strategic alternative during SWOT analysis;

4) organization of value chains and business processes of the company. This restriction is essential when choosing a strategic alternative, since the successful implementation of the investment project requires coordinated interaction between the entities.

The selection of strategic alternatives should be based on criteria:

1) optimality of the solution. The solution is called optimal if it provides an extremum, that is, the maximum or minimum of the selection criterion for an individual decision, or satisfies the principle of matching in a group choice. Full information about the problem or situation is needed.

2) the effectiveness of the solution. Effective are solutions that require less cost to implement them in relation to the expected result.

3) the effectiveness of the decision. The solution is effective in comparison with others, which provide a greater degree of achievement of the goals.

These criteria should be specified as a set of indicators that will vary according to the objective of the strategy. The set of indicators can be reduced to integral, taking into account the selected weights of indicators.

Strategic alternatives increase the production capacity of mining and metallurgical company is an investment project for the extraction and enrichment of mineral raw materials, which has an impact on the subsequent stages of the production chain of the company (metallurgical redistribution).

Therefore, to assess the impact of the strategic alternatives, both generally accepted indicators of economic efficiency of investment projects and additional indicators are used (Egerev et al. 2003). It should meet the criteria of optimality, efficiency, effectiveness; ensure compliance of activities with the principles of sustainable development and process approach. Taking into account the principles of the strategic alternatives concept, indicators are developed in three directions: economic, environmental and social, for each direction three indicators corresponding to the criteria of efficiency (input), optimality (process), and effectiveness (output) are selected.

The economic indicators characterizing the efficiency of attracting and using resources (that is, entering the process of efficient production with an increase in production capacity) include capital productivity, material intensity of resources, capital intensity. Other indicators of efficiency of use of fixed and current assets.

For mining and metallurgical production is a very informative indicator of economic efficiency is the ratio of capital costs to revenue. This is due to the fact that investments in the construction of processing plant are very much (for example, investments in the construction of Nezhinsky processing plant in 2018-2019 amount to 950 million euros, the volume of investments in the construction of a new mining and processing plant at the Apatite-nepheline ore Deposit of Oleniy Ruchey is 11.5 billion rubles). Especially high capital intensity is characterized by an underground method of production due to the need to create expensive production facilities and communications directly in the bowels. The main factors that determine the capital intensity in the mining industry: the share of open-pit mining, technological progress, intensification and concentration of production, completeness and complexity of the use of reserves, mining and technical conditions of development, social and environmental measures. Therefore, the selected indicator measures the intensity of investment.

The criterion of optimality among the economic indicators are the most significant: the coefficient of profitability, return on equity, return on invested capital, return on production assets, etc. liquidity ratio, maneuverability, self-sufficiency ratio. The preferred indicator is the level of costs per ruble of commodity products. This is the most generalized indicator of the cost of production, reflecting its direct relationship with profit.

Looking at economic indicators from the perspective of performance, the key ones are sales volume, sales revenue, economic value added and other absolute indicators. Selected economic indicator: economic value added takes into account the cost of capital invested and reflects the size of the economic profit of the enterprise, net of taxes and investment in capital.

Among the environmental factors for the degree of efficiency of use of resources can be used: the number of products that can be reused at the end of life or restored, the percentage of re-used products, the amount of waste per unit of output per year, the number of hazardous substances per unit of output, year. The total amount of waste disposed outside the territory of the enterprise, etc. From a set of possible indicators of ecological efficiency of the enterprise, the key is chosen: the ratio of the area of the broken lands to volume of products. This indicator has the greatest weight, because when developing the field in an open way, large areas of land are exposed, which negatively affects the landscape, hydrogeology, the fertile layer of the earth is disturbed.

Indicators of optimality in the environmental sphere should be based on the legislation of the Russian Federation, which provides for payment for negative environmental impact. The types of negative environmental impact include: emissions of pollutants and other substances into the air; discharges of pollutants and other substances and microorganisms into surface water bodies, underground water bodies and catchments; pollution of subsoil, soils; disposal of production and consumption waste; environmental pollution by noise, heat, electromagnetic, ionizing and other physical effects; other types of negative impact on the environment.

Currently, the fee for negative environmental impact is charged only for the following types of negative impact:

o air emissions of pollutants from stationary and mobile facilities;

o discharges of pollutants into surface and underground water bodies;

o disposal of production and consumption waste.

Thus, the selected indicator will take into account the ratio of the total cost of environmental impact to the volume of production.

The effectiveness in the environmental sphere should be assessed by the level of non-waste and low-waste mining and metallurgical production in the following forms:

- complete processing of waste, both underground and open-pit mining;

In mining and metallurgical production it is necessary to introduce a set of measures aimed at environmental protection. It:

- processing and involvement of liquid, gaseous and solid waste, reduction of discharges and emissions of harmful substances with wastewater and waste gases;

- as building materials for roads, wall blocks and mines, it is possible to use multi-tonnage dump solid waste of concentrating and mining production;

- increasing the efficiency of newly created and existing processes to capture the by-components that are released from waste water and waste gases;

- full use of all Ferroalloy and blast furnace slag, as well as the establishment of processing of steel waste;

- extensive introduction of dry methods for cleaning gases from dust debris for the entire metallurgical production;

- reduction of water consumption, as well as reduction of wastewater through the subsequent development and introduction of anhydrous processes and drainage systems for water supply;
- introduction of cleaning equipment and devices for control of various environmental pollution factors at the enterprise;

increased use of microelectronics, for potential savings of energy and materials, and monitoring of waste products and their active contraction.

Thus, among the environmental indicators, the developed system includes: in terms of efficiency – the ratio of the area of disturbed land allotment to the annual volume of production, optimality – the ratio of the amount of environmental payments/t to the annual volume of production, efficiency – the ratio of the total waste of the enterprise/year to the annual volume of production.

A number of indicators characterizes the social sphere: the ratio of consumer prices and wages of the industry, the level of staff turnover, the level of wages in the industry and the economy as a whole, the level of professional growth of employees.

An important indicator is the average wage index in the region of presence to the average wage in the company. This indicator is one of the main in the study of the main socio-economic indicators of the standard of living of the population, and as a result affects the motivation of employees.

The indicators of optimality (efficiency of the production process) include various types of productivity. The level of labor productivity in the mining industry is determined by the worker and one employee of industrial and production personnel (hereinafter -PPP). Since the system of indicators is developed for investment projects in the field of ore mining, the most important indicator for the assessment is the ratio of annual production to the number of PPP.

Among the performance indicators in the field of social responsibility, it is advisable to consider the indicator of the impact of the enterprise on stakeholders at the local, Federal, regional levels, the degree of labor satisfaction of staff, the degree of change in the reputation of the company. The existing relationship between the performance of mining and metallurgical companies and the state tax policy determines the choice of tax sustainability as a component of the economic stability of the organization. It is proposed to use the calculation of tax intensity as the ratio of accrued taxes/year to investment in the project.

Thus, among the social indicators selected: efficiency -the level of the ratio of s\n in the region to the average s/n in the mining enterprise.

4 RESULTS

Comparing variants of strategic alternatives for the company «PMH»

Approbation of the proposed approach to the choice of strategic alternatives is performed for the company of mining and metallurgical complex. The company "Industrial and metallurgical holding" is aimed at the implementation of the strategy of expansion of production capacity, as the company is in need of its own raw materials for the production cycle (ore mining and production of iron ore concentrate). In the conditions of the concentrated market, the shortage of own raw materials at the moment leads to incomplete loading of production capacities, and in the near future the company needs compensation for the outgoing production capacities for the extraction of iron ore raw materials.

In the design version of «PMH» to increase the degree of availability of its own raw materials, it is planned to increase the production of ore by underground method at the plant due to the reconstruction of the company. Provided by the opening of three new vertical shafts and development of reserves of new operational floor.

A possible SA is the acquisition of a license for the development of iron ore deposits.

The investment analysis showed that the indicators of economic efficiency of projects are comparable. According to the volume of production and terms of development, the project of acquisition of a production license is preferable. It allows providing the design annual capacity

Table 1. Key indicators characterizing the sustainable development of the mining and metallurgical company and meeting the criteria of efficiency, optimality, effectiveness.

Criterion	Economic performance	Ecological index	Social indicators
Resource efficiency	The measure of the intensity of investment: $\frac{\text{Capital expenditures}}{V_{yr}}$, (RUB/RUB) Where capital costs - long-term investments, RUB; Vyr - revenue, RUB/	$\frac{\text{S of disturbed land)}}{\text{Volume of production per year}}$, (Ha/t)	$\frac{\text{Average salary in the region where}}{\text{average salary for mining company}}$, (RUB/RUB)
Optimality of the production process	The level of costs per ruble of commodity products: $\frac{S/s}{V_{yr}}$, (RUB/RUB) Where S/s – the cost of production of rubles.; Vyr – revenue, RUB	$\frac{\sum \text{environmental charges/t)}}{\text{Volume of production per year}}$, (RUB/t)	Labor productivity per 1 production worker in the project $HRV = \frac{\text{Production volume}}{\text{Number of personnel}}$, (ton/person)
Performance of the company	EVA	$\frac{\text{The volume crushi}}{\text{Volume production}}$, ton/ton	Tax intensity = $\frac{\text{Amount of potential taxes}}{\text{Investments paid}}$, (RUB/RUB)

of 21 million tons, in contrast to the reconstruction project – 7.7 million tons/year. The reconstruction option ensures the division's own demand for concentrate, and the alternative project will additionally allow the company to sell the surplus of finished products on the market. At the same time, the projected growth rates of markets and prices are important. At the same time, the size of the investment is preferable to the reconstruction option, because the ratio of the amount of debt of the object of study to the EBITDA indicator is large, respectively; the size of the investment for the company is acceptable strategic alternative to increase the capacity of its own plant.

Thus, the clear choice for two of alternatives impossible to do, then the calculations of the indicators to which are applied the comparison-based scoring. The score "1" received an alternative with the best value of the indicator, the score "0" was set to the alternative, the indicator of which was worse when compared quantitatively. This approach has the following drawback: it is not taken into account that the values of the indicators can be close to each other and only estimates 0 and 1 are set.

Table 2 presents key economic indicators when comparing alternatives.

When comparing the environmental performance should pay attention that, the option of reconstruction of the facilities of the enterprise have already applied to non-waste production.

Table 2. Key economic indicators when comparing alternatives.

Criterion	Economic performance			
	Own production	Point	Acquisition of mineral deposits	Point
Resource efficiency	0,4 RUB/RUB	1	1,4 RUB/RUB	0
Optimality of the production process	0,04 RUB/RUB	1	0,3 RUB/RUB	0
Performance of the company	29%	0	35%	1
Total points for alternative	2		1	

Table 3. Key environmental indicators when compared to the alternatives.

| Criterion | Ecological index | | | |
	Own production	Point	Acquisition of mineral deposits	Point
Resource efficiency	0,008 thousand tons	1	0,03 thousand tons	0
Optimality of the production process	33,35 RUB/thousand tons	1	166,6 RUB/thousand tons	0
Effectiveness	20,7 ton/thousand tons	1	24,7 ton/thousand tons	0
Total points for alternative	3		0	

Unique stowing facility provides complete waste disposal and processing of production in the mine, closing the cycle of the production process (extraction of ore – beneficiation – processing of waste). In this case, the enrichment tailings of the plant are used for the production of the filling mixture. Table 3 presents key environmental indicators when comparing alternatives.

Table 4 presents key social indicators when comparing alternatives.

Table 4. Key social indicators when comparing alternatives.

| Criterion | Social indicators | | | |
	Own production	Point	Acquisition of mineral deposits	Point
Resource efficiency	0,8 RUB/RUB	0	0,6 RUB/month	1
Optimality of the production process	5217 ton/person	0	6857 ton/person	1
Effectiveness	0,8 RUB/RUB	1	0,2 RUB/RUB	0
Total points for alternative	1		2	

5 SUMMARY

The performed calculations indicate that the economic and environmental indicators of the priority is the project to increase capacity in its own production. However, according to the selected social indicators, the best is an alternative project. The final choice of strategic alternatives should take into account the established general goal of the company, which, as a rule, is focused on the priority development of one sphere of strategic alternatives (economic, ecological, social) and implemented by management.

1. The limitations of the system of performance indicators of investment projects, including net present value, internal rate of return, payback period and return on investment index for comparison of strategic alternatives to increase the production capacity of mining and metallurgical companies.

2. A methodological approach to the additional evaluation of strategic alternatives based on the principles of optimality, efficiency, effectiveness in conjunction with the concept of sustainable development and proposed a system of indicators.

3. The approbation of the developed methodological approach to the choice of strategic alternatives is carried out and the choice of a strategic alternative to increase the production capacity of the company "Industrial and metallurgical holding" is justified.

REFERENCES

Arnold G., Davies M. Value-Based Management: Context and Application. John Wiley & Sons: Chichester, UK., 2000 30.History value based management. URL:Valuebasedmanagement.net/faq_history_value_based_management.html

Pierce J. A., Robinson R. B. Strategic management: Strategy Formulation and Implementation. Third ed.Irwin, Homewood, 1988.

Ponomarenko, T., Sergeev, I. Quantitative methods for assessing levels of vertical integration as a basis for determining the economic and organizational sustainability of an industrial corporation. Indian Journal of Science and Technology, 2016, 9 (20).

Ponomarenko, T.V., Fedoseev, S.V., Korotkiy, S.V., Belitskaya, N.A. Managing the implementation of strategic projects in the industrial holding. Indian Journal of Science and Technology, 2016, 9 (14).

Rue L. W., Holland P. G. Strategic Management: Concepts and Experiences. Second ed. N. Y Mac Graw Hill.One thousand nine hundred eighty nine.

Cartright R. Strategies for Hypergrowth. Capstone Publishing, Oxford, 2002.

I. Ansoff. New corporate strategy. Peter, St. Petersburg, 1999.

Bowman K. fundamentals of strategic management./Per. with English. ed L. G. Zayceva, M. I. Sokolova. — Moscow: Banks and exchanges, UNITY, 2007. — 175 p.

Volkov D. L. Theory of value-oriented management: financial and accounting aspects.- 2nd ed. - D. L. Volkov; graduate school of management St. Petersburg state University. - SPb. Publishing house: "Higher school of management; Published. house S.-peterb. State. UN-TA, 2008. - 320, p. 19.

Egerev I. A. business Value: the Art of management. M.: Case, 2003, p. 45-47.

Efremov V. S. Strategy of the business. Concepts and methods of planning./Studies'. benefit. — M.: Ed. "Finpress", 2008. — 192 sec.

Ibrahimov R. cost Management as a management system.//Corporate management.URL:cfin.ru/management/finance/valman/vbmassystem.shtml.

Information on the practice of forming clusters and priority development zones in the Belgorod region and development prospects: [electronic resource of the Federation Council Committee for Federation Affairs and regional policy]. - URL:www.komfed.ru

Collis David J., Montgomery Cynthia A. Corporate strategy. Resource approach. Moscow: ZAO "Olympus-Business", 2007.

Copeland, T Koller, J. Murrin. The cost of companies: assessment and management. - Moscow: ZAO "OLYMPUS-BUSINESS", 2002, p. 576.

MBA course on strategic management/ed. by A. Faye, R. rendella. 4-e Izd. M.: Alpina Business Books, 2007.

Meskon M. H., albert M., Hedouri F. Fundamentals of management, 3rd edition. M.: LLC "I. D. z "Williams", 2008.

Pearce II, J., Robinson R. Strategic management. 12 ed. SPb.: Peter, 2013.

Popov S. A. Strategic management. Modular program for managers "organization development Management" Module 4. M.: "Publishing house "Infra-M", 1993.

Porter M. international competition. M.: International relations, 1993.

Porter M. Competitive advantage: How to achieve high results and ensure its sustainability. — M.: Alpina Business Books, 2005.

Robbins S., Coulter M. Management. Eighth edition. M.: Publishing house "Williams", 2007.

Segal-horn S., P. Quintas Competition based on capacity: proc. benefit. kN. 4. Zhukovsky: MIM LINK, 2001.

Skoch A.V. Synergetic effect of crater-forming investments: methods of quantitative and qualitative assessment//Management in Russia and abroad. – 2008. - №3. - P. - 23-30.

Tarasova J. N. Features of the assessment carried out in the process of managing the value of the company. Proceedings of the all-Russian scientific and practical conference of the faculty and graduate students in the field of economic Sciences "Modern problems of Economics and management in the service sector", 2014.

Thompson Jr., A. A., Strickland III A. J. Strategic management: concepts and situations for analysis. 12 ed. M.: Publishing house "Williams", 2009.

Topical Issues of Rational Use of Natural Resources 2019 – Litvinenko (Ed)
© 2020 Taylor & Francis Group, London, ISBN 978-0-367-85720-2

Impact of market prices of carbon dioxide -within planetary boundaries- to the extractive industry

A.M. Gómez Cuartas
TU Bergakademie, Freiberg, Germany
Montanuniversität, Leoben, Austria

R. Biastoch
TU Bergakademie, Freiberg, Germany

ABSTRACT: This paper analyses the hypothetical economic impacts on revenues of 16 of the largest producers of copper, aluminium, gold and iron, if CO_2 emissions were appropriately costed. For this, the author compiled prices of CO_2 per ton prices from academic studies that are given under a scenario in which global temperature may not rise more than 2°C, as well as self-reported annual CO_2 emissions of the mining companies. Total costs of emissions were subtracted from the revenues of each assessed firm. The results show, that costs of CO_2 emissions per ton of copper, aluminium, iron and gold vary highly; on the one hand due to the differences in mining method and processing, but also due to the large scale of prices across the different studies. It becomes apparent that an appropriate costing of released carbon emissions would significantly diminish certain companies' revenues. The applied methodology may allow for a more in-depth analysis within the mining industry or other industries and can inform targeted environmental policy making and taxation in the future.

1 INTRODUCTION

The environment has been seriously threatened by human activity over the last 300 years. Since the late 1900s, experts made efforts to assess the economic impacts of such adverse environmental pressures. For instance, in the 50s, authors advocated for the consideration of environmental impacts of economic activities, not only based on a cost-benefit approach, but *"as social costs resulting from the structure and incentives in free market capitalism"* (Ciriacy-Wantrup, 1952; Kapp, 1961 In: Nadeau, 2015). Geogerscu-Roegen (1971) claimed that *"nature plays an important role in the economic process as well as in the formation of economic value"*. Albeit such efforts, till date, economic activities are measured preferably based on economic utility, efficient resource allocation, or human welfare without considering *"that welfare also depends to a large extent on ecosystem services"* (Daly & Farley, 2004).

Experts found that a mean to ensure a sustainable future is through the regulation of human activities by global thresholds or planetary boundaries, such as the establishment of a cap on allowed ppm emissions of CO_2. Rockström et al (2009) analysed the levels of anthropogenic perturbation with the purpose of identifying thresholds that shall not be trespassed in order to prevent undesirable shifts in environment dynamics. The authors defined nine planetary boundaries, seven of them are able to be quantified (Table 1). The level of uncertainty of each planetary boundary compromises the Earth Systems stability, thus the capacity to remain *"low- a safe operating space for a societal development"* (Steffen et al, 2015).

Planetary boundaries are perceived as a relatively new concept that has been introduced into the global political agenda along with resolutions or pacts like the Paris

Table 1. Planetary boundaries' thresholds. Source: Rockström et al (2009).

Planetary Boundary	Limit Value
Climate change: CO_2 concentration in the atmosphere	< 350 ppm and/or maximum change of +1 W m^{-2} in radioactive forcing.
Ocean acidification	Mean surface seawater saturation state with respect to aragonite >80% of pre-industrial levels.
Stratospheric ozone depletion	<5% reduction in O_2 concentration from pre-industrial level of 290 Dobson Units.
Biogeochemical flows: Nitrogen and phosphorus cycle	Limit industrial and agricultural fixation of N2 to 35 Tg N yr^{-1}. Annual phosphorus inflow to oceans shall not to exceed 10 times the natural background weathering of phosphorus.
Global freshwater use	<4000 km^3 yr^{-1} of consumptive use of runoff resources.
Land system change	<15% of the ice-free land surface under cropland.
Biosphere integrity: Biological diversity	Annual rate of <10 extinctions per million species.
Atmospheric aerosol loading	No global quantification
Novel entities	No global quantification

Agreement and the Sustainable Development Goals (SDGs), however compliance is not yet mandatory (Steffen et al, 2015). Nevertheless, 193 countries have ratified to incorporate the SDGs into their national legislation and to develop an action plan, which includes the engagement of firms and companies. Economists suggest that applying a cost to carbon pollution is an incentive for countries to reduce CO_2 emissions (United Nations, 2015). There are different methodologies to calculate CO_2 per ton prices, the most well-known is the Emissions Trading Scheme (ETS) developed and applied in the European Union, which has also been implemented in Australia, New Zealand, Kazakhstan and Canada (The World Bank, 2019a). Countries such as Colombia, Chile, Mexico, Brazil, as well as some of the US states, are considering the implementation of the ETS or a carbon tax (The World Bank, 2019a). According to the conducted literature review, the most commonly used methodologies are the social cost of carbon, marginal abatement costs, the Dynamic Integrated Climate-Economy (DICE) model, and other Integrated Assessment Models (IAMs) (Table 2).

As shown in Table 2, costs of CO_2/ton vary widely in literature, with ranges from 5 USD CO_2/ton to 260 USD CO_2/ton. This does not only depend on the applied valuation methodology, but also on the global environmental scenario or planetary boundary, time horizon, and currency considered in the calculations. The Environmental Protection Agency (2014) estimates that in 2010, ca. 25% of global anthropogenic greenhouse gas emissions (GHG) are attributable to electricity and heat production from burning fossil fuels, while all other industrial activities are accountable for ca. 21% of global anthropogenic GHG emissions. Tost et al (2018) estimated that global CO_2 emissions for the year 2016 were ca. 36 Gt and that less than 1% of these were contributed by the mining of bauxite, copper, gold and iron. These findings imply that the metals extractive industry should be held accountable for the costs of a total of ca. 0,36 Gt of CO_2 emissions.

This paper focuses on climate change as a relevant planetary boundary for the extractive industries. The author adopts the long-term development goals of the Paris Agreement that aims to tackle climate change by limiting global temperature to rise in the 21st century by more than 2°C (United Nations, 2015).

2 METHODOLOGY

Literature review and data compilation was conducted to identify costs of CO_2/ton under the mentioned threshold (Table 2), likewise carbon emissions of 16 of the Large-

Table 2. Summary of literature costs for CO_2/ton under a 2°C threshold scenarios.

Price CO2/ton			Time Horizon	Valuation Methodology	Source
Min	Max	Unit			
5	20	USD/t CO_2	2020	Marginal abatement costs	ECD, 2017
14	31	GBP/t CO_2	2020	Marginal Abatement Costs	Department of Energy and climate change, 2009
18	250	USD_{2010}/t CO_2	2020	Integrated Models	Rogelj et al, 2014
24	50	USD/t CO_2	2020	Marginal abatement costs	CDP, 2018
25	50	EUR/tCO_2	2020	Expert interviews	Canfin et al, 2016
30	90	GBP/t CO_2	2020	Marginal Abatement Costs	Department of Energy and climate change, 2009
30	50	USD/t CO_2	2020	Marginal abatement costs	IEA, 2015
40	80	USD/t CO_2	2020	Social Cost of Carbon	CPLC, 2017
-	40	USD/t CO_2	2020	Social cost of Carbon (IAM)	DDP, 2017
-	264	USD_{2005}/t CO_2	2015	IAM (DIECE)	Nordhaus, 2007
	60	USD_{2005}/t CO_2	2020	IAM (DIECE)	Nordhaus, 2013

Scale Mining (LSM) companies that extract iron, gold, copper and aluminium, which excluding gold, represent ca. 96% of all metals mined worldwide, in terms of bulk tonnage (Tost et al, 2018).

No commonly accepted carbon valuation methodology exists yet, therefore the author compiled different costs of CO_2 from 11 influential studies with a 2°C threshold boundary. Costs of CO_2/ton obtained from marginal abatement costs- and social cost of carbon methodologies display more conservative ranges than the values which result from integrated models. In comparison with the average cost of CO_2/ton (ca. 5 USD), in which carbon was traded in the EU ETS in the year 2016 (Sandbag, 2019), the hypothetical prices given in literature are higher. This can be due to the over-allocation of allowances, as well as the volatility of policies and global crisis.

In order to make pricing scenarios comparable, all costs were converted to 2016 USD base year prices (The Balance, 2019) and currencies were converted to USD using World Bank exchange rates (World Bank, 2019b). While not explicitly stated as such in the source, the price quoted in Nordhaus (2007, 2013) is assumed to be a maximum value, a minimum value is not given in this source. Unless specified differently, USD values are considered to be from the year of publication of each study. For this paper, an average of the studies' adjusted minimum and maximum values was formed which accrued to a minimum price of USD 18.97 per CO_2/ton and a maximum price of USD 100.39 per CO_2/ton.

The impact on revenues of the largest producers of copper, aluminium (bauxite ore), iron and gold are assessed by multiplying the corresponding average of the obtained minimum and maximum prices of CO_2/ton to the amount of companies' self-reported emissions in the year 2016 (Table 3). Such theoretical costs are subtracted from revenues of each company, in order to identify the relative additional costs of CO_2 emissions.

The four aluminium producers considered in this study account for approximately 39% of global anthropogenic carbon outputs for the year 2016 (Tost et al, 2018). Among those companies, thus primarily engaged in the extraction of Copper bear a share of ca. 33% of such outputs, Gold producers 19% and Iron producers approximately 33% (Tost et al, 2018).

Table 3. Summary of self-reported revenues and emissions in the year 2016 from LSM companies that extract aluminium (bauxite ore), copper, gold and iron.

Companies	Revenues (USD)	Emission (t)	Source
Aluminium (bauxite ore)			
Rio Tinto	$ 33,781,000,000	17,500,000	Rio Tinto, 2016
Alcoa	$ 9,318,000,000	24,300,000	Alcoa, 2016
Chalco	$ 25,616,038,410	68,000,000	Chalco, 2017
Hydro	$ 9,756,309,524	13,000,000	Hydro, 2016
Copper			
Codelco	$ 11,537,000,000	5,000,000	Codelco, 2016
Freeport-McMoRan	$ 14,830,000,000	10,400,000	Freeport-McMoRan, 2016
Glencore	$ 42,142,000,000	3,510,000	Glencore, 2016
BHP	$ 30,900,000,000	6,900,000	BHP, 2016
Southern Copper	$ 5,379,000,000	5,390,000	Southern Copper Corp, 2016
Gold			
Barrick	$ 8,558,000,000	3,646,671	Barrick, 2016
Newmont	$ 6,711,000,000	4,300,000	Newmont, 2016
Anglo Gold Ashanti	$ 4,254,000,000	4,062,000	Anglo Gold Ashanti, 2016
Goldcorp	$ 3,510,000,000	1,143,616	Goldcorp, 2016
Kinross	$ 3,472,000,000	1,568,000	kinross, 2016
Iron			
Vale	$ 27,488,000,000	14,000,000	Vale, 2016
FMG (Fortescue Metal Group)	$ 7,100,000,000	1,760,000	FMG, 2016

3 RESULTS

All reviewed companies reported on their emissions and revenues in annual reports. In terms of total outputs, the LSM that extract bauxite ore are accountable for the largest CO_2 emissions to the atmosphere with an overall of ca. 123 Mt, followed by copper producers with ca. 31 Mt. Iron producers contribute with ca. 16 Mt and gold producers with 15 Mt respectively. Accordingly, a pricing scheme for carbon emissions would have the highest negative impact on aluminium producers' revenues (Table 4). Notably, copper mining companies' revenues were not only the highest in 2016 for this sample, but their total CO_2 emissions were also only about a fourth of those from bauxite ore extraction.

Comparing companies' total production outputs with CO_2 emissions may serve as an indicator for mineral specific taxation. The data shows that the carbon emission levels are strongly dependent on the commodity, with copper having the highest emissions per ton produced, then bauxite, and with much lower emissions per output iron (see Figure 1). The output of gold is measured in ounces and is shown separately in Figure 2.

The two graphs also illustrate that significant differences exist in the efficiency of production per ton measured in CO_2 output. Since the major source for emissions in the mineral industry stems from processing and the associated power consumption, these findings suggest that efficiencies can be improved for low-performing companies. However, the absence of a clear regulation for reporting of CO_2 emissions makes such values difficult to compare (Tost et al, 2018). While those companies listed on international exchanges are required to disclose their emissions, there is no defined methodology in doing so. This becomes particularly apparent for the CO_2 emissions reported by the Chinese state-owned Aluminium Corporation of China (Chalco). Chalco's sustainability report does not distinguish CO_2 emissions by commodity type. While its

Table 4. Impact of CO_2/ton prices to extractive companies' revenues.

Companies	Revenues [M USD]	CO2 Emis-sion [t]	min. CO2 [M USD]*	max. CO2 [M USD]**	min. share of revenues %	max. share of revenues %
Aluminium (Bauxite ore)						
Rio Tinto	$33,781	17,500,000	$332	$1,750	0.98%	5.20%
Alcoa	$ 9,318	24,300,000	$461	$2,430	4.95%	26.18%
Chalco	$21,527	68,000,000	$1,290	$6,800	5.99%	31.71%
Hydro	$ 9,756	13,000,000	$247	$1,300	2.53%	13.38%
Copper						
Codelco	$11,537	5000,000	$95	$500	0.82%	4.35%
Freeport-McMoRan	$14,830	10,400,000	$197	$1,040	1.33%	7.04%
Glencore	$42,142	3,510,000	$67	$351	0.16%	0.84%
BHP	$30,900	6,900,000	$131	$690	0.42%	2.24%
Southern Copper	$ 5,379	5,390,000	$102	$539	1.90%	10.06%
Gold						
Barrick	$ 8,558	3,646,671	$69	$366	0.81%	4.28%
Newmont	$ 6,711	4,300,000	$82	$430	1.22%	6.43%
Anglo Gold Ashanti	$ 4,254	4,062,000	$77	$406	1.81%	9.59%
Goldcorp	$ 3,510	1,143,616	$22	$114	0.62%	3.27%
Kinross	$ 3,472	1,568,000	$30	$157	0.86%	4.53%
Iron						
Vale	$27,488	14,000,000	$266	$1,400	0.97%	5.12%
FMG (Fortescue Metal Group)	$ 7,100	1,760,000	$33	$176	0.47%	2.49%

* Assuming a min. price of CO_2/ton of USD 18.97
** Assuming a max. price of CO_2/ton of USD 100.39

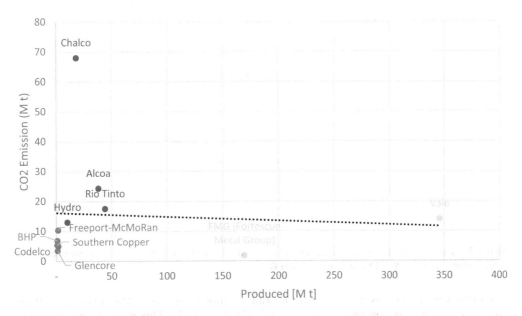

Figure 1. Relationship between amount of emissions of CO_2 per produced ton of commodity. Source: Companies' annual reports.

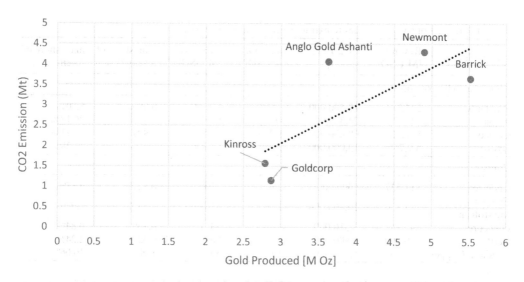

Figure 2. Relationship between amount of emissions of CO_2 per produced ounce of gold. Source: Companies' annual reports.

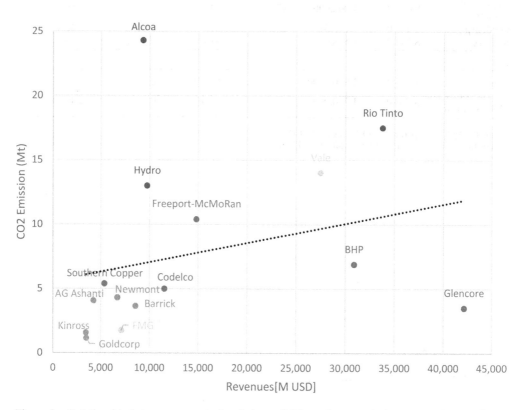

Figure 3. Relationship between amount of emissions of CO_2 and companies' revenues. Source: Companies' annual reports, WMD and Tost et al (2018).

economic key activity is the exploration and mining of bauxite, also other resources are exploited by this company, such as coal (Chalco, 2017). This may explain the outlier of Chalco.

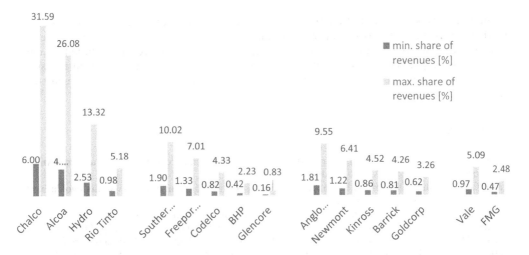

Figure 4. Impact of costs of CO_2/ton to companies' revenues, assuming a minimum cost of USD 18,97 CO_2/t and a maximum cost of USD 100,39 CO_2/t.

This study further analysed the impact on companies' profitability. This depends largely on the pricing scenario, the quantity of emissions, and the revenues of the firm. The differences in pricing scenarios have been discussed above. If the same costs for emissions are applied to all companies, the extent to which the firm's profitability is affected depends on amount of emissions and revenues. Figure 3 shows that significant differences in both revenues and emissions exist across companies, independent of which mineral they produce. Copper producers tend to have the most revenues compared to their CO_2 emissions, with Glencore and BHP performing significantly better than other companies. The next best ratio is by iron producers, followed by gold producers. Bauxite producers perform the worst, with all companies below the overall average. Notably, Chalco was excluded in the graph, due to the distortion its' extreme values of carbon emissions of 68 Mt.

Coherent with these findings, costs under different pricing scenarios for CO_2 emissions turn out differently. In the minimum pricing scenario (USD 18,97 CO_2/t), the costs for emissions accrue to an average share of each companies' revenue of 1.62% and 8.54% under the maximum pricing scheme (100,39 CO_2/t). These averages however mask the very significant effect on the worst-performing companies, which, for example, under the maximum pricing scheme accrue to up to 31.71% of revenues of Chalco or 26.18% of those of Alcoa. In comparison, the share of revenue from such additional taxes of the best performing company Glencore is affected by less than 1% in both pricing scenarios (see Figure 4, Table 4).

4 DISCUSSION

The limitations of this study include the relatively small sample of firms which only allow to draw limited conclusions regarding other producers or the industry as a whole. These companies however represent a significant share for each of the minerals produced and mineral production overall. For a comprehensive analysis, small and medium-sized enterprises should be included. Another issue is that no revenue or profit values by commodity were assessed, but only for the firm as a whole. Most of the companies in this sample engage in the production of more than one mineral. Another discussed shortcoming is the lack of a consistently applied methodology to account for self-reported CO_2 emissions by companies.

This research can inform the appropriate design for taxation of GHG emissions in the mining sector. Understanding the large differences of adverse effects of such a tax across different mineral sectors, but also among the producers is important to design a methodology which rewards the best performers, while setting strong punitive mechanisms to enforce the reduction of emissions of the worst polluters. In this context, this research should be continued to develop a better understanding of the source for significant differences in emissions among producers of the same commodity. The effect on revenues must also be compared to that of other industries.

REFERENCES

Alcoa. 2016. Annual Report 2016. Retrieved from: http://www.annualreports.com/HostedData/Annual ReportArchive/a/NYSE_AA_2016.pdf.

Anglo Gold Ashanti. 2016. Annual Financial Statements. Retrieved from: https://www.anglogoldashanti. com/investors/annual-reports/#2016.

Baiocchi, G. 2012. Methods on dimensions of ecological economics. Ecological Economics 75: 1-9.

Barrick. 2016. 2016 Year-end Report and Fourth Quarter Results. Retrieved from: https://barrick.q4cdn. com/788666289/files/doc_financials/quarterly/2016/q4/Barrick-2016-Q4-Year-End-Report.pdf.

BHP. 2016. BHP Billiton Annual Report 2016. Retrieved from: https://www.bhp.com/media-and-insights/reports-and-presentations/2016/09/2016-annual-reporting-suite.

Canfin, P., Grandjean, A & Mestrallet, G. 2016. Mission Report-Proposal for aligning carbon prices with the Paris Agreement. 15 P.

Carbon Pricing Leadership Coalition (CPLC). 2017. Report of the High-Level Commission on Carbon Prices. 69 P.

CDP. 2018. Carbon Pricing Corridors: The Market View 2018. Carbon Pricing Leadership Coalition. 39 P.

Chalco. 2016. 2016 Annual Results Announcement. Retrieved from: http://www.chalco.com.cn/chalcoen/ rootfiles/2017/03/23/1490229844504018-1490229844506299.pdf.

Codelco. 2016. Annual Report 2016. Retrieved from: https://www.codelco.com/memoria2016/en/.

DDP. 2017. Carbon prices in national deep decarbonization pathways Insights from the Deep Decarbonization pathways project (DDP). 13 P.

Daly, H.E & Farley, J. 2004. Ecological Economics: Principles and Practice. Island Press, Washington, D.C.

Department of Energy and Climate Change. 2009. Carbon Valuation in UK Policy Appraisal: A Revised Approach. Climate change Economic Department of Ebergy and Climate Change. 128 P.

European Comission Decisison (ECD). 2017. Climate action, environment, resource efficiency and raw materials. 102 P.

FMG. 2016. Annual Report 2016. Retrieved from: http://www.annualreports.com/HostedData/Annual ReportArchive/F/ASX_FMG_2016.pdf.

Freeport-McMoRan. 2016. 2016 Annual report. Retrieved from: http://www.annualreports.com/Hosted Data/AnnualReportArchive/f/NYSE_FCX_2016.pdf.

Georgescu-Roegen, N. 1971. The Entropy Law and the Economic Process. Cambridge, MA: Harvard University Press.

Glencore. 2016. Annual rpeort 2016. Retrieved from: https://www.glencore.com/dam/jcr:79d87b60-d53a-4f1a-9dbe-4d523f27de83/GLEN-2016-Annual-Report.pdf.

Goldcorp. 2016. Annual Report 2016. Retrieved from: https://s22.q4cdn.com/653477107/files/doc_finan cials/2016/2016-Annual-Report.pdf.

Hydro. 2016. Annual Report 2016. Retrieved from: https://www.miningdataonline.com/reports/annual/ NorskHydro-2016-annual-report.pdf.

International Energy Agency (IEA). 2015. Energy Technology Perspective: Mobilising Innovation to Accelerate Climate Action. 418 P.

Kinross. 2016. Annual Report 2016. Retrieved from: http://www.annualreports.com/HostedData/Annual ReportArchive/k/TSX_K_2016.pdf.

Newmont. 2016. 2016 Annual Report and Form 10-k. Retrieved from: https://s1.q4cdn.com/ 259923520/files/doc_financials/annual/2016/Newmont-2016-Annual-Report-Bookmarked-PDF-for-website.pdf.

Nodeau, R.L. 2014. The unfinished journey of ecological economics. Ecological Economics 109:101-108.

Nordhaus, W. 2007. The Challenge of Global Warming: Economic Models and Environmental Policy. Yale University. 153 P.

Nordhaus, W. 2013. DICE 2013R: Introduction and User's Manual. Second Edition. 102 P.

Rio Tinto. 2016. 2016 Annual report. Retrieved from: https://www.riotinto.com/documents/RT_2016_An nual_report.pdf.

Rockström, J., Steffen. W, Noone, K., Persson, Å., Chapin, F.S., Lambin, E., Lenton, T.M, Scheffer, M., Folke, C., Schellnhuber, H., Nykvist, B., De Wit, C.A., Hughes, T., Van der Leeuw, S., Rodhe, H., Sörlin, S., Snyder, P.K., Costanza, R., Svedin, U., Falkenmark, M., Karlberg, L., Corell, R.W., Fabry, V.J., Hansen, J., Walker, B., Liverman, D., Richardson, K., Crutzen, P & Foley, J. 2009. Planetary boundaries: exploring the safe operating space for humanity. Ecology and Society 14(2): 32.

Rogelj, D., Shindell, K., Jiang, S., Fifita, P., Forster, V., Ginzburg, C., Handa, H., Kheshgi, S., Kobayashi, E., Kriegler, L., Mundaca, R., Séférian, M.V & Vilariño. 2018, Mitigation pathways compatible with 1.5°C in the context of sustainable development. In: Global warming of 1.5°C. An IPCC Special Report on the impacts of global warming of 1.5°C above pre-industrial levels and related global greenhouse gas emission pathways, in the context of strengthening the global response to the threat of climate change, sustainable development, and efforts to eradicate poverty [V. Masson-Delmotte, P. Zhai, H.O. Pörtner, D. Roberts, J. Skea, P.R. Shukla, A. Pirani, W. Moufouma-Okia, C. Péan, R. Pidcock, S. Connors, J.B.R. Matthews, Y. Chen, X. Zhou, M. I. Gomis, E. Lonnoy, T. Maycock, M. Tignor, T. Waterfield (eds.)]. In Press.

Sandbang. 2019. Carbon Price Viewer. Retrieved form: https://sandbag.org.uk/carbon-price-viewer/.

Southern Copper Corp. 2016. Annual Report 2016. Retrieved from: http://www.annualreports.com/Hos tedData/AnnualReportArchive/s/NASDAQ_SCCO_2016.pdf.

Steffen, W., Richardson, K., Rockström, J., Cornell, S.E., Fetze, I., Bennett, E.M., Biggs, R., Caprtenter, S.R., De Vries, W., De Wit, C.A., Folke, C., Gerten, D., Heinke, J., Mace, G.M., Persson, L.M., Ramanathan, V., Reyers, B & Sörlin, S. 2015. Planetary Boundaries: Guiding human development on a changing planet. Science (344) 3223.

The Balance. 2019. US Inflation Rate by year from 1929 to 2020: How bad is inflation? Past, Present, Future. Retrieved from: https://www.thebalance.com/u-s-inflation-rate-history-by-year-and-forecast-3306093.

The Environmental Protection Agency. 2014. Global Greenhouse Gas Emissions Data. Retrieved from: https://www.epa.gov/ghgemissions/global-greenhouse-gas-emissions-data.

The World Bank, 2019a. Carbon Pricing Dashboard. Retrieved from: https://carbonpricingdashboard. worldbank.org/map_data.

The World Bank. 2019b. Official exchange rate (LCU per US$, period average). Retrieved from: https:// data.worldbank.org/indicator/pa.nus.fcrf.

Tost, M., Bayer, B., Hitch M., Lutter, S., Moser, P & Feiel, S. 2018. Metal Mining's Environmental Pressures: A Review and Updated Estimates on CO2 Emissions, Water Use, and Land Requirements. Sustainability 10:2–14.

United Nations, 2015. Paris Agreement. Retrieved from: https://unfccc.int/files/essential_background/con vention/application/pdf/english_paris_agreement.pdf.

Vale. 2016. Annual Report: Form 20-F. Retrieved from: http://www.vale.com/EN/investors/information-market/annual-reports/Pages/default.aspx.

Topical Issues of Rational Use of Natural Resources 2019 – Litvinenko (Ed)
© *2020 Taylor & Francis Group, London, ISBN 978-0-367-85720-2*

Economical efficiency analysis of a company based on VBM indicators

K.D. Deripasko & M.A. Khalikova
Ufa State Petroleum Technological University, Ufa, Russia

ABSTRACT: The paper analyzes the VBM concept indicators, such as economic value added (EVA), added cash flow (CVA), shareholder value added (SVA). The main models for efficiency estimation were highlighted, the advantages and disadvantages are indicated. On the basis of VBM indicators, an assessment of the efficiency of the oil and gas company has been carried out. It has been suggested that the company needs to use VBM indicators together with a model of sustainable economic growth for more successful management of the company.

1 INTRODUCTION

The goal of the research paper is appliance of VBM indicators in efficiency estimation in the case of Russian oil and gas companies. The problem of choosing indicators for estimation of company's efficiency on operating and strategic levels is rather relevant these days. According to the goal research objectives are highlighted: analysis of the cur-rent state and tendencies in the oil and gas industry; determination of indicators that impact value and efficiency of a company; study of traditionally used efficiency indicators, their advantages and disadvantages; comparison of company's accounting and financial models; company efficiency analysis based on VBM indicators; connection between sustainable economic growth model and VBM indicators.

There is a noticeable interest of economists and top management of large companies to the concept of VBM. Many scientists worldwide, such as T. Copeland, A. Rappaport, M. Scott, G. Stewart, D. L. Volkov, V. E. Esipov and others, have studied the ways to control the cost management of a company.

Oil and gas industry is one of the main sources of state budget replenishment in Russia, it provides economic and energy security to the country and remains a section, attractive for investments. Vertically-integrated oil companies are mostly interested in an increase of their productive efficiency and their value on the market. In this case, choosing indicators for estimation of company's efficiency, both on operating and strategic levels, is absolutely essential.

Nowadays, Russian oil and gas companies work in increasing competition among international companies. The rising share of hard to recover reserves among all reserves is also troubling. Moreover, many companies face the effects of oil price volatility, political instability and changes in taxation. Great amount of analytical companies forecast rising energy consumption all over the world, making satisfying this rising demand one of the main priorities for those producing energy sources. At the same time the energy sector is expected to change in support of renewable energy. Shelf and hard to recover reserves will take the leading role in oil production. Also, digitalization is certainly one of the more promising trends when it comes to improvements in efficiency.

All oil and gas companies set strategic goals, which may be separated into goals in the business-segment, in exploration & production, refining & petrochemicals, marketing & distribution, corporate governance. However, two of the most essential goals for the company is rising profitability and consistent value creation. In this case, factors that influenced company value must be determined. External factors such as geopolitical instability, oil price volatility and rising competition definitely have an impact on company value. These factors can't be changed but management of the company should take them into consideration while planning. Internal factors such as resource base, operational plan, financial policy, investment decisions, process utilization rate and other might be controlled for rising efficiency.

2 MATERIALS AND METODS

Russian companies today use the traditional analytical model for efficiency estimation of a company. Traditional accounting oriented metrics for vertically-integrated oil companies like EBITDA, earnings per share, return on average capital employed, focus on past performance and as a result they are not correlated with actual value creation. The main advantages of accounting model are simplicity and wide application.

The concept of VBM is a projection of financial analytical model. VBM assumes focus on the future, which means it is essential to catch signals of approaching changes and to react on them as early as possible. The main disadvantages of this concept are a noticeable complexity in calculation and a necessity of adaptation for Russian reality.

Value Based Management (VBM) is the management philosophy and approach that enables and supports maximum value creation in organizations, typically the maximization of shareholder value. The main difference between traditionally used indicators and VBM indicators is that the latter signalizes the changes of values to the shareholders and measures how successful the company is at achieving its strategic goals.

A full analysis of the main VBM-indicators, such as Economic Value Added (EVA), Cash Value Added (CVA) and Shareholder Value Added (SVA) has been conducted in this paper.

2.1 Economic value added

Economic Value Added (EVA®) - an indicator of economic profit, which is calculated as the difference between net operating profit after taxes and fees for all capital invested in a company, taking into account special amendments to profits and capital. This indicator,

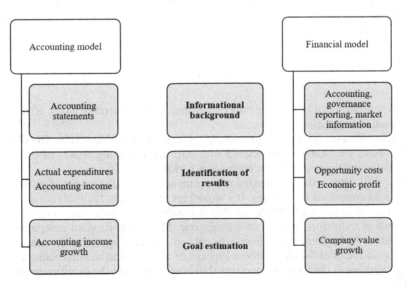

Figure 1. Comparison of accounting and financial models.

developed and patented by the American company Stern Stewart & Co. The calculation procedure is presented in Equation 1:

$$EVA = NOPAT - IC * WACC \qquad (1)$$

where NOPAT = net operating profit after tax; IC = invested capital; WACC = weight average cost of capital.

It is worth noting that when calculating EVA indicator, it is worth making certain adjustments. There are a large number of possible adjustments for calculating the EVA, but it is advisable to use only those that will have a significant impact on the value of this indicator. So, when calculating the economic value added for an oil company, it is necessary to take into account the costs of research and development, exploration and development of non-operating wells.

2.2 Cash value added

Added cash value (CVA) - residual cash flows generated by investments in the organization. CVA is net cash flow minus capital expenditures, calculated using the Equation 2:

$$CVA = CBIj - NA_0 * WACC \qquad (2)$$

where CBIj = cash flow before interest; NA0 = net assets assessed at original cost.

CVA calculation is based on company cash flow. Therefore, the model results do not depend on the accounting policies and accounting standards adopted by the company. Another obvious advantage of the indicator is that it can be used not only by public, but also by private companies. Due to the fairly transparent ideas of the company, the cost of personnel training in the implementation of the approach is minimal. When using the CVA approach, it may be difficult to use specific adjustments if the expected cash income is irregular.

2.3 Shareholder value added

Shareholder value added (SVA) is defined as the actual increment of share capital. Added value for shareholders appears when the profitability of a company's new investments exceeds the weighted average cost of capital. The SVA method is developed by Alfred Rappaport. The indicator is calculated by the Equation 3:

$$SVA_n = \left[\frac{\Delta EBI}{WACC} * \frac{1}{(1 + WACC)^{n-1}} \right] - \left[\frac{\Delta In}{(1 + WACC)^{n-1}} \right] \qquad (3)$$

where ΔEBI = influence of effects depends on earnings before interests; ΔIn = effect connected with additional investments.

The indicator focuses on directly measuring the change in business value for shareholders. At the same time, the change is associated with both the operating results of operations and investment decisions made during the evaluation period.

One of the difficulties with value-based management is the large amount of indicators developed by various consulting companies. It's important to weight in their advantages, disadvantages and areas of implication. Based on that it is safe to conclude that companies must use whole systems of indicators, instead of a single one. Moreover, Russian companies may face some difficulties in implementing these concepts in the existing systems, these include: an over-reliance on the established traditional means of analysis and management, a blind commitment to thinking at the level of operational management, and the insufficient level of managerial skills.

3 RESULTS

«Bashneft» is a vertically-integrated oil and gas company that produces and refines oil and gas, as well as sells petroleum and petrochemicals products on domestic market and on export. The company plays a huge role in the development of the Republic of Bashkortostan and remains its main socio-economic foundation.

In the research work the main key performance indicators were analyzed. In 2018, oil production at license areas totaled 18.90 million toe, which is 8.25% lower than 2017. The decrease is due to the production restrictions imposed by the agreement of the OPEC + countries. Natural gas production amounted to 828.07 million m3, associated petroleum gas - 30.50 million m3, which is lower by 6.20% and 3.48%, respectively, compared with 2017. At the end of 2018, the company showed an increase in profit indicators, despite the existing restrictions on the volume of production. Sales revenue in 2018 amounted to 755.44 billion rubles, which is 35.25% higher than in 2017. The company's EBITDA in-creased by 15.61% and amounted to 175.52 billion rubles. The growth of indicators is the result of the work of management to improve the efficiency of commercial activities, optimization of the structure and volume of production of petroleum products, as well as reducing administrative costs.

To conclude, in 2017-2018 despite high volatility of oil prices, changes in tax legislation, regulation of prices for oil products in the domestic market and impact from restriction under OPEC+ Agreement, the company demonstrated successful production and financial results.

The main goal of the company is to increase the profitability of the business and, essentially, it's value. The value of the company is influenced by both external factors (indirect and direct impact) and internal factors. The company cannot influence external factors, but it must be taken into account when planning its activities. By managing internal factors, the company will ensure the effective functioning and growth of the company's value.

It is very important to plan and forecast company activity based on strategic goals. That is why VBM indicators are chosen for efficiency estimation through the lens of goals achievement.

Based on theoretical aspects, chosen VBM indicators for PAO ANK "Bashneft" were calculated. Dynamic of indicators presented in Figure 2.

The dynamics of all considered indicators of economic efficiency of PAO ANK "Bashneft" is positive, which indicates the efficiency of the enterprise, the growth of its value and attractiveness for shareholders. In 2018, the value of economic value added amounted to 80.3 billion rubles. which is 34.3% more than in 2017. Cash value added amounted to 113.5 billion rubles, which is 32.4% more compared with the previous year. In 2016, the shareholder value added was negative. However, there has

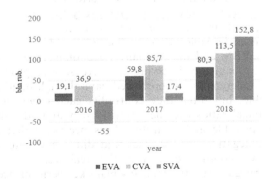

Figure 2. Dynamic of VBM indicators for PAO ANK "Bashneft".

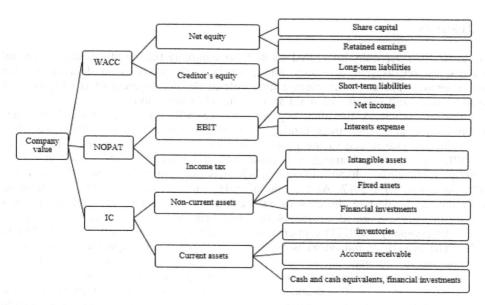

Figure 3. Value driver tree.

been a significant increase in 2018 and the figure has reached its highest value and amounted to 152.8 billion rubles.

After VBM indicators are calculated, we can design a tree of drivers that influence the change in company's value. It will help to determine the «bottleneck» and to identify which indicators must be controlled in order to improve the efficiency.

To sum up, depending on the detalization of the indicators, we can plan activities at both the strategic and operational levels. The main indicators that need to be managed are the weighted average capital rate, the amount of capital invested and the amount of profit.

4 DISCUSSIONS

Since development is one of the main goals of any company, it is assumed that the sales volumes of the studied company will increase. However, this growth should not be achieved at any cost, otherwise the company may not strengthen, but weaken its position until the state of bankruptcy, as world practice shows.

In order to assess the growth of the company, maintain operational control over the level of achievement of the planned values of performance indicators and develop measures to improve economic efficiency, it is necessary to use the Van Horn model of sustainable economic growth in connection with and economic value added indicator. Using this approach, it is possible to trace the relationship between the company's strategy, the concept of VBM and the company's economic growth rates.

Forming its plans, the company is trying to solve several problems: sales growth, production flexibility, moderate use of borrowed capital, high dividends. Each of these tasks in itself is valuable, but often they contradict each other. And when this happens, one or more financial ratios become unsatisfactory and adjustments are required, which are not always possible and desirable. It is necessary that the growth of the company did not have negative consequences and did not violate its financial condition, and moreover, it was coordinated with market realities.

Focusing on the set strategic goals, the company plans key performance indicators. Next, it is necessary to calculate VBM indicators in order to assess how the planned indicators will

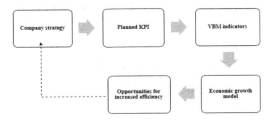

Figure 4. Model for estimating efficiency based on VBM indicators and economic growth model.

affect the growth of the company's value. In order for the growth of the company to be sustainable, it is necessary to build a model of the company's sustainable economic growth. Further, in the process of operation, it is essential to identify deviations from the planned indicators and accordingly develop the necessary measures aimed at improving efficiency.

To conclude, using VBM indicators along with economic growth model allows to determine the company's performance; to determine the economic growth rate; to identifying reserves available for improving economic efficiency; to consider the measures.

5 CONCLUSION

The research work reviewed the theoretical aspects of assessing the economic efficiency of the company. Today, the traditional approach to evaluating efficiency is most widely used; however, large oil and gas companies should consider using a financial assessment model. The mapping of this model is the concept of value-based management, the essence of which is to focus on the strategic goal of increasing the value of the company.

Selection of economic efficiency indicators depends on the goals of research. Shareholder value added indicator is commonly used for evaluating return from investment projects. It`s better to use Cash value added indicator for dynamics observation and long-term forecasting. Economic value added indicator is usually used for forecasting and surveys in short-term.

The results of calculation leads to the conclusion that PAO ANK "Bashneft" is creating value, what successfully illustrates high efficiency of the company and attractiveness for shareholders.

The concept of value-based management is currently most relevant for analyzing the performance of an oil and gas company. Based on selected VBM indicators, it is possible to build a model of sustainable economic growth for any oil company, which will determine the degree of achievement of strategic goals and will correctly identify reserves available for improving economic efficiency.

REFERENCES

Copeland T., Koller T., Murrin J. 1995. Valuation: Measuring and Managing the Value of Companies. New York: John Wiley & Sons.
Drucker P.F. 2015. The practice of management. Moscow: Mann, Ivanov and Ferber.
Ivashkovskaya I. 2004. The value based management: challenges to the Russian managers. *Russian Management Journal*. Vol. 6.
Morin R. A., Jarrell S. L. 2001. Driving Shareholder Value Value-Building Techniques for Creating Shareholder Wealth. New York: Mc Graw-Hill.
Rappaport A. 1986. Creating Shareholder Value: The New Standard for Business Performance. New York: Free Press.
Stewart B. The Quest for Value: A Guide for Senior Managers. Harper Business: N. Y., 1991.
Volkov D. L. 2004. Managing value of companies and the problem of choice of an adequate model of evaluating. Saint-Petersburg: Vestnik S.-Petersburg U.
Van Horne, James C. 2001. Financial Market Rates and Flows, 6thed. Upper Saddle River, New Jersey: Prentice Hall.

An approach to assessment of sustainability of the large-scale Russian liquefied natural gas project

O.O. Evseeva & A.E. Cherepovitsyn
St-Petersburg Mining University, Saint-Petersburg, Russia

ABSTRACT: The relevance of this work is due to the need to diversify the directions of export of Russian gas through the development of the industry of liquefied natural gas (LNG). The specificity of LNG projects is such that due to their high capital intensity and low prices for LNG in the world market, such projects are mainly focused on the achievement of external effects. The authors of the article defined the concept of sustainable development in the context of project management, analyzed the specifics of Russian LNG projects, defined the concept of sustainability of the LNG project, proposed a list of indicators to assess the sustainability of LNG projects taking into account the interests of stakeholders.

1 INTRODUCTION

One of the key drivers in the development of the global energy market is the technology of transporting and storing gas in a liquefied state. Although the history of active use of liquefied natural gas (LNG) in the process of diversification of energy consumption in the world has only about half a century, the pace of development of the industry is high.

Russia is the largest producer of natural gas, but its position in the global LNG production is currently weak. Since Russia's interest in the development of new gas markets in modern conditions is quite high, the intensification of LNG production, including through the construction of new large-tonnage LNG assets, is an important strategic objective. To solve it, it is necessary not only to accumulate the resources and efforts of the parties concerned, but also to form an effective management system for large-scale LNG projects, including the development of a specific approach to their assessment.

LNG technology is an effective way to sell natural gas, which allows not only optimizing the cost of transporting this energy source, but also expanding the geography of sales. The LNG industry is today the fastest growing sector in the field of energy, its growth rate is about 7.6% per year, which is almost two times higher than this figure for natural gas and three times for oil (Braginsky, O.B. 2006, E&Y. 2013).

Currently, there are a sufficient number of promising market niches in the global LNG market, due to the presence of existing and under construction regasification terminals. Shell predicts that in the coming years, the demand for LNG will increase dramatically (mainly due to the demand for LNG in China, where the goal of reducing fuel that pollutes the atmosphere), while the growth rate of supply for this type of fuel is slower, despite a large amount under construction factories (Shell. 2019).

In accordance with the Energy Strategy of Russia until 2035, it is planned to increase the export of Russian LNG to 100 million tons, to build new LNG production facilities, to expand its presence in the global LNG market.

Today in the country function only two the export-oriented large-capacity the LNG production plant - Sakhalin-II with design capacity of 9.6 million t. and 16.5 million t Yamal LNG. Besides the existing large-capacity plants, to realization four more LNG projects are planned: Far East LNG, Pechora LNG, Baltic LNG and Arktik LNG-2. In case of successful implementation of the stated large-capacity LNG projects in the specified time frames, the outputs in the country will increase every year, and by 2025 the share of Russia in the world LNG market can already make more than 15%.

2 LITERATURE REVIEW

Projects of a LNG assets construction are unique. Infrastructure LNG - projects are hi-tech, capital-intensive and are characterized by high risk level, besides, often such projects are implemented in remote and undeveloped regions that can be in addition complicated by need of providing projects foreign the equipment, materials, technologies. Project works are, as a rule, carried out by a large number of contractors and subcontractors, the great number of suppliers and service companies and also financial structures is involved in projects. Before management company there is a difficult task of coordination of a large number of elements of the design system directed to achievement of the general result.

For allocation of specific characteristics of projects of creation of LNG - assets it should be noted key features of the industry in which they are implemented. On the basis of results of the analysis of a global trend in the LNG sector, it is possible to note the following characteristics:

–Difficult chain of creation of value of a LNG product. Depending on the chosen business model, stages of creation of value can include processes of investigation/arrangement of field/ gas production, preparation, separation of gas/liquefaction, storage, transportation, LNG regasification. Thus, the sector of LNG accumulates in itself a set of branch tasks (Evseeva, O.O. & Cherepovitsyn, A.E. 2017).

–Need of maintenance of high competitiveness in the world LNG market demands continuous technological development that emphasizes the high importance of innovative capacity of the LNG-sector.

–Adverse ecological effect on the environment in areas of construction and operation of assets. The applied technologies have to guarantee full safety of work.

–Influence of the LNG industry on processes of globalization in a power complex of the world. The flexibility and mobility of LNG allows to carry out entry into the new markets for the remote countries. The limited number of suppliers of technologies for liquefaction promotes development of the international relations between the manufacturing companies.

The specifics of the industry cause need of introduction in activity of the companies participating in development of the LNG industry, specific design approach of which are characteristic:

–LNG - projects render multiplicative effect on development of the adjacent and accompanying industries, such as gas processing, shipbuilding, mechanical engineering, metallurgy and other.

–Great number of participants of the project that demands application of the advanced methods of management sent for approval of interests, ensuring effective interaction and achievement of the maximum synergetic effect;

–Implementation of LNG - projects often demands development of the international cooperation that is caused by high capital intensity and knowledge intensity of creation of hi-tech capacities and also need of decrease in market risks.

–High degree of uncertainty throughout all life cycle of the project, capable to render both positive impact on implementation of the project, and negative. So regular monitoring of an environment of the project and the analysis of factors of its external and internal environment is necessary.

–The LNG projects has significant effect as on achievement of commercial and strategic objectives of the company, and on acceleration of social and economic development of the region in which the project is implemented.

To summarise, it is possible to draw a conclusion that LNG effects has the expressed external effects. In present conditions when the prices of LNG interfere with payback of the majority of the capital-intensive LNG projects, their realization in - much depends on the state support in a type of tax benefits and the state investments into infrastructure (Sigra Group. 2014). At the same time a main objective of providing the state support is maximizing external effects of projects.

3 METHODOLOGY

The aim of this study is to develop an approach to assessing the effectiveness of LNG projects, taking into account the consequences of such projects in the external environment. To achieve the goal, the following tasks were set and solved: 1) Define the concept of sustainability of the LNG project; 2) Analyze the external environment of the LNG project with the identification of key stakeholders and their areas of interest; 3) Propose a system of indicators for assessing the sustainability of LNG projects On the basis of stakeholders ' interests; 4) Test the proposed approach on the example of assessing already implemented LNG projects. To solve these problems, an integrated approach was used using methods of synthesis, analogy, grouping, comparison, as well as tools for strategic analysis, investment evaluation and socio-economic forecasting.

4 DISCUSSION AND RESULTS

The project's focus on achieving not only internal commercial performance indicators, but also external ones is closely related to the concept of sustainable development, which means achieving the company's business goals while respecting the interests of stakeholders. The theory of sustainable development is based on the so-called triune outcome - the economy, society and ecology. Hence the concept of sustainability of the project as a balanced approach, in which the created asset should maximize not only economic value, but also social and environmental benefits. The implementation of this approach involves the identification of elements that ensure the achievement of sustainability, i.e. points of balance of interests. These elements are mechanisms that provide long-term prospects for the project and include both financial drivers and non-financial drivers.

The concept of stakeholder management, which originally emerged in strategic management at the end of the last century, is today one of the key areas of knowledge in project management. At the same time, the influence of stakeholders on the outcome and success of the project is higher than in corporate governance, since the economic activity in projects is more intense (Grabar, V.V. & Salmakov, M.M. 2014). Forming a portfolio of projects, programs and individual projects, taking into account aspects of sustainability, is an important methodological and practical task. The Project Management Institute asserts that sustainability in project management is a new global business model that allows you to achieve sustainable development goals in every phase of project implementation (Gutierrez, M. 2014).

The popularity of the idea of sustainable development in the business environment is confirmed by the growth in the volume of responsible investment, an increase in the number of published reports in the field of social responsibility and sustainable development, as well as the growing popularity of standards and guidelines for sustainable development. Based on the basic principles of the concept of sustainable development, as applied to the LNG industry, it can be represented as follows: (Figure 1)

The need to take into account the sustainability indicators of LNG projects at the planning stage is related to the fact that sustainable development is part of the business strategies of companies implementing such projects. In addition, the Russian experience in the implementation of LNG projects already demonstrates an example of the influence of individual groups of stakeholders on the progress of the project. For example, the launch of the implementation of the first Russian LNG project Sakhalin-2 was delayed due to complaints from indigenous peoples. In particular, among the concerns of local communities were: violation of the conditions of spawning of salmon due to the crossing of several hundred rivers by pipelines; uncertainty with the possibility of full

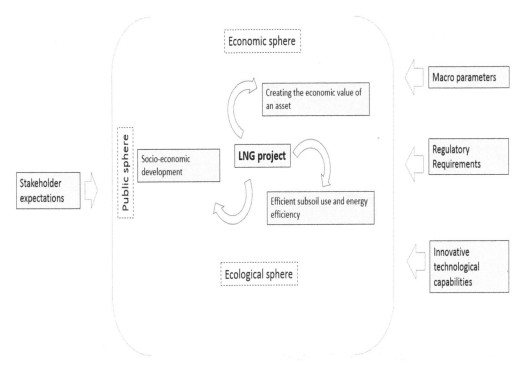

Figure 1. LNG sustainability concept.

compensation for damage in the event of an oil spill; the dumping of industrial waste into the productive points of the Okhotsk Sea, the threat of the "Red Book" Okhotsk-Korean gray whale population (100 individuals).

According to international practice, sustainability assessment is based on a set of indicators, and there is no generally accepted methodology. The process of methodological support for assessing sustainability consists of the following steps:

1. Determination of areas of indicators by areas of interest of key stakeholders
2. Formation of the list of indicators
3. Establishment of indicators
4. The establishment of weight characteristics.

The initial stage in the development of the methodology is to identify the main stakeholders, since the project sustainability indicators should correspond to their interests. Based on the analysis of the experience of LNG projects in Russia and in the world, the main categories of stakeholders were identified. Further, through the author's checklist tool, a quantitative assessment of each stakeholder was carried out (Ilinova A. et al. 2018). It was concluded that the greatest attention in the planning and implementation of LNG projects should be focused on business (industrial companies and investors), government agencies and society (Figure 2).

Project sustainability indicators should be in line with the expectations of key LNG project stakeholders, as represent a quantitative expression of their interests. An analysis of the expectations of the main categories of stakeholders identified in the first stage was carried out. The results of the analysis are presented in Table 1.

In accordance with the identified expectations, a list of indicators was formed in accordance with the interests of stakeholders – economic, social and environmental. Next, a survey of experts of economic and oil and gas profile was conducted, during which each indicator was evaluated with a score from 1 to 10. Further, on the basis of

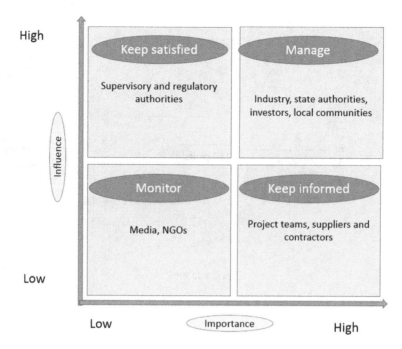

Figure 2. Matrix of key LNG project stakeholders.

Table 1. Expectations of the main stakeholders of LNG projects.

Stakeholder	Interests
Companies	Compliance with project deadlines and budgets, product quality, high contracting of products, capacity utilization, high performance, no accidents, high yield of the useful product, synergy effect for other company assets.
State	Socio-economic development of the country, development of industries, budgetary and tax revenues, import substitution, diversification of gas export supplies, strengthening the country's image in the global arena, development of remote and offshore hydrocarbon fields, development of human capital, GDP growth.
Society	Job creation, creation/modernization of social infrastructure, minimum impact on the environment, gasification
Investors	Return on investment, quick payback, responsible investment

the principle of balance of economic, public and environmental spheres, the weight distributed proportionally for each subgroup was set. The results are presented in Table 2.

The applicability of the proposed system of indicators was tested in assessing the sustainability of two already implemented LNG projects - Sakhalin-2 and Yamal LNG. The results of calculations based on data from open sources showed that from the point of view of maximizing the sustainability indicators, the Yamal LNG project is the most effective (Table 3). This project, possessing a higher production capacity, allowed the formation of a high-tech infrastructure framework in the region, including facilities of both transport and logistics infrastructure, and social. In addition, the first Russian natural gas liquefaction technology was developed for the needs of this project.

Table 2. Scorecard for assessing the sustainability of LNG projects.

Area of interest	Efficiency group	Indicator	Type	Average expert point (1-10)	Weight
Economic	Managerial	Capacity	max	5,33	2,28
		Capacity utilization	max	8,00	3,42
		The degree of product contraction	max	8,00	3,42
		Production losses	min	4,33	1,85
		Impact on other (including unallocated) projects	max	8,00	3,42
		Construction time	min	5,33	2,28
	Investment	Payback period	min	8,33	7,72
		NPV	max	3,33	2,60
		Return on investment	max	9,67	8,95
Public	Macroeconomic	GDP	max	4,67	1,17
		Creation of regional infrastructure	max	8,67	2,17
		State revenue	max	7,67	1,92
		Gasification of Russian regions	max	6,67	1,67
		The share of Russian equipment and technology in the project assets	max	5,67	1,42
	Geopolitical	Increase gas production	max	8,00	2,74
		Increasing global LNG market share	max	8,33	2,85
		Entry into new markets	max	8,00	2,74
	Social	Creation of social infrastructure in the region	max	9,67	3,31
		Contribution to the preservation of the traditional lifestyle of indigenous peoples	max	3,67	1,26
		Job creation	max	7,33	2,51
		Employment of the local population in the project	max	3,67	1,26
	Innovative	Creation of fundamentally new technologies in mining	max	5,67	1,65
		Creation of fundamentally new technologies in liquefaction	max	9,00	2,62
		Creation of fundamentally new technologies in the sea transportation	max	9,00	2,62
		Targeted training	max	5,00	1,45
Ecological	Ecological	Greenhouse gas emissions	min	8,67	8,33
		Accidents with environmental consequences	min	8,33	8,01
		Energy intensity of technology	min	7,33	7,05
		Gas flaring	min	5,00	4,81
		Volumes of refueling of domestic courts	max	5,33	5,13

Table 3. Calculation results of the integral stability indicator for implemented Russian LNG projects.

Value by area of assessment	Yamal LNG	Sakhalin-2
Economic	28,8	25,72
Public	25,42	20,51
Ecologic	20,19	20,19
Integral evaluation	74,2	66,4

5 CONCLUSIONS

The activities of the industrial sector has a significant impact on the development of society. Business affects the economy through job creation, payment of taxes, as well as stimulation of related industries, the environment through control over the technologies used and investment in environmental protection, the social component of sustainable development by ensuring equal access to jobs, as well as corporate social programs of various kinds. The implementation of the concept of sustainable development at the company level requires appropriate improvements in the project management system. One of the key aspects in this area is the development and implementation of an approach to assessing the sustainability of project initiatives.

LNG projects are large-scale and involve a complex value chain of the finished product, with the company having the ability to influence the achievement of corporate sustainable development goals at every stage of the production cycle. The specificity of LNG projects implies a greater focus on achieving external effects, which are determined by the interests of key stakeholders. The main stakeholders of LNG projects are industrial companies, investors, the state and society.

The sustainability model of the LNG project is based on the economic value of the asset, rational subsoil use and energy efficiency, as well as the socio-economic development of the region of presence and the country as a whole. The assessment of the effectiveness of the LNG project, taking into account the concept of sustainability, should include indicators of economic, public and environmental efficiency.

The proposed approach based on a system of indicators allows for rapid assessment of the effectiveness of projects based on the theory of stakeholders and compare projects with each other.

REFERENCES

Braginsky, O.B. 2006. Liquefied natural gas: a new factor of the world energy market. Moscow: The Russian Academy of Sciences.

E&Y 2013. The world market of LNG New demand + new supply = new prices? Available online: http://www.ey.com/Publication/vwLUAssets/Global-Oil-and-Gas-LNG-RU/%24FILE/Global-Oil-and-Gas-LNG-RU.pdf (accessed on 01. 03.2019).

Evseeva, O.O. & Cherepovitsyn, A.E. 2017. Comparative effectiveness analysis of Russian and foreign liquefied natural gas projects. Economic and technological conditions of development in extractive industries. Gliwice: Wydawnictwa Politechniki Śląskiej, pp. 73-83.

Grabar, V.V. & Salmakov, M.M. 2014. Analysis of stakeholders of the project: methodology, methodology, tools. ARS ADMINISTRANDI 2014, 2, 36-44.

Gutierrez, M. 2014. Applying PRiSM Methodology in the Canadian Construction Sector. PM World J. 2014, 3.

Ilinova A., Cherepovitsyn A. & Evseeva O. 2018. Stakeholder Management: An Approach in CCS Projects. Resources 2018, 7, 83; doi:10.3390/resources7040083.

Shell LNG outlook 2019. Available online: https://www.shell.com/energy-and-innovation/natural-gas/liquefied-natural-gas-lng/lng-outlook-2019.html (accessed on 01.03.2019).

Sigra Group. 2014. Government Support to Upstream Oil & Gas in Russia. How Subsidies Influence the Yamal LNG and Prirazlomnoe Projects. Available online: http://www.iisd.org/gsi/sites/default/files/ffs_awc_russia_yamalprirazlomnoe_en.pdf (accessed on 01.03.2019).

Topical Issues of Rational Use of Natural Resources 2019 – Litvinenko (Ed)
© 2020 Taylor & Francis Group, London, ISBN 978-0-367-85720-2

Reusing underground mine space of closing mines

A.I. Grincheko & H.E. Golovneva
Donetsk National Technical University, Ukraine

ABSTRACT: The article addresses opportunity of the reuse, appearing in the process of closing coal-mining enterprises. It represents approaches of cost minimizing due to closing mines in the reusing closed mine object based on tentative example.

1 INTRODUCTION

Relevance of this work. Closing coal-mining enterprises in the Donbass is time-consuming process because of the specific technologies of coal getting in this region. In 2016, the Council of Ministers of the DPR came to a decision to close 22 unprofitable mines. As of 1.07.2017, 18 mines were liquidated, as of 1.11.2018, 3 unprofitable mines were shutted down too. However, upon the huge volume of unexercised space, leaving after liquidation, the question arises about the consideration of a certain type of resource. Notwithstanding, a huge structure of questions of what should be taken into account is the main problem, appearing during the mine liquidation.

The aim of this work is to indicate the possibilities of reusing underground space, consider the most promising areas of its reuse, as well as the construction of an economic and mathematical model to choose the best option for the reuse of mining enterprises.

V.A. Bezpflug and M.K. Durnin are researches of this sphere. The use of coalmine methane and various methods for its production are observed in a work of the previously mentioned authors, "Comparative economic evaluation of various technologies for utilization of coal mine methane" (Sorenkov V.M., Nedoluzhko V.N., Begicheva et al. 2012). Besides that, A.S. Kuznetsov in his work explained the advantages of using coal-water fuel (Kuznetsov A.S. et al. 2012).

The process of legal mine liquidation goes through several stages, namely:

1. Development of justifications for the feasibility of further mine exploitation and planning costs associated with its further use.
2. Implementation of design decisions.
3. Adoption of measures for employment and social protection of the dismissed, at the closure of the mine, personnel.
4. Creation of new jobs in regions where mine closure is envisaged.

We need elaborate a certain strategy of the mine liquidation: - to develop a socio-economic and hydrogeological forecast of the region, - justify the gradual closure of the mines, linking it with investment and financing opportunities, - to improve the regulatory framework, and study environmental problems of the mine liquidation (Panishko A.I. et al.2009).

Based on the above analysis, the process of liquidation of unprofitable mining enterprises can be characterized as having no the end in the future. It provokes a sufficiently high level of costs for maintaining the mine in the period of expansion work, reconstruction, construction of water drainage at the mining enterprise, overcoming social and economic consequences (costs of free coal supply, social safety net) (Levkin Yu. M. et al. 2004).

Physical closure provides for the dismantling of equipment, backfilling of trunks, ensuring drainage, the implementation of measures related to the elimination of unsuitable buildings

and mine workings. The most labor-intensive things are coal loading bunkers, overhead structures, buildings of main ventilation fans, lifting machines and boiler rooms with monolithic
reinforced concrete foundations. The last one should be destroyed to the level of minus
0.2-0.3 m from the ground level (Makarov A.A., Shevtsov N.R. et al.2002).

According to the intendended purpose, underground structures are conventionally divided
into several main groups:

- Transport and hydraulic tunnels; structure underground system; power plants (mainly
  hydro); basic warehouses and refrigerators;
- Medical institutions, military facilities;
- Industrial enterprises;
- Tanks for the disposal of hazardous industrial waste;
- Oil and gas storage facilities;
- Tanks for drinking water;
- Objects of municipal facilities (pedestrian crossings, garages, collectors, etc.) (Panishko
  A.I. et al. 2009).

There are certain difficulties in the reuse of mine workings in coalmines, which are largely
because the use of development systems with roof collapse or laying them with empty rock
does not provide the preservation of the waste space. As a rule, it is difficult to use the capital
mine workings (transport, ventilation, near-barrel yards) due to their insufficient cross-
sections. In addition, it is hard to place efficient production flow lines in an extensive network
of workings of small cross-section. In technological processes in gas mines, operations associated with high temperature (welding, soldering, etc.), should be excluded. There are high costs
for the maintenance of mine equipment and the creation of safe working conditions (maintenance workings, organization of drainage, ventilation, power supply, the operation of lifting
equipment, etc.).

Foreign experience in solving the problems of the mine's liquidation is achieved through
economic, legal, and innovation-technical steps.

The implementation of economic and legal steps is carried out by changing the regulatory
framework, based on objective economic laws and mechanisms (Panishko A.I. et al. 2009).

Therefore, in Poland, they made changes to the law on subsoil, which provided for
mining companies to pay a special environmental fund - 10% with open pit and 3% with
underground mining. This allowed to accumulate funds for the planned liquidation of
mining enterprises.

In Germany, the practice of accumulating funds for the needs of liquidating mining enterprises designates the need for special licensing. At the same time, not only fines for environmental damage are paid, but also a license is purchased for carrying out activities that are
potentially dangerous for the environment.

Thus, based on foreign experience, taking into account the unfavorable economic conditions in the Donbass, we can recommend tax-subsidy regimes that allow you to accumulate
funds necessary for socially acceptable closure of mines. The trend in the application of
innovative technical measures is reduced mainly to the spread of the rational use of waste
underground space.

The most promising areas to reuse underground space of coalmines are:

- underground warehouses and storage facilities;
- peak reserve underground pumped storage power plants and hydro power plants;
- underground wind power.

2 ECONOMIC-MATHEMATICAL MODEL

An economic-mathematical model for the disposal of a mining enterprise has been developed
to select the optimal reuse option.

$$C_1 + C_2 + C_3 \rightarrow min \qquad (1)$$

$$k_1(k_2C_2 + C_3) \leq R$$

$$C_1 + k_1(k_2C_2 + C_3) \leq L$$

$$\frac{0.53(0.57C_1) + 0.53(0.57C_2) + 0.53(0.57C_3)}{n} \geq \frac{0.53(0.57\Sigma C_i^0)}{N}$$

$$\Sigma C_i^0 \geq 0$$

$$\Sigma C_i \leq \Sigma C_i^0$$

$$\Sigma C_i \leq R$$

Where C1, C2, C3 - the costs of designing, developing and operating a new production;

R - revenues from the exploitation of new mastered production;

L - the costs associated with overcoming the effects of "wet" mine;

S - costs associated with overcoming the social consequences of the mine's liquidation;

N, n - the number of workers in accordance with redundant workers at the closure of the mine and employed workers to a new production;

α, β - shares in the costs of mastering and operating a new production attributable to the wages of workers;

k1 – coefficient, which is taking into account the increase in costs in connection with the geological conditions and the life extension of the underground structure;

k2 – coefficient, which is taking into account the increase in operating costs in connection with the increase in the depth of work.

Consider an example of using this model with the following conditional data. Based on the condition, the company allocated $ 250,000 for the introduction of new production. The planning department calculated the following values presented in Table 1.

Let's find an optimal solution of the conditional problem using the Excel function "Search for solution". We see that the amount of expenses is $ 279,320. The componentwise representation of expenses, in it turns, is represented by the following indicators:

Table 1. Conditional data for solving the optimization problem.

№	Denomination	Indicator name	Value, thousand. $
1	2	3	4
2	C1	The cost of designing a new production	30.00
3	C2	The cost of mastering a new production	120.85
4	C3	The cost of operating a new production	128.47
5	R	Revenues from the exploitation of new production	1,000.00
6	L	Costs associated with overcoming the effects of "wet" mine	1,000.00
7	n	The number of employees taken to the new works	50.00
8	N	The number of workers laid off due to the closure of the mine	50.00
9	S	Costs associated with overcoming the social consequences of the mine's liquidation	84.38

* conditional data has been formed on the basis of research of Donetsk institute for design organization of mine construction and enterprises of the construction industry, which was reflected in the research work 'Development of proposals on the possibility of using underground spaces in connection with the closure of mines'.

Optimized cost by elements	C1	C2	C3
Materials cost	$ -	$ 10,00	$ 15,00
Miscellaneous expenses	$ 30,00	$ -	$ -
Manufacturing expenses	$ -	$ 50,00	$ 50,00
Labour expenses	$ -	$ 35,00	$ 37,00
Depreciation charge	$ -	$ 15,00	$ 15,00
Unified social tax	$ -	$ 10,85	$ 11,47
Sum total, prime cost	$ 30,00	$ 120,85	$ 128,47

Figure 1. The componentwise representation of expenses before optimization.

Profitability of expenses will be equal to $ 280, profitability of revenue - $ 3580, which, in turn, diagram demonstrates us that $ 1,000 of income accounted for $ 280 in expenses and $ 1,000 in expenses — $ 3,580 in income.

It is necessary optimally to redistribute the amount of expenses that the company is prepared to incur with minimal losses. With the help of the function "Search for solution", we get the following values:

The solution of an optimization issue showed that it would be more reasonably to use mixed method with drawing on contract organizations in implementation of new contracting organizations to preserve the optimality of the plan, but with a decrease for expenses. You can consider the option of re-using materials to save the use of material resources. It is necessary to increase the productivity of workers engaged in the development and operation of a new production to reduce the value of the wage parameter. Hence, if the salary is dropping, the unified social tax will be falling too.

We also see that because of solving the optimization model, the profitability of expenses decreased by $ 30 from each thousand revenues, and the profitability of revenue rose by $ 420 from each thousand expenses, which undoubtedly indicates that the optimization of expenses on this model favorably affects the redistribution the amount of the planned expenses.

In Table 2 we can see that all specified constraints have been accomplished, and the absence of the given gradient says that when a parameter is included in the plan that the optimization model does not offer, to a large extent, the value of the objective function (in this case, the sum of expenses) will not be changed.

The financial statement, which is shown in Table 3, proves the results of research by many scientists who dealt with the problems of the coal-mining enterprises elimination in particular, the most significant and substantial component in the total costs will be the parameters, which are responsible for eliminating the socio-economic aspect.

Optimized cost by elements	C1	C2	C3
Materials cost	$ -	$ 7,70	$ -
Miscellaneous expenses	$ 30,00	$ 1,32	$ 0,47
Manufacturing expenses	$ -	$ 47,56	$ 47,42
Labour expenses	$ -	$ 31,66	$ 33,64
Depreciation charge	$ -	$ 15,00	$ 15,00
Unified social tax	$ -	$ 9,81	$ 10,43
Sum total, prime cost	$ 30,00	$ 113,05	$ 106,95

Figure 2. The componentwise representation of cost after optimization.

618

Table 2. Sustainability report.

Box	Name	The final value	Reduced gradient
1	2	3	4
$K$15	Material cost C2	7,696131868	0
$L$15	Material cost C3	0	0
$K$16	Miscellaneous expenses C2	1,319243077	0
$L$16	Miscellaneous expenses C3	0,46907931	0
$K$17	Manufacturing expenses C2	47,56181722	0
$L$17	Manufacturing expenses C3	47,42031796	0
$K$18	Labour expenses C2	31,656455	0
$L$18	Labour expenses C2	33,63622406	0

Table 3. Results report.

Box	Name	Meaning of the box	Formula	Condition	Access
1	2	3	4	5	6
$E$10	Parameter non-negativity conditions 1	30,00	$E$10>=$G$10	No binding	30,00
$E$11	Parameter non-negativity conditions 2	113,05	$E$11>=$G$11	No binding	113,05
$E$12	Parameter non-negativity conditions 3	106,95	$E$12>=$G$12	No binding	106,95
$E$13	Conditions whereby the optimized parameters will be less than the planned one 1	30,00	$E$13<=$G$13	Binding	0
$E$14	Conditions whereby the optimized parameters will be less than the planned one 2	113,05	$E$14<=$G$14	Binding	0
$E$15	Conditions whereby the optimized parameters will be less than the planned one 3	106,95	$E$15>=$G$15	Binding	0
$E$16	Production breakeven	250,00	$E$16<=$G$16	No binding	750,000001
$E$7	Comparative breakeven	337,53	$E$7<=$G$7	No binding	662,4743495
$E$8	Comparative cost value, linking with the choice of liquidation method	367,53	$E$8>=$G$8	No binding	632,4743495
$E$9	Social and economic aspect to attract redundant workers in the time of liquidation to new ways of works	1,51	$E$9>=$G$9	Binding	0,00
$F$3	Choice criterion	$250,00	$F$3=250	Binding	0
$K$18	Labour expenses C2	$31,66	$K$18>=$C$31	No binding	$1,66
$L$18	Labour expenses C3	$33,64	$L$18>=$C$31	No binding	$3,64

Thus, the performance review of the mine workings in the closed coalmines to assess its reuse, it will be carried out in several stages in the following sequence:

- select options for reuse of mine workings,
- considerate the mine network of workings and the selection of those that are technically suitable for the chosen direction of use,
- study the impact of the surrounding geological environment on the selected workings, their assessment of stability, depth, increase in service life,

- study of the state of mine workings in terms of the need for their repair and re-equipment for new production, taking into account the chosen direction of reuse,
- economic assessment of the chosen direction and place (selected in the process of analysis of mine workings) of reusing underground space in the coal mine according to the "cost minimization" principle . That is, the project self-sufficiency, in order to justify the "dry" or combined preservation of the mine and save part of the fund of jobs.

World experience in the use of underground space, which does not participate in coal mining, suggests that the introduction of such technologies is possible in the countries of Eastern Europe and Asia. Mine workings, which have no prospects for possible further exploitation for their intended purpose, can be used as a storage of household and industrial non-toxic waste.

The analysis, forecast and evaluation of a part of the mine survey monitoring system of a coal mining enterprise, which can be used later for multi-purpose, are necessary requirements (Golovneva, E.E. et al. 2002).

As a result, we can conclude about the technical complexity and, often, the economic inexpediency of locating production of industrial products in mining. Although there are contradictions, regarding the mine reuse, that have not been solved at the moment, we propose the following concept of reusing underground space, which consists of several steps presented in Figure 3:

- At the first stage - inventory and systematization of all mine workings are performed;
- At the second stage - the selection of workings promising for the placement of the national economy objects in them is realized;

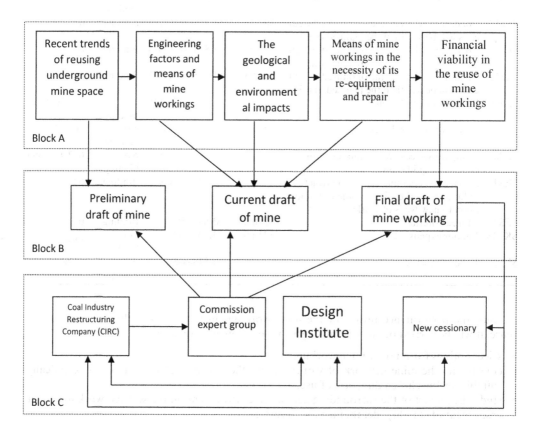

Figure 3.   Flow chart of the strategy to reuse underground space.

- At the third stage - the workings are classified according to the main directions of use and are assessed according to the specific requirements of the objects placed in such workings.

The presented economic-mathematical model allows choosing the optimal variant of reusing underground space of a closing mine according to the presented strategy of determining the optimal plan.

It should be noted that the use of this model should not be authoritative when you decide to reuse of underground space. This strategy performs only a supporting function for the planning section and technical departments.

Together with the use of an optimization model, which has been built on the presented strategy, you can get quite good results on the profitability of using the new production.

3 CONCLUSIONS

Based on the above, we can make conclusion that the closure of the mine is an expensive, lengthy and difficult process with many attendant factors. Alongside this, it is implementable to receive economic benefits from a closed mine, allowing to minimize the cost to liquidate an enterprise. For example, it is for the production of electricity or for the creation of special storage facilities. Using the developed economic-mathematical model in completed with the Excel function "Search for solution", it is viable to calculate the economic feasibility and choose the best option to reuse the underground space of a liquidated mine. Reports on solution of the optimization model will help to analyze the resulting model and improve it. According to the results of our calculations, using the underground space of mines as warehouses for materials of construction industries and the burial of mining waste is the appropriate way for Donetsk region. This is due to a very large development depth of 700-1000 m.

REFERENCES

Bezfflug V.A., Durnin M/K., Comparative economic evaluation of various technologies for utilization of coal mine methane//Coal. – 2007. - №12, pp.59-60.

Chukanov S.Yu., Substantiation and development of technological solutions for the use of underground space of closed coal mines in the Prokopyevsk-Kiselevsky district of Kuzbass for waste disposal [Electron.resource]: official site - Electron.dan. - Access mode: https://www.dissercat.com/content/obosnova nie-i-razrabotka-tekhnologicheskikh-reshenii-po-ispolzovaniyu-podzemnogo-prostranstv - Title. From the screen.

Gaiman L.M., Underground structures/Great Soviet Encyclopedia [Electron.resource]: official site - Electron.dan. - Access mode: https://gufo.me/dict/bse/- Title. From the screen.

Golovneva, E.E., Improvement of organizational and regulatory documents in the context of the restructuring of the coal industry in Ukraine [Electron.resource]: official site - Electron.dan. - Access mode: http://masters.donntu.org/2002/ggeo/golovneva/diss/refer.htm - Title. From the screen.

Kuznetsov A.S., Why is not water coal fuel used in Ukraine?//Coal of Ukraine. - 2012. - №3, p.40-43.

Levkin Yu. M. Factors determining the feasibility of multipurpose use of the underground space in coalmines//GIAB. 2004. №2.

Makarov A.A., Shevtsov N.R., New technologies for liquidation of objects in closed coal mines//Coal of Ukraine. - 2002. - №12, p.28-31.

Malkin A.S., Myasnikov V.V., Agafonov V.V. Actual problems of designing and using objects when developing the underground space of the subsoil//GIAB. - 2013. - №11, p.5-9.

Panishko A.I., Problems of coal industry enterprises liquidation and ways of its solution//Coal of Ukraine.– 2009. - №12, p.3-5.

Sorenkov V.M., Nedoluzhko V.N., Begicheva T.V. The issue of the mines elimination in the Central region of the Donbass//Coal Industry of Ukraine. - 2012. - №2, p.31-35.

Topical Issues of Rational Use of Natural Resources 2019 – Litvinenko (Ed)
© 2020 Taylor & Francis Group, London, ISBN 978-0-367-85720-2

Development of the Donbass mines potential on the basis of underground coal gasification

I.V. Kochura, M.V. Chaika & V.A. Shapoval
Donetsk National Technical University, Donetsk, Ukraine

ABSTRACT: The analysis of mining and geological and technical conditions of the closed coal-mining enterprises of Donbass was carried out on the subject of using the technology of underground coal gasification. Its essence and methods, advantages and disadvantages of using are considered. The comparative description of the experience of applying the technology of underground gasification of coal in different countries has been carried out. The issues of the economic efficiency of the technology of underground coal gasification and the receipt of a number of benefits from its use are considered.

1 INTRODUCTION

Currently in the world a significant number of non-working coal seams remain out of balance and are not used. The renewal of production activities of mines that are in the process of closure, for the final extraction of these seams by conventional methods is not economically effective. The most promising solution for the development of the economic potential of such coal-mining enterprises is the implementation of well mining methods, which include underground gasification of coal. Today it is a competitive technology, as a result of which enterprises can provide themselves with their own heat and electricity, as well as produce gasoline, diesel and aviation fuel, electricity and many different chemicals.

1.1 *Literature review*

In modern conditions, such scientists as Gridin S.V., Zhukov E.M., Kondyrev B.I., Kropotov Yu.I., Luginin I.A., Lazarenko S.N., Kreynin Ye.V., Makagon Yu.V., Ryabchinin OM., Raimzhanov B.R., Saltykov I.M., Chizhik Yu.I., Yankovsky M.A. and others were concerned with the prospects of using underground coal gasification. Many of them considered the experience and effectiveness of this technology in different countries (Russia, China, USA, Germany, Uzbekistan and others), which differ in mining and geological and other conditions (Zhukov et al. 2016, Kondyrev et al. 2005, Kreinin et al. 2009, Raimzhanov et al. 2008, Schilling et al. 1986). In the research (Belov et al. 2009, Gridin et al. 2013, Yankovsky et al. 2011) alternative sources of energy in terms of energy and resource shortage, as well as the general process of producing synthesis gas by the method of underground gasification and possible prospects for its further use in Ukraine, are proposed.

The analysis of publications and materials of already implemented projects and programs showed that the economic and social aspects of using the technology of underground coal gasification are not sufficiently developed. This is especially true of Donbass coal deposits with difficult mining and geological conditions, as well as a lot of political and economic problems caused by the uncertainty of the external environment.

1.2 *Object*

The purpose of the study is the grounds of the potential of using the underground gasification of coal technology at coal mines of Donbass in the future.

The goal requires the following tasks:

- consideration of the essence of the technology of underground gasification of coal, features of its use, advantages and disadvantages;
- analysis of world experience in the use of underground coal gasification technology;
- analysis of the mining and geological and mining conditions of the closed coal-mining enterprises of Donbass for the application of this technology;
- consideration of economic efficiency issues and obtaining a number of benefits from the use of technology in the coal-mining enterprises of Donbass.

2 THE GROUNDS FOR PROBLEM

2.1 *Theoretical approaches to the problem*

The idea of underground coal gasification (UCG) belongs to D.I. Mendeleev. "It is enough to set fire to coal underground, turn it into luminous, or generator, or water gas and take it through pipes made of paper, impregnated with resin and entwined with wire,"- was the first official mention of the underground gasification of coal idea in 1888 (Mendeleev 1939). The development of this technology was investigated by the English chemist William Ramsay (1912) and the first attempts to realize the coal gasification were made in England. The technology of underground gasification of coal was further developed in the USSR in the 1930s, where I. E. Korobchansky, V. A. Matveev, V. P. Scaff, and D. I. Filippov worked on this issue.

The essence of the technology is in the transformation of coal into gas directly at the place of its occurrence without previous extraction. First, prepare suitable areas in the coal seam for gasification, and then gasify it. Preparation can be done by underground mining with the help of mines and mine workings, and surface - with the help of specially oriented wells. According to underground mining method (Figure 1) the gasification panel is prepared by conducting an air supply and gas exhaust shafts (1,2), two inclined mine workings (3,4) and a burning road through the coal seam (5). Firewood is laid in the burning road, which is ignited by incendiary cartridges. As the result burning spreads across the coal surface in the gasification channel (6). In this scheme, the number 7 indicates the slag zone.

Air is blown through one of the inclined workings, and the resulting gas is blown to the surface. In the process of burning coal, the firing face moves upwards along the coal seam uprising.

Currently, most countries use surface preparation of sites for coal gasification, which was also used in the USSR. Its essence is in the fact that the mine workings are replaced by wells directed from the surface to the seam. Horizontal seams are opened by a grid of vertical wells 25–30 meters apart. To begin gasification, the first 3-4 rows of wells are drilled. The subsequent drilling of wells is done in the process of seam gasification.

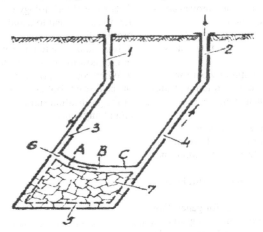

Figure 1.   Technological scheme of the flow method of underground coal gasification.

Depending on the production technology and the heat of combustion of the produced gas, it is possible to obtain a different composition of the product, which also has different uses (Table 1).

The technology of underground coal gasification has a number of advantages and disadvantages, which are presented in Table 2.

Table 1. The purpose of gas depending on the heat of combustion.

Gas	Heat of combustion, kJ/nm^3	Purpose
Generator gas	3800 – 4600	Electric and thermal energy
Synthesis gas	10900 – 12600	For chemical technology
Reducing gas	12600 – 16800	For metallurgical and industries machine-building
City gas	16800 – 21000	Fuel/consumer gas
Synthetic natural gas	25000 – 38000	For long distance transportation

Table 2. Advantages and disadvantages of underground coal gasification technology.

Advantages	Disadvantages
1. The ability to extract the reserves when the method is inadvisable for any reason.	1. Relatively low heat of combustion of the mining produced gas.
2. Relatively small amount of underground work.	2. The difficulty of controlling the spread of the gasification front.
3. It is not necessary for additional preparation of fuel for the consumer.	3. The strong dependence of performance indicators on geological and hydrogeological conditions, as well as the difficulty of its quantitative description.
4. Processing of coal into other types of fuel (liquid and gaseous) is directly at the place of its occurrence, which makes it more convenient technological raw materials.	4. The change of product gas output and its quality depending on the methods of the of gasification.
5. Dangerous underground labor of workers is eliminated compared to underground mining	5. Technical difficulties associated with the creation of channels in the coal seam between 8. two wells.
6. The formation of waste gangue is eliminated.	6. The application of technology at great depths is associated with an increase in the number of wells and it increases costs of gas.
7. The top layer of the earth is not removed compared to the open mining method.	7. The use of technology in thin seam (less than 1.5 m) causes heat loss and a decrease in production profitability.
8.The possibility of using coal with different ash and high sulfur content.	8. It is impossible to control the possible pollution of groundwater by the products of burning coal.
9. Processing is carried out without violation of the biosphere reducing the opening, extraction, transportation and storage environmental consequences associated with coal.	9. When coal is burned, subsidence and collapse of the coal seam roof occur, and landslides in the area of the gasification station can occur at any place and at any time.
10. Creating new workplaces in the region.	
11. The possibility of creating mineral raw materials due to the technogenic transfor mation of natural hydrocarbon deposits	
12. Used in everyday life, replacing natural gas, having a lower cost.	
13. It is used as an energy resource for generating electricity both at the underground coal gasification station itself and at power stations located nearby	
14. Wide opportunities to automate the process	

2.2 World experience of usage of underground coal gasification

The technology has much more advantages than disadvantages, therefore it is of great interest all over the world. The experience of gasification technology took place in different countries: the USSR, Belgium, England, USA, Russia, France, Poland, Czechoslovakia and others. Comparative characteristics of the conditions and output parameters of the use of underground coal gasification technology for different countries are given in Table 3.

Practical application of technology was first implemented in the USSR. In the early 1960s, five «Podzemgaz» experimental industrial stations were operated in the USSR (Shatskaya at Mosbass, Angrenskaya in Central Asia, Lisichanskuyu in Donbass, Uzhno-Abinskaya in Kuzbass). The results of the application of the technology showed that the quality of theproduced gas was rather low due to the imperfection of the technology. The best technical and economic indicators were achieved in 1964-1966. at Angren station «Podzemgaz» with a capacity of 1.4 billion m^3 of gas/year and Uzhno-Abinskaya station «Podzemgaz» (Kuzbass) with a capacity of 0.45 billion m^3/year. The Uzhno-Abinskaya station was operated for forty years and was closed in 1996 due to physical deterioration of equipment. So far, only the Angren station in Uzbekistan is operating, which in 2007 was acquired by Linc Energy, the world leader of underground coal gasification.

Outside the USSR, the first experiments of underground gasification were carried out in the United States and in Italy. In 1970-1980 several pilot projects were launched in the United States (HoeCreek, Hanna, CarbonCounty, Centralia), which resulted in the pollution of groundwater with benzene and phenols due to too high blast pressure.

At the beginning of the 21st century, interest in underground gasification of coal grew in China, Kazakhstan, the USA, India, both Koreas, Australia, Belgium, Spain.

Now China is on the first place in the world in terms of the number of operating industrial plants of underground gasification and in terms of the volume of natural gas substitutes which is produced for the energy and chemical industries. The impetus to the development of this technology was given by the problem of resource saving and the complete extraction of coal reserves. The new program of Chinese authorities suggests the creation of nine large plants using this technology, the total capacity of which will be more than 37 billion m^3/year of gas. And in the future, another thirty such plants are supposed to be built. Also in China, they intend to close about two thousand coal mines in order to improve the state of the environment and increase the level of industrial safety.

Ukraine, which has large reserves of coal, also considered underground gasification as one of the promising direction for realizing the potential of coal, the extraction of which is not available through the traditional mine method. Donetsk Fuel Energy Company (DFEC), a Ukrainian company, and Linc Energy, an Australian company, have announced a partnership to implement a number of projects to convert thermal power facilities to the use of water-coal fuel, as well as to build coal gasification plants using Shell technologies that China has bought. In 2012, Naftogaz of Ukraine signed a credit agreement with the State Development Bank of China for the amount of $ 3.66 billion under state guarantees. The implementation of this project would allow Ukraine to save up to 4 billion m^3 of natural gas annually, create more than two thousand new workplaces, provide a market for coal produced in the country in the amount of up to 10 million tons/year. However, due to the political and economic crisis in Ukraine, these projects were not yet destined to be realized.

The economic indicators of the projects are quite different. This happens because the gasified coals, the characteristics of seams, the technologies used are very different from each other.

So in the Chinese pilot project of the company ENN, the cost price of synthesis gas is estimated at $39-73 per thousand cubic meters when recalculating its calorific value for traditional natural gas. This is the lowest figure for current gasification plants.

In the USA, the cost is $ 65-295 per thousand cubic meters.

Table 3. Comparative characteristics of the conditions and output parameters of the use of underground coal gasification technology by country.

Ni/o	Country	Thickness of coal seam, m	Angle of incidence of coal seam, degrees; coal seam position	Depth of exstraction, m	Method of preparation underground gas generator	Heat of combustion of coal, MJ/kg	Gas output per year, billion m³	Capital investment, million dollars	Cost price, $ per 1000 cubic meters	Stage of research
1	2	3	4	5	6	7	8	9	10	11
1.	The USSR Moscow region (1958)	2-4	near horizontal	45-60	well method (flow)	11.3-14.7	0.66 (project)	19.58-35.5 (for the whole project)	no data	Experienced
	Kuzbass (1955-1989)	2-8.5	35-56	50-300	—"—	31.5	0.45		2.38-8.5	Industrial
	Angren (1964-66)	4-20	near horizontal	120-220	—"—	16.7	1.4		2.02-7	Industrial.
	Dneprobass (1957-1960)	3.5	near horizontal	60	—"—	25.5-29.2	-		no data	Experienced
	Donbass, (1934-35 гг.)	0.5-1	near horizontal and inclined 38-60	60-250	mine method	33.4	-		no data	Experienced
2	England (1949-56)	1.5-1.8	near horizontal	500-600	well method	15	0.07	55	no data	Experienced
3	USA (Hannah, 1976) (1973-88)	8-9	6-9	84	well and mine method	21.3	0.00004-0.00011	24	65-295	Experienced and Industrial.
		2-11	near horizontal and large slope	30-270			0.0014-0.006			
4	France (1979)	1.2	large slope	1170	well method	32-35.5	7.5 (project)	37	no data	Experienced
		1.8; 9-20		880						
5	Belgium-FRG (1979-1987)	1-2	near horizontal	860	well method	20-30.8	0.5	25	no data	Experienced

1	2	3	4	5	6	7	8	9	10	11
6	China (2003)	2-6	5-15	350-600	well and mine method	31-34	0.6-0.9	25-33	39-73	Industrial
7	Australia (since 2003)	10	5-15	120-400	well method	23.8-33.8	0.9	17,5	60-80	Industrial
8	Ukraine Lviv and Volyn field (2007)	0.2-0.5	horizontal and inclined	300-650	well method	27.8-36.3	0.5	22-28	no data	Project
9	South Korea	1.8-6	near horizontal	80-190	well method	11.4-12.2	0.75	20	60-100	Experienced
10	Russia Kuzbass Lankovskoye, Uralskoye and Shkotovskoye coal deposits (2009)	2-10 5-7	35-45 3-12	100-400 50-400	well method	31.5 18.8-19.6	1-2 100-120 MW 0.5-1.8	6-31 12-39	60-70 12-15	Industrial Industrial

The cheapest gas in the project of GasTech, which in 2007 began gasification of seams with a thickness of more than 10 m at a depth of 152-610 m. The most expensive synthesis gas is from relatively thin layers.

As can be seen from Table 3, not all experimental tests are brought to industrial implementation. From individual test experiments to industrial deployment, the distance is quite large.

All of the above studies, as well as experimental work and work on an industrial scale in enterprises, were carried out using various technologies that were developed for various conditions and were aimed at various purposes. Analysis of existing technologies, mining and geological conditions and technical and economic characteristics will help to determine possibility of development of the technology for the conditions of the Donetsk coal region.

2.3 *Analysis of potential of the closed coal-mining enterprises in the Donetsk region*

Currently, there are eighteen coal-mining enterprises operating in the Donetsk region, the extraction of which so far covers the domestic needs of operating industrial enterprises. At the same time, nineteen mines with significant reserves of coal were closed. This is especially true of the State Enterprise «Artemugol» and the State Enterprise «Ordzhonikidzeugol». This is the central region of Donbass, where about 60 coal seams are concentrated. Even today, it is of interest to use these reserves in perspective. However, innovative methods of extracting reserves are needed because of the very difficult mining and geological conditions of the Donbass fields. It must differ from those used methods by greater economic efficiency, safety and other benefits. The technology of underground gasification of coal can just become a promising direction of development, at least for the numerous closed coal mining enterprises with the remaining reserves. In order to get positive results of the use of this technology, reliable mining and geological, mining and other conditions of the developed field are necessary. The study analyzed the mining and geological conditions and other indicators of the closed coal-mining enterprises of the Donetsk region from the point of view of the possibility of their further development (Table 4).

Gasification of coal deposits is considered appropriate if the conditions are met - "Time criteria for the suitability of coal deposits for underground coal gasification" - 1986. These include: reserves and mark of coal, thickness, depth and angle of the coal seam and other characteristics. Nevertheless, according to the results of the study (Fedorova 2018, Schilling et al. 1986), it can be concluded that only the geological factor (thickness and depth of the seam) plays a significant role in increasing the technical (combustion heat of the generator gas) and economic (increasing the distance between wells) efficiency of the process of underground coal gasification.

Let us discuss some of the factors influencing the process of underground coal gasification. Coal reserves should provide it with at least 30 years of operation and for coal is 10 million tons, and 30 million tons for brown coal. In our case, the remaining industrial reserves of coal are 750 million tons with an annual expenditure of about 7-8 million tons of coal. This is at least 100 years of use. In this case, even in the case of the resumption of the underground mining method, not all reserves can be extracted. But underground gasification allows the use of seams, which are quite difficult to extracted.

For the production of synthesis gas, it is preferable to use such marks of coal with the lagest contents of volatile matter: B, D, DG and G. It will allow to get more gas.

The minimum thickness of the seams, below which the heat loss increases, is 1.5-2 m. Another factor influencing the efficiency of underground coal gasification is the depth of the seam. Experimental work was carried out mainly at not big depths of about 100-400 m. For many European countries, the problem of applying the method of underground gasification for the exploitation of deposits located at great depths of about 1000 m and more is actual. As the underground coal gasification goes to deeper horizons, the requirements for drilling accuracy increase and their number increases as well. Ultimately, the cost of gas production increases. On the other hand, the use of UCG for deep seams significantly reduces the environmental consequences for human settlements.

Table 4. Mining and geological and technical conditions of closed coal mines DPR.

No	SE	Coal mines	Depth of extraction, m	Average coal seam thickness, m	Angle of incidence of coal seam, degrees (°)	Balance reserves, million tons	Mark of coal	Heat of combustion of coal, MJ/kg	Volatile matter, %
1	2	3	4	5	6	7	8	9	10
1	SE «Artemugol»	mine named after M.I. Kalinina	1080	0.83	51-53	20.6	T	30.6- 36.6	12
2		mine named after K.A. Rumyantsev	1090	0.9	56-60	37.2	OS	35.4-36.8	15
3		mine named after A.I. Gayevogo	975	0.4-2.8	58-63	21.3	OS	35.4-36.8	15
4		mine named after V.I. Lenina	1190	0.98	42-60	66.1	K	35.2- 36.4	20
5	SE «Ordzhonikidzeugol»	mine named after K. Marks	1000	0.84	67-72	30.2	OS	35.4-36.8	15
6		mine «Enakievskaya»	477	0.93	43-55	14.6	T	30.6- 36.6	12
7		mine «Poltavskaya»	477	0.92-1.15	40-60	10.7	T	30.6- 36.6	12
8		mine «Bulavinskaya»	530	0.86	54-63	18.8	T	30.6- 36.6	12
9		mine «Olkhovatskaya»	546	0.79	55-60	36.4	T	30.6- 36.6	12
10		mine «Uglegorskaya»	820	1.05	60-65	20.9	T	30.6- 36.6	12
11	SE «DFEC»	mine «Octyabrskiy Rudnik»	1148/911	1.29 - 1.55	10-12	80	G	33.1- 36.0	36
12		mine named after E.T. Abakumova	840	1.08-1.3	8-9	80	DG	32.5-34.5	35-40
13		mine «Trudovskaya»	900	1.61	11-13	70	D, DG	31.4- 33.5	35-45
14		mine «Udarnik» SE«Torez-anthracite»	670	0.9-1.1	5-10	8.4	A	33.4- 36.5	<8
15	SE «Makeevugol»	Chaikino mine	1100	1.58-1.85	9-14	117	Zh	34.7- 36.4	30
16		mine «Severnaya»	715	1.15	9-13	20	OS	35.4-36.8	15
17		mine named after V.M. Bazhanova	1300	1.61-1.91	4-15	80	K	35.2- 36.4	20
18		mine «Butovskaya»	1110	1.15-2.10	4-15	18	G	33.1- 36.0	36
		Total				750.2			

The angle of incidence of coal seam in some coal deposits is considered the main danger of the spontaneous combustion of coal. It is known that underground burning under certain conditions can extend into the depth of a coal seam, cover a significant area of it and continue for thousands of years.

The mines of SE «Artemugol» and SE «Ordzhonikidzeugol» developed low-thickness seams (0.8-1.05 m), the angle of incidence of some reached 72 degrees, the depth of development was from 477 to 1090 m, while the plan was for some seams 1800 m. All seams are supercategory on gas, dangerous for sudden emissions of coal and gas and some of them are prone to spontaneous combustion.

As for the remaining eight mines belonging to SE «DFEC», «Makeevugol» and «Torezan-tratsit», these are seams of a gentle fall to 15 degrees with an average power of 1.5 m and more. The development was carried out at great depths from 900 to 1300 meters. Mines are supercategories of gas, dangerous due to sudden emissions. Such conditions cause the complexity of the classic mine extraction, the final extraction of such seams by traditional methods is unprofitable and dangerous. The introduction of special mining methods in this situation may be an alternative way to generate energy. However, the industrial introduction of underground gasification of deep-seated and thin seams will be hindered by some difficulties, but in view of the benefits that it can provide, many countries are conducting studies of the industrial implementation of this technology for difficult seam conditions.

Figure 2 shows the economic, social and environmental benefits that the technology of underground coal gasification for the Donetsk region can provide.

3 RESULTS

Design and construction of the mine takes 8-12 years and requires a huge amount of investment. Creating underground coal gasification station will take 2-3 times less, capital and labor intensity will also be 2-3 times lower than that of the underground coal mining enterprise.

To produce 1 billion cubic meters of gas, it is necessary to gasify 2-2.4 million tons of coal (Gridin). In our case, 750 million tons of coal (Table 4) will make it possible to

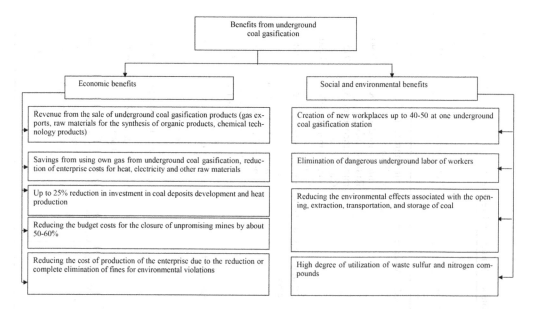

Figure 2. Benefits from the implementation of underground coal gasification projects.

obtain from 312.5 to 375 billion cubic meters of gas. In terms of a year, from 8 to 10 million tons of coal can save up to 4-5 billion cubic meters of natural gas. In accordance with the calculations of various experts (Kreinin, Grabskaya), a substitute for natural gas can be obtained from underground gasification gas at a cost of 60 to 180 dollars per 1000 cubic meters, depending on the geological conditions and the technology. Natural gas is supplied to enterprises of the republic at a price of $ 203 per 1000 cubic meters (13168 rubles), that is, savings can range from $ 23 to $ 133 per 1000 cubic meters. For comparison, for Ukraine, the weighted average price of natural gas for March 2019, based on the results of electronic exchange trading on the Ukrainian Energy Exchange, was $ 283 per 1000 cubic meters. Russia supplies Donbass more than 2.4 billion cubic meters of natural gas per year, so the annual savings will be from 55 to 319 million dollars. The remaining volumes of gas up to 2.5 billion cubic meters can be exported at a price of 203 dollars per 1000 cubic meters, while the gross annual income can range from 57 to 300 million dollars.

There is a possibility of obtaining electricity through coal gasification, but it can be 10–20% more expensive than traditional energy of heat power stations. However, using it for the needs of the enterprise will reduce the cost of production. For example, some Donbass mines when using its own electricity, the cost of 1 ton of coal can be reduced by 15%.

In the metallurgical industry underground gasification of coal gas can be used as a reducing agent for iron ore in blast furnace production and in direct iron production processes. This saves expensive and scarce coke and natural gas.

On the basis of the underground gasification of coal from the resin and gas it is possible to obtain raw materials for the synthesis of organic products, and if necessary, final products: carbon black, polymers, monomers, surfactants, dyes, solvents for varnishes and paints, as well as artificial liquid fuel.

In recent years, the approach to fuel use has changed dramatically. In the first place are not only the economic benefits of its use, but also environmental and social.

The implementation of projects to transfer restructured mines to the category of gas producers will reduce the budgetary costs of closing unpromising mines by about 50-60%, as well as reduce social tensions during their closure, creating an additional 40-50 jobs per position. The elimination of underground labor will help reduce the level of professional diseases, injuries, as well as the number of accidents at coal-mining enterprises of Donbass.

Despite the disadvantages of the UCG technology associated with the harmful effect on the ecology (Table 2), it has significantly less environmental impact than traditional mining methods, which is presented in Table 5.

In addition to improving the environmental situation, it will also reduce the cost of production by reducing or fully eliminating fines for environmental violations in the form of environmental taxes.

Table 5. The impact of UCG use on the environment.

Ecological disturbances	Impact of UCG
The creation of waste gangue	prevented (5–6 tons/ton of coal)
Land alienation	prevented (10-20 ha/million tons of coal)
Coal dust ejection	excluded (0.3-15 kg/ton of fuel equivalent)
The emission of ash	excluded (13.7-17.2 kg/ton of fuel equivalent)
Emission of sulfur dioxide	decreases from 6-9 to 1.6 kg ton of fuel equivalent
Nitrogen oxide emissions	decreases from 2-2.5 to 1-1.5 kg/ton of fuel equivalent
The discharge of suspended substances into wastewater	decreases from 0.452 to 0.044 kg/ton of fuel equivalent

4 CONCLUSIONS

Thus, the development of unconventional mining technology, which involves the use of coal without underground work, has great potential. The experience of using underground gasification of coal technology in different countries, both at the experimental level and at the level of practical application, allows us to conclude that this technology can be used in perspective in the Donbas areas where there are coal-bearing areas with hard-to-extract reserves. The implementation of underground coal gasification projects will make it possible to fully supply the republic with gas for both industrial and domestic needs, as well as export the surplus; reduce the cost of production of enterprises through the use of its own gas, heat and electricity, as well as by reducing or completely eliminating the amount of fines for environmental violations; reduce budget costs for the closure of unpromising mines; will contribute to the creation of new jobs, eliminate dangerous underground work of workers; reduce the environmental consequences associated with the opening, extraction, transportation, and storage of coal.

REFERENCES

Belov, A.V. & Grebenuk, I.V. 2009. Prospects for the chemical processing of gas underground coal gasification. *Mining industry*: 36. Research and Production Company «Gemos Limited».

Fedorova, M.A. 2018. Justification of innovative design solutions for the rational development of the potential of coal-gas fields based on LUGEC: Dis. cand. econ Sciences: 25.00.21: Moscow.

Gridin, S.V. & Vertela, S.A. 2013. Analysis of the prospects and methods of using gas-generating gas in order to develop energy-efficient solutions to save energy. *Energy saving. Energy. Energy audit*: 31-40.

Kondyrev, B.I. & Belov, A.V. 2005. The experience of underground coal gasification in the People's Republic of China. *Mining information and analytical bulletin (scientific and technical journal)*: 286-289.

Kreinin, E.V. & Grabskaya, E.P. Underground gasification of coal seams as the most efficient alternative to clean coal technology in fuel energy. – URL: https://cyberleninka.ru/article/n/podzemnaya-gazifikat siya-ugolnyh-plastov-kak-naibolee-effektivnyy-variant-ekologicheski-chistoy-ugolnoy-tehnologii-v-toplivnoy.pdf - Access Date: 26.03.2019.

Kreinin, E.V. 2009. Technical and economic prospects of underground gasification of coal. *Mining Informational Analytical Bulletin (scientific and technical journal)*: 347-352.

Kreinin, E.V. Unconventional hydrocarbon sources. New technologies for their development. – URL: https://books.google.com.ua/books - Access Date: 26.03.2019.

Lazarenko, S.N. & Kreinin, E.V. 1994. Underground gasification of coal in Kuzbass: present and future. Novosibirsk: VO «Science».

Mendeleev, D.I. 1939. Works. V.1. L .: Literary heritage.

Raimzhanov, B.R., Saltykov, I.M. & Yakubov, S.I. 2008. Underground gasification of coal: historical information and problems. *Gorny Vestnik of Uzbekistan*: 154-160.

Schilling, G.D., Bonn, B., Kraus, W. 1986. Coal gasification: Mining-raw-energy. M .: Nedra.

Yankovsky, M.A., Makogon, Yu.V., Gubatenko, M.I. & Ryabchin, O.M. 2011. Alternatives to natural gas in Ukraine in condition of energy and resource deficiency: industrial technologies. Donetsk: DonNU.

Zhukov, E.M., Kropotov, Yu.I., Luginin, I.A. & Chizhik, Yu.I. 2016. Prospects for the use of underground gasification in old industrial areas of Kuzbass. *Young scientist*; 146–148.

Topical Issues of Rational Use of Natural Resources 2019 – Litvinenko (Ed)
© 2020 Taylor & Francis Group, London, ISBN 978-0-367-85720-2

Value chain of the refining and petrochemical industries: assessment and strategy integration

P.E. Rezkin
Deputy Dean of the Faculty of Finance and Economics, Chairman of the Council of Young Scientists, Polotsk State University, Republic of Belarus

ABSTRACT: The modern concept of value chains is considered. The author developed and adapted theoretical concepts of the value chain concept to the conditions of the national economy of the Republic of Belarus; developed a methodical approach to the analysis of value chains in the oil refining and petrochemical industries, taking into account technology and peculiarities of price formation for oil and oil products; determined the method of identifying the most productive links in the value chain in the refining and petrochemical industries based on the added value indicator; formulated two types of strategies for integrating value chains in the studied industry.

1 INTRODUCTION

The aggravation of competition between manufacturers leads to a constant search for innovations that provide the long-term competitive advantages of an economic entity, as well as its stable functioning in a dynamically changing external environment. The performance of these entities in the refining and petrochemical industries is determined by many different factors, including a balanced value chain.

In modern conditions actualized the problem of constructing a value chain, which, on the one hand, makes it possible to improve the economic entity's own economic efficiency and on the other - does not burden the implementation of it is not characteristic activities and does not reduce the effect of specialization.

The concept of value chains is based on the idea of M. Porter about corporate value chains, as well as on the previously developed concept of supply chains in logistics. As a result of the use of outsourcing by modern industrial companies, a distributed production model has been formed, in which individual technological operations are localized in various regions of the world, which has made it possible to reduce total costs and increase the flexibility of the production process. Thus, the final product is created within the global value chains, where each country specializing in certain technological operations contributes to the value added of the product. The key actors in the industrial sector of the world economic system are not national economies, industries and companies, but global value chains, which are an intermediate form between centrally managed structures and the market.

2 STATEMENT OF THE PROBLEM

The aim of the study is to develop theoretical and methodological guidelines, specific methodologies and practical recommendations on the formation of strategies for integrating value chains, as well as an assessment of their macroeconomic and microeconomic efficiency in the oil refining and petrochemical industries.

3 DESCRIPTION OF THE CURRENT SITUATION

Fundamental provisions of the concept of value chains were formulated by M. Porter (Porter 1985), as well as continued in the writings of G. Gereffy (Gereffi 2001), R. Kaplinsky & M. Morris (Kaplinsky 2002). Foreign scientists-economists P. Grant (Grant 2008), A. Thompson & A. Strickland (Thompson et al. 1998), J. Schank & V. Govindarajan (Schank. et al. 1999.), M. Rother & J. Shuk (Rother et al. 2005), as well as Russian scientists M. Melnik & V. Kogdenko (Melnik 2010), D. Saveliev (Saveliev 2010), S. Tolkachev (Tolkachev 2016), D. Tatarkin & O. Bryantseva & V. Dyubanov (Tatarkin et al. 2014), O. Yuldasheva & O. Yudin (Yuldasheva et al. 2012), L. Maslennikova (Maslennikova 2001.), V. Repin (Repin 2014), R. Khasanov (Khasanov 2009), T. Andreeva (Andreeva 2013) and others made significant contributions to the development of the theoretical and methodological foundations for the formation and evaluation of the effectiveness of value chains.

Practical research in the field of product value chains in certain sectors of the Russian economy is reflected in the scientific studies of S. Avdasheva & I. Budanov & V. Golikova & A. Yakovlev (Avdasheva et al. 2005), N. Rubtsova (Rubtsova 2012), A. Yudaev (Yudaev 2011), T. Andreeva (Andreeva 2013).

Belarusian scientists V. Baynev (Baynev et al. 2015), N. Bogdan (Bogdan 2011), A. Bykov (Bykov 2014.), V. Gusakov (Gusakov 2011), R. Ivut (Ivut et al. 2016), L. Nekhorosheva (Nekhorosheva 2014), T. Kasayeva (Kasaeva et al. 2012), I. Poleschuk (Poleshchuk 2011.), P. Rezkin (Rezkin 2016), D. Rutko (Rutko 2014), V. Fateev (Fateev 1985), V. Shimov (Shimov 2010.), O. Shimova (Shimova et al. 2013), V. Shutilin (Shutilin 2015), G. Yasheva (Yasheva 2009), etc. are also involved in issues of managing value chains and supply chains, as well as network structures in industry.

At the same time, a number of problems associated with the formation of an effective value chain in specific sectors of the national economy are due to insufficient theoretical and methodological development.

There remain debatable questions about the assessment of the effectiveness of value chains and the choice of key activities in their formation.

4 RESEARCH RESULTS

Building an effective value chain, in the opinion of most domestic and foreign scientists and economists, is the most important factor in increasing the competitiveness of business entities and, as a result, the country's economic growth. Taking into account the theoretical research, it can be argued that the value chain consists of consecutive links - activities that are aimed at creating the value of the final product, i.e. product of value to the buyer, and are performed in a chain of economic entities or their units. The main purpose of these activities is to increase the efficiency of the entire value chain or its individual link and maximize long-term profit.

As part of the concept of value chains, the function of participants in the entire value chain is revealed in relations with the external environment (suppliers, consumers) - the transformation of the flow of incoming resources into a valuable product for the consumer. A value chain is a system of interrelated activities, the links between which allow to identify alternative ways of implementation, which can later lead to competitive advantages.

The study summarized and supplemented the classification of value chains for a number of characteristics (Rezkin 2016):

- type of management (managed by the manufacturer or consumer);
- the dominant link (with the dominant supplier/consumer/intermediary/manufacturer);
- the place of the entity in the value chain (supplier, manufacturer, distribution channel, consumer);
- interaction intensity (with high or low interaction intensity);
- the duration of interactions (non-recurring (short-term) or repetitive (long-term) value chain);
- coordination (coordinated, polycentric);

- method of construction and detailing (standard (according to M. Porter), unique);
- type of integration (integrated up, integrated down, fully integrated, specialized);
- scale (individual product, division, individual business entity, industry, global);
- territorial basis (local, regional, national, international);
- ownership (state, private, mixed).

The classification of value chains allowed generalizing and systematizing the theoretical principles of the concept of value chains, as reflected in the works of domestic and foreign scientists, as well as deeper revealing the essence of the category "value chain".

The results of a comprehensive analysis of value chains predetermine the further development of the business entity and industry, as well as contribute to the proper evaluation of competitive strategies in order to increase the efficiency of their activities.

The study of scientific papers in the field of research on the concept of value chains has shown that all existing methods of analysis can be grouped into five approaches (Rezkin P. E. 2016):

1) assessment of the value chain of an economic entity regarding the efficiency and optimality of building a business process system;

2) assessment of the competitiveness of the value chain;

3) assessment of the chain in terms of the cost approach;

4) assessment of the chain from the standpoint of differentiation;

5) assessment of the consumer value of the proposal of a business entity as a result of the functioning of the value chain.

Based on the analysis of existing approaches to the assessment of value chains, it can be concluded that there is no universal methodology for assessing chains of this kind.

The study proposed the use of an integrated approach that will take into account various aspects of the listed approaches to assessing value chains, which will improve the quality and effectiveness of the analysis of value chains, conduct a more complete analysis of the activities of an economic entity and formulate a strategy that ensures its high level of competitiveness and sustainable economic growth.

It is proposed to use value added instead of the currently most common profit as an indicator for assessing the performance of value chains.

The "value added" indicator has a special practical significance in forming value chains and evaluating its effectiveness (Rezkin 2016):

- characterizes the overall efficiency of resource use and takes into account the economic interests of internal (owners, employees) and external (state) participants in the value chain;

- allows you to determine the exact contribution of the employee to the creation of new value in the framework of the value chain and a particular link in particular;

- it makes it possible to improve the system of labor remuneration of an employee through the establishment of the dependence of the size of his remuneration on the value of the added value created (observance of the rule of faster growth of labor productivity over the rate of wage growth);

- allows you to determine and, if necessary, change the share of external contractors (suppliers of material resources) in the value of the product;

- makes it possible to determine the dependence of one link (enterprise) on another within the framework of one value chain;

- allows you to determine the controlling link in the value chain by structural division of its participants (suppliers, producers, consumers);

- prepares a basis for studying the connections between the links of the value chain and their organizational and economic relations in the process of formation and functioning of the value chain in order to increase the efficiency of activities, competitiveness and sustainability of each individual link, as well as identify sources and ways to increase the value of the product (product services) for the end user;

- allows you to resolve issues regarding the construction of a new or improvement of the existing value chain.

A synthesis of methodological approaches to the determination of value added confirms the need and importance of calculating this indicator. Obviously, the added value for each individual economic entity will be different, but its formation in vertically integrated structures will be much more efficient.

It has been proven that a properly built, productive value chain is the basis for a stable and efficient functioning of an economic entity, industry and national economy as a whole. This concept is one of the key tools in overcoming the existing crisis phenomena in the economy.

The next stage of the research is the application of the concept of value chains to the oil refining and petrochemical industries.

The author conducted a comprehensive analysis of the global oil market and identified trends in its development. Selected factors affecting the price of oil, including: the movement of capital, the demand and supply of oil, geopolitics and natural factors. Assessing the status and trends in the oil refining and petrochemical industries revealed the possibility of applying the concept of value chains in the studied industry, which allowed us to build an industry value chain with the identification of key links, as well as to determine the proportions of the distribution of material flow and the value added value along the chain. The research results allowed to develop the author's methodology for identifying the most productive links in the value chain in the refining and petrochemical industries.

Today, the world oil market is structurally close to the seller's oligopoly. According to the data for 2018, oil buyers are 118 countries of the world community, while major importers - only 10 with purchases of more than 50 million tons (Report on the activities of the Belarusian state concern for oil and chemistry 2018).

Analysis of the world market for oil and oil products has allowed us to build a global oil-producing network based on the flow of crude oil. This network includes four phases:

1) prospecting and mining (ascending phase);
2) transportation and storage (intermediate);
3) processing, sales and consumption (downward);
4) return to the environment (reverse).

The identified features of the material flow in the global oil producing network serve as the basis for building the industry value chain in the oil refining and petrochemical industries (Figure 1).

The built value chain has become the starting point in assessing the state and trends of the oil refining and petrochemical industries in Belarus. In Bearus, a relatively high level of external openness is observed in the studied industry, which is due to the high export orientation and dependence on oil imports.

Calculations showed that the creation of 1 BYN of gross value added in the studied industry requires 1.01 BYN of import costs. The excess of the level of export orientation over the level of import dependence in the oil refining and petrochemical industries indicates a high efficiency of foreign trade, but the value of imports actualizes interest in the problem of import substitution in Belarus.

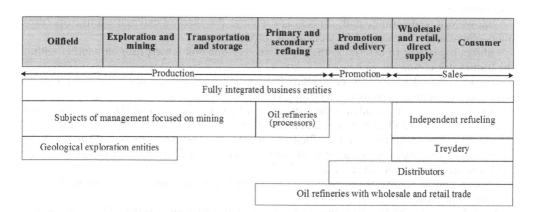

Oilfield	Exploration and mining	Transportation and storage	Primary and secondary refining	Promotion and delivery	Wholesale and retail, direct supply	Consumer
←————————————Production————————————→				←—Promotion—→	←————————Sales————————→	
Fully integrated business entities						
Subjects of management focused on mining			Oil refineries (processors)		Independent refueling	
Geological exploration entities					Treydery	
				Distributors		
			Oil refineries with wholesale and retail trade			

Figure 1. Sectoral value chain in the oil refining and petrochemical industries (author's elaboration).

Analyzing the geographic focus of foreign trade in oil and oil products, it can be noted that the main sales markets for petroleum products according to data for 2018 are Ukraine (25.0%), Great Britain (34.9%) and the Netherlands (13.0%); crude oil - Germany (100%). As for imports, Russia is the main supplier of both oil and oil products for this year (100%) (Report on the activities of the Belarusian state concern for oil and chemistry. 2018).

An important result of the study is the application of the concept of value chains to the oil refining and petrochemical industries. The author has built a valid value chain in the oil refining and petrochemical industry of the Republic of Belarus, the key links of which were primary and secondary refining. Based on the study of the process units of value chains built technological schemes and directions of spending of the incoming raw material - oil. For the first time in the Republic of Belarus, a pattern of distribution of the material flow, as well as the value added across the value chain (Figure 2), has been built.

In order to analyze the existing value chain, the author developed a method for identifying the most productive links in the refining and petrochemical industries, consisting of five consecutive steps.

Step 1. Build a functioning value chain. Identification of key links in the value chain and their interrelation.

Step 2. Constructing the material flow along the value chain – determining the proportion of incoming raw materials (oil) in the finished product.

Step 3. Calculate the value added for each individual link in the value chain. Determination of the share of the value added of each link in the total value added of the value chain.

Step 4. Calculate the coefficient of productivity of the added value of the i-th link in the value chain (I_i^{AV}). This indicator was introduced by us for the purpose of calculating the efficiency of functioning of the links in the value chain.

$$I_i^{AV} = \frac{AV_i}{IRM_{i.}}, \tag{1}$$

where AV_i – the proportion of the added value of the i-th link in the total value added of the value chain; IRM_i – the share of incoming raw materials (oil) in the produced product of the ith link in the value chain.

Stage 5. The ranking of activities according to the coefficients of the value added efficiency of the i-th link in the value chain is made (I_i^{AV}). Based on the method of calculating this coefficient, as well as its economic essence, it is easy to see that the coefficient value is maximized in the case of greater efficiency ($I_i^{AV} \rightarrow max$), that is, the larger the coefficient value, the more effectively the link of the value chain functions

For the calculations, the author used data from the Belarusian State Concern on Oil and Chemistry, which are in closed access and are a commercial secret of enterprises of the studied industries. For this reason, the study presents only the results, without intermediate calculations.

In the course of the study, the features of the functioning of the value chains in the oil refining and petrochemical industries of the Republic of Belarus were identified, which served as the basis for the formation of development strategies for this industry. It can be stated that the formation of gross value added in the studied industry is largely due to the trade in primary refining products (77.56% of the total gross value added of the industry), while its prevailing part was formed due to export (40.45% of the total gross added value).

In general, the primary oil refining, taking into account sales, provides about 85.42% of the gross value added of the Belarusian petrochemical industry versus 14.03% of the secondary oil refining.

At the same time, most of the gross value added of the secondary oil refining is formed by the scope of production and sale for further industrial use (more than 13.94% of the total gross value added, including 4.45% due to export deliveries), and the sale of consumer goods is only a small part of the gross value added of the industry (0.09%). Based on the distribution of gross value added and material flow throughout the value chain of the industry, the

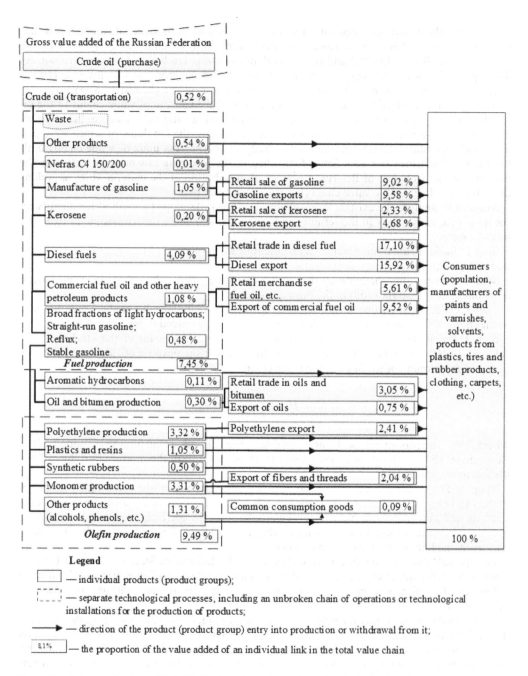

Figure 2. Scheme of value added distribution throughout the value chain in the oil refining and petrochemical industries of the Republic of Belarus (author's elaboration).

coefficients of productivity of the value added of each link in the value chain are calculated and ranked. The calculations were made separately for the production and sales of the products of the refining and petrochemical industries (Table 1).

According to the results of the study, two directions of development of the studied industry can be identified:

Table 1. Ranking of the chains of the value chain of the refining and petrochemical industries by coefficient of productivity for 2018 (fragment, author's elaboration).

Rank	Option 1 (production) Link value chain	I_i^{AV}	Option 2 (retail and export) Link value chain	I_i^{AV}
1st	Fibers and threads	12,64	Retail trade in oils	2,29
2nd	Alcohols, phenols, etc.	10,45	Export of fibers and threads	2,28
3rd	Methyl acrylate, sulfate ammonium, etc.	8,51	Kerosene export	2,20
4th	Polyethylene	7,52	Export of AI-92 gasoline	2,08
5th	Synthetic rubbers	3,17	Retail sale of kerosene	1,70
6th	Plastics and resins	2,62	Retail sale of other gasoline	1,55
7th	Waxes, dyes, etc.	2,55	Polyethylene export	1,53
8th	Orthoxylol	0,91	Retail sale of gasoline AI-92	1,27
9th	Gasoline AI-95	0,76	Export of AI-95 gasoline	1,21
10th	Diesel fuel	0,68	Retail trade in gasoline Normal-80	1,12
11th	Tulol	0,67	Retail trade in diesel fuel	1,03

1) the development of primary refining in order to sell final products in the domestic and foreign markets;

2) the development of secondary refining in order to manufacture products for further industrial use.

Analyzing these areas, it can be argued that the most promising is the development of primary refining through its deepening, which is due to lower costs when correlated with the value added compared to the second direction, as well as the special interest of the state in primary refining from the position of value added distribution in the form export duties and taxes.

The main paths of development should be energy saving, increasing the depth of oil refining, i.e. processing with the greatest return, production of higher quality products that meet modern environmental requirements, the needs of consumers and the state.

On the basis of generalization, development and adaptation to the realities of the Belarusian economy of the theoretical positions of the concept of value chains, the author has developed a flowchart of the formation of an effective value chain (Figure 3).

Based on the developed flowchart, the author proposed strategies for integrating value chains in the studied industry, namely:

- change of value chains by extending it, which is expressed in the form of reverse ("integration back") and (or) forward-going integration ("integration forward");

- changing value chains through its development (modernization).

The author also developed a system of indicators for assessing the effectiveness and efficiency of functioning of individual links in the value chain (Table 2).

The approbation of this system of indicators was carried out by the author on the example of the oil refining and petrochemical industry of the Republic of Belarus.

5 CONCLUSIONS

This task was successfully solved by the author.

1. The theory of the concept of value chains has been developed and adapted to the conditions of the national economy of Belarus, the use of the value added indicator in the assessment of vertically integrated corporations is justified.

2. A methodical approach to the analysis of value chains in the oil refining and petrochemical industries has been determined, taking into account the technological process and the peculiarities of the formation of prices for oil and oil products.

3. A methodology has been developed for identifying the most productive links in the value chain in the oil refining and petrochemical industries, in which the indicator of value added

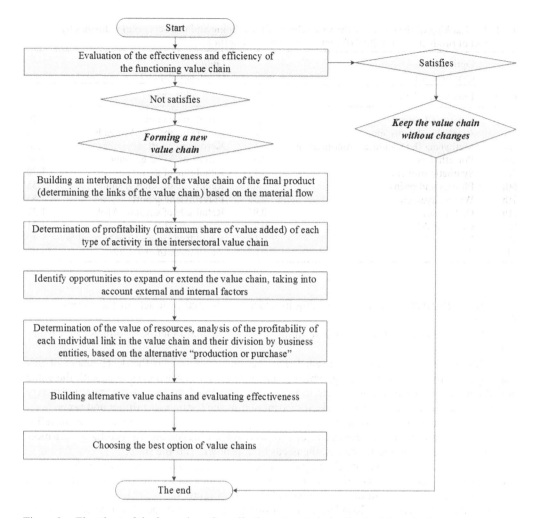

Figure 3. Flowchart of the formation of an effective value chain (author's elaboration).

generated within each process has been used for the first time. The method allows to select the final and intermediate products with the highest level of value added per physical unit of raw materials used. The author has proposed a system of indicators for assessing the effectiveness of individual links in the value chains, which can be used to form a competitive strategy for enterprises and increase competitiveness. This system includes indicators of the structure of value added, indicators of the dynamics of value added, resource efficiency indicators.

4. Differentiated strategies for integrating the value chains of the oil refining and petro-chemical industries of the Republic of Belarus are proposed. A distinctive feature of strategies is to focus on maximizing value added

Practical application of the research results allows building an effective value chain for interacting economic entities, thereby increasing their efficiency and competitiveness. The results of the reasearch were introduced into the educational process of the Polotsk State University, tested and used in the activities of Belarusian enterprises of the oil refining and petro-chemical industries (Naftan OJSC, Mozyr Oil Refinery OJSC, Steklovolokno OJSC and others), which is confirmed by the relevant acts of implementation in the production activities.

Table 2. The system of indicators for assessing the effectiveness and efficiency of functioning of individual links in the value chain (author's elaboration).

Name of the indicator	Formula for calculating
1. Indicators of the structure of value added,%	
1.1. Proportion of wages (employee income) in the total value added	$\frac{W}{VA}$
1.2. Proportion of depreciation in the total value added	$\frac{DC}{VA}$
1.3. Share of profit from sales in the total value added	$\frac{PS}{VA}$
1.4. Proportion of income of owners in the total value added	$\frac{IO}{VA}$
1.5. Proportion of state revenues in the total value added	$\frac{SR}{MC}$
2. Indicators of the dynamics of value added,%	
2.1. Rate of value added	$\frac{VA_1}{VA_0} \cdot 100\% - 100\%$
2.2. Rate of wage growth (employee income)	$\frac{W_1}{W_0} \cdot 100\% - 100\%$
2.3. Growth rate of incomes of owners	$\frac{IO_1}{IO_0} \cdot 100\% - 100\%$
2.4. Growth rate of depreciation	$\frac{DC_1}{DC_0} \cdot 100\% - 100\%$
2.5. Growth rate of sales profit	$\frac{PS_1}{PS_0} \cdot 100\% - 100\%$
2.6. Growth rate of state revenue	$\frac{SR_1}{SR_0} \cdot 100\% - 100\%$
3. Indicators of resource efficiency	
3.1. Value added per monetary unit of material costs	$\frac{VA}{MC}$
3.2. Value added per monetary unit of sales revenue	$\frac{VA}{SRV}$
3.3. Value added per monetary unit profit from sales	$\frac{VA}{PS}$
3.4. Value added per monetary unit of fixed assets	$\frac{VA}{AVFA}$
3.5. Value added per employee	$\frac{VA}{ANE}$
3.6. Ratio of growth rates of labor productivity and wages	$\frac{R_{LP}}{R_W}$ where: $R_W = \frac{W_1}{W_0} \cdot 100\%$ $R_{LP} = \frac{LP_1}{LP_0} \cdot 100\%$
3.7. Increase in value added per monetary unit of investment costs	$\frac{\Delta VA}{IC}$
3.8. Increase in value added per monetary unit of investment costs	$\frac{\Delta VA}{INC}$

Note. 0, 1 - base and reporting periods, respectively; W - wages; VA - value added; DC - depreciation charges; PS - profit from sales; IO - the income of the owners; SR - state revenues; MC - material costs; SRV - sales revenue; AVFA - average annual value of fixed assets; ANE - the average number of employees; LP - labor productivity; ΔVA - absolute increase in value added; IC - investment costs; INC - innovative costs.

REFERENCES

Andreeva, T.V., 2013 *The chain of value creation of a product: the formation and evaluation of efficiency.* Moscow. INFRA-M.

Avdasheva, S.B. & Budanov, I. A. & Golikova, V. V. & Yakovlev, A. A. 2005. Modernization of Russian enterprises in value chains (using the example of the pipe and furniture industry in Russia). *Economic Journal of the Higher School of Economics* 3: 361–377.

Baynev, V. F. & Vinnik V. T. 2015. Vertical integration as a condition for the modernization of the national economy. *State regulation of the economy and improving the efficiency of business entities.* Minsk. Academy of Management under the President of the Republic of Belarus: 240–243.

Bogdan, N. 2011 Innovative development of Belarus in the context of European integration. *Science and Innovation* 5: 38–40.

Bykov, A. A. 2014. On the causes and possible consequences of the decline in world oil prices. *Belarus economic journal* 4: 4–16.

Fateev, V. S. 1985. *Investigation of the level and effectiveness of the territorial concentration of production in the industries of the Belarusian economic region.* Institute of Economics of the Academy of Sciences of the BSSR. Minsk.

Gereffi, G. 2001. Shifting Governance Structures in Global Commodity Chains, with Special Reference to the Internet. *The American Behavioral Scientist* 10: 1616–1637.

Grant, R. M. 2008. Modern Strategic Analysis. St. Petersburg. Peter, 2008.

Gusakov, V. 2011 Conditions and factors for the effectiveness of cooperative-integration associations. *Agricultural Economics* 3: 2–6.

Ivut, R. B. & Krasnova, I. I. & Kisel, T. R. 2016. *Inventory management*. Minsk. BNTU.

Kaplinsky, R. 2002. The Globalization of Product Markets and Immiserising Growth: Lessons from the South African Furniture Industry. *World Development* 7: 1159–1177.

Kasaeva, T. V. & Tsynkovich O. G. 2012. Estimation of the intensification of economic growth in terms of gross value added. *Herald of the Vitebsk State Technological University* 23: 160–171.

Khasanov, R. Kh. 2009. A Methodology for Assessing Enterprise Competitiveness. *Modern competition* 4: 99–109.

Maslennikova, L. A. 2001. *Forming an economic mechanism for managing an organization based on the concept of a value chain*. Moscow.

Melnik, M. V. 2010. The concept of economic analysis focused on the assessment of the value chain. *Economic analysis: theory and practice* 7: 2–9.

Nekhorosheva, L. N. 2014. The concept of the formation and development of innovation-industrial clusters in the context of the industrial policy of the Republic of Belarus: problems and directions for implementation. *Clusters and sectoral agreements, as an example of market cooperation*: 75–97.

Poleshchuk, I. I. 2011. Development Strategy for Commodity Distribution Networks and International Logistics Systems. *The integration of science, education and law - a strategy for the development of an innovative economy. International scientific and practical conference. Ekaterinburg,* January 25–26, 2011 *Ural State Economic University:* 57–61.

Porter, M. E. 1985. How Information Gives You Competitive Advantage. *Harvard Bussiness Review* 85 (July–August): 149–160.

Repin, V. V. 2014. *Business Processes. Modeling, implementation, management*. Moscow. Mann, Ivanov and Ferber.

Report on the activities of the Belarusian state concern for oil and chemistry. 2018. Minsk.

Rezkin, P. E. 2016. Application of the concept of value chains in the fuel and petrochemical industries of the Republic of Belarus. *Management problems. Series A and B* 4 (61): p. 80–86.

Rother, M., & Shuk, J., 2005. *Learn to see business processes. The practice of building value stream maps.* Moscow. Alpina Business Books.

Rubtsova, N. V. 2012. The value chain of a tourist product. *Regional economics: theory and practice* 40 (271): 46–53.

Rutko, D. F. 2014. *Clusters in the European Union: the mechanism of formation and development trends.* Minsk. Belarusian State Economic University.

Saveliev, D. A. 2010. Analysis of the value chain of aircraft manufacturing companies of the world and Russia. *Transportation in Russia* 4: 26–31.

Schank, J. K. & Govindrajan V. 1999. *Strategic cost management*. St. Petersburg. CJSC "Business Micro".

Shimov, V. N. 2010. Directions of the structural transformation of the industrial complex of the country in the context of global trends. *Scientific works of the Belarusian State Economic University* 3: 3–10.

Shimova, O. S. & Reutyonok, T. A. 2013. Features of the development of a system of environmental and economic indicators for the production strategy of the enterprise. *Scientific works of the Belarusian State Economic University* 6: 448–455.

Shutilin, V. Yu. 2015. Import as a source of technological modernization of the machine-building complex of the country: a comparative analysis, opportunities and problems. *Scientific works of the Belarusian State Economic University* 8: 456–467.

Tatarkin, A. I. & Bryantseva, O. S. & Dyubanov, V. G. 2014. Assessing the impact of new technologies on changing value chains when processing zinc-containing technogenic raw materials. *The region's economy* 4 (40): 178–187.

Thompson Jr., A. A. & Strickland, A. J. 1998. *Strategic Management. The art of developing and implementing strategies*. Moscow. UNITY.

Tolkachev, S. A. 2016. Methodological Foundations for the Analysis of Transformation of Global Value Chains During Neo-Industrialization. *Russia's economic revival* 3 (49): 57–65.

Yasheva, G. A. 2009. *Formation and implementation of a cluster approach in managing the competitiveness of light industry enterprises of the Republic of Belarus*. Minsk.

Yudaev, A. V. 2011. *Interfirm Interaction Management Based on the Concept of Value Chains (on the example of the pharmaceutical market)*. Moscow. Moscow Finance and Law Academy.

Yuldasheva, O. U. & Yudin, O. I. 2012. Modeling the value chain. *Problems of modern economy* 1: 218–222.

Topical Issues of Rational Use of Natural Resources 2019 – Litvinenko (Ed)

CUMR projects: Possibilities for implementation on Arctic territories

V. Solovyova & A. Cherepovitsyn
St. Petersburg Mining University, Saint Petersburg, Russian Federation

ABSTRACT: To date, issues related to the complex use of mineral resources (CUMR) are becoming increasingly important especially for Arctic zone. This article deals with economic evaluation of the possibility for implementation of CUMR projects on Arctic territories. The study presents features of projects of complex use of mineral raw materials and effects of its implementation. Criteria for assessing prospects of these projects are formed. Potential mechanisms for support of the CUMR projects implementation are discussed. Much attention is given to analysis and practical implementation of support mechanisms. The study makes a number of recommendations aimed at promotion of implementation of the complex use of mineral raw materials projects.

1 INTRODUCTION

Efficient development of the resource potential of the Arctic zone is one of important strategic priority for Russia. Over 80% of national mineral reserves are located in the Russian Arctic, including one third of the world's reserves of nickel and platinoids, cobalt (15% of world reserves), diamonds, 90% of tin, gold, mica, apatite and other types of raw materials (Larichkin et al. 2012).

Most part of mineral deposits situated in Arctic are complex: contain two or more components. Ores of non-ferrous metals, as a rule, contain molybdenum, copper, silver, gold, rare-earth metals, tin, etc. The iron ores contain titanium, phosphorus, zinc, vanadium, nickel and other elements. In general, about 70 components are extracted from complex ores. Therefore, there is a great potential within complex use of mineral resources (Kaplunov et al. 2011, Larichkin 2011).

Projects of complex use of mineral raw materials have a critical feature: focus on long-term ecological, social and innovative results. At the same time, projects, the purpose of which is to increase the level of complex use of mineral base, are notable for a high level of knowledge and capital intensity. The need for development unique mining and refining technologies leads to significant costs in the stage of research and development and in implementation phases. Projects of complex use of mineral raw materials have also specific characteristics that should be taken into account. So these projects are characterized by high level of technological and economic risks. It leads to necessity to search potential mechanisms for support of complex use of mineral raw material projects' implementation.

2 LITERATURE REVIEW

The existing literature on the theoretical aspects of use of mineral resources is extensive and focuses particularly on analysis of the prospects and problems of CUMR within development of Russian mineral resources base (Zuckerman 1995; Larichkin 2008; Vinogradov et al. 2011; Vukolov 2016). To date, several studies have investigated possibilities and necessity of CUMR on Arctic territories (Larichkin et al. 2012; Trubetskoy et al. 2015).

Much of the current literature on complex use of mineral raw materials pays particular attention to refinement of approaches to the economic evaluation of extracting individual components out of feedstock (Marinina & Vasiliev 2009; Lobanov & Noskov 2013; Mirzoyeva 2013).

Numerous studies have attempted to explain institutional and organizational problems connected with the rational use and protection of mineral resources (Nevskaya & Ligotskiy 2013; Nevskaya & Marinina 2017). To date, several studies have investigated the opportunities of stimulation of investment environmental protection activity (Solovyova 2008; Yashalova 2011).

Despite the range of existing papers devoted to different aspects of complex use of mineral raw materials, there are no publications devoted to economic evaluation of CUMR projects from the point of view of their features and possibility of using modern mechanisms for their support.

3 METHODOLOGY

The study uses qualitative analysis in order to gain insights into main directions within environmental management and theoretical aspects of complex use of mineral resources. Key features of the development of the Arctic and existing ecological problems were studied on the basis of scientific works of Russian researchers (Krapivin et al. 2015; Bolsunovskaya et al. 2015).

Effects of implementation of CUMR projects with use of decomposition method were formed. The methods of strategic analysis were used to create the criteria of assessing prospects of CUMR projects. In order to identify the possibilities of applying support mechanisms and to assess their effectiveness, the methods of economical evaluation of projects were used. The Net present value (NPV) method was used to determine the current value of all future cash flows generated by a project. The method "Profitability Index (PI)" was used to measure a proposed project's costs and benefits by dividing the projected capital inflow by the investment (*Mining Project Evaluation*; Boadway 2000). The payback period (PP) was used to determine the period of time required to recoup the funds expended in an investment. It was also used the method of comparative analysis for determining the most effective alternative.

The legal basis of the study is formed by analyzing the Federal Law "On Subsoil", the "Strategy of development of mineral resources base of the Russian Federation till 2035", the Program "Socio-economic development of the Arctic zone of the Russian Federation for the period until 2025" et al. The methods collectively used in the study made it possible to ensure the reliability and validity of the conclusions.

4 RESULTS

The relevance of such problem as complex use of mineral raw materials on Arctic territories is increasing now due to the growth of ecological risks, gradual depletion of mineral and raw materials and necessity of meeting the important challenges within environmental management.

The development of the Arctic territories is associated with the need to solve significant problems in economic and social development of the region. At the same time, effective development of Arctic territories is one of the most essential strategic priority for our country (Larichkin 2011).

An important feature of projects of complex use of mineral raw materials is that in addition to commercial and budgetary efficiency, they also focus on obtaining long-term macroeconomic, geopolitical, innovative, social and environmental effects. Macroeconomic effects related to increase in investment of regional economy, growth of macroeconomic indicators. The importance of social and environmental effects is dictated by the possibility of reduction of ecological risks and improvement the quality of life of the population. Innovative effects are connected with increase in innovative activity of Russian mining companies and with development of national science in CUMR direction. Geopolitical effects are also important due to the possibility of strengthening national positions in the global market of mineral raw materials.

These types of effects were chosen as a basic because they represent expected achievements in different areas most comprehensively. So macroeconomic effects include also microeconomic aspects, innovative effects include scientific and technological aspects. Figure 1 presents an overview of estimated effects of implementation of CUMR projects on Arctic territories.

The interests of the State in matters of complex use of mineral raw materials related to the long-term effects of expanding production of minerals and metals, reproduction of the resource potential and intensification of the use of environmental protection technologies.

However, there is a problem of lack of interest in implementation of the complex use of mineral raw materials projects at the side of subsoil users. This fact can be explained by complicity of such projects. CUMR projects are notable for a high level of knowledge and capital intensity. The need for development unique mining and refining technologies leads to significant costs in the stage of research and development and in implementation phases. Moreover, there is a high level of uncertainty in macroeconomic environment. There may be no demand for some types of minerals in the market or current costs may not be covered under certain market conditions. For example, the cost of extracting gallium from polymetallic ores exceeds the price of this product.

The production of a new type of product requires an extensive analysis of the market, an assessment of the possibilities of implementing a particular type of product, the formation of an effective marketing policy and a strategy for entering new market segments, the use of planning tools and long-term forecasting.

The following features of CUMR projects were highlighted as those of particular importance:

- High level of complicity and capital intensity of technologies for extraction of valuable components
- The difficulty in determining the optimal composition of extracted components
- The need to use differentiated economic assessment
- Uncertainty in establishing real market needs for individual components of mineral raw materials
- Difficulties in justifying and implementing a business diversification strategy
- Presence of specific risks of implementation of CUMR projects

The projects of complex use of mineral raw materials should comply with certain criteria. To determine the perspectives of individual projects following criteria have been formed:

1. Location (availability of ports, railways, transshipment points);
2. Geologic conditions (optimal occurrence depth, high degree of extraction);

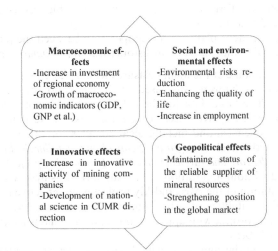

Figure 1. Effects of implementation of CUMR projects.

3. Categories of reserves (reserves of associated components of categories A, B, C1. If their distribution is uneven, the category of reserves can be reduced to C2);
4. Raw material quality (Compliance with the requirements of limiting grade of ore - in accordance with the "Methodical recommendations for the integrated study of deposits and the calculation of reserves of associated minerals and components");
5. Competitiveness of mineral raw materials (the strategic importance of raw materials for the national economy; favorable market conditions);
6. The availability of technology (possibility of extracting associated products based on existing enrichment technologies);
7. Economic efficiency (cost-effectiveness, NPV>0);
8. Socio-economic efficiency (positive contribution to the economic development of the regions; development of related industries (technological, chemical and other areas), implementation of social and infrastructure programs);
9. Environmental friendliness (compliance of implemented projects with the established standards in the field of environmental safety).

Thus, only the availability of technology does not suggest the possibility of implementation of such projects. Despite of the strategic importance of certain types of mineral raw materials, development of complex deposits is frequently unprofitable because of high capital costs, high operating costs et al. This relates in particular to extraction of rare earth metals, that requires the building of special processing complexes. As an example, Afrikandskoe titanium-magnetite ore deposit is a complex deposit located in the Murmansk region. The total resources are estimated at 70 million tons of titanium, 860 thousand tons of rare-earth metals, 580 thousand tons of niobium and tantalum. Implementation of this project allows to provide the national industries, especially military-industrial complex, with rare earth metals (*Integrated chemical and metallurgical complex for the production of titanium dioxide, rare and rare earth metals. Afrikandskiy project* 2016). However, the project is not worthwhile in current conditions from an economic point of view.

According to the "Strategy of development of mineral resources base of the Russian Federation till 2035" the economic incentive model is planned to use, especially for the development of new deposits, where there is no infrastructure, deposits with hard-to-recover reserves et al. However, there are no special economic mechanisms to support projects of complex use of mineral raw materials. It requires analyzing the possibilities of existing instruments and tax regimes. Following support mechanisms were chosen:

- Regional investment project (RIP) - investment project aimed at production of goods on the territory of one of the Russian regions
- Advanced Development Territories - zones with favorable tax conditions, simplified administrative procedures and a number of other privileges
- Special investment contract - platform for cooperation between the public party (federal, regional and municipal authorities) and investors (*Ministry of industry and trade Russia*)

The table below illustrates some of the main measures within chosen support mechanisms.
The use of these support mechanisms was based on the evaluation of the effectiveness of the "Afrikandskiy" project. The basic data for the calculations were obtained according to key recommendations represented by Kola Science Centre Russian Academy of Science within formation of Integrated chemical and metallurgical complex for the production of titanium dioxide, rare and rare earth metals (*Integrated chemical and metallurgical complex for the production of titanium dioxide, rare and rare earth metals. Afrikandskiy project* 2016). The baseline scenario was calculated without application of any support measures. According to the baseline scenario, implementation of the Afrikandskiy project is not effective now that makes relevant the discussion of possibilities of using modern support measures. The calculation of economic efficiency of this project with application support mechanisms was made with using measures represented in Table 1.

The results obtained from the comparative analysis of applied mechanisms are presented in Table 2.

Table 1. Characteristic of support mechanisms *(Tax measures of state support for investors* 2017*)*.

Name of mechanism	Measures (based on mechanisms, that offered in the Murmansk region)
Regional investment project (RIP)	- The federal part of profits tax rate for a participant of a Regional investment project is 0%; - The regional part of profits tax rate for a participant cannot be more than 10% during first 5 years - Property tax reduction
Advanced Development Territories	- The federal part of profits tax rate is 0% during first 5 years - The regional part of profits tax rate for a participant is no more than 5% during first 5 years from the moment of receiving profit, no more than 12% in the next 5 years - Exemption from property tax and land tax - Reduction in the rate for insurance premiums to 7.6% - The possibility of attracting federal and regional funding for the preparation of a production site (up to 80% of the cost of infrastructure development)
Special investment contract	- The federal part of profits tax rate is 0% during first 5 years - The regional part of profits tax rate for a participant is 12,5% during first 5 years - Property tax reduction

Table 2. The results of comparative analysis of applied measures (Afrikandskiy project).

Indicator	Baseline scenario	Regional investment project	Advanced Development Territories	Special investment contract
NPV, mln rub	-1711.62	203.23	2180.66	-31.07
IRR, %	11.30	12.64	14.05	12.48
PI	0.91	1.01	1.11	1.00
PP, years	>20	20	16	>20

According to the obtained results, the greatest effect for subsoil users is formed due to use support measures within the framework of the tax-legal regime of Advanced Development Territories – NPV = 2180.66 million rubles, payback period -16 years.

Regional investment project's support measures also have a positive effect on the final financial results of the project, thereby increasing its commercial efficiency and investment attractiveness. Due to applying support measures of Special Investment Contract (SPIC), the overall effectiveness of the project remained negative (-31.07 million rubles).

In general, it is necessary to form the integrated approach to stimulate the implementation of CUMR projects. It is important not only to provide financial resources, but also to have an effective tax policy mechanism and to develop forms of interaction between private and public partners. Therefore, it should be offered administrative, financial and budgetary mechanisms – Figure 2.

Modern approaches to environmental management should be based on the principles of consistency and comprehensiveness, which implies the creation of a unified system of inter-action between the technological, industrial and economic spheres in the field of solving environmental problems. The forecasting of future needs in CUMR technologies and setting strategic priorities in this sphere can be carried out with using of mechanisms of foresight-research (Popper 2017).

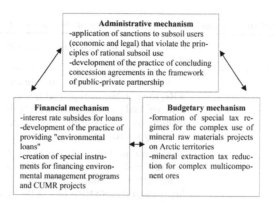

Figure 2. Effects of implementation of CUMR projects.

The development of scientific and technological bases in the CUMR direction should be realized due to use cluster approach. For example, Afrikandskiy project can be included in Kola Chemical Technology Cluster, designed to ensure the strategic sectors of the Russian economy by "critical materials" (Kalinnikov 2014).

It is also important to discuss the possibility of applying mechanisms of "green economy". They include such instruments as introduction of sustainable production and consumption models, promotion of business development in the field of "green" technologies and formation of a sustainable innovation infrastructure (Reilyy 2012).

5 DISCUSSION

The findings of this study suggest that the complex use of mineral raw materials is significant direction within environmental management.

The formed criteria can become a basis for the creation of a list of high-priority CUMR projects of Arctic territories in order to clarification of the "Strategy of development of mineral resources base of the Russian Federation till 2035".

The results of this study indicate that there are state mechanisms, which usage allows increasing the economic efficiency of CUMR projects. According to the results obtained, the greatest effect for subsoil users is formed due to use support measures within the framework of the tax-legal regime of Advanced Development Territories. At the same time, there is a necessity of creation special regimes to support such projects owning to their features. New mechanisms should be formed according to combination of administrative, financial and budgetary instruments. One of the key issue remains the possibility of severance tax rate reductions for complex multicomponent ores. Measures proposed aim to capacity-building within the CUMR direction. The use of economic support mechanisms allows increasing of investment attractiveness of projects of complex use of mineral raw materials and achieving important effects in innovative, social and environmental spheres.

Such modern mechanisms as foresight-research, instruments of "green economy" and the cluster approach can complement the system of interaction between the technological, industrial and economic spheres in the field of solving environmental problems.

The findings of this study have a number of important implications for future practice.

6 CONCLUSION

The Arctic region of the Russian Federation is important for the whole country because of its unique mineral raw material base. Efficient development of the resource potential of the Arctic zone is one of the strategic priority for our country. However, now there are such trends as the growth of ecological risks, gradual depletion of mineral and raw materials base et al. These leads to necessity of meeting the important challenges within environmental management. That is why the relevance of the problem of complex use of mineral raw materials is obvious.

This study has shown that the CUMR projects focus on obtaining long-term macroeconomic, geopolitical, innovative, social and environmental effects. The implementation of such projects allows both to increase in investment of regional economy and reduce existing environmental risks. The investigation has identified the features of projects of complex use of mineral raw materials and criteria for assessing prospects of these projects.

The investigation of the state support mechanisms and assessment of possibility of their applying, have shown that the use of measures of Advanced Development Territories and Special in-vestment contract allows to increase the economic efficiency and investment attractiveness of CUMR projects.

The study has made also a number of recommendations aimed at promotion of implementation of complex use of mineral raw materials projects. They include administrative, budgetary, financial measures and such modern mechanisms as foresight-research, instruments of "green economy" and the cluster approach. The results achieved will form the basis for creation of managerial and organizational arrangement within development of the complex use of mineral raw materials concept on Arctic territories.

REFERENCES

Boadway, R. 2000. The economic evaluation of projects. Kingston: Queen's University.

Bolsunovskaya Y.A. & Bolsunovskaya L.M. 2015. Ecological risk analysis as a key factor in environ-mental safety system development in the Arctic region of the Russian Federation *IOP Conf. Series: Earth and Environmental Science* 1 (24): 012003.

Integrated chemical and metallurgical complex for the production of titanium dioxide, rare and rare earth metals. Afrikandskiy project. 2016. viewed 13.04.2019. http://www.osatom.ru/mediafiles/u/files/News/Atomexpo%202016/2016.05.30_01_Afrikanda_Trenin.pdf.

Kalinnikov, V.T., Nikolaev, A.I. & Gerasimova, L.G. 2014. Kola chemical-technological cluster for solving the problems of economy and ecology of the Russian Arctic *North and the market: forming the economic order* 3 (40): 21-24.

Kaplunov, D.R.& Radchenko, D.N. 2011. Justification of the full cycle of integrated development of mineral resources in the development of solid mineral deposits Mining informational and analytical bulletin (scientific and technical journal) 447-455.

Krapivin, V.F., Varotsos, C.A. & Soldatov, V.Y. 2015. The Arctic Environmental Problems In: New Ecoinformatics Tools in Environmental Science *Environmental Earth Sciences* Springer, Cham.

Larichkin, F.D. 2008. Mineral Resources in the Russian Economy *Journal of Mining Institute* 179: 9-13.

Larichkin, F.D. 2011. Evolution and formation of the modern paradigm (model) complex use of mineral raw materials *Herald of the Kola Science Centre of the Russian Academy of Sciences* 4 (11): 8-14.

Larichkin, F.D. 2011.Problems of study and rational development of mineral resources of the North and the Arctic *Kola Scientific Center of the Russian Academy of Sciences* 15-23.

Larichkin, F.D., Cherepovitsyn, A.E. & Fadeev, A.M. 2012. Problems of study and development of mineral resources of the Arctic region *Arctic: Ecology and Economy* 1 (5): 8-15.

Lobanov, N.Y. & Noskov, V.A. 2013. Economic efficiency of complex use of mineral deposits *Journal of Mining Institute* 201: 59-63.

Marinina, O.A. & Vasiliev, Y.N. 2009. Economic efficacy estimate of minerals multiple use for the purpose of efficient nature management *Journal of Mining Institute* 184: 235-239.

Methodological Recommendations on the Integrated Study of Deposits and Assessment of Reserves of Associated Minerals and Components: Recommended by the Protocol of Russia's Ministry of Natural Resources of April 03 2007 11- 17/0044-pr.

Mining Project Evaluation. SRK Conculting's International Newsletter. viewed 20.04.2019. http://www.srk.co.uk/sites/default/files/SRKNews52_MiningProjectEvaluation.pdf.

Ministry of industry and trade Russia. viewed 02.04.2019. http://minpromtorg.gov.ru/en/public_ministry/open_data/.

Mirzoyeva, A.R. 2013. Evaluation of the economic efficiency of the integrated use of raw materials *Economic analysis: theory and practice* 3 (336): 51-60.

Nevskaya, M.A. & Ligotsky D.N. 2013. Organizational and economic problems of rational subsoil use and ways to solve them in modern conditions *Journal of Mining Institute* 201: 51-60.

Nevskaya, M.A. & Marinina O.A. 2017. Stimulating innovative transformations for the sustainable development of Russia's mineral and raw materials sector *Naukovedenie.* 6: 1-13.

Popper, R. 2007. Methodology: Common Foresight Practices & Tools *International Handbook on Foresight and Science Policy: Theory and Practice.* Edward Elgar.

Reilyy, J. 2012. Green growth and the efficient use of natural resources *Energy Economics* 34: 585-593.

Smulders, S., Toman, C. & Withagen C. 2014. Growth theory and 'green growth' *Oxford Re-view of Economics Policy* 3 (30): 423-446.

Solovyova, E.A. 2008. Methodical foundations and principles for the formation of incentive taxation for subsoil use *Journal of Mining Institute* 179: 14-19.

Tax measures of state support for investors. KPMG-2017. viewed 28.04.2019. https://www.messefrankfurt.com/content/dam/automechanikarussia/TIAF/Presentations/TIAF2017/KPMG_Bourliand.pdf

Trubetskoy, K.N., Galchenko Y.P., Kalabin G.V. & A.N. Poshlyakov. 2015. Geotechnological paradigm of development of integrated mineral resources development in the Arctic zone of Russia *Arctic: ecology and economy* 3 (19): 54-65.

Vinogradov, AN, Glushchenko, Y.G. 2011. Mineral and Raw Material Potential of the North-West and Problems of its Rational Use *Journal of Mining Institute* 191: 107-112.

Vukolov, A.N. 2016. Problems and prospects of the use of technogenic mineral raw materials *International research journal* 6 (48): 130-131.

Yashalova, N.N. 2011. Organization of stimulation of investment environmental protection activities in the region *Regional Economics: Theory and Practice* 37 (220): 66-67.

Zukerman, V.A. 1995. Problems of the complex use of the mineral raw materials of the Kola peninsula *Mining informational and analytical bulletin (scientific and technical journal)* 181-188.

Topical Issues of Rational Use of Natural Resources 2019 – Litvinenko (Ed)
© 2020 Taylor & Francis Group, London, ISBN 978-0-367-85720-2

Comprehensive overview of Lean Management tools used in the project to build a miniature production line „AGH Lean Line"

K. Styk & P. Bogacz
AGH University of Science and Technology, Krakow, Poland

ABSTRACT: The world is changing, modifications, innovations and improvements are constantly progressing. The approach to organisation management is also changing. Nowadays, classic business and organization management methodologies are being abandoned and more and more often modern concepts are being implemented. The implementation of such a new system in a production company is not an easy task. Very often there is resistance and misunderstanding on the part of production workers, but also of workers at higher levels. Sponsored by currently one of the most popular methodologies implemented in manufacturing companies and gaining popularity in non-production environments is Lean Management. It is a concept that has developed on the basis of TPS (Toyota Production System) principles and tools. However, introducing changes in the functioning and approach of employees is extremely complicated, it requires understanding of the philosophy of this methodology, getting to know the basic principles and developing the habit of self-discipline and self-improvement. These aspects are all the more difficult to achieve, the longer the life of the unit and (statistically) the older the staff. Student Scientific Circle Management functioning at the Faculty of Mining and Geoengineering AGH University of Science and Technology in Kraków (Poland), recognizing the need to educate and understand the basics of Lean Management methodology, with the support of its Alma Mater decided to build a tool allowing for effective learning of elementary LM methods. The project „AGH LeanLine"; consists in building a comprehensive Lean Manufacturing system based on RFID and Andon technology, as well as Lego Mindstorms, which is another development element in the Quality and Production Engineering Laboratory (Lean Lab) of the Faculty of Mining and Geoengineering of AGH University of Science and Technology. This article describes the conditions and diagram of the process of building the AGH LeanLine project and a description of individual methods used in the simulation game of production of a specific finished product (including Value Stream Mapping, 5S, PokaYoke).

1 INTRODUCTION

The civilization of our world is changing, what happens thanks to modern technologies and innovative solutions. The approach to the management of organisations is also changing. Today there is a shift away from classic management methodologies based on a typically structural, linear and later behavioural approach, towards methodologies with higher levels of flexibility, but also for learners. One of them is the Lean Manufacturing methodology, which since the turn of the 20th and 21st century has revolutionized thinking about the management of a production company. This is the world's most widely used production management concept (Fertsch et al. 2003). Its implementation is carried out in various industries, such as automotive, FMCG, household appliances or light industry.

Students of the AGH University of Science and Technology in Cracow, who are also members of the Student Research Group „Management" (hereinafter referred to as SRGM AGH), they participate in technological trips every year, visiting several European companies from ous variindustries every year. During their visits they gain the opportunity to observe the current situation of production companies and current trends in the area of production management. The observations made by both students and university lecturers can be summarised in the following sentence - in order to keep the company in good shape nowadays, to achieve strategic and operational goals, as well as to catch up with the ubiquitous competition, it is necessary to constantly introduce changes and improve processes, regardless of whether it is a large corporation or a small local entrepreneur. This is one of the main opportunities for building and implementing key success factors. Lean Manufacturing meets these challenges.

Sponsored by Despite the wide application of Lean Manufacturing, specialists point to the lack of comprehensive and modern education in flexible, change oriented production and quality management, claiming that the recruitment of employees with appropriate qualifications is extremely complicated. According to the MotoBarometer survey conducted in 2018 by Exact Systems, 2/3 of automotive companies in Poland indicate a lack of personnel with a major education in production and quality management and, as a result, an insufficient number of candidates joining the recruitment process. This phenomenon generates demand for experienced specialists in the field of process optimization. This is also confirmed by the data presented on the platforms collecting job advertisements. Statistics of 30. 09. 2018 shows that on that day as many as 9805 active job offers were available in Poland in the scope of Lean Manufacturing and derivatives (e. g.: optimization engineer, production engineer, process engineer).

At the same time, a situation is observed in the market in which young people completing their studies have a difficult beginning of their professional career due to a lack of practical experience. This is confirmed by the research conducted by the Polish Central Statistical Office on the labour market in Poland. The lack of experience itself may result not only from too short practices in a given industry (obligatory practices during education in engineering faculties) during studies, but also from lack of developed skills of practical action. The structure of the school and university education system is based on traditional forms, such as lectures or auditorial exercises, during which a large number of theories are passed on. This is not sufficient for the beneficiary, as it does not involve attention and does not favour the memorisation of information and does not allow for its processing. Workshops and laboratories that allow participants to get involved and use theoretical content in practice are devoted far too little time - about 2/3 less time compared to traditional forms.

For four years now, the Student's Research Group „Management" (SRGM) has been carrying out projects aimed at filling in the gaps and deficits mentioned above. They are implemented on the basis of Lean Manufacturing methods (hereinafter referred to as LM interchangeably), which helps beneficiaries to understand complex issues in an attractive and accessible way. One of them is the educational workshop „Minute to the effect", which implements the methods and tools of the Toyota Production System to the everyday school and home activities of students. Following the concept of continuous improvement, the members of the SRGM take up new challenges in subsequent editions of the project (2019 is the fifth edition of the project), with each stage more complicated and complex. All changes are aimed at introducing new tools and methods of their implementation, constantly improving the level of education in the area of process optimization. The following article presents the AGH University of Science and Technology LeanLine as a modern, effective way of teaching Lean Manufacturing, comprehensive but also attractive from the point of view of the didactic tools used.

2 DEVELOPMENT OF MODERN MANAGEMENT METHODS

2.1 The template file

Although management and organization began to function as separate fields of science only at the beginning of the 20th century, they have existed in an unstructured way since ancient times.

Classical management methods are intuitive. The key processes, functions and managerial skills they indicate are still valid and used where appropriate. However, they have not been sufficient for several decades now, being once created for small organizations, with a low level of complexity and not having to change so often due to the requirements of the environment. Nowadays, individuals are often extremely complex and constantly changing. The concepts proposed in the classical approach were created as universal and nowadays they do not fit into specific types of organizations. The research carried out at that time focused on the whole enterprise, without noticing the important role of the entity which now plays a key role in the functioning of the organization. For man is a resource, not a tool of work.

Behavioural approach, created in the 1930s, contributed to the development of the field of employee motivation and the arrangement of interpersonal processes in organizations. The employee was no longer one of the modes of the big machine, but became an important determinant of the development of enterprises. It should be remembered, however, that the human factor is too unpredictable to be used on a large scale, given the high complexity of processes and the complexity of the organization's functioning.

Using quantitative management since the 1940s, managers have gained additional methods, techniques and procedures that have supported the decision-making process. They have helped to better understand the complexity of the processes, in particular those related to planning and subsequent control. However, the models included in this theory are subject to a certain risk because they do not take into account individual behaviours and attitudes. A variable may differ significantly from reality or be unfeasible, causing a disruption of the result.

The behavioural, classical and quantitative theories presented above are theoretically mutually exclusive. However, from a broader perspective, they can complement each other. They need a common denominator, which is the so-called integrating management, functioning in the form of two approaches: systemic and situational (Griffin et al. 2017, Janik et al. 2018). The system approach is to treat an organisation as a system that is a related set of elements functioning as a whole. The second integrating approach is the situational approach, which says that universal theories are not applicable, because every organization is unique (Szafrańska et al. 2011).

At the end of the 20th century, due to the dynamic development of the organization, the interest in management theory and practice increased significantly. Research into new management solutions has therefore been launched. They were often born from the bottom up, at the employee level, and then perceived as accurate, developed, refined and developed in a comprehensive manner. Figure 1 shows a summary of management methods used in the theory and practice of enterprises today.

The concepts, methods and tools are briefly described below

1) Benchmarking - is a process of systematic comparison of your company with another organisation operating in the same field, which is the best on a given scale (Szafrańska et al. 2011).

2) Controlling - has an extremely wide range of terms and definitions. Looking for the simplest definition one can say that controlling is a system of strategic and operational support for the management of the entire enterprise (computer-aided) (Bieńskowska, Kral et al. 2011).

3) Customer Relationship Management - is treated as a new approach to marketing issues (using one-to-one contacts), because on this level the most important for the company is to acquire and maintain a valuable customer (Ciurla et al. 2011).

4) Integrated IT systems supporting ERP management - in simple terms, ERP systems - are „comprehensive, standard, integrated software packages supporting a wide range of business processes carried out in a given organization" (Kandora et al. 2011).

5) Lean Manufacturing, i. e. lean management - a methodology based on lean management of organization's resources and lean production. (Aleksandrowicz et al. 2016).

6) Outsourcing - separating individual processes, tasks and functions to an external contractor (Małkus et al. 2018).

7) Business Process Reengineering - or reengineering - is another of the management concepts based on „fundamental rethinking and radical re-design of company processes, leading to dramatic (breakthrough) improvement". (Hopej, Kral et al. 2011).

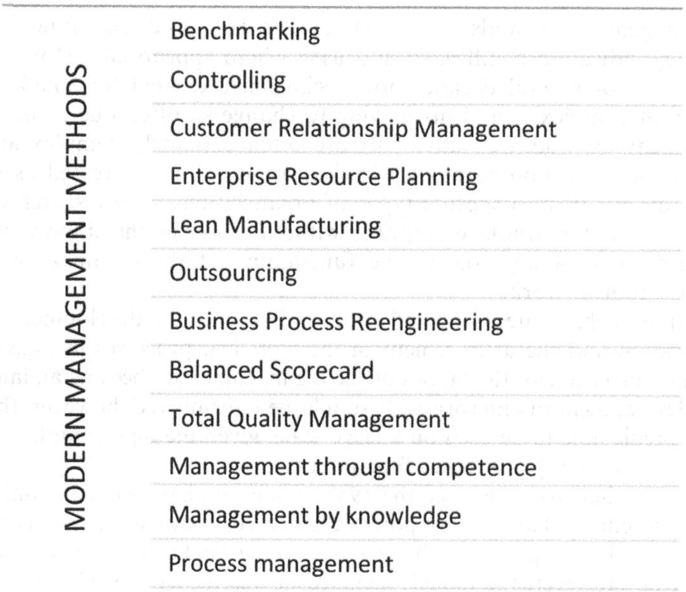

MODERN MANAGEMENT METHODS

Benchmarking

Controlling

Customer Relationship Management

Enterprise Resource Planning

Lean Manufacturing

Outsourcing

Business Process Reengineering

Balanced Scorecard

Total Quality Management

Management through competence

Management by knowledge

Process management

Figure 1. Summary of modern management methods Source: own calculations based on (Hopej, Kral et al. 2011).

8) Balanced Scorecard - a tool for presentation and detailing the strategy of the organization, de facto a document, designed to highlight and focus on the key processes of the company's functioning, resulting directly from the mission and vision of the organization (Hopej, Kamiński et al. 2011).

9) Total Quality Management (TQM) - is a pro-quality approach to the way of management in the company, in all individual tasks. Its primary objective is to achieve long-term and lasting benefits for both the company and the customer, employees and the environment (Centrum Jakości et al. 2018).

10) Management through competence - is one of the elements of human resources management. It is a management method aimed at defining the competence needs of the organization and such an arrangement of employees in the processes that specific competences are available in a given place at the right time (Bieńkowska, Brol et al. 2011).

11) Management by knowledge - is related to the knowledge possessed by the enterprise (know-how); it is characterized by terminological inconsistency, but it should always be cyclical and systematized, so that it brings certain benefits to the organization. Its main objective is to focus on the processes of acquiring, transferring and using knowledge. Knowledge management should be closely linked to the business processes of a given unit and use available technology to provide support for particular processes (Kłak et al. 2010).

12) Process management - a set of activities, activities performed sequentially, with defined inputs, which aim at generating a specific result, i. e. a product or service with a value level defined by the customer (Bitkowska et al. 2013). The main objectives of this methodology include, among others, improvement of processes in the organization through their control, elimination of restrictions between groups, adaptation of business processes to strategic plans and customer requirements, improvement of the organization's effectiveness (Korsan-Przywara et al. 2013).

3 METHODS LEAN MANUFACTURING

Lean Manufacturing, called lean management, is a methodology based on creating added value for the customer, common sense use of organizational resources and effective production management. Her mother's roots and working culture are in the Toyota manufacturing plant in Japan. Sakichi Toyoda has developed a unique system that combines the advantages of mass production and craftsmanship in order to match the size and efficiency of production to American concerns, referring to completely different products and conditions of production. This is how Toyota Production Systems came into being, a methodology based on eliminating waste and unnecessary processes, while maintaining the best quality of products and services and low costs, called Toyota Production Systems (Aleksandrowicz et al. 2016, Chyła et al. 2018, Lean enterprise Institute Polska et al. 2018).

To start implementing the Lean Manufacturing methodology, to understand the philosophy and principles of the Toyota Production System (TPS). Only when you understand its meaning and goals can you start working with basic methods such as 5S, Total Productive Maintenance (TPM), One Point Lesson (OPL) and production levelling with Yamazumi, Flow Chart, Value Stream Mapping (VSM) (Faron et al. 2011).

All TPS elements have been formed into a Toyota House, which is a collection of methods and tools for lean business management. Its foundation is standardization, and earlier understanding the concept of added value in processes (VA) and 5S - maintaining order in the workplace, which directly affects the workers themselves. An important element is also TPM, responsible for machine management and the initial part of the Kaizen system, indicating the role of the employee and the possibility of his participation in the process. All these factors of production try to standardize in the foundation of Lean Manufacturing. The first pillar is represented by the Just-in-time concept, which consists mainly in eliminating waste from processes and delivering a specific product to the right place at the right time. This pillar uses such tools as One Piece Flow and SMED, measures the cycle and tact time to get information about meeting customer needs, introduces Kanban to manage the flow of information, and then Heijunka and the Dairy Way. The second pillar of the Toyota House is the Yiddock, i. e. the quality built into the workstation. At this stage, quality management tools are used, such as 5Why, Poka Yoke, data consolidation, SPC and MSA, or Training Within Industry (TWI). The heart of the Toyota House is continuous improvement, based on the development of Kaizen systematics. All these elements, worked out in small steps and maintained at a high level, allow us to achieve the construction of a roof, which is identified with the objectives of the highest quality, at the lowest cost, in the shortest time, with maximum safety, while maintaining a positive morale among employees and customer (Bogacz et al. 2018, Lean enterprise Institute Polska et al. 2018).

According to Lean Enterprise Institute Polska, the introduction of Lean Manufacturing methodology allows for a real improvement of many production indicators (results presented in 2010), such as (Lean enterprise Institute Polska et al. 2018):
• performance ratio - increase of 66%,
• machine utilisation rate (OEE) - an increase of 59%,
• inventory ratio of work in progress - 80% reduction,
• availability of space for production - increase by 61%,
• time index from raw material to finished product (Lead Time) - reduction by 70%,
• changeover time index - reduction by 96%.

In addition, it is not only production indicators that are improving, because scientists in research show that the morale of employees, who are more committed and praise more efficient communication in the organization, is also changing. It should not be overlooked that companies implementing Lean Manufacturing record increased competitiveness on the market and improved quality of offered products, adjusting them to current consumer needs (Faron et al. 2010, Lean enterprise Institute Polska et al. 2018). All this makes Lean Manufacturing the most popular production and quality management methodology in the world.

Looking at the many benefits of the Japanese corporate governance concept, it seems that these are quite simple and common sense rules that are very easy to implement. This is somewhat true, however, many companies selectively use particular methods and tools, reducing the whole concept to occasional projects. In such cases, the implementation of the methodology is also associated with drawbacks. A concept that is simple and bottom-up, if misinterpreted, can transform into rationalisation, causing stress and loss of motivation for employees. As a result of the improvements made and the reduction in operation times, there is a need to lay off workers with lower qualifications. Therefore, it should be remembered that the process of Lean Manufacturing implementation is a long-term one, requiring a common approach of the entire organization and the consequences of the action (Chyła et al. 2018, Faron et al. 2010, Lean enterprise Institute Polska et al. 2018).

Lean Management has developed based on the principles, methods and tools of the Toyota Production System (TPS), which are described below (Dietrich et al. 2018, Hamrol et al. 2002 Zapłata et al. 2009).

3.1 *Fundamentals of lean manufacturing*

This section briefly presents the essence of the various tools that underpin the foundations of the Toyota House, which are aimed at building customer value, eliminating losses and widely understood standardization.

3.1.1 *5S – method of place management*
Workstation organization method, based on five steps: sorting, systematics, cleaning, standardization and self-discipline. 5S is the best opportunity to directly involve employees in the process of change, because it is conducted at every workplace and it forces to maintain order in accordance with established standards.

3.1.2 *TPM – method of machinery and equipment management*
It is a method used to conduct proper maintenance of machines, based on the construction of a solid company by maximizing the use of the production system. It provides a stable system for preventing loss of production and eliminating incompatibilities, failures and accidents.

3.1.3 *Visual management - a method of systematising communication*
It is a management method designed to present information in a very visible way to both employees and management so that the current state of the operation and future objectives are understood by everyone. Otherwise - a method of managing visual information, based on visual communication tools and the 5M method (Manpower, Machine, Materials, Methods, Measurements). Visual signs allow for quick identification of the position of tools and components, constant monitoring of the condition of machines and the quantity of materials, as well as increase work safety and ensure easy assimilation of information.

3.1.4 *Standardisation - a method for systematising work*
A standard can be defined as the most effective way to perform a given process in terms of time, cost and quality. Standardization is used to control production processes, leading to proper planning and monitoring of the performance of all processes, making it easier to manage them and seek further improvements. In the framework of standardization, Yamazumi system is very often included, which allows to level the production line in order to achieve equal occupancy of individual production positions.

3.2 *Production pillar - just in time*

In the analyzed subchapter the tools and methods included in the production pillar of the Toyota House are presented. Just in time is a system that puts the customer first - he is to receive products of the required quality and at the best time.

3.2.1 *Systematic pull - suction rather than pushing the product*
This is the principle of material flow management and information, which consists in replenishing only those stocks that have already been used. The main assumption of the system is to deliver the exact quantity, at the right place, at the right time. It consists in supplementing them only when the resources are used. The process uses a system of"sucking"; the materials (or information) that are needed in the production process. It allows to reduce the level of

stocks and minimize production batches, which in turn results in higher quality of products. The Pull Systematics is mainly driven by real demand, guaranteeing a specified level of production and thus reducing costs.

3.2.2 Systematics of supermarkets - management of production warehouses
Supermarkets are specially prepared places for temporary storage of products, where parts are available for subsequent processes. They are used between processes that have different cycle times, or in situations where there are other problems in the production processes or in the organization itself that do not allow for a continuous flow.

3.2.3 Kanban
Kanban is a visual production control system that allows you to determine what, when and how much to produce or deliver. It functions primarily as a signal, meaning the need to place an order, and on this basis to supplement the shortage of warehouse space. However, this is not the only way to use and form the Kanban system. It can function as a card or even a container.

3.2.4 Heijunka – management of intra-production logistics
Heijunka is used to level production over a defined period of time, ensuring an even distribution of manpower, materials and ergonomic movements. With its help, it is possible to effectively satisfy customer requirements while minimizing costs, maintaining the lowest possible level of inventory and low production lead time.

3.3 Quality pillar – Jidoka

Jidoka means quality built into the workplace. Its main goal is to never allow a defective product to pass to another station. Achieving embedded quality is understood as a change of approach to a problem that creates opportunities for improvement in the process.

3.3.1 Statistical Process Control (SPC), including metrology (MSA)
The SPC uses data collected directly from processes to verify deviations from the set standards. With the help of statistical analysis of the source data, the SPC gives you the opportunity to react before non-compliance occurs on the basis of studies, forecasts and trends. This is a method that examines the level of process stability. Its sub-element, also treated as an independent element, is the MSA, i. e. the process of analyzing the measurement system. It is based on Shewhart's control cards and a process capability analysis. The Shewhart control card is a graphical representation of the variation in the results of a process. The graph on the control card always has a horizontal centerline, which indicates the desired value. Depending on the product specification, it also has horizontal lines representing the upper and lower limit of results. Such a chart makes it possible to verify the stability of the process and therefore to plan preventive actions before the occurrence of discrepancies.

3.3.2 Poka Yoke
Poka-Yoke is a set of devices that do not allow the error to occur or to quickly detect it (even before the defect occurs). There are introduced regulatory (regulatory functions) or organisational (setting functions) solutions preventing mistakes and product control at the workplace (directly by the operator - this & quot; control"; is a part of the operation and does not take up additional time by the operator). Control functions include control; control methods (process stop) and warning methods (audible or light alarms). On the other hand, organizational functions include contact methods (sensors, formats), methods of set values (checking the number of movements), methods of necessary step (clocking the process), methods of the necessary step (clocking the process).

3.4 Kaizen – heart of the system

Kaizen, i. e. the concept of continuous improvement, is understood as the involvement of all employees regardless of the level, with the main focus on operational employees. It is based on a constant search for ideas to improve all areas of the organisation. The system is based on the small steps method, obtaining more effective results through continuous work and so-called small improvements than through revolutionary transformations.

4 „AGH LEAN LINE" AN INNOVATIVE TEACHING PROJECT TO TEACH LEAN MANUFACTURING METHODOLOGY

The „AGH LeanLine" project, created by the AGH University of Science and Technology (SRG Management AGH), is a comprehensive response to the needs and expectations of the final beneficiaries of the project, i. e. students, and thus the broadly understood socio-economic environment. It assumes the development of personal competences, professional qualifications, gaining experience and preparing students for entering the labour market, in its part concerning production and quality management, broadly presented from the deficit side in earlier parts of this article.

The project consists in building a comprehensive system for teaching Lean Manufacturing methods and tools on the basis of modern technologies: barcode scanners, Vorne 800XL and LEGO Mindstorms EV3 boards, but first of all on the basis of an innovative production system and approach to didactic processes. The main task carried out within the project implementation is to design and build an automated production line based on structures made of LEGO Mindstorms blocks, based on technical, process and organizational solutions used in real enterprises, used to simulate the production process and its improvement in accordance with the principles of Lean Manufacturing. The resulting structures will be used to conduct workshops based on them, during which participants will learn and use in practice knowledge about production optimization. These elements of the production line will allow for the practical application of methods and tools derived from the concept of Lean Manufacturing and for controlling the effectiveness of their use depending on various variables of the production process. The use of robots (made of LEGO Mindstorms blocks) and various sensors providing data from the production process in real time will allow to simulate the real conditions in production companies, where employees work every day not only with people, but also with machines. And with the Vorne 800XL, you can set the necessary performance indicators for your production line in real time, both in terms of productivity and financial performance. Traditional simulation games, based on subsequent attempts to improve the production process, are usually limited to the following steps:
• warehouse organisation - segregation of materials and components, designation of specific locations to improve visual identification,
• improvement of the system of production stands according to the course of the production process,
• determination of storage bays for finished semi-finished products for individual workstations,
• training of employees in operations performed on particular positions,
• development of templates, templates and instructions facilitating work on activities requiring precision,
• checking the times of each operation and trying to balance their load in order to eliminate bottlenecks that limit the group's performance.

The above tasks are practical in terms of design and optimisation of production. However, in order to better illustrate the actual conditions in manufacturing companies, it is necessary to remember about machines and equipment, which are an indispensable part of them, especially in the era of Industry 4. 0 revolution. Employees responsible for the organization of work in factories must focus not only on group work, training of employees and the visual aspect of the workplace, but also, or perhaps above all, on the optimization of the process and the elimination of restrictions resulting from e. g. incomplete working hours of machines.

5 PRINCIPLES, METHODS AND TOOLS OF LEAN MANUFACTURING IMPLEMENTED EDUCATIONALLY INTO THE PROJECT „AGH LEAN LINE"

In the project of creating a simulation of the production process (the diagram of the proposed arrangement of positions is presented in Figure 3, whose task is to educate participants in the field of principles, methods and tools of Lean Manufacturing, the following solutions have been applied (they are shown in relation to the construction of the Lean Manufacturing House, presented in Chapter 2).

In our simulation, the product flow will take place according to the presented scheme „from left to right", it will include positions such as: warehouseman, logistics specialist, employee and customer.

5.1 Lean Manufacturing foundaments

5.1.1 5S – method of place management
In case of a project, the construction of each production station will be based on the 5S method. The participants of the simulation will receive a set of tools enabling the implementation of the method, e. g. in the form of shadow tables for production orders (examples of such tools are presented in Figure 3).

5.1.2 TPM – method of machinery and equipment management
In the simulation described above, in the next stage of the project, robots will be used to automate the production line. These are machines and equipment, so it will be necessary to use the

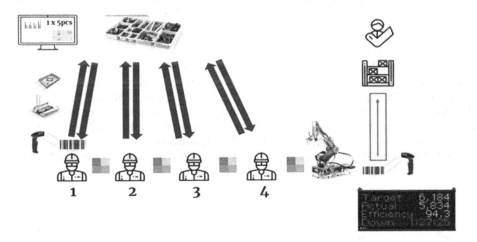

Figure 2.   Diagram of the proposed composition of the production site Source: AGH Lean Line project.

Figure 3.   On the left container for the segregation of LEGO brake pads in a raw material warehouse; on the right containers for raw materials transported by trolleys Source: AGH Lean Line.

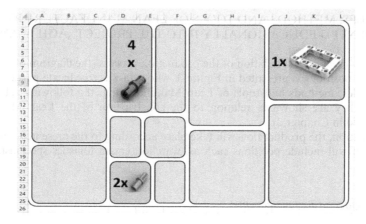

Figure 4. A proposal for a sheet that automatically visualizes the amount of required materials and their location. Source: AGH Lean Line project.

TPM method. Three types of cards will be built: autonomous maintenance for robots, EWO and A3 for possible failures.

5.1.3 *Visual management - a method of systematising communication*
Visual Management will be used to speed up the decision-making process in the project. This will be done mainly by introducing visual signs such as: arrows and descriptions, Andon whiteboard, instructions, kanban system. The selected VM tools to be used in the project are shown in Figure 4.

5.1.4 *Standardisation - a method for systematising work*
Standardisation will be based primarily on the creation of work instructions (activity instructions), containing a list and manner of performing activities, work instructions (containing a description of the organisation of a given position) and a product template sheet (containing a description and specification of a given product). These instructions will be developed based on observations and measurements made during the simulated production process. As a result of data collection and analysis, participants and/or organisers of the simulation will be able to work out optimal parameters for the given process. An example of a product benchmark card is shown in Figure 5.

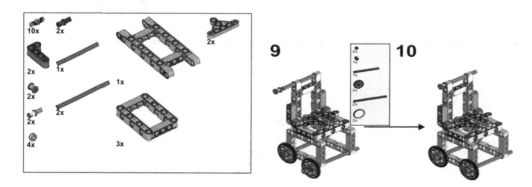

Figure 5. On the left - BOM card project - demand for raw materials for 1 piece of finished product; on the right - Extract from the product card for the production workstation Source: AGH Lean Line.

5.2 Production pillar – just in time

5.2.1 Systematic pull - suction rather than pushing the product
In the presented project, the production will be based on the pull system. The main step allowing the introduction of the above mentioned system is the use of supermarkets and the determination of the maximum number of pieces at individual production stations. As well as the application of the following methods to produce a suction production process.

5.2.2 Systematics of supermarkets - management of production warehouses
In the case of the presented project, supermarkets will be used as dedicated storage areas for specific variants of finished products.

5.2.3 Kanban
The project will use electronic Kanban cards. An example of such a card is presented in Figure 6. This card contains the code of the required material and its quantity, the weight of the order and the order number to be processed.

5.2.4 Heijunka – management of intra-production logistics
Heijunka jest is a kind of Kanban box, which allows us to see the flow of production throughout the process. An example of a structure is shown in Figure 7.

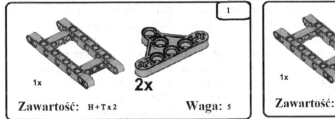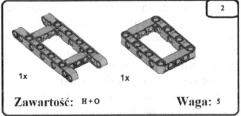

Figure 6. Examples of material order cards Source: AGH Lean Line project.

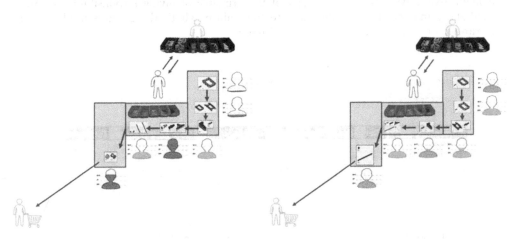

Figure 7. Visualisation of the analysis of the measurement of time of operations at workstations - production levelling. The upper diagram shows the loads before the introduction of the system, the bottom one after the introduction of the system Source: AGH Lean Line project.

5.3 Quality pillar – Jidoka

The project assumes, in accordance with the principles of Lean Manufacturing, the possibility of committing quality errors, while delegating responsibility for them to the employees.

5.3.1 Statistical process control (SPC), including metrology (MSA)

In the presented project, the role of control cards will be played by specially prepared Excel sheets (author's EPOK system, not described in this article), where data from the process will be collected, among others, about the quality or time obtained on individual activities. Then the data will be analyzed in order to develop standards for the activity and the whole process. Visualization of a sample EPOK spreadsheet is shown in Figure 8.

5.3.2 Poka Yoke

The project will provide tools and materials for the use of this method, but concrete solutions and combinations will have to be developed by the workshop participants, with the possibility for them to use them, of course. Poka Yoke will create the opportunity to reduce or prevent defects, which will result in high quality at every stage of the process. An example of using this tool is the introduction of formats limiting the transition of an element from a position to a position, which will prevent the erroneous product from moving to a further position or the application of instructions and separating the appropriate (exact) amount of raw material for a given position (only as many blocks as are part of a single finished product).

5.4 Kaizen – heart of the system

In the project, the system is implemented by leaving the participants the possibility to analyze and improve the current state of the processes. This enables us to effectively develop innovative solutions and methods of their implementation using the available materials and tools. The simulation will be designed in such a way that there are quality errors that can be eliminated after improvements are made. However, the participants themselves have to work on improvements. They will not receive instructions or ready-made solutions in the form of manuals. The participants of the simulation will be different groups of people, having different knowledge, creativity and approach to problems, therefore the worked out solutions will be different from each other, and as a result will give different solutions. The layout of the workshop room will also change, as will the distribution of positions depending on the ideas developed by individual groups, and therefore we cannot plan the simulation proces „rigidly", because it is simply impossible and ineffective, especially in the context of the principle of giving responsibility to operational staff, and therefore, in this case, to the trainee. That's the quintessential Kaizen. An example of a Kaizen card is shown in Figure 9.

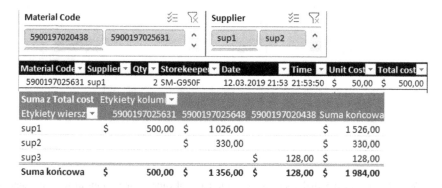

Figure 8. Visualization of a sample EPOK spreadsheet from Raw Materials Source: AGH LeanLine project.

KAIZEN BASE

Team name: _____

LP.	KAIZEN/ IMPROVEMENT	SIMULATION NUMBER	PERFORMANCE REVIEW (1-5)	NOTES
1	Job training of employees			
2	Development of a system for efficient detection and repair of quality errors			
3	Study and analysis of the cycle time of individual positions			
4	Levelling of production			
5	Poka-Yoke solutions			
6	Planning of purchases and supplies of raw materials			
7	Determining the minimum conditions needed for proper production flow - eliminating time losses and bottlenecks			
8	Workplace organisation and visual management - Creation of shelving areas for WiP and finished products			
9	Organisation of the in-house warehouse (Collection of containers)			
10	Improving the way in which raw material purchase orders are collected and realised			
11				
12				

SCALE MARKINGS: 1-VERY SLIGHT; 2-POOR; 3-MEDIUM; 4-FINE GOOD; 5-VERY GOOD

Figure 9. Kaizen database containing a list of possible improvements, simulation number, evaluation of effectiveness and comments. (The list can be extended with suggestions from participants) Source: Oleanpiada AGH project.

5.5 *Analysis and monitoring of the AGH LeanLine production process*

In the project, thanks to the use of Vorne 800XL and EPOK system, it is possible to collect information on the flow from individual production stations, which, after proper analysis, will allow for the transfer of activities, tasks or materials between individual stations, so that the load is even (removal of bottlenecks).

Figure 10 shows the layout of the stands during the first round of the simulation, when the simplest and seemingly most logical layout of the room during the simulation and the production of the entire product on individual stands is used. The logistician (7) takes the raw materials from the external warehouse (right, top position), then delivers them to the internal warehouse, from where the appropriate raw materials are collected directly to the workstations. After the production of the product is finished by the stations, they are transferred to the last station, from where they are transferred to the customer. Logist (7) after receiving the next order takes the raw materials from the external warehouse again - the next round starts.

Figure 11 is a presentation of a simulation after changing the arrangement of workstations on the production line - this time we do not carry out unit production at individual

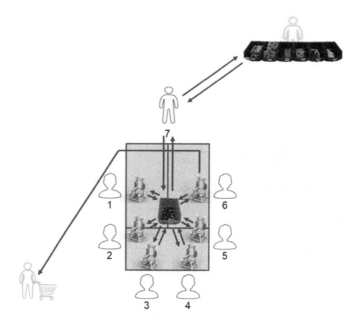

Figure 10.   Organization of stands during the simulation - unit production Source: AGH Lean Line.

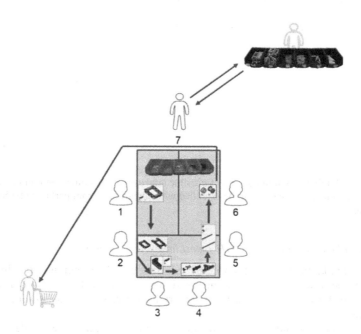

Figure 11.   Organization of stands during the simulation - product flow Source: AGH Lean Line project.

workstations, but we introduce streamlining - - product flow through the process and special-ization of workstations (each of the 1-6 employees has a specific activity to perform). The logistician (7) takes the raw materials from the external warehouse, delivers them to the internal warehouse located centrally to the entire production line, from where the raw

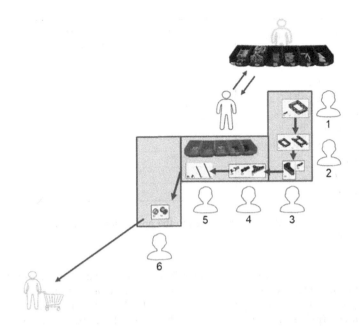

Figure 12.   Room organization during simulation - linear product flow Source: AGH Lean Line Project.

materials are then distributed to the individual workstations. The production cycle begins at position 1 and ends at position 6, which then passes the product to the customer.

The introduction of a solution for specialization of workstations allows to shorten the production time thanks to faster performance of particular activities. It also allows to implement further LM tools, methods and rules, e. g. 5S, VM or workstation instructions.

Another improvement of the product flow through the process is presented in Figure 12 - this time there has been a change in the way the arrangement of workstations has been organised - a linear arrangement of workstations has been applied and the location of the internal warehouse has been relocated. This shortened the communication routes between the external warehouse and logistics; logistics and production stands; the last stand and the customer.

The use of this modernized system allows for the introduction of further LM tools and methods, e. g. pull system, kanban or heijunka.

Introducing changes and improvements during the production process has a measurable effect in the form of costs. During each simulation round, data on a given cycle is collected: sales revenue, other revenues, raw material costs, inventory costs, labour costs, transport costs and other costs. An example of a data collection sheet is presented in Figure 13. The aim is, of course, to maximise production, minimise costs and ensure the highest quality of the finished product. The presentation of the results after the simulations is completed allows the workshop participants to see the benefits of implementing tools, principles and methods of „lean production”.

6 SUMMARY

Analyzing statistics on current trends, knowledge gained during classes at AGH University of Science and Technology in the field of Production Management and Engineering and Lean Manufacturing specialization, as well as experience gained over the years during visits to industrial plants, one can clearly see tendencies to the growing demand for specialists in the field of management and optimization of production. As the Lean Manufacturing methodology is the most developing concept of process improvement in the world, it is worth noting

FINANCIAL STATEMENT	SIMULATION ROUND no …				
	Team 1	Team 2	Team 3	Team 4	Team 5
Revenues from sales					
Other revenues					
TOTAL REVENUE					
Costs of raw materials					
Labour cost					
Transport costs [to customer]					
Other costs					
TOTAL COSTS					
Profit (R-C)					

Figure 13. Sheet for counting the profit generated in the production process in one simulation Source: AGH Lean Line project.

the form of education in this direction, both for future and current employees who use it. It is through an appropriate level of awareness of the applied principles, methods and optimization tools that it is possible to develop appropriate standards of conduct leading, as a result, to the development of enterprises. Through projects such as „AGH LeanLine", employees can learn about practical ways of using theoretical knowledge and test the ideas to be implemented in reality in workshops without incurring additional costs. So it can be done effectively and without risk.

The main task of the „AGH Lean Line" project is to educate both students associated in the Student Management Scientific Circle, who must acquire the necessary knowledge to design the whole process, as well as participants of the simulation, i. e. other students and students, i. e. future employees, who must learn the definition and manner of applying individual principles, methods and tools of Lean Manufacturing in the conducted production process. SRG Management is currently working on the final version of the simulation game, where it will be possible to apply all the above mentioned elements of Lean Manufacturing. At present, elements such as electronic kanban cards, EPOK system, formats for input materials, limiting formats (Poka Yoke) have been developed. Students developed the first basic simulation to produce chairs according to the customer's order. To this first test simulation, tools such as workplace standardization, production leveling, flow optimization, layout reorganization and Lean style logistics organization were introduced. The entire training project is carried out in accordance with the principles and stages of building the Lean Manufacturing House, which ensures comprehensiveness and teaches the system of proceedings. The workshop offered by AGH University of Science and Technology students is an extremely interactive tool, which is conducive to effective learning. As we know, 29% of the population is visual, i. e. learners through image memorisation, and 37% are;best learned through direct engagement related to movement and emotions'; and 34% are listeners, i. e. learners through the assimilation of auditory stimuli (Wiadomości gazeta et al. 2019) . Current and future generations of employees are exposed to a huge amount of stimuli every day and their learning process changes - the form has to be attractive and interactive. The AGH University of Science and Technology project developed by SRG Managemant AGH has these features.

REFERENCES

Aleksandrowicz J. 2016. Tools of Lean methodology in the processes of urban public transport improvement, Buses: technology, operation, transport systems, no 12, 1726–1730.
Bieńkowska A., Brol M. W. 2011. Management by competence. In Hopej M. & Kral Z (eds), *Modern methods of management in theory and practice*. Wrocław: Politechnika Wrocławska.
Bieńkowska A., Kral Z., Zabłocka-Kluczka A. 2011. Controlling. In Hopej M. & Kral Z (eds), Modern methods of management in theory and practice. Wrocław: Politechnika Wrocławska.
Bitkowska A. 2013 Process management in modern organizations, Warszawa: Difin.

Bogacz P. 2018. Teaching materials for the "Quality Management" course, Kraków.

Centrum Jakości 2018.07.29.TQM - Total Quality Management Comprehensive management through quality, https://centrum.jakosci.pl/podstawy-jakosci,zarzadzanie-przez-jakosc.html.

Chyła D., Szybalska K. 2018.07.29.Lean management, https://mfiles.pl/pl/index.php/Lean_management.

Ciurla M., Nowak M. 2011. Customer Relatmionship Management. In Hopej M. & Kral Z (eds), *Modern methods of management in theory and practice*. Wrocław: Politechnika Wrocławska.

Dietrich M., 2018.06.06. Quality management models, http://semafor.euke.sk/zbornik2007/pdf/dietrich2.pdf.

Faron A. 2011. Lean management, in Hopej M. & Kral Z (eds), *Modern methods of management in theory and practice*. Wrocław: Politechnika Wrocławska.

Fertsch M. (eds.) 2003. Production logistics, Institute of Logistics and Warehousing, Poznań.

Griffin R.W. 2017. Fundamentals of organisation management, Wydawnictwo Naukowe PWN, Wyd. 3, Warszawa.

Hamrol A, Mantura W. 2002. Quality management. Theory and practice, Wyd. 3, Wydawnictwo Naukowe PWN, Warszawa.

Hopej M., Kamiński R.: Reengineering. In Hopej M. & Kral Z (eds), *Modern methods of management in theory and practice*. Wrocław: Politechnika Wrocławska.

Hopej M., Kral Z. 2011. Modern methods of management in theory and practice. Wrocław: Politechnika Wrocławska.

Janik A., Marzec E. 2018.07.29.Management theory. https://mfiles.pl/pl/index.php-/Teoria_zarz%C4%85dzania.

Kandora M. 2011. Enterprise Resource Planning, In Hopej M. & Kral Z (eds), *Modern methods of management in theory and practice*. Wrocław: Politechnika Wrocławska.

Kłak M. 2010. Knowledge management in a modern enterprise. Kielce: Wydawnictwo Wyższej Szkoły Ekonomii i Prawa im. prof. Edwarda Lipińskiego w Kielcach.

Korsan-Przywara P., Zgrzywa-Ziemak A. 2013. Process controlling in organization management. In *Governance and Finance* no 4.

Lean Enterprise Institute Polska. 2018.07.29.Lean on production – Lean Manufacturing, https://lean.org.pl/lean-w-produkcji/.

Małkus T., Polis P. 2018.07.29.Outsourcing, https://mfiles.pl/pl/index.php/-Outsourcing.

Szafrańska A., Szyran-Resiak A.: Benchmarking. In Hopej M. & Kral Z (eds), *Modern methods of management in theory and practice*. Wrocław: Politechnika Wrocławska.

Wiadomości gazeta.pl – portal internetowy 2019.04.13. http://wiadomosci.gazeta.pl/wiadomosci/1,156046,6506256,Wzrokowcy_czy_sluchowcy___kto_ma_lepiej_.html, dostęp na13.04.2019

Zapłata S. 2009. Quality management in the company. Evaluation and determinants of effectiveness, Wyd. 1, Wolter Kluwer SA.

Topical Issues of Rational Use of Natural Resources 2019 – Litvinenko (Ed)
© 2020 Taylor & Francis Group, London, ISBN 978-0-367-85720-2

Analysis of the project selection technique in the process of a portfolio building for oil and gas company (on the example of GPN-Development LLC)

A.O. Vasilchenko
St. Petersburg Mining University, St. Petersburg, Russia

ABSTRACT: The article is devoted to the portfolio management in an oil and gas company, taking into account the peculiarities of the oil and gas sector. The existing methods of project ranking used by company GPN-Development in the process of a portfolio building and the degree of influence of expert opinion on the results of project selection are analyzed. Based on the analysis, problems arising from the selection of projects in the oil and gas company portfolio are highlighted, and methods for their partial solution are proposed.

1 INTRODUCTION

Current market conditions place high demands on competitiveness of existing enterprises. In order to meet the requirements of a continuously changing external environment, an increasing number of organizations are beginning to use a project approach in their activities, and business of large companies with widely diversified activities is based on project portfolio management. However, portfolio management continues to evolve, which determines the high relevance of issues related to this area of project management.

With the beginning of active implementation of portfolio management in activities of companies, it became necessary to standardize it. In 2005, the Standard for Portfolio management was issued by the Project Management Institute. The Standard defines a portfolio as a "collection of projects and/or programs and other work that are grouped together to facilitate the effective management of that work to meet strategic business objectives" (Ross & Shaltry, 2005) and describes portfolio management in terms of process approach. The process of portfolio management was considered in more detail by individual researchers, whose proposed definitions, models of formation and management of a portfolio, methods for evaluating its effectiveness are analyzed and summarized in modern scientific works (Borovyh, 2016; Simionova et al. 2017). Special attention in these studies is paid to the problem of forming a portfolio of projects, which is associated with complexity of this process and imperfection of existing methods for its implementation. According to the results of a study conducted in 2000 by marketing professors at Canadian McMaster University (Cooper et al. 2000), the imperfection of methods and models for managing a portfolio of projects is one of the main causes of problems arising in the process of portfolio management. A number of works studied by the author (Perebatova et al. 2017; Solovyeva & Galtyaev, 2017; Shtefan & Ornatskij, 2015) is devoted to the analysis of methods of ranking projects in the formation of portfolio, their advantages and disadvantages, as well as possibilities of their modification. However, the author noted the lack of scientific work carried out taking into account the specifics of industries, in particular - the oil and gas industry. The study conducted by British audit and consulting company Ernst & Young (EYGM. 2016) describes the processes of portfolio management taking into account characteristics of the industry, but there are no attempts to highlight existing shortcomings of the portfolio management of oil and gas companies. In this regard, the need to study in practice oil and gas company features of selecting projects in its portfolio and to identify existing problems of portfolio management was revealed.

The author hypothesized that methods for selecting projects in a portfolio used in oil and gas companies are strongly influenced by expert opinion, which is associated with the peculiarities of the oil and gas sector and complexity of the task of forming an optimal portfolio of projects. According to the author, improvement of the existing expert-analytical methods will increase the objectivity and accuracy of obtained results, as well as the effectiveness of the process of project portfolio formation.

2 METHODS

Based on the analysis and systematization of the primary information obtained in the company GPN-Development, the author highlighted the features of the oil and gas industry, which affect requirements for the portfolio of projects of oil and gas company and the method of its formation. In the process of further research, an analytical study of the method of project prioritization used in the company GPN-Development and the degree of influence of expert assessments on its results was conducted. Based on the analysis of existing methods for ranking projects their advantages and disadvantages, the relative accuracy of obtained results, as well as the possible reasons for the difference in solutions obtained by different methods were determined empirically. At the final stage of the work, the obtained results were summarized and on their basis main problems faced by GPN-Development in the formation of the project portfolio were identified and possible directions for their partial solution were proposed.

3 RESULTS AND DISCUSSION

Portfolio management includes an assessment of the effectiveness of projects that can potentially be included in the portfolio, portfolio formation, planning the implementation of the portfolio projects, resource allocation between projects and operational portfolio management. One of the most difficult and demanding tasks within the framework of project portfolio management is portfolio formation, since the direction of further development of the company and the effectiveness of its activities depend on the decisions made at this stage. The ultimate goal of this process is to form an optimal project portfolio that maximizes targeted economic indicators and creates additional value, taking into account company's strategic goals, existing constraints on financial, human and other resources, as well as existing interdependencies between projects.

The formation of a portfolio of projects in oil and gas companies is complicated by a number of features related to the specifics of this industrial sector.

Most of the projects in the oil and gas industry are related to the development of fields, which determines the presence of a license-defined time frame to start the project, as well as the dependence of the feasibility of its implementation on the location and geological characteristics of the field, the availability of infrastructure for the transportation of hydrocarbons.

One of the important aspects that should also be taken into account when selecting oil and gas projects is the presence of industry risks that increase the likelihood of failure to achieve the project goals. Project's totals are strongly influenced by the degree of geological exploration of developed field, since the discrepancy between the value of actual reserves and the initial estimates may reduce the economic efficiency of the project. For example, in 2015, as a result of the non-confirmation of the forecasted reserves of shale gas in Poland, companies such as Marathon, Talisman, Exxon and Eni refused to invest in the project due to lower production profitability (Kostylev & Skopina, 2015). Besides, it is necessary to take into account technological features of a field, since the wrong choice of a development options can lead to a drop in field productivity, as it happened during the development of the unique Samotlor oil and gas condensate field. In addition, the actual sales revenue may differ significantly from planned values due to the volatility of exchange rates and oil prices: only in the last 5 years, oil prices have ranged from $ 34.74 to $ 112.36 per barrel. These circumstances determine high importance of the risk level of projects as one of criteria for their evaluation and selection.

The specifics of the industry also have an impact on current objectives and strategic goals of an oil and gas company, which should be achieved as a result of the implementation of its project portfolio. For example, for GPN-Development LLC, in addition to achieving high economic indicators, an important task is the realization of obligations to the parent company Gazprom Neft PJSC and the state. In this regard, one of the key strategic goals of the company until 2025, which should be implemented in the process of managing its project portfolio, is to increase the annual production volume to 100 million tons of oil equivalent.

Thus, in the process of selecting projects in the portfolio of an oil and gas company, experts need to take into account a large number of requirements for a portfolio of projects due to the specifics of the industry. Based on this, the process of forming a portfolio of projects in the company GPN-Development LLC was divided into several stages.

At the first stage, projects that do not meet strategic goals of the company are eliminated. Then, remaining projects are evaluated and prioritized on the basis of their quantitative estimates (economic indicators and production volume indicators) using economic and mathematical modeling. At the final stage, experts form a portfolio of projects with the highest priority of implementation, analyze and balance it.

The central process in this chain, on the basis of which the project portfolio is formed, is the process of prioritizing (ranking) of projects using the methods of economic and mathematical modeling, so it is advisable to conduct its more detailed study and analysis.

As the main criteria for evaluating projects using an economic-mathematical model, company's experts identified such indicators as the volume of production in 2020 and 2025 (Q_{2020}, Q_{2025}), the net present value of a project (NPV), the amount of free cash flow (FCF), the profitability index (PI), the capital expenditures of a project (CAPEX). For the selection of projects according to these criteria, a scoring method for evaluating alternatives was chosen. By summing up the project scores for the selected indicators, normalized to the specific weight of the criteria, which reflects the degree of their influence on the achievement of company's strategic goals, the multicomponent integrated indicator (MII) is formed:

$$MII = 0,1 \cdot Q_{2020} + 0,1 \cdot Q_{2025} + 0,2 \cdot NPV + 0,2 \cdot FCF + 0,2 \cdot PI + 0,2 \cdot CAPEX \quad (1)$$

The obtained MII values allow to quantify the significance of the project for the company and to determine the priority of its implementation. This method is quite simple and can be used to evaluate projects in any industry subject to a change in the composition of criteria in accordance with the specifics of a particular industry.

However, this method of ranking projects has a number of significant drawbacks. First, the formation of MII involves selection of criteria and determination of their significance based on expert assessments, which increases the likelihood of inaccurate results. For example, equivalence for the company of such economic indicators as PI and CAPEX can be questioned, since the company currently does not have a lack of financial resources, and therefore, the amount of investment in the project with sufficient profitability should not have a big impact on the final investment decision. In addition, in the MII formula, production volume indicators are assigned half the weight compared to economic indicators, which does not correlate with the current strategy of GPN-Development LLC, focused on the increasing volume of production. The task of correct evaluation of the specific weights of the criteria is also complicated by the use of interrelated indicators of economic efficiency (such as NPV, FCF, CAPEX, PI). Besides, it should be noted that since calculation of the MII uses a limited set of quantitative indicators, such financial, economic and managerial indicators as payback period, the urgency of project implementation and project riskiness at the stage of selecting components in the portfolio of GPN-Development are not taken into account.

In order to analyze the impact of possible experts' mistakes on the results of project selection, 15 current projects of GPN-Development LLC were ranked according to the available data on

Table 1. Financial and economic indicators of the projects of GPN-Development LLC.

Project	NPV, mln rub	CAPEX, mln rub	PI, unit shares	FCF, mln rub	HC production in 2020, ktoe	HC production in 2025, ktoe
Project 1	16 395	-3 124	9,38	26 075	2 994	1 411
Project 2	159 554	-35 889	6,79	312 405	14 812	13 548
Project 3	66 812	-25 543	4,08	124 632	1 994	1 787
Project 4	4 804	-3 143	3,60	3 434	242	122
Project 5	19 274	-20 437	2,11	47 520	813	666
Project 6	856	-3 882	1,32	7 666	93	86
Project 7	6 203	-11 291	1,83	23 345	481	347
Project 8	33 563	-53 849	2,04	131 514	902	4 099
Project 9	4 645	-13 763	1,61	22 630	249	1 068
Project 10	-7 977	-51 421	0,77	51 496	0	3 129
Project 11	4 427	-4 222	2,58	20 227	0	255
Project 12	14 989	-55 890	1,43	96 209	2 347	5 161
Project 13	3 981	-43 809	1,14	45 962	193	936
Project 14	-12 236	-29 404	0,20	-14 307	0	581
Project 15	-3 052	-17 592	0,61	5 057	5	318

their financial and economic indicators presented in Table 1 (names of the projects are not specified in connection with the protection of company's interests).

The calculations were carried out in the following variants:

1) according to the initial formula of the MII;
2) according to modified formula of the MII with a change in the specific weights of criteria;
3) according to modified formula of the MII with a change in the number of criteria.

During the first calculation, the projects were scored for each of the indicators (production volume in 2020 and 2025, net present value, free cash flow, profitability index, project capital expenditures), and the total value of the MII was calculated according to the initial formula.

In the second calculation, the MII formula was used with the adjustment of the specific weights of production volume and capital expenditures:

$$MII = 0,15 \cdot Q_{2020} + 0,15 \cdot Q_{2025} + 0,2 \cdot NPV + 0,2 \cdot FCF + 0,2 \cdot PI + 0,1 \cdot CAPEX \quad (2)$$

In the third calculation, free cash flow and capital expenditures were excluded from the formula of the MII simultaneously with an increase in the weights of net present value and profitability index:

$$MII = 0,15 \cdot Q_{2020} + 0,15 \cdot Q_{2025} + 0,35 \cdot NPV + 0,35 \cdot PI \quad (3)$$

The results of all calculations are presented in Table 2.

As the analysis shows, the priority of projects, with the exception of the three leading ones, which indicators significantly exceed the indicators of other projects, depended on the composition of evaluation criteria and the specific weight assigned to each of the criteria in the overall indicator. Thus, the use of subjective expert opinions in the calculation of the MII reduces the probability of obtaining reliable results and may lead to the loss of effective options.

In order to analyze the impact of a selected ranking method on obtained results, the prioritization of the GPN-Development projects was carried out by the method of pairwise comparison, the rank method, and the hierarchy analysis method using the criteria proposed by the experts of GPN-Development LLC in the initial formula of the MII.

Table 2. The results of the prioritization of projects for options calculations.

Project	The result of the 1-st calculation	The result of the 2-nd calculation	The result of the 3-rd calculation
Project 1	2	2	2
Project 2	1	1	1
Project 3	3	3	3
Project 4	4	6	6
Project 5	5	7	7
Project 6	10	11	12
Project 7	8	9	9
Project 8	7	4	4
Project 9	9	10	10
Project 10	15	13	13
Project 11	6	8	8
Project 12	11	5	5
Project 13	13	12	11
Project 14	14	15	15
Project 15	12	14	14

The method of pairwise comparison assumes the prioritization of projects using a matrix, the cells of which can acquire one of two values: if the project by row is more significant than the project by column, the corresponding cell of the matrix is set to one, otherwise - zero. Then all values in the rows are summarized and on the basis of obtained values each project is given an appropriate rank (the maximum cumulative value corresponds to the highest priority).

When using the rank method, projects are ranked by each of the criteria, followed by averaging, and the highest priority is given to the project with the highest average rank.

The method of hierarchy analysis involves more complex calculations than in previous methods and is implemented in several stages. At the first stage, the specific weights of the criteria are calculated by quantifying their importance relative to each other, calculating the geometric mean for each criterion and rationing the geometric mean values obtained for each of the criteria by the sum of this values. Then in the process of pairwise comparison of projects the ratios of their indicators for each criterion are calculated and the weight priorities of projects for each of the criteria are calculated using the average geometric value. At the final stage, project priorities are set taking into account the entire set of criteria by summing up the project priorities for each of the criteria weighted by the specific weights of the criteria.

The results of project prioritization obtained using these ranking methods are shown in Figure 1 (the project of the 1st rank has the highest priority, the project of the 15th rank has the lowest priority).

The analysis of the graph shows that the ranking results differ depending on the chosen method and in some cases this difference can be quite significant.

It should be noted that for such projects as Project 8, Project 10 and Project 12 the priority assessment obtained using the MII is significantly underestimated compared to the results of calculations by other methods. This discrepancy may be related to the system of scoring projects according to the criteria in the process of calculating the MII, which involves dividing the interval between the maximum and minimum values of the indicator for projects into equal intervals, each of which corresponds to a certain score. In this regard, in the case of a significant excess of the indicator value of one of the projects over the indicators of the other projects, there arises the difficulty of differentiating the remaining projects. This phenomenon indicates inaccuracy of the MII and the risk of losing effective options in the case of reliance in the process of projects selection solely on the results of calculations by this method.

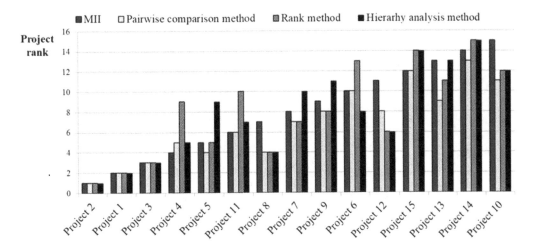

Figure 1.   The results of ranking projects by different methods.

The advantages of the pairwise comparison method are the simplicity for understanding and the absence of the need for complex mathematical calculations. However, this method is convenient only for single-criteria evaluation of projects, when one of indicators (NPV, PI, FCF, etc.) is selected as the main criterion for ranking. And the results of a multi-criteria evaluation by this method are completely dependent on the expert's intuitive-willed decisions. In addition, the project priority estimates for this method are integer values, and therefore as a result of the ranking some projects may be assigned the same rank, which complicates the selection process.

When using the rank method, there is also a frequent discrepancy of the ranking results with the results obtained by other methods (for projects 1, 3, 4, 6, 11). This can be explained by the neglect of the significance of the criteria when ranking projects by this method, which is its significant drawback.

The hierarchy analysis method, despite the complexity of calculations in comparison with other methods, may be more convenient to use, since it allows an expert to establish specific weights by determining the significance of criteria relative to each other. Allowing pairwise comparison of evaluation criteria and projects, as well as taking into account the importance of individual criteria, this method combines the advantages of both the scoring method, on which the MII is based, and the pairwise comparison method. However, this method can also lead to inaccurate results.

The analysis allows us to identify a number of problems in the formation of the portfolio of projects in the company GPN-Development.

The first problem is related to the fact that all the considered methods of economic and mathematical modeling are based on expert estimates. This leads to the dependence of the results of the selection of projects on the competence, awareness and integrity of the expert, increases the likelihood of an erroneous decision and the possibility of manipulating calculations to obtain the desired results. To solve this problem, the specialists of GPN-Development LLC and independent researchers are striving to develop a mathematical model that would allow to determine the projects to form an optimal portfolio and thus reduce the negative impact of expert's subjectivity on the selection results.

However, the achievement of this goal is hampered by the second difficulty: the multidimensionality and versatility of project evaluation in the process of their selection in the portfolio of GPN-Development LLC. In addition to the achieved production volume and economic indicators, it is also necessary to take into account many other aspects due to the specifics of the industry, including the urgency of project implementation; the interdependence between projects and the possibility of achieving a synergistic effect through their simultaneous

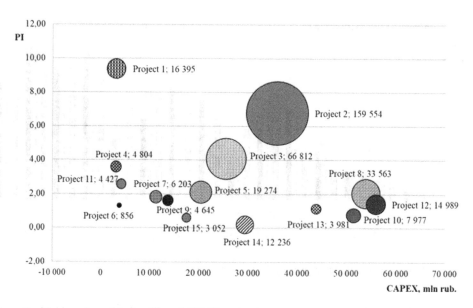

Figure 2.   Presentation of the projects of GPN-Development LLC on a bubble chart.

implementation; the availability of development options within individual projects; risks of various types and potential opportunities for their reduction. Due to the fact that all of the existing conditions of implementation and requirements for the projects cannot be accurately reflected in the mathematical model, it is not possible to exclude the use of expert analytical methods in the project selection process.

The third problem arises from the two previous ones and is related to the complexity of expert evaluation of projects and making final decisions on the inclusion of projects in the portfolio under the conditions of multidimensional requirements for the final portfolio. This problem is complicated by the underdevelopment of the company's knowledge and information management system in relation to the project portfolio management. As modern researches show (Lavuschenko et al. 2016; Ponomarenko & Khaertdinova, 2015), the introduction of an effective knowledge and information management system allows to facilitate the decision-making process and significantly improve the quality of decisions made through the accumulation and analysis of existing management experience and the formation of new knowledge.

In author's opinion, the solution of existing problems in order to improve the accuracy and objectivity of the results of project selection should be the main objective of portfolio management at GPN-Development.

The author offers the following recommendations to reduce the severity of these problems:

- involving not one, but several competent experts to the task of forming a project portfolio to increase the objectivity of the assessments;
- providing experts with the most complete list of portfolio requirements;
- implementation of a knowledge and information management system for project portfolio management;
- raising the awareness of experts by regularly providing information on changes in the external and internal environment affecting project performance and requirements for the final portfolio;
- expansion of the number of criteria used in the process of economic and mathematical modeling through the inclusion of qualitative indicators of projects in their composition;
- careful choice of selection criteria for projects included in the mathematical model, taking into account the possibility of their adequate assessment, compliance with the portfolio requirements and significance for the company;

- conducting the primary selection of projects using several different economic and mathematical methods;
- use in the process of deciding on the inclusion of projects in the portfolio of project prioritization results based on economic and mathematical methods as an additional, auxiliary source of information, the correctness of which can be questioned;
- use of methods for graphical representation of information on projects, such as bubble charts (Figure 2), to simplify decision-making by experts.

4 CONCLUSION

Thus, according to the results of the study, the following conclusions can be drawn.

1) When managing a portfolio of oil and gas projects, it is necessary to take into account the specifics of the oil and gas sector, which determine the great importance of expert-analytical methods in the process of portfolio formation.
2) The results of the prioritization of projects using the methods of economic and mathematical modeling depend on the chosen ranking method, as well as the composition of the criteria for evaluation and the specific weight assigned to each of the criteria in the overall indicator, and therefore also subject to the influence of subjective expert opinion.
3) In order to improve the accuracy of project selection results, it is recommended to involve several competent experts in the task of portfolio formation, to select the project evaluation criteria covering the various portfolio requirements carefully, to carry out the primary ranking of projects using several different economic and mathematical methods and to use all relevant information on projects to make the final decision.

REFERENCES

Borovyh, A.A. 2016. The concept of modeling a portfolio of projects for the development of integrated economic systems. *Bulletin of the Volga University. V.N. Tatishcheva* 1(23): 19–29.

Cooper, R.G., Edget, S.J. & Kleinschmidt, El.J. 2000. New problems, new solutions: making portfolio management more effective. *Research-Technology Management* 2(43): 18–33.

Kostylev, A.O. & Skopina, L.V. 2015. The growth of uncertainty factors in investment decision-making in the oil and gas industry. *Interexpo Geo-Siberia* 2(12): 121–126.

Lavuschenko, V.P., Garipov, A.K., Ponomarenko, T.V., Khaertdinova, D.Z. 2016. Project-based approach to the knowledge management in vertically integrated oil companies (on the example of Tatneft PJSC). *Neftyanoe Khozyaystvo - Oil Industry* (1): 20–22.

Perebatova, E.A., Alfyorov, A.A. & Grishina, N.V. 2017. Application of expert assessment methods for ranking investment projects//*Russian transport business* 5(14): 17–18.

Ponomarenko, T.V. & Khaertdinova, D.Z. 2015. Corporate systems of the knowledge management in the practices of integrated companies. *Mediterranean Journal of Social Sciences* 6(3S3): 203–212.

Ross, D.W. & Shaltry, P.E. 2005. *The Standart For Portfolio management*, Pennsylvania: Project Management Institute.

Shtefan, M.A. & Ornatskij, A.A. 2015. Heuristic methods in the evaluation of investment projects. *Finance and credit* 5(629): 51–63.

Simionova, N.E., Krivosheev, D.N. & Krivosheeva, I.N. 2017. Project portfolio: problems of planning and performance evaluation. *Bulletin of the UGNTU. Science, education, economy* 2(20): 16–21.

Solovyeva, I.A. & Galtyaev, A.V. 2017. Development of a multi-criteria model for selection and ranking of projects in the formation of the company's investment program. *Science of science* 2(16): 12–14.

Portfolio management in oil and gas: Building and preserving optionally [Electronic resource]//EYGM, 2016. URL: http://www.ey.com/oilandgas/capitalprojects

Environment protection and sustainable nature management

Topical Issues of Rational Use of Natural Resources 2019 – Litvinenko (Ed)
© 2020 Taylor & Francis Group, London, ISBN 978-0-367-85720-2

Phytoremediation of the salt mine water by common reed

A. Brodská & J. Nováková
VSB –Technical University of Ostrava, Czech Republic

ABSTRACT: Salt mining waters are characteristic for deep mining in the Ostrava - Karviná district, located in the Czech Republic. Due to their high mineralization and associated salinity, they may cause salt stress in some organisms. However, there are halophytic plants that are adapted to high salinity and they can withstand the salt stress.Common reed is classified as a halophytic plant and after long-term research it was confirmed to have phytoremediation ability. Phytoremediation is a technology that uses green plants to remove, accumulation or decompose contaminants in the environment found in groundwater, surface water, sediments soil, sludge. Phytoremediation processes in salt mine waters were experimentally tested in laboratory conditions on hydroponically grown reeds. The reduction of the salt concentration in the water samples from the hydroponic system, together with other chemical parameters, was monitored. For comparison, the experiment was also performed on common reed hydroponics taken from the mine Lazy, that are adapted to salt mine water, and common reed hydroponics taken from pond reeds, that are not loaded with high salt concentrations. It was prove by research that the common reed is able to work as root zone wastewater treatment plant for salt waters (not only salt mine waters) and it is able to withstand salt stress.

1 INTRODUCTION

Exploitation of mineral resources, particularly in the Ostrava-Karviná district, has a significant impact on the environment, not only during mining, but also after its finish in the post-mining landscape. The main environmental burdens include the issue of mine water management, their pumping and regulated discharge to recipients (Grmela, 1999).

The problematic mine water is the water produced during deep coal mining, the so-called salt mine water. Salt mine water consists of dissolved anorganic salts (RAS), chlorides and sodium. In addition to Cl^- and Na^+, sulphates, bicarbonates, alkali metals (Ca, K, Mg), metals (Fe), toxic metals (Cd), strategic metals (Mn) and suspended solids (NL) occur in salt mine waters (Dvorský et al., 2006).

These mine waters with high mineralization and high content of chlorides and sulphates are toxic to most plants because they bioaccumulate and negatively affect the environment. (Chytrý, 2007). Morphological changes such as plant growth are observable. However, plants called halophytes can grow, reproduce and survive even in places with high to extreme salt content (Flowers et al., 2008). The common reed (*Phragmites australis*, known for its resistance and perseverance) belongs to halophytic plants (Haslam, 1969).

Characteristic capabilities of reed are to cope with fluctuations in pH and temperature and tolerance to pollution. It can also resists salt stress, that is why it also occurs in salt habitats. Thus, it tolerates brackish conditions as well as oligotrophic and eutrophic aquatic environment and requires a nutrient-rich environment for life. For example, in mineral springs, salt lakes or even seas (Turkan, 2011).

Phytoremediation is one of the possibilities how to reduce salinity. It is a gentle, natural process that uses green plants to remove water and soil contaminants and it has aesthetic importance in nature. The principle of phytoremediation is the transformation of substances through metabolic detoxification processes and subsequent deposition in plant bodies. Some toxic substances cannot be transformed by the plant, so they only accumulate in the body (in this case the plants are treated as hazardous waste after harvesting) (Kadlíková, 2010). Halophytic plants were proven to be effective in reducing salinity by phytoremediation in soil and water. For example, Israeli scientists have found that Bassia indica, ranked among halophytes, can reduce salinity in water by 20-60% and deposit salts in leaves (Shelef et al., 2012).

Phytoremediation is used as a wastewater treatment technology. One of the working examples is the root zone wastewater treatment plant, consisting mainly of common reed (*Phragmites australis*).

2 MATERIALS & METHODS

The plants for an experiment, in which phytoremediation of the common reed was monitored, were selected from two different locations. The first 3 samples of reed plants were taken from the sludge tank site of the Lazy mine. These samples were adapted to salt stress for a long time. Another three samples of reed plants were taken from the non-salty locality Ostrava - Svinov, specifically from the eutrophic pond Rojek, which did not show significant values of the conductivity, that are specific for dissolved inorganic salts.

All samples were placed in hydroponic systems with a precise description on June 30, 2017 (see Figure 1). Plants in hydroponic systems have been named according to location:

- reed plants from mine Lazy (sample L1, L2, L3)
- reed plants from pond Rojek, Ostrava Svinov (sample R1, R2, R3)

Samples L2 and R2 were used as blank samples, other samples were exposed to salt stress. In order to form a hydroponic system, suitable containers, a substrate, and a nutrient solution for proper plant growth and development are needed. Selection of the right habitat, which does not affect the conditions of the experiment and at the same time ensures the plants basic conditions for development and growth, is very important.

Therefore, plant pots were placed out under the canopy so that subsequent analyzes were not affected by precipitation, while the reeds had a natural climate and sufficient solar energy

Figure 1.   Common reed in hydroponic system, Brodská 2018.

for photosynthesis processes. Hydroponics remained under the roof all summer and half of autumn and they survived the winter in the unheated room.

Frost-resistant and non-flammable clay-like material was chosen as a growing substrate. Because plants do not have a nutrient supply in the hydroponic system, nutrients must be supplied to them via nutrient solution.

Hydroponex has been selected as the nutrient solution due to its suitable composition (free of chlorides), essential nutrients, including stimulants and trace elements, which contribute to healthy and rapid plant growth.

One dose of nutrient solution was prepared by mixing one spoon attached with 1 liter of distilled water. Because of the failure in previous experiment, when feeding hydroponics with the nutrient solution at the interval of every 14 days lead to death of plants, the nutrient solution was now added to hydroponics only three times during the experiment, namely on 10 July 2017, 14 August 2017 and 30 October 2017.

Hydroponics were watered also with salt solution, in addition to the nutrient solution,. The saline solution in the experiment replaced salt mine water and was prepared from NaCl and distilled water at four concentrations of 0.1%, 0.2%, 1% and 2%.

One hydroponic system with a reed from the Lazy mine (L2) and one from the Rojek pond (R2) served as blank samples. This means that they were only watered with distilled water and nutrient solution. The other two L1 and R1 hydroponic systems were first watered with 0.2% saline solution followed by 2% saline solution. Hydroponics L3 and R3 were watered with saline solutions of 0.1 and 1% concentrations.

The Table 1 below shows the days of watering hydroponics with salt solutions. Watering with salt and nutrient solutions depended on the actual amount of water in the pot, which was monitored by a dipstick showing the minimum and maximum water in the pot.

In hydroponic systems, in situ and ex situ analyzes of phytoremediation (reductions of salt in water in pots) of common reed were investigated. Dissolved oxygen, pH, conductivity, and salinity were monitored in water samples from pots. However, the analysis of chlorides in water samples taken from hydroponics was the most important for the experiment.

Measurements of the above parameters took place from July 12, 2017 to February 5, 2018. Measurements of dissolved oxygen, pH, conductivity and salinity parameters were taken in situ every Monday morning at 7 am, by immersing the appropriate probes directly in the pot. For the determination of dissolved oxygen, the Greisinger type GMH 3630 oximeter was used, and the pot temperature was also measured. Dissolved oxygen was monitored in the water samples to control the proper functioning of the roots, which need oxygen for the intake of nutrients and breathing. PH measurement was performed using a pH 3310/set 2 pH meter. Conductivity and salinity of potted water, caused by salt solution watering, was measured by the WTW Condi 330i conductometer.

Chloride analysis was performed in a specialized laboratory ex situ after prior measurement of parameters in situ. Water samples were taken every 14 days from the pots using a pipette to the sample boxes in which they were transferred to the laboratory. Here, the water samples were filtered with filter paper and the solution was mixed with the solution in the cuvette due

Table 1. Watering hydroponics by salt solution, Brodská 2018.

Watering hydroponics by salt solution (SS)	Hydroponics		
	L1, R1	L2, R2	L3, R3
24.07.2017	Salt solution 0,2 %	Distilled water	Salt solution 0,1 %
07.08.2017	Salt solution 2 %	Distilled water	Salt solution 1 %
25.09.2017	Salt solution 0,2 %	Distilled water	Salt solution 0,1 %
09.10.2017	Salt solution 2 %	Distilled water	Salt solution 1 %
27.11. 2017	Salt solution 2 %	Distilled water	Salt solution 1 %

to precision micropipettes. The cuvettes used for chloride determination are specifically designed to analyze chloride in a Hach spectrophotometer called the UV - VIS DR6000 spectrophotometer. The greater the saline concentration, the greater the amount of chlorides in the hydroponics water samples. Chlorides were analyzed in water samples to detect possible reduction of salts in samples due to phytoremediation.

3 RESULTS

From July 12, 2017 to February 5, 2018, measurement results were evaluated both in situ and ex situ. Unfortunately, due to extreme drought during the summer, there was a large evaporation of water from the pots, which made it impossible to measure the parameters of L1 (1x), L2 (3x) and L3 (1x) hydroponics. This problem was related only to hydroponics with reeds taken from the Lazy Mine, while hydroponics from the Rojek pond showed no problems every week (32x).

Measured values of dissolved oxygen in water samples from hydroponics are reported in mg/l and the temperature is measured in ° C. In all six samples taken from hydroponics, the dissolved oxygen readings fluctuate during the experiment. The average dissolved oxygen value was about 6.06 mg/(see Table 2).

Since it was measured in the early morning hours, the measured temperature did not exceed 25 ° C even in summer. The average temperature measured by the oximeter during the whole period was 10.55 ° C. The minimum temperature of 3 ° C was measured on 18 December 2017 for all three hydroponics water samples from the Lazy u mine and the maximum 24.8 ° C on 31 August 2017 for the water sample from hydroponics R2.

The graph (Figure 2) below shows the measured pH values for all six hydroponics water samples. It can be seen from the Table 3 that the pH is around a neutral value of 7.

The highest pH values were found in hydroponics L2 and R2, or blank samples that were not watered with saline.

The highest pH values were found in hydroponics L2 and R2, or blank samples that were not watered with saline.

Figure 3 shows the conductivity of individual water samples from hydroponics. Although water samples from all hydroponics are from two different territories, we can see on the graph that the conductivity does not differ significantly. All water samples from hydroponics were measured, although so-called blank samples (L2, R2) were not watered with salt solution. Therefore, sample L2 showed the lowest conductivity value as shown in the Table 4.

Conductivity of water samples from the Lazy mine was in most cases lower than that of water samples from the Rojek pond. The omitted graph lines again indicate omission of conductivity and salinity measurements due to the large evaporation of water in the pot.

Salinity as well as conductivity was measured with a conductometer and reported without units. After immersing the conductometer probe in the pots and switching the button on the instrument, the salinity value was displayed.

The graph (Figure 4) shows the salinity values during the experiment, when the water samples from L1 and R1 hydroponics were the highest. We can read the measured salinity values from the table. The average of all measurements (including L2, R2, which had zero salinity) was 3.7.

Table 2. Dissolved oxygen values, Brodská 2018.

Dissolved oxygen values		Hydroponics/Measurement Date
Minimum	2,71 mg/l	R3/18. 9. 2017
Maximum	11,5 mg/l	L2/23. 10. 2017
Average from all measurements	6,06 mg/l	

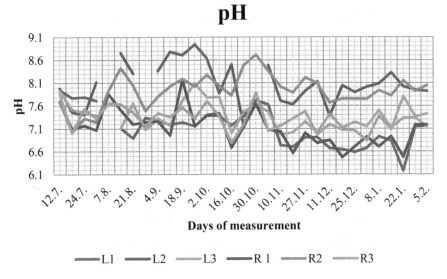

Figure 2. Chart pH, Brodská 2018.

Table 3. Dissolved oxygen values, Brodská 2018.

Dissolved pH values		Hydroponics/ Measurement Date
Minimum	6,176	L1/22. 1. 2018
Maximum	8,945	L2/25. 9. 2017
Average from all measurements	7,500	

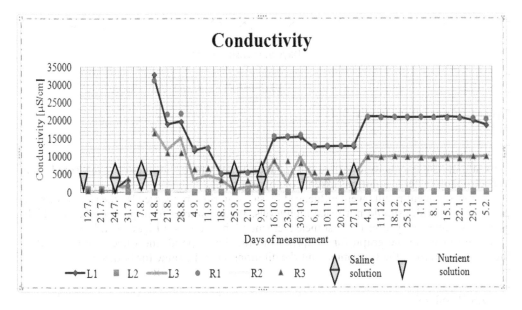

Figure 3. Chart conductivity, Brodská 2018.

Table 4. Conduktivity values, Brodská 2018.

Conductivity values		Hydroponics/ Measurement Date
Minimum	0,3 µS/cm	L2/22. 1. 2018
Maximum	32 800 µS/cm	L/25. 9. 2017
Average	7136 µS/cm	

Table 5. Salinity values, Brodská 2018.

Salinity values		Hydroponics/ Measurement Date
Minimum	0	L2, R2
Maximum	20,3	L1/14.8. 2017
Average	3,7	

Salinity

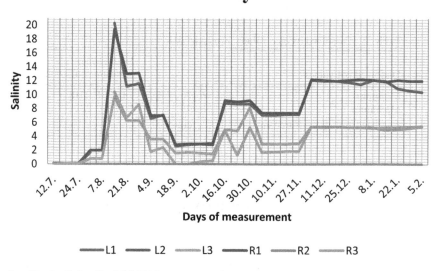

Figure 4. Chart salinity, Brodská 2018.

Ex situ measurements were carried out in laboratories where chlorides were measured using a spectrophotometer in solutions taken from hydroponics.

The measurement results were influenced by salt watering, but were also measured for samples where chlorides should not reach high levels. Water samples from pots reached the highest values immediately after the first measurement of chlorides (see Figure 5 and Table 6).

As we can see in the graph for samples from the Rojek pond, the values are higher than those measured from the Lazy mine, but the dilutions were the same for both samples.

4 DISCUSSION

Because salt mine water taken from a previous Lazy mine experiment had a different chloride concentration every time, depending on mining, saline mine water was replaced by salt

684

Chlorides

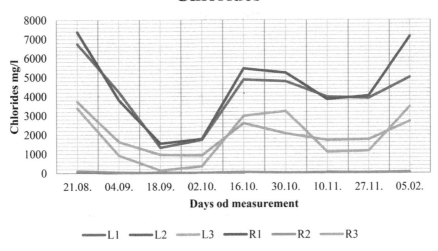

Figure 5. Chlorides, Brodská 2018.

Table 6. Chlorides value, Brodská 2018.

Chlorides values		Hydroponics/ Measurement Date
Minimum	4,04	L2/04. 09. 2017
Maximum	7367 mg/l	R1/21.8. 2017
Average	2153,3 mg/l	

solutions. The hydroponic system was chosen for the experiment to improve water analysis accuracy. Hydroponics lack the microorganisms that occur in the soil and bind to the roots, also nutrient intake works differently. The soil substrate, full of minerals and nutrients, has been replaced by ceramsite and nutrient solution. The distilled water, in which both the nutrient solution and the saline solution were dissolved, did not change the results, it was demineralized water.

The difference in cultivation is also the availability of oxygen for plants. In some cases, oxygen must be supplied artificially by the apparatus for hydroponics. In all measurements, oxygen was judged to be sufficient due to the openings that were formed among the ceramsite. Comparable values of dissolved oxygen were measured for all six samples. Dissolved oxygen values depend on temperature. In the experiment, the relationship between these parameters was not confirmed. The temperature difference was only one degree between the lowest and the highest dissolved oxygen values. The saline solution had no effect on the dissolved oxygen values.

The pH values measured for hydroponics corresponded to the salt mine water collected in the Lazy mine, where the pH was also around neutral 7. As already mentioned for the proper functioning of the hydroponic system, the pH should be slightly acidic (5.5 to 6.5), but the reeds benefited even at neutral pH during the experiment (Benton J., 2004).

In a hydroponic system, conductivity plays an important role, because by measuring it, we can adapt nutrient delivery to support the growth and development of the plant grown in this way. The increase in conductivity and salinity was associated with saline delivery. When the conductivity increased, the salinity also increased. The supply of nutrients in the form of nutrient solution influenced the conductivity. The exception was the so-called blank samples, which were not supplied with saline, and therefore zero salinity was measured.

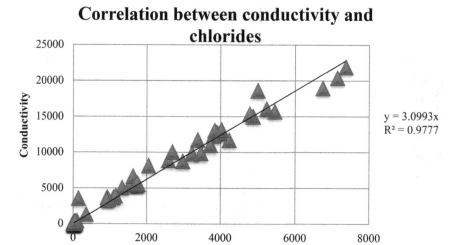

Figure 6.   Correlation between conductivity and chlorides, Brodská 2018.

The result of the research was also a correlation between conductivity and chloride content in solutions, which is shown in the graph. The high chloride concentration increased the conductivity in the solutions.

Phytoremediation was confirmed by decreased chloride concentration in water samples from hydroponics, when it dropped visibly after a short period of time (see Chlorides Chart, Conductivity). The reed behaving like halophyte resisted stressful conditions. The natural value of chlorides in surface waters (250 mg/l) was exceeded 29 times.

Plants taken from the Lazy Mine were more tolerant to salt stress and had new additions during the experiment. The reeds, whose original habitat was a non-salted pond Rojek, were suffering because of salt watering. After a month, their above-ground part burned and the stem gradually dried up. However, the roots were still functional. For the dormancy period, the plants were cut and in spring only the more resistant L1, L2 and L3 reeds, which have survived to date, have grown.

5   CONCLUSIONS

In the experiment, the phytoremediation capabilities of the common reed were demonstrated when salinity decreased in hydroponics solutions. The common reed is already widely used for water scrubbing and is an integral part of root sewage treatment plants. The research revealed that this plant can also reduce the salinity of salty waters, not just mine water, in addition to wastewater contamination.

Proof that reed can accumulate salts and thus reduce salinity is the measurement of conductivity, salinity and especially chloride analysis. All these parameters were closely related, in particular the higher chloride concentration increased the conductivity.

The overall appearance of plants in relation to salt stress was also monitored. Saline-adapted reeds turned out to be more responsive to salt stress. Although the reeds from the Rojek pond have successfully passed the experiment, they have not grown up after a period of dormancy.

I continue in this experiment and one of the other goals is to determine the location of salt accumulation in the plant.

REFERENCES

Benton J. a JR. 2004. Hydroponics a Practical Guide for the Soilless Grower. 2nd ed. Hoboken: CRC Press,. ISBN 9781420037708.

Chytrý, M. ed. 2007. Vegetace České republiky: *Vegetation of the Czech Republic*. Praha: Academia, ISBN 9788020014627.

Dvorský J., Malucha, P., Grmela A, A., Rapantová, N. 2006. Ostravskokarvinský detrit - spodnobádenská bazální klastika české části Hornoslezské pánve. Monografie. Montanex, Ostrava. ISBN 80-7225-231-3.

Floweras, T., J. a T. D. Colmer. 2008. Salinity tolerance in halophytes. New Phytologist 179(4), 945–963 [cit. 2019-02-22]. DOI: 10.1111/j.1469-8137.2008.02531.x. ISSN 0028646x.

Grmela, A. 1999. Problematika důlních vod a ochrana kvality povrchových vod při jejich vypouštění. In: Sborník vědeckých prací VŠB – TU Ostrava. Řada hornicko – geologická,. Ostrava: ISSN 0474-8476- Roč. 45

Haslam S. M. 1969. The Development and Emergence of Buds in Phragmites communis Trin., *Annals of Botany*, Volume 33, Issue 2, Pages 289–301, https://doi.org/10.1093/oxfordjournals.aob.a084283.

Kadlíková L. 2010. Fytoremediace aneb rostliny které čistí půdu. *Příroda.cz* [online]. [cit. 2019-03-14]. https://www.priroda.cz/clanky.php?detail=1242

Shelef, O., A. Gross a S. Rachmilevitch. 2012.The use of Bassia indica for salt phytoremediation in constructed wetlands. Water Research [online]. 46(13), 3967–3976 [cit. 2019-03-14]. DOI: 10.1016/j.watres.2012.05.020. ISSN 00431354. http://linkinghub.elsevier.com/retrieve/pii/S0043135412003478

Turkan I.. 2011. Plant responses to drought and salinity stress developments in a post-genomic era. *Oxford: Academic*. ISBN 9780123876928.

Topical Issues of Rational Use of Natural Resources 2019 – Litvinenko (Ed)
© 2020 Taylor & Francis Group, London, ISBN 978-0-367-85720-2

Comparison of the basic principles of state monitoring of water objects in Russia and the European Union

N.A. Busko
State Marine Technical University, St.Petersburg, Russian Federation

ABSTRACT: Acquaintance with the principles of state monitoring of water objects in different countries will help to identify the positive aspects and borrow these principles for the best func-tioning of the monitoring systems in these countries.

The systems of state monitoring of water objects in Russia and in Europe are developed enough and are widely used at the moment. The state monitoring of water objects in Russia and in the European Union is very similar by most criteria.

There are several differences between the monitoring systems in question, but these differences don't allow us to conclude that one system or another is worse. However, it became possible to introduce one of the principles of state monitoring of water objects in Europe into the Russian monitoring system.

1 INTRODUCTION

State monitoring of water objects is an integral part of environmental protection.

Each country has a system for monitoring water bodies, regulated by law. All of these principles will reveal the positive aspects and borrow these principles to ensure the best functioning of the monitoring system.

The purpose of this article is a comparative analysis of the main provisions on state monitoring of water bodies in the legislation of the Russian Federation and the European Union with the aim of identifying the positive aspects of monitoring in the Russian Federation and the EU and further mutual implementation of the identified best practices.

Tasks:

1) Determination of the main legislative documents regulating the state water monitoring in the Russian Federation and the EU;
2) Identification of the main provisions on the state monitoring of water objects from previously defined sources;
3) Comparison of the principles of state monitoring in Russia and the EU, based on the analysis of the above provisions.

2 PROVISIONS ON STATE MONITORING OF WATER OBJECTS IN THE RUSSIAN FEDERATION AND THE EUROPEAN UNION

2.1 *State monitoring of water objects in the Russian Federation*

State monitoring of water objects is a system of observations, assessment and forecast of changes in the state of water objects.

The tasks of the state monitoring of water objects:

1) Identification of negative impacts on water objects, forecasting the development of negative processes that may affect the quality and condition of water objects, development and implementation of measures to prevent the consequences of these processes;
2) Assessment of the effectiveness of measures for the protection of water objects;
3) Information support in the field of use and protection of water bodies, including state supervision over the use and protection of water objects.

The state monitoring of water objects includes the monitoring of surface water objects, the state of the seabed and coast, water protection zones, monitoring of groundwater, as well as monitoring of water management systems and hydraulic structures.

In the process of conducting state monitoring of water objects, observations are made of quantitative and qualitative indicators of the state of water resources, as well as of the regime for their use; collecting, processing and storing information and entering this information into the state water register; assessment and forecasting of changes in indicators of water resources is being carried out.

State monitoring of water objects is divided into levels: local, regional, territorial and federal.

The organization and implementation of state monitoring of water bodies is carried out by the Ministry of Environment of Russia together with the Russian Federal Service for Hydrometeorology and Environmental Monitoring and other authorized state bodies in the field of environmental protection.

Entering information obtained as a result of observations into the State Water Register, perform out by the Federal Agency for Water Resources of the Russian Federation.

2.2 *Monitoring of surface water quality, groundwater quality and protected areas of the European Union*

Monitoring of European water objects is carried out by numerous regional and national authorities. The development of monitoring programs in accordance with the Framework Water Directive is the most effective way to obtain information about the quality of water in the European Union and the harmful effects affecting water bodies.

The objectives of the monitoring are:

1) Identification of areas with serious environmental problems;
2) Identification, assessment and prediction of environmental threats at the regional and global levels;
3) Provision of information on the state of water resources;
4) Carrying out of measures to improve the ecological status of water bodies.

Monitoring of water objects quality in Europe is done in relation to surface water, bottom sediments and shores, groundwater and for protected areas.

In accordance with the Framework Water Directive, Member States need to regularly assess the status of their waters. To this purpose, the European Environment Agency (EEA), together with the Member States, has established a unified monitoring network, Eionet Water.

The organization and implementation of water monitoring in Europe is conducted by the European Commission, which includes the Directorate General for the Environment, the Joint Research Center and Eurostat. The Commission should evaluate progress in monitoring at regular intervals and inform the European Parliament, the Council and the public about the results of its assessments.

Information on the conduct and results of monitoring is published in the European Water Agency's Environment Information System on Water (WISE).

3 COMPARISON OF THE PRINCIPLES OF STATE MONITORING IN THE RUSSIAN FEDERATION AND IN THE EUROPEAN UNION

It is difficult not to notice the obvious similarities between monitoring water objects in Russia and in the European Union. This similarity lies in many aspects, first of all - the purpose of monitoring. They can be formulated differently, but the essence is identical.

The process of conducting state monitoring of water objects in the Russian Federation and the EU is very similar - it is the observation of the quantitative and qualitative characteristics of the state of water resources, the mode of their use, as well as the assessment and forecasting of changes in the state of water resources.

Databases for storing information about condition of water resources are identical - In Russia, this is the State Water Register, and in the EU, the Information System on water resources in Europe. The objects of monitoring are also similar, with the exception of a few - the Water Framework Directive of Europe does not provide for monitoring of hydraulic structures and water management systems. Its organization and implementation is handled by the Directorate General for Energy and public non-governmental organizations.

In the EU, the European Environment Agency is responsible for collecting information through the Eionet Water integrated monitoring network, whose stations are distributed throughout the territory of all EU member states. In Russia, first of all, monitoring is divided into levels, and at each level, monitoring is carried out by authorized bodies of the Ministry of Natural Resources of Russia and the Federal Service for Hydrometeorology and Environmental Monitoring. In principle, there is a similarity here, except that in the European Union monitoring is not divided into levels.

As for the agencies that are engaged in the implementation of state monitoring of water objects, the systems are very similar. Although in Europe the role of the Ministry of Natural Resources and Environment is played by three organizations that are part of the European Commission, but in fact these organizations perform the same functions. This fact can be regarded as a similarity, and as a difference.

Also, the comparison should take into account the positive and negative sides of each of the systems. In the Russian system, the division into levels can be a weakness. In the process of monitoring at the regional level today, small cities and numerous settlements, the vast majority of diffuse sources of pollution, which do not allow for a complete picture of the state of water bodies, remain practically not covered by the network of observations. The introduction of one of the principles of the European system of state monitoring can solve this problem. Formation of public non-governmental organizations and their admission to monitoring would eliminate "blind zones".

The mutual introduction of principles of state monitoring systems is necessary not only for the best functioning of these systems, but also for interaction in monitoring in zones with transboundary water bodies. Using of joint principles, as well as the development of joint criteria are necessary for assessing the quality indicators of transboundary basins and coordinating measures water management.

4 CONCLUSION

Objectively assessing the state monitoring systems of water objects in Russia and Europe, we can conclude that they have more similarities than differences. Nevertheless, these differences are present:

- The Water Framework Directive of Europe does not provide for monitoring of hydraulic structures and water management systems;
- State monitoring is not divided into levels in the European Union;
- A slight difference in the bodies involved in monitoring water bodies in the Russian Federation and the EU;

It is difficult to say whether these differences make monitoring in Russia more effective than in Europe and vice versa. In the course of the analysis of the provisions on state monitoring of water objects, it can be argued that both in Russia and in the European Union the systems are sufficiently developed, can be widely used and successfully applied at the moment. The only adjustment can be made to the Russian monitoring system of water objects – formation of public non-governmental organizations which will control the small rivers and reservoirs located on them that are not check at the regional level under the current system.

REFERENCES

Directive 2000/60/EC of the European Parliament and of the Council of 23 October 2000 establishing a framework for Community action in the field of water policy.

Establishing a dynamic system of surface water quality regulation: Guidance for Countries of Eastern Europe, Caucasus and Central Asia

European Commission https://ec.europa.eu/commission/index_en

European Environment Agency https://www.eea.europa.eu/UDC 711.27

Rasskazova N.S., Bobylev A.V. Some problems of regional environmental monitoring and modern solutions

Recommendation 1668 (2004) Management of water resources in Europe «Water Code of the Russian Federation» dated 03.06.2006N 74-FZ (as amended on 12. 27.2018)

Topical Issues of Rational Use of Natural Resources 2019 – Litvinenko (Ed)
© 2020 Taylor & Francis Group, London, ISBN 978-0-367-85720-2

Application of a modular-type construction in thermochemical facilities

M.A. Fedorova, V.A. Sasarov & I.N. Novikov
FSBEI HE «P. A. Solovyov Rybinsk State Aviation Technical University» (RSATU), Rybinsk, Yaroslavl region, Russia

ABSTRACT: The article raises the issue of waste recycling. A modular scheme of a thermochemical facility for recycling of hydrocarbon-containing waste with producing of a valuable product is proposed. A module for preparation for reprocessing of solid household waste, coal dust and woodworking waste, a gasification module and a module for producing of a valuable product – hot water, electrical energy, liquid fuel – are considered. In the article cost price of an obtained valuable product is presented, and also a conclusion is made concerning profitability of a facility for recycling of one or other raw material with generation of a valuable product.

1 INTRODUCTION

In today's world the problem of an uncontrolled growth of amounts of waste is acute. The pace of increase of the quantity of waste exceeds its recycling, decontamination and utilization. In Russian Federation more than 7 billion tons of household, agricultural, industrial and other kinds of waste are being generated annually. The amount of production of solid household waste (SHW) in localities of Russian Federation comprises 150 million cubic meters (30 million tons) a year. The major part of the SHW is being stored in landfills of different types and numerous dumps. About 85 % of waste gets to landfills, only 5 % is being reprocessed, and about 10 % of waste is being lost during its transportation. Accumulation of waste causes enormous ecological, economical and social harm (Konstantinov, 2000).

Negative impact of waste also manifests in growth of levels of sickness among people, worsening of their living conditions, decrease of productivity of natural resources. Specific pathologies occur, caused by a chronic influence of small concentrations of technogenic contaminants, which get into an organism from environment and remain in it, and a gradual accumulation of such substances is taking place (Konstantinov, 2000).

A solution to the SHW handling issue is being implemented in three directions:

– Recirculation – turning waste into resources suitable for secondary use, new products or semi-finished items for re-use. Recirculation of waste demands organization of its collecting and sorting, availability of economical encouragement for its recycling, and also existence of an informational system about sources of secondary resources.
– Waste disposal is being organized through its placement to dumps. But this leads to a whole range of negative implications for environment.
– Waste is most commonly being eliminated via incineration. But this way also doesn't help to resolve the problem of non-harmful waste, both for economy and environment. During the incineration process hazardous materials are being produced. As a result of incineration of 1 ton of solid household waste more than 330 kg of slag, about 30 kg of fly ash and up to 6 thousand cubic meters of flue gas are generated. Their composition includes hydrogen fluoride and chloride, sulphur dioxide, nitrogen and carbon oxides and toxic hydrocarbons.

Recyclable resources are also being eliminated along with other trash. (Zyryanova et al, 2011; Kireyenko, 2014)

Apart from SHW, about 420 million tons of coal are being extracted annually in Russia by mining enterprises, and large amounts of it (about 3 % from the total obtained coal) are turned into waste in a form of coal dust, which is combustible, and currently considerable sums are being spent on its utilization.

Moreover, woodworking facilities produce sawdust and chips as waste, which accumulate in significant amounts and also require an effective utilization.

Due to that, the issue of recycling of solid household waste, coal dust, woodworking waste and other hydrocarbon-containing industrial waste in order to prevent a negative impact on ecology, economy and the social sphere becomes relevant (Konstantinov, 2000).

At the same time, hydrocarbon-containing substances can serve as fuel for generation of a valuable product.

2 METHODOLOGY

One of the methods of utilization of solid hydrocarbon-containing waste is proposed in a form of a gasification with production of a synthesis gas. It can be used not only as fuel, but also as a chemical raw material for various synthesizes depending on applied gasifying agents (air, steam, gas mixtures).

Nowadays both methods and facilities for recycling of hydrocarbon-containing waste with generation of a valuable product are available, but most of them consider one type of raw material and one or two kinds of a resulting valuable product. For example, obtaining of hydrocarbon oil from plastic waste (Fukuda et al, 1989), and also pyrolysis of unwashed and combined plastic waste with generation of useful hydrocarbon liquids and other valuable materials (Scheirs & Kaminsky, 2006). This reduces their universality and usability.

In order to increase usability and universality, facilities for thermal recycling of a hydrocarbon-containing raw material can be of a modular type. Some units of the facilities are the same for reprocessing of different kinds of waste. For instance, all types of facilities, which employ the gasification principle, require a thermochemical reactor, devices for loading of a raw material and discharging of slag, applicators, raw material accumulating bunkers, etc. in their construction. Modular-type facilities are able to generate various valuable products (hot water, liquid fuel, electrical energy). For this purpose necessary modules should be added to or removed from the design of the facility. A configuration of the facility is chosen depending on the feedstock and the resulting valuable product.

The goal is to propose modular-type construction methods of hydrocarbon-containing substances gasification facilities family with additional modular blocks for generation of a valuable product. On Figure 1 composite modules of facilities for recycling of different kinds of waste with generation of a valuable product are presented.

The facility may include the following modules: 1 – accumulating bunker for coal dust, 2 – accumulating bunker for SHW, 3 – accumulating bunker for woodworking waste, 4 – dryer, 5 – transporter, 6 – grinder, 7 – loading device with a doser, 8 – thermochemical re-actor, 9 – condenser, 10 – afterburning chamber, 11 – scrubber, 12 – heat exchanger, 13 – gas-powered engine, 14 – smoke extractor, 15 – slag discharger. The facility consists of three blocks: I – the raw material preparation block, II – the gasification block, III – the valuable product generation block.

3 EXPERIMENTAL

Let's consider the process of recycling of a hydrocarbon-containing raw material in the modular-type facility on the example of SHW reprocessing with generation of, for in-stance, hot water.

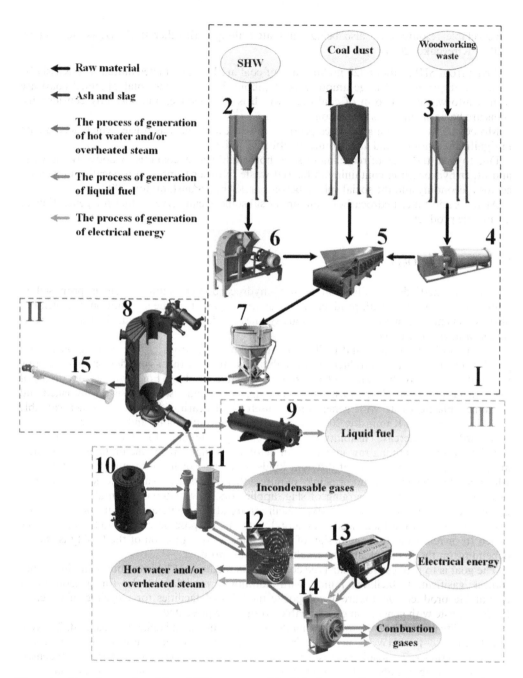

Figure 1.   The scheme of modules of different types of facilities for recycling of hydrocarbon-containing raw materials.

Waste is stored in the SHW accumulating bunker 2. From there it's being taken to the grinder 6, where it's being shredded in order to speed up processes in the thermochemical reactor. Then the raw material is put onto the transporter 5 to be taken into the loading device 7. From there the waste is forwarded to the thermochemical reactor, in which the gasification process is taking place.

On Figure 2 an example of a design of the thermochemical reactor is presented.

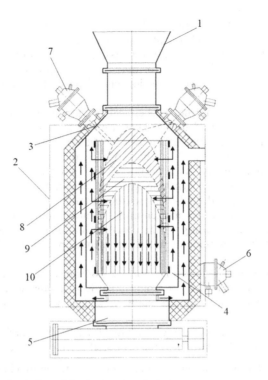

Figure 2.   The example of a design of the thermochemical reactor.

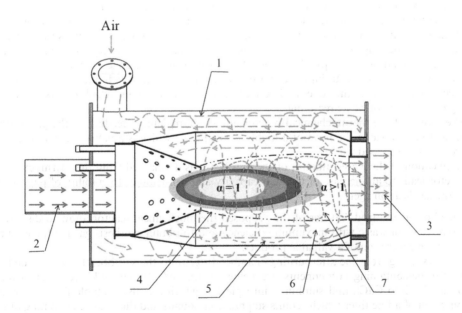

Figure 3.   The combustion chamber of a whirling countercurrent type: 1 – air; 2 – flammable gas; 3 – combustion products; 4 – flows separation border; 5 – outer whirl; 6 – whirling chamber; 7 – inner whirl; α – air excess coefficient.

The thermochemical reactor is comprised of: 1 – loading device, 2 – gasification chamber with openings 3 and 4 for release of gaseous products – a gasifying agent (blowing by air, overheated steam, combustion products), 5 – receiving container for recycled waste (ash, slug), 6, 7 – gasifying agent forming and feeding device, which is constructed in a form of one or a number of combustion chambers or burner units (Novikov et al, 2011).

Three zones are formed in the thermochemical reactor. The heating zone 8, where the recycled substance is being heated due to heat transfer from the gasifying agent which is flowing directly towards it. In the drying zone 9 the substance is losing water which is being taken away as water vapor, which then forms a part of the synthesis gas. The pyrolysis and gasification zone 10. As a result of the pyrolysis, organic elements of the recycled substance break into the synthesis gas and solid matters. Solid matters are solid non-combustible remaining and coal-like – carbon matter. While moving forward in the gasification chamber, the coal-like matter and the non-combustible elements are being exposed to further heating. In doing so the coal-like matter is breaking into the synthesis gas, and non-combustible components are being melted, producing slag or ash (Novikov et al, 2011).

Ash generated in the thermochemical reactor is being transported to the slag discharger 15. Analyses of slag and ash remnants, taken from similar functioning facilities, correspond to the hazard class 4 according to laboratories' data, which means they can be stored in SHW landfills.

Application of a thermochemical reactor is a must for effective recycling of various types of waste (Bilgesu et al, 2006). The features of the proposed thermo-chemical reactor are:

- Bottom extraction of a pyrolysis gas is implemented in the thermochemical reactor. Due to this, there is almost no resin in it, consequently, the risk of formation of deposits on gas pipelines' walls is reduced, and also a maxi-mum concentration of flammable gases and the temperature close to maxi-mal are achieved.
- Bottom extraction of a pyrolysis gas is implemented in the thermochemical reactor.
- The design of the thermochemical reactor allows to provide feeding of a gasification agent to the waste through its upper end face surface and/or through its entire side surface, which leads to the heating zone becoming the outer in relation to the drying zone, which is also external in relation to the pyrolysis zone. Such arrangement of zones significantly expands the surface area of heat input and the gasification agent, which intensifies heat-mass exchange processes in the zones and accelerates chemical reactions, which results in increasing productivity of waste recycling.
- The design of the thermochemical reactor allows to implement feeding of the gasification agent to the waste in the form of a highly twisted flow, which is able to cover a maximum area of the processed waste.
- Application of the whole amount of the pyrolysis gas discharged from the thermochemical reactor leads to an increase of the heat capacity and con-tributes to stability of parameters of the gasification agent.

The above-mentioned features of the proposed thermochemical reactor allow to dispose of different types of waste with maximum efficiency and to obtain synthesis gas for further processing with generation of a valuable product.

Gases escaping from the afterburning chamber 10 get into the scrubber 11, where sulphur- and chlorine-containing components are being neutralized. "Lime milk" (Ca(OH)2) turns chlorine (Cl) into CaCl, and sulphur – into plaster, which are also suitable for further use. Application of a fine filter which retains sulphur compounds and dust particles coming out of the scrubber 11 is possible. In the heat exchanger 12, where combustion gases transfer heat energy to water, hot water and/or overheated steam is being generated. The amount of over-heated steam and hot water can be varied according to needs of a consumer. After purification and work execution the combustion gases are being emitted into the atmosphere through the smoke extractor 14. Thanks to the purification system the combustion gases are obtained which are pure enough to meet European standards, which is confirmed by measurements taken on similar active facilities.

In contrast to other facilities for gasification and afterburning of synthesis gas with generation of heat as a valuable product, for instance, a multi-purpose system for waste recycling (Alves et al, 2011), implementation of the combustion chamber of a whirling countercurrent type:

– contributes to breakdown of hydrocarbons and dioxins;
– contributes to breakdown of hydrocarbons and dioxins;
– constrains synthesis of nitrogen oxides;
– allows to introduce reagents which bond harmful components of dioxins with the formation of neutral compounds;
– helps to speed up chemical reactions.

All of this have a positive effect on eco-friendliness of the proposed modular facility.

A configuration of the facility should be changed for recycling of other kinds of waste or generation of another valuable product (see Figure 1).

For reprocessing of coal dust the accumulating bunker 1, the transporter 5 and the loading device 7 are applied in the raw material preparation block.

For recycling of woodworking waste the accumulating bunker for this kind of waste 3, the dryer 4, as sawdust has an excessive humidity, the transporter 5 and the loading device 7 are applied.

The gasification block design can vary depending on the purpose of the facility, but the principle of the gasification process remains the same.

Liquid fuel is obtained in the valuable product generation block as a result of the following processes. The synthesis gas from the thermochemical reactor 8 is directed to the liquid fuel condenser 9, where the synthesis gas is divided into a liquid fraction and incondensable gases. The condensed fraction can be used as a separate fuel product. Other-wise, it can be additionally processed with resulting consumable goods: methanol, dimethylether or motor fuels applying traditional oil and gas industries devices. Incondensable gases are forwarded to the scrubber 11 for further purification and the subsequent cooling in the heat exchanger 12. After that the purified and cooled incondensable gas is flowing into the gas-powered engine 13. The produced electricity is used for functioning of the facility as a whole, and excesses can be directed to consumers. By varying temperatures of the process in the thermochemical reactor, it's possible to obtain different proportions of liquid fuel and incondensable gases depending on demands of one or other enterprise. After purification and work execution the gas is being emitted into the atmosphere through the smoke extractor 14.

If only electrical energy is needed, then the following processes are taking place in the valuable product generation block. The synthesis gas obtained in the thermochemical reactor 1 is directed to the scrubber 11, where it's being purified. Then the gas gets into the heat exchanger 12 for producing of hot water. After that the gas is forwarded to the gas-powered engine 13 for generation of electrical energy. After purification and work execution the gas is being emitted into the atmosphere through the smoke extractor 14.

Based on the above-mentioned, it becomes obvious that SHW, coal dust, sawdust and other hydrocarbon-containing substances aren't only waste, but can also become a source of energy and a raw material, which can be used to produce a valuable product.

But it's still necessary to be aware of expenditures for a recycling technological process and/or reprocessing of hydrocarbon-containing waste, which will vary in each individual case depending on a raw material, production volume of the facility and obtained valuable product. That's why the further goal was to determine how profitable such facilities would be. For this purpose a payback period was set at 5 years. Consequently, if the facility pays off in less than 5 years, let's assume it's profitable, and if in more than 5 years – unprofitable, as after 5 years additional investments in maintenance of the facility's equipment will be required. For each individual case a payback period may vary.

4 RESULTS

On the basis of a market analysis an average cost of the modular equipment, which has been used for calculations, has been determined.

Results of the analysis and the calculations are summarized in Table 1 and Table 2.

In Table 1 the cost of the valuable product and the profitability of the facilities are shown depending on amounts of recycled waste.

It can be concluded from Table 1 that the facility for recycling of coal dust with generation of a valuable product becomes profitable with a lower capacity if compared with facilities for reprocessing of woodworking waste and SHW. Moreover, a common pattern of all facilities for recycling of hydrocarbon-containing waste is that profitability with generation of hot water and liquid fuel is gained on a lower capacity than with producing of electrical energy. It can be seen that the profitability of the facilities increases with the growth of the capacity.

In Table 2 the cost of the equipment depending on the capacity for different configurations of the facility is shown.

Table 1. The cost of the valuable product and the profitability of the facilities based on the capacity.

Capacity	Coal dust			Sawdust			SHW		
	Hot water	Electricity	Liquid fuel	Hot water	Electricity	Liquid fuel	Hot water	Electricity	Liquid fuel
1650 kg/day	77,8 rub/m³	5,7 rub/kW	13,4 rub/L	134,7 rub/m³	10,4 rub/kW	24,1 rub/L	148,9 rub/m³	10,8 rub/kW	25,5 rub/L
3 t/day	50,4 rub/m³	3,7 rub/kW	8,9 rub/L	87,2 rub/m³	6,8 rub/kW	16,2 rub/L	98 rub/m³	7,2 rub/kW	17,3 rub/L
6 t/day	29,7 rub/m³	2,22 rub/kW	5,4 rub/L	51,3 rub/m³	4 rub/kW	9,8 rub/L	58.6 rub/m³	4,3 rub/kW	10,6 rub/L
10 t/day	26,6 rub/m³	2 rub/kW	4,9 rub/L	46 rub/m³	3,6 rub/kW	8,8 rub/L	52,1 rub/m³	3,9 rub/kW	9,6 rub/L
25 t/day	19 rub/m³	1,4 rub/kW	3,4 rub/L	32,8 rub/m³	2,6 rub/kW	6,3 rub/L	38,5 rub/m³	2,8 rub/kW	6,8 rub/L
50 t/day	14,7 rub/m³	1,1 rub/kW	2,6 rub/L	25,3 rub/m³	1,9 rub/kW	4,9 rub/L	29,7 rub/m³	2,1 rub/kW	5,3 rub/L
100 t/day	10,5 rub/m³	0,8 rub/kW	1,8 rub/L	18,1 rub/m³	1,4 rub/kW	3,4 rub/L	21,3 rub/m³	1,51 rub/kW	3,7 rub/L
150 t/day	7,6 rub/m³	0,6 rub/kW	1,3 rub/L	13 rub/m³	1 rub/kW	2,5 rub/L	15,3 rub/m³	1,1 rub/kW	2,7 rub/L
300 t/day	4,2 rub/m³	0,3 rub/kW	0,72 rub/L	7,2 rub/m³	0,57 rub/kW	1,3 rub/L	8,6 rub/m³	0,61 rub/kW	1,45 rub/L

Non-profitable (a payback period is more than 5 years)

A payback period is about 5 years

Profitable (a payback period is less than 5 years)

Table 2. The cost of the equipment based on the capacity.

Ca-pacity	Coal dust			Sawdust			SHW		
	Hot water mln rubles	Electricity mln rubles	Liquid fuel mln rubles	Hot water mln rubles	Electricity mln rubles	Liquid fuel mln rubles	Hot water mln rubles	Electricity mln rubles	Liquid fuel mln rubles
1650 kg/day	1,5	1,7	2	1,5	1,7	2	1,7	1,9	2,2
3 t/day	4,5	5,1	6	4,5	5,1	6	5,1	5,7	6,6
6 t/day	8	9,1	10,7	8	9,1	10,7	9,1	10,1	11,7
10 t/day	12,5	14,2	16,7	12,5	14,2	14,2	14,2	15,8	18,3
25 t/day	22,5	25,5	30	22,5	25,5	30	25,5	28,5	33
50 t/day	36	40,8	48	36	40,8	48	40,8	45,6	52,8
100 t/day	48	54,4	59,2	48	54,4	59,2	54,4	58,6	65,6
150 t/day	60	68	74	60	68	68	68	73,2	82
300 t/day	85	96,3	100,3	85	96,3	100,3	96,3	102	107,6

5 CONCLUSIONS

It's possible to end this article with a conclusion that waste recycling through gasification process in a modular-type thermochemical facility (Figure 1) can be accompanied by generation of a valuable product, which can become a source of energy for various technological processes.

A significant obstacle to wide implementation of waste as raw material for obtaining of a valuable product is a relative sophistication of the facilities (necessity of additional equipment) and lesser, if compared with natural resources, calorific value. But these dis-advantages are insignificant versus the positive points which waste recycling demonstrates: it's not only a low cost of electrical and thermal energy, but of liquid fuel as well, and also purer emissions in comparison with simple incineration and a high efficiency of the facilities. By reprocessing waste, in particular coal dust, sawdust and SHW, by recycling waste in a modular-type thermochemical facility (Figure 1) we aren't only reducing emissions, but also obtaining a valuable product (hot water, overheated steam, electrical energy, liquid fuel), which are always in demand.

Recycling of the hydrocarbon-containing waste can be not loss-making, but rather profitable on condition that its technological process is effectively organized, for instance, in application of the modular scheme and implementation of serial industrial equipment. In many cases development of hydrocarbon-containing waste recycling facilities doesn't re-quire designing of new ones but rather utilization of already existent modules.

Profitability of the facility which is based on the proposed modular-type structural layout (Figure 1), depends directly on the feedstock, the generated valuable product and the capacity (Table 1). It's necessary to consider these factors in order to make profit from waste recycling.

REFERENCES

Bilgesu, A. Y.; Kocak, M. C.; Karaduman. 2006. Waste Plastic Pyrolysis in Free-Fall Reactors. Feedstock Recycling and Pyrolysis of Waste Plastics, 2006: 605-623.

Fukuda T., Saito K., Suzuki S., Sato H., Hirota T. 1989. Processing for producing hydrocarbon oils from plastic waste. Patent Number US 4851601, 1989.

Kireyenko V.P. 2014. Environmental management economics. A teaching method. Minsk: SIMST BSU.

Konstantinov V.M. 2000. Protection of nature. A study guide for students in higher teacher education. M.: "Academia" publishing center.

Mário Luis Alves Ramalho Gomes. 2011. Waste to liquid hydrocarbon refinery system. Patent Number US20110158858A1, 2011.

Novikov N.N., Ustinova I.S, Novikov I.N. 2011. Technical specifications (TS) № 1. Waste recycling facility. - Intr. 2011-10-16. Rybinsk: Ltd "New energy".

Scheirs J., Kaminsky W. 2006. Feedstock Recycling and Pyrolysis of Waste Plastics: Converting Waste Plastics into Diesel and Other Fuels. John Wiley & Sons, Ltd.

Zyryanova U.P., Kuznetsov V.V., Lazarev V.N. 2011. Environmental management and environmental protection economics. A study guide. USTU typography.

Topical Issues of Rational Use of Natural Resources 2019 – Litvinenko (Ed)
© 2020 Taylor & Francis Group, London, ISBN 978-0-367-85720-2

Study of the equilibrium parameters and laboratory experiments of heavy metal(Cu^{2+}) adsorption from aqueous solution by nanoclay

S. Hasani & M. Ziaii
Petroleum and Geophysics Engineering, School of Mining, Shahrood University of Technology, Shahrood, Iran

F. Doulati Ardejani
School of Mining, College of Engineering, University of Tehran, Tehran, Iran

M.E Olya
Department of Environment Research, Institute for Color Science and Technology, Tehran, Iran

ABSTRACT: In this article, adsorption mechanism of heavy metal(copper) from aqueous solution onto nano-montmorillonite was investigated. The effect of several operational parameters such as adsorbent dosage, pH, agitation speed and temperature was studied. The laboratory experiments showed that the copper(Cu^{2+}) adsorption was dependent on adsorbent dosage, pH and temperature. And also equilibrium isotherms were analyzed by Langmuir, Freundlich and Temkin adsorption models. It was observed that the equilibrium data fitted well with the Freundlich adsorption model. In addition, the adsorption capacity, Q_0, was calculated 37.04 $mg.g^{-1}$ for adsorption of copper on nano-montmorillonte. Thermodynamic parameters suggest that the adsorption of Cu^{2+} on nano-montmorillonite is spontaneous and endothermic.

1   INTRODUCTION

Abandoned mines and mine wastes include heavy metals like copper, lead, cadmium, iron, aluminum, zinc and manganese which are the causes of surface and groundwater contamination. These days, water pollution by heavy metals is considered as a global and critical environmental issue because of their toxicity and carcinogenicity (Atasoy and Bilgic, 2018; Ma et al., 2018). The main applied techniques of heavy metals removal from wastewaters are including precipitation, ion exchange, membrane filtration, electrolytic methods, solvent extraction and adsorption (Fakhre and Ibrahim, 2018; Li et al., 2017). Currently, researchers are trying to find environmentally friendly and cost-effective wastewater treatment methods in order to remove heavy metals from aqueous solutions (Ahmed and Ahmaruzzaman, 2016; Park et al., 2016; Uddin, 2017).

Among the numerous methods, adsorption has been considered as an effective and economic wastewater treatment (Ali et al., 2016). Recently, a number of methods have been studied in order to develop low-cost and renewable adsorbents for the adsorption of various toxic metal contaminants. Numerous materials have been applied as adsorbents, chitosan (Li et al., 2015), agricultural waste material (Acharya et al., 2018), polymer-based nanocomposites (Zhao et al., 2018), and natural materials (Bhattacharyya and Gupta, 2008).

Clay and clay composite materials are used as environment-friendly adsorbents for toxic metal removal from wastewater due to their high surface area and high ion exchange capacity (Gu et al., 2018; Han et al., 2019; Hokkanen et al., 2018). Montmorillonite is one of the most plentiful minerals in nature (Zhu et al., 2018), and also it is mostly used for wastewater treatment because of its high adsorption capacity, beneficial surface properties and high cation exchangeability (Chu et al., 2019; Elsherbiny et al., 2018). So, raw and modified montmorillonite have been applied as adsorbents for numerous metals such as Cd^{2+}(Liu et al., 2018), Pb^{2+} and Cu^{2+}(Chu et al., 2019), Cr^{6+}(Cai et al., 2017), Ni^{2+}(Vieira et al., 2018).

In this study, adsorption mechanism of copper(Cu^{2+}) from aqueous solution by nano-montmorillonite as a natural adsorbent was investigated. The optimum conditions of copper (Cu^{2+}) adsorption from simulated water were determined. In addition, the adsorbate distribution between the liquid phase and solid phase is assessed by different isotherm models. Also, in order to figure out the nature of copper(Cu^{2+}) adsorption onto nanoclay, the thermodynamic parameters were determined.

2 MATERIALS AND METHODS

2.1 Adsorption isotherms

The isotherm experiments were performed by changing initial dye concentrations (C_0) from 5 to 100 mg.l^{-1} and mixing 0.25 g nano-montmorillonite in 200 ml solution. In addition, the experiments were done with an agitating speed of 30 rpm at a constant temperature 25±1°C for 15 minutes in order to achieve equilibrium conditions. Also, the initial pH of solution was determined 4.67. The amounts of Cu^{2+} adsorbed per unit weight of nanoclay (q_e) were calculated as following equation (Hasani et al., 2017):

$$q_e = \frac{(C_0 - C_e).V}{M} \tag{1}$$

Where C_0 = the initial concentration; C_e = the equilibrium concentration; V = the volume of the solution; and M = the mass of the dried adsorbent.

In order to determine the adsorption mechanism, the adsorption isotherms and their specific parameters are investigated (Castro et al., 2018). So, Langmuir, Freundlich, and Temkin isothermal models were used to indicate the adsorbate distribution between the liquid phase and solid phase.

2.2 Langmuir isotherm

In order to describe single-solute systems, the Langmuir isotherm has been usually applied. The Langmuir model assumes that adsorption takes place at homogeneous sites in the adsorbent and the adsorption of adsorbate on the surface of adsorbent is monolayer [3]. The Langmuir isotherm is expressed as follows (Ardejani et al., 2008):

$$q_e = \frac{Q_0 K_L C_e}{1 + K_L C_e} \tag{2}$$

Where Q_0 = the maximum adsorption capacity of adsorbent; and K_L = the Langmuir isotherm constant.

The separation factor, R_L, shows the shape of isotherm (Lei et al., 2018) which is expressed as Eq. (3):

$$R_L = \frac{1}{1 + K_L C_0} \tag{3}$$

$R_L > 1$ shows unfavorable adsorption, while $0 < R_L < 1$ shows favorable adsorption. Also, $R_L = 1$ and $R_L = 0$ show linear adsorption and irreversible adsorption, respectively.

2.3 Freundlich isotherm

In this research, Freundlich isotherm was used to describe heterogeneous systems. All adsorption sites within the adsorbent are heterogeneous and also the adsorption is multilayer adsorption based on Freundlich isotherm assumption. The Freundlich isotherm equation is given as (An et al., 2017):

$$q_e = K_F C_e^{1/n} \qquad (4)$$

Where K_F = the Freundkich constant; and n = the heterogeneity factor.

2.4 Temkin isotherm

The Temkin isotherm assumes that the heat of adsorption of adsorbate molecules on the surface of adsorbent particles would reduce linearly due to the adsorbent-adsorbate interactions (Araújo et al., 2018). Temkin isotherm can be expressed as following equation (Bordoloi et al., 2017):

$$q_e = \frac{RTln(K_T C_e)}{b} \qquad (5)$$

Where K_T = the equilibrium binding constant; b = the Temkin isotherm constant; R = the gas constant; and T = the temperature.

2.5 Thermodynamic study

In order to evaluate thermodynamic parameters, batch experiments were conducted by adding 0.25 g of nano-montmorillonite to 200 ml of 10 mg.l^{-1} metal ion solution at different temperatures (298 K, 318 K and 338 K) for 15 minutes.

The nature of copper(Cu^{2+}) adsorption onto nanoclay is recognized by thermodynamic parameters including changes in standard entropy (ΔS^0), standard enthalpy (ΔH^0) and standard Gibb's free energy (ΔG^0). The negative Gibbs free energy change represents favorable adsorption mechanism (Manirethan et al., 2018). The thermodynamic parameters are calculated as follows:

$$\Delta G^0 = -RTlnK_d \qquad (6)$$

$$K_d = \frac{q_e}{C_e} \qquad (7)$$

$$lnK_d = \frac{\Delta S^0}{R} - \frac{\Delta H^0}{RT} \qquad (8)$$

3   RESULTS AND DISCUSSION

3.1 Effect of nanoclay dosage

The effect of adsorbent dosage was investigated by adding 0.04-0.35 g of nano-montmorillonite to 200ml of 10 mg.l^{-1} solution. From Figure 1, it is perceived that the percentage of copper removal increased from 42.66 to 87.05% due to the increase in surface area and the availability of more adsorption sites. In addition, the adsorption capacity decreased because of the adsorption sites remaining unsaturated (Iriel et al., 2018). Also, increase in the

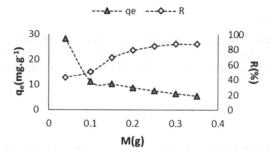

Figure 1. Effect of adsorbent dosage on copper adsorption (Condition: pH=4.67, T=25±1°C, C_0=10 mg.l^{-1}, equilibrium time=15 min and stirring speed=200rpm).

amount of nanoclay from 0.25 g to 0.35 g did not have effect on copper removal from aqueous solution. So, 0.25 g of nano-monrmorillonite was considered as the optimum amount of adsorbent.

3.2 *Effect of pH*

The surface charge of the adsorbent can be affected by the pH of solution (Hasani et al., 2017). So, the initial pH of solution was changed from 2 to 6 in order to investigate the influence of the pH on copper(Cu^{2+}) adsorption. As Figure 2, the adsorption of copper(Cu^{2+}) increased with growth of pH due to decrease in H^+ ions and increase in the available adsorption sites (Nguyen et al., 2018). According to Figure 2, increase in pH of solution from 4 to 6 did not affect copper(Cu^{2+}) removal from aqueous solution. So, the natural pH of solution(pH=4.67) was considered as the optimum value.

3.3 *Effect of stirring speed*

To study the effect of stirring speed on copper(Cu^{2+}) adsorption mechanism, the stirring speed varied from 30 to 200 rpm. Adsorption experiments were done at optimum values of pH and adsorbent dosage for equilibrium time(15 min). Figure 3 shows that the stirring speed did not have significant effect on copper(Cu^{2+}) adsorption from aqueous solution. So, the stirring speed of 30rpm was considered as the optimum value.

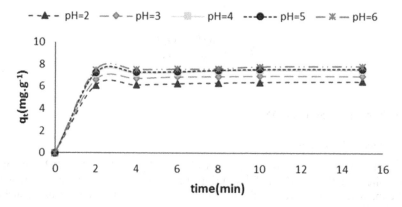

Figure 2. Effect of pH on copper adsorption (Condition: M=0.25g, T=25±1°C, C_0=10 mg.l^{-1} and stirring speed=200rpm).

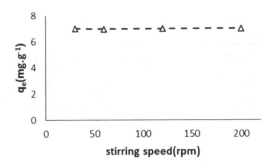

Figure 3. Effect of stirring speed on copper adsorption (Condition: M=0.25g, T=25±1°C, C_0=10mg.l^{-1}, pH=4.67 and equilibrium time=15 min).

3.4 Effect of temperature

The relationship between the amount of copper(Cu^{2+}) adsorption and solution temperature was studied at 298, 318 and 338 K at a pH value of 4.67 and 0.25 g adsorbent dosage to determine the enthalpy and entropy changes. As Figure 4, the amount of copper(Cu^{2+}) adsorption increased from 6.90 to 8.13 mg.g^{-1} with an increase in temperature because a greater number of molecules gain adequate energy to undergo an interaction with active sites on nano-montmorillonite surface (Al-Harahsheh et al., 2015).

3.5 Adsorption isotherms

Table 1 and R^2 values indicate that the experimental adsorption data of Cu^{2+} had the best agreement with Freundlich model. So, the adsorption of Cu^{2+} onto nano-montmorillonite is multilayer (An et al., 2017). And also, the value of n shows lesser heterogeneity and favorable adsorption because n value is between 1 and 10 (Lei et al., 2018).

According to the R_L value which is shown in Table 1, the Cu^{2+} adsorption onto the adsorbent is favorable and the value of Q_0 showed the high adsorption capacity of nano-montmorillonite.

3.6 Thermodynamic study

Thermodynamic parameters of copper(Cu^{2+}) adsorption were calculated at 298, 318 and 338 K (Table 2) using Eqs. (6), (7) and (8). As can be seen, ΔG^0 values were negative at all the temperatures. So, the adsorption of copper(Cu^{2+}) onto nanoclay was feasible and spontaneous. Also, the positive values of ΔH^0 and ΔS^0 show the endothermic nature of copper(Cu^{2+})

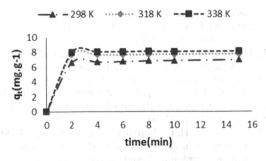

Figure 4. Effect of temperature on copper adsorption (Condition: M=0.25g, stirring speed=30rpm, C_0=10 mg.l^{-1} and pH=4.67).

Table 1. Parameters of the Freundkich, Langmuir and Temkin isotherms of Cu^{2+} adsorption.

Langmuir isotherm				Freundlich isotherm			Temkin isotherm		
Q_0	K_L	R_L	R^2	K_F	n	R^2	K_1	K_2	R^2
37.04	0.09	0.22	0.96	5.72	2.39	0.99	1.01	429	0.93

Table 2. Thermodynamic parameters of Cu^{2+} adsorption.

Metal	$\Delta G\left(\frac{kj}{mol}\right)$			$\Delta H\left(\frac{kj}{mol}\right)$	$\Delta S\left(\frac{J}{mol.K}\right)$	R^2
Cu^{2+}	298	318	338			
	-3.53	-4.73	-5.93	14.37	60.05	0.9751

adsorption and an increased randomness of the adsorbed Cu^{2+} ions on nano-momtmorillonite surface, respectively (Mahmood-ul-Hassan et al., 2018).

4 CONCLUSION

In this study, the adsorption of copper(Cu^{2+}) by nano-montmorillonite was investigated. The laboratory experiments showed that the copper(Cu^{2+}) adsorption was dependent on adsorbent dosage, pH and temperature. In addition, it was observed that the equilibrium data fitted well with the Freundlich adsorption model and the adsorption process is multilayer. Also, the adsorption capacity, Q_0, was calculated 37.04 mg.g^{-1} for adsorption of copper on nano-montmorillonte. Thermodynamic parameters suggest that the adsorption of copper on nanoclay is spontaneous and endothermic. According to the high adsorption capacity and thermodynamic consistency, nano-montmorillonite can be used as an effective adsorbent for Cu^{2+} removal from wastewater.

REFERENCES

Acharya, J., Kumar, U., and Rafi, P.M., 2018, Removal of Heavy metal ions from wastewater by chemically modified agricultural waste material as potential adsorbent-a review: International Journal of Current Engineering and Technology, v. 8, no. 3, p. 526–530.

Ahmed, M.J.K., and Ahmaruzzaman, M., 2016, A review on potential usage of industrial waste materials for binding heavy metal ions from aqueous solutions: Journal of Water Process Engineering, v. 10, p. 39–47.

Al-Harahsheh, M.S., Al Zboon, K., Al-Makhadmeh, L., Hararah, M., and Mahasneh, M., 2015, Fly ash based geopolymer for heavy metal removal: A case study on copper removal: Journal of Environmental Chemical Engineering, v. 3, no. 3, p. 1669–1677.

Ali, R.M., Hamad, H.A., Hussein, M.M., and Malash, G.F., 2016, Potential of using green adsorbent of heavy metal removal from aqueous solutions: adsorption kinetics, isotherm, thermodynamic, mechanism and economic analysis: Ecological Engineering, v. 91, p. 317–332.

An, F.-Q., Wu, R.-Y., Li, M., Hu, T.-P., Gao, J.-F., and Yuan, Z.-G., 2017, Adsorption of heavy metal ions by iminodiacetic acid functionalized D301 resin: kinetics, isotherms and thermodynamics: Reactive and Functional Polymers, v. 118, p. 42–50.

Araújo, C. S., Almeida, I. L., Rezende, H. C., Marcionilio, S. M., Léon, J. J., and de Matos, T. N., 2018, Elucidation of mechanism involved in adsorption of Pb (II) onto lobeira fruit (Solanum lycocarpum) using Langmuir, Freundlich and Temkin isotherms: Microchemical Journal, v. 137, p. 348–354.

Ardejani, F. D., Badii, K., Limaee, N. Y., Shafaei, S. Z., and Mirhabibi, A., 2008, Adsorption of Direct Red 80 dye from aqueous solution onto almond shells: Effect of pH, initial concentration and shell type: Journal of hazardous materials, v. 151, no. 2-3, p. 730–737.

Atasoy, A., and Bilgic, B., 2018, Adsorption of copper and zinc ions from aqueous solutions using montmorillonite and bauxite as low-cost adsorbents: Mine Water and the Environment, v. 37, no. 1, p. 205–210.

Bhattacharyya, K. G., and Gupta, S. S., 2008, Adsorption of a few heavy metals on natural and modified kaolinite and montmorillonite: a review: Advances in colloid and interface science, v. 140, no. 2, p. 114–131.

Bordoloi, N., Goswami, R., Kumar, M., and Kataki, R., 2017, Biosorption of Co (II) from aqueous solution using algal biochar: kinetics and isotherm studies: Bioresource technology, v. 244, p. 1465–1469.

Cai, J., Lei, M., Zhang, Q., He, J.-R., Chen, T., Liu, S., Fu, S.-H., Li, T.-T., Liu, G., and Fei, P., 2017, Electrospun composite nanofiber mats of Cellulose@ Organically modified montmorillonite for heavy metal ion removal: Design, characterization, evaluation of absorption performance: Composites Part A: Applied Science and Manufacturing, v. 92, p. 10–16.

Castro, L., Blázquez, M. L., González, F., Muñoz, J. A., and Ballester, A., 2018, Heavy metal adsorption using biogenic iron compounds: Hydrometallurgy, v. 179, p. 44–51.

Chu, Y., Khan, M. A., Wang, F., Xia, M., Lei, W., and Zhu, S., 2019, Kinetics and equilibrium isotherms of adsorption of Pb (II) and Cu (II) onto raw and arginine-modified montmorillonite: Advanced Powder Technology.

Elsherbiny, A. S., El-Hefnawy, M. E., and Gemeay, A. H., 2018, Adsorption efficiency of polyaspartate-montmorillonite composite towards the removal of Pb (II) and Cd (II) from aqueous solution: Journal of Polymers and the Environment, v. 26, no. 2, p. 411–422.

Fakhre, N. A., and Ibrahim, B. M., 2018, The use of new chemically modified cellulose for heavy metal ion adsorption: Journal of hazardous materials, v. 343, p. 324–331.

Gu, S., Kang, X., Wang, L., Lichtfouse, E., and Wang, C., 2018, Clay mineral adsorbents for heavy metal removal from wastewater: a review: Environmental Chemistry Letters, p. 1–26.

Han, H., Rafiq, M. K., Zhou, T., Xu, R., Mašek, O., and Li, X., 2019, A critical review of clay-based composites with enhanced adsorption performance for metal and organic pollutants: Journal of Hazardous Materials.

Hasani, S., Ardejani, F. D., and Olya, M. E., 2017, Equilibrium and kinetic studies of azo dye (Basic Red 18) adsorption onto montmorillonite: Numerical simulation and laboratory experiments: Korean Journal of Chemical Engineering, v. 34, no. 8, p. 2265–2274.

Hokkanen, S., Bhatnagar, A., Srivastava, V., Suorsa, V., and Sillanpää, M., 2018, Removal of Cd2+, Ni2+ and PO43− from aqueous solution by hydroxyapatite-bentonite clay-nanocellulose composite: International journal of biological macromolecules, v. 118, p. 903–912.

Iriel, A., Bruneel, S. P., Schenone, N., and Cirelli, A. F., 2018, The removal of fluoride from aqueous solution by a lateritic soil adsorption: kinetic and equilibrium studies: Ecotoxicology and environmental safety, v. 149, p. 166–172.

Lei, C., Yan, B., Chen, T., and Xiao, X.-M., 2018, Preparation and adsorption characteristics for heavy metals of active silicon adsorbent from leaching residue of lead-zinc tailings: Environmental Science and Pollution Research, v. 25, no. 21, p. 21233–21242.

Li, S., Wang, W., Liang, F., and Zhang, W.-x., 2017, Heavy metal removal using nanoscale zero-valent iron (nZVI): Theory and application: Journal of hazardous materials, v. 322, p. 163–171.

Li, X., Zhou, H., Wu, W., Wei, S., Xu, Y., and Kuang, Y., 2015, Studies of heavy metal ion adsorption on Chitosan/Sulfydryl-functionalized graphene oxide composites: Journal of colloid and interface science, v. 448, p. 389–397.

Liu, W., Zhao, C., Wang, S., Niu, L., Wang, Y., Liang, S., and Cui, Z., 2018, Adsorption of cadmium ions from aqueous solutions using nano-montmorillonite: kinetics, isotherm and mechanism evaluations: Research on Chemical Intermediates, v. 44, no. 3, p. 1441–1458.

Ma, J., Liu, Y., Ali, O., Wei, Y., Zhang, S., Zhang, Y., Cai, T., Liu, C., and Luo, S., 2018, Fast adsorption of heavy metal ions by waste cotton fabrics based double network hydrogel and influencing factors insight: Journal of hazardous materials, v. 344, p. 1034–1042.

Mahmood-ul-Hassan, M., Yasin, M., Yousra, M., Ahmad, R., and Sarwar, S., 2018, Kinetics, isotherms, and thermodynamic studies of lead, chromium, and cadmium bio-adsorption from aqueous solution onto Picea smithiana sawdust: Environmental Science and Pollution Research, p. 1–9.

Manirethan, V., Raval, K., Rajan, R., Thaira, H., and Balakrishnan, R. M., 2018, Kinetic and thermodynamic studies on the adsorption of heavy metals from aqueous solution by melanin nanopigment obtained from marine source: Pseudomonas stutzeri: Journal of environmental management, v. 214, p. 315–324.

Nguyen, T. C., Loganathan, P., Nguyen, T. V., Kandasamy, J., Naidu, R., and Vigneswaran, S., 2018, Adsorptive removal of five heavy metals from water using blast furnace slag and fly ash: Environmental Science and Pollution Research, v. 25, no. 21, p. 20430–20438.

Park, J.-H., Ok, Y. S., Kim, S.-H., Cho, J.-S., Heo, J.-S., Delaune, R. D., and Seo, D.-C., 2016, Competitive adsorption of heavy metals onto sesame straw biochar in aqueous solutions: Chemosphere, v. 142, p. 77–83.

Uddin, M. K., 2017, A review on the adsorption of heavy metals by clay minerals, with special focus on the past decade: Chemical Engineering Journal, v. 308, p. 438–462.

Vieira, R. M., Vilela, P. B., Becegato, V. A., and Paulino, A. T., 2018, Chitosan-based hydrogel and chitosan/acid-activated montmorillonite composite hydrogel for the adsorption and removal of Pb+ 2 and Ni+ 2 ions accommodated in aqueous solutions: Journal of Environmental Chemical Engineering, v. 6, no. 2, p. 2713–2723.

Zhao, G., Huang, X., Tang, Z., Huang, Q., Niu, F., and Wang, X., 2018, Polymer-based nanocomposites for heavy metal ions removal from aqueous solution: a review: Polymer Chemistry, v. 9, no. 26, p. 3562–3582.

Zhu, H., Xiao, X., Guo, Z., Han, X., Liang, Y., Zhang, Y., and Zhou, C., 2018, Adsorption of vanadium (V) on natural kaolinite and montmorillonite: Characteristics and mechanism: Applied Clay Science, v. 161, p. 310–316.

Topical Issues of Rational Use of Natural Resources 2019 – Litvinenko (Ed)
© 2020 Taylor & Francis Group, London, ISBN 978-0-367-85720-2

Technological aspects of reducing ecological risks from industrial wastewater pollution

A.S. Kitaeva, V.F. Skorohodov & R.M. Nikitin
Mining Institute of the Kola Science Centre, Apatity, Russia

ABSTRACT: A technology was developed which meets modern requirements for wastewater treatment from multicomponent pollutions as a result of coagulation, sorption and flotation in activated water dispersion of air by accumulating contaminants in a multiphase system in a single volume, based on a flotation machine constructed in the Mining Institute of the Kola Science Centre of the RAS. The technology gives a necessary purification degree independently on pollutions at the input and doesn't require preliminary sedimentation of suspended matters. Industrial testing indicates its high competitive ability compare to other known methods on wastewater treatment.

1 INTRODUCTION

As a result of the mining enterprises work the large volumes of the wastewater, containing harmful impurities, are dumped into the natural water bodies. This entails increasing the ecological load on the environment and the additional costs of the industrial enterprises for the penal sanctions, as well as the huge costs of the environmental protection. The volume of water consumed by mining enterprises for mines is 6-10 million m^3 per year, quarries – 1-7 million m^3 per year, concentration plants – 10-300 million m^3 per year.

Until now, the problem of the wastewater treatment of the mining enterprises is one of the most topical. This fact is confirmed by the almost unchanged volume of the discharged polluted wastewater from the mining in the surface water bodies in 2014 – 813.2 million m^3, 2015 – 839.1 million m^3, 2016 – 801.3 million m^3 (National environmental report, 2017). The costs of the environmental protection, aimed at the wastewater selection and treatment of the mining and processing enterprises occupy a leading position in relation to enterprises of other activities. Thus, in the production of the food products, including potables and tobacco, 1,705.8 million rubles were spent in 2016 on the wastewater selection and treatment, and in the extraction of the mineral resources – 17,762.3 million rubles (National environmental report, 2017).

Due to the uniqueness and wealth of the mineral resources, the Kola Peninsula has a highly developed industrial potential. Powerful mining and processing and metallurgical enterprises are located in the Murmansk Region, a nuclear power plant has been built, cities and towns have been established. The main minerals in the peninsula are apatite-nepheline (the Khibiny and Lovozero tundras deposits), iron (Olenegorsk and Kovdor deposits) and copper-nickel ores (the Pechenga and Monchegorsk deposits groups). The largest reserves of the rare-earth metals are concentrated in the depth of the Lovozero deposit. The mineral extraction and processing of the Kola Peninsula by the powerful mining and processing enterprises led to a significant anthropogenic load on the surface waters of the territory under consideration. Thousands tons of the mineral salts, suspended solids, biogenic elements, hundreds tons of the heavy metals enter the lakes with the wastes of the mining complex, in addition there is aerotechnogenic pollution of the territory due to the transfer some of these substances through the atmosphere (Rumjanceva et al, 2015).

The main causes of the aquatic ecosystem pollution during the mineral resources extraction and processing are:

1. A huge amount of the industrial wastewater.
2. Complex chemical composition of the wastewater.
3. Low effectiveness of the applied methods for treatment.

The main methods of the wastewater treatment include the removal of the suspended solids by precipitating them in open tank, which area is calculated from the sedimentation time of the coagulated impurities and suspended particles (Jahanshahi et al, 2018). Depending on this parameter and the volume of treated water, the areas of the tanks reach significant values, which sometimes make it difficult to use any methods of the wastewater treatment because of the high costs required to create treatment facilities. The restrictions are imposed, when using filters made of synthetic and natural materials. One of these restrictions is the volume of water to be purified (Pearce, 2017). In addition, large volumes of waste to be disposed of are generated by the sorption purification methods (De Gisi, 2016). Therefore, the considerable attention is paid to the issues of the wastewater treatment processes intensification.

The wastewater treatment of the industrial enterprises should be accompanied by the achievement of the necessary requirements for its composition for reuse, or comply with the MPC when discharging treated water into the open water bodies. The replacing natural fresh water consumed for the technological needs with treated wastewater will solve the problem of the scarcity and prevent the depletion of their reserves. Thus. the development of new highly efficient and cost-effective technological and technical solutions for the wastewater treatment is one of the critical tasks facing the enterprises of the mining complex.

2 THEORETICAL STUDIES

At the first stage, the task was to form aggregates, including components present from the treated waters. The removal of these components from treated water will achieve a high degree of the purification. As a rule, petroleum products, finely dispersed substances, which are represented by fine mineral phase and dissolved chemical compounds, are present in the wastewater. The system under consideration is distinguished by aggregative and kinetic stability. Water is not clarified by the simple settling. The aggregative stability is due to the ability of the smallest particles to have their own double electric layer of the ions and salvation shells, and kinetic stability is caused by the ability of the particles to resist the gravity. Thus, each particle of a finely dispersed phase is subjected to the action of two opposite forces: gravity and diffusion, which impedes it. The vision is aggravated by an increase in the viscosity of the system, which in turn is associated with the increased influence of the shear forces. To bring the presented system out of balance it is necessary to enlarge the particles by their aggregation, reducing their surface charge, so the gravity of the formed aggregates predominates over the remaining forces. Various coagulants are used for this purpose. The second part of the dispersed system is represented by petroleum products. They can be carried to the colloidal part of the considered system. These particles should be easily a part of micelles, without having a solubilizing effect on them due to relatively small concentrations. Many researchers have noted the formation of associations from the mineral constituents and organic substances. With an increase in the hydrophobicity degree of the mineral suspensions, the ability to form micelles in the presence of the surface-active reagents should increase. This results in a change in the hydrophobicity of the solid phase in this case and the complex composition of the micelle in the presence of petroleum products in the water.

Consequently, one of the solutions for the effective wastewater treatment is the choice of the coagulation conditions, i.e. this may include the selection of a non-deficient coagulant that is unable to accumulate in the treated water and the conditions choice for the use of a coagulant (Prakash et al, 2014; Chebakova, 2001).

The wastewater treatment by the contaminants coagulation with the following removal of the formed floccules is one of main methods due to its cheapness, comparative simplicity of equipment, and possibility of the effective removal of obtained hardly soluble compounds with their disposal. The study uses the ferrous sulfate as a coagulant, which has a variety of advantages. One of the main ones is the use of $Fe_2(SO_4)_3 \cdot 7H_2O$ in a wide range of the pH values.

The formation of the ferrum hydroxides sols can be considered as three interrelated stages: hydrolysis, a rise of the nuclei solid phase, and the growth of nuclei and their transformation into the particles, forming a micro heterogeneous system by coagulation of the primary particles. At the hydrolysis stage, the coordination of water molecules with metal ions forms hydro complexes of the form of $Fe(H_2O)_6OH^{2+}$. For the heavy metal hydroxides, the characteristic form in the presence of the sulfate ions is compounds of the type of $MeSO_4$ and $Me(oh)_2H_2O$. In the dilute solutions the balance offsets sideward the formation of the simple hydrates. The supermicellar structures of the hydrated ferrum hydroxides are formed as a result of the coagulation processes. They have a huge active surface on which the various water contaminants can be sorbed. The formed floccules differ with an extremely extended surface and consist of the three-dimensional cells, inside which water is contained; it determines their deposition rate. The formation of the ferrum hydroxides is a difficult multifactorial process, depending on the occurrence conditions. In the presence of the extraneous metal cations, the structure of hydroxides, in addition to the sorbed elements, may also include a metal cation, forming a common metal-ferrum-hydroxide complex of the $Me_xFe_{3-x}O_4$ form.

The coagulation process can be divided down into stages: the first stage is hydrolysis of the injected coagulant, the second is the floccules nucleation and flocculation; and the third stage includes the floccules enlargement, their aging and sedimentation.

The analysis of the theoretical calculation of coagulation conditions shows their possible use in the certain systems with significant restrictions, which dramatically reduces their practical applicability to real systems. Therefore, to simplify the task, the optimum pH value was established experimentally and is 9-9.5, which satisfies the conditions of the ferrum hydrolysis and coagulation.

At the first research stage it was established that the input of a fatty acid collector into the purified water allows intensifying the coagulation process. When it is entered, a more rapid formation of hardly soluble ferrum compounds with the collector anion occurs, which accelerates the formation of the nuclei during coagulation. Floccules formed in the presence of the ferrum hydroxide and fatty acid collectors are removed from the effluents being purified by flotation.

For the successful flotation, the gas phase dispersivity and the total gas content are of a decisive importance. It is known that in the existing pneumatic machines the regulation of gas phase dispersion and gas saturation is limited. For example, a mechanical increase in air consumption for all designs of aerators leads to the formation of not only useful fine bubbles, but also the appearance of large bubbles in the pulp, violating the hydrodynamics of the process. At that, the structural changes in air bubbles are associated both with coalition phenomena and with the mechanics of bubble formation and gas outflow through pores of various sizes.

3 WASTEWATER TREATMENT TECHNOLOGY

To solve the problem of the reducing environmental pollution in the discharged waters, the Mining Institute of the Kola Science Centre of the Russian Academy of Science has developed a technology of the wastewater treatment, based on a synergistic effect when concentrating pollution in a multiphase system (Stepannikova et al, 2016). This technology includes coagulation, sorption and flotation in an activated water dispersion of air (AWDA) (Figure 1). The basic device for the implementation of a new method for wastewater treatment is a flotation machine for the extraction of the contaminants in AWDA (Figure 2). The operating principle of the proposed device for the wastewater treatment is based on the ability of the activated gas bubbles to interact with the hydrophobic flocculating pollutants presented in industrial wastewater.

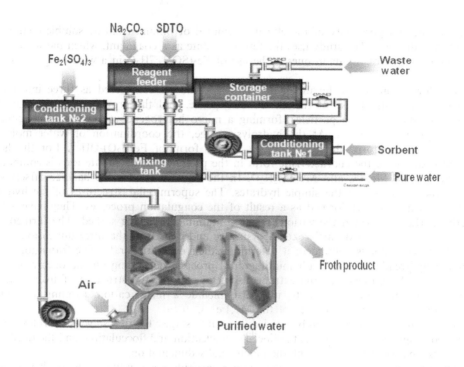

Figure 1. The wastewater treatment technology developed in the Mining Institute.

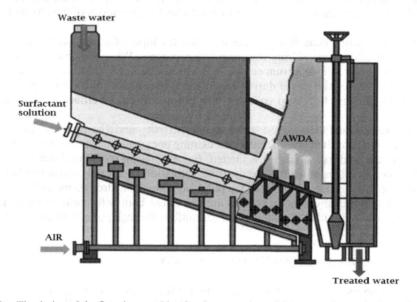

Figure 2. The design of the flotation machine for the extraction of the contaminants in AWDA.

The advantages of the proposed technology include:

1. Small compared to other methods, the volume of waste to be disposed of;
2. The use of non-toxic reagents, which has a positive effect on the quality of purified water;

3. The proposed technology allows processing large volumes of treated water without significant capital expenditures, since does not require the construction of large waste water treatment plants.

This wastewater treatment technology has been applied at a number of the mining enterprises of the Kola Peninsula, such as OJSC Kovdor MPP, LLC Lovozero MPP, OJSC
Apatite.

The wastewater of these enterprises are characterized by the presence of such contaminants
as copper, manganese, calcium, ferrum, fluorine, sulfate ions, petroleum products, suspended
solids, BOD, COD. The Lake Lovozero has been exposed to fluorine-containing waste water
from the Karnasurt mine for over 60 years. The mineral salts, fluorine and suspended substances are dumped into the lake as part of the wastewater. And the long-term inflow of wastewater into the Lake Kovdor led to an increase in the total salinity of its water, accompanied
by a change in its ionic composition (Petrova et al, 2011). The lake water corresponds to the
class of sulphates, which account for more than half of the total content of anions, the predominant cation is calcium. Along with the growth of salinity in the lake Kovdor water dramatically increased the content of the heavy metals such as strontium, aluminum, iron, zinc,
copper.

Table 1 and 2 present the results of the wastewater treatment at the Kovdorsky GOK and
Lovozersky GOK. According to these results, it is possible to see a decrease in the concentration of contaminants below the maximum permissible concentrations for all controlled
components.

Table 1. The test results of the wastewater treatment technology at the Kovdorsky GOK.

Analyzed elements	Units of measurement	Waste water	Purified water
pH	-	8.5	7.4
Suspended solids	mg/dm^3	8	0
NH_4^+	mg/dm^3	0.2	0.19
NO_2^-	mg/dm^3	0.15	0.02
NO_3^-	mg/dm^3	2.72	0.29
Cl^-	mg/dm^3	7.73	6.3
SO_4^{2-}	mg/dm^3	261	5.9
Ca^{2+}	mg/dm^3	48.1	3
Mg^{2+}	mg/dm^3	31.6	1.8
Petroleum products	mg/dm^3	0.66	0.025
PO_4	mg/dm^3	0.58	0.02

Table 2. The test results of the wastewater technology at the Lovozersky GOK.

Analyzed elements	Units of measurement	Waste water	Purified water
pH	-	8.5	7.4
Suspended solids	mg/dm^3	5.8	1.6
NH_4^+	mg/dm^3	0.157	0.067
NO_2^-	mg/dm^3	0.198	0.02
NO_3^-	mg/dm^3	1.85	0.32
Cl^-	mg/dm^3	2.38	1.53
SO_4^{2-}	mg/dm^3	7.6	2.71
Cu^{2+}	mg/dm^3	0.005	0.002
Mn^{2+}	mg/dm^3	0.1	0.01
Petroleum products	mg/dm^3	0.17	0.01
PO_4	mg/dm^3	0.47	0.11
F^-	mg/dm^3	7.46	0.63

4 CONCLUSION

On the basis of the studies performed and the results obtained, it can be concluded that this technology of the wastewater treatment for discharge into open water bodies allows achieving high performance and effectively removing multicomponent impurities without significant capital expenditures with the required high degree of the purification. It should be noted that the advantages of the developed technology also include eliminating the need for preliminary sedimentation of suspended solids and coagulated dissolved impurities and reducing the amount of waste to be disposed of, which in turn makes the proposed method more effective compared to other known methods.

The developed technology of the wastewater treatment meets modern requirements. This technology allows send up to 80% of the treated water in the circulating water supply, which reduces the consumption of clean water and the volume of the wastewater discharged into the natural water bodies. The implementation of the wastewater treatment technology allows reducing the volume of the discharged water by 5.7 million m^3/year (on the example of the Kovdor GOK), reducing the amount of penalties for exceeding the MPC discharged into open reservoirs of the wastewater, improving the quality of water in these reservoirs and providing favorable conditions for their use for fisheries.

REFERENCES

Chebakova, I.B. 2001. The wastewater treatment: textbook for university students. [Omsk: OmGTU, p. 84. In Russian.

De Gisi, S. 2016. Characteristics and adsorption capacities of low-cost sorbents for wastewater treatment: A review, *Sustainable Materials and Technologies*, V. 9, pp. 10–40.

Jahanshahi, M. & Taghizadeh, M.M. 2018. Pre-sedimentation tank effects on water treatment unit operation, *Environmental quality*. №28, pp. 35–42.

National report on the state and environmental protection of the Russian Federation, 2017.

Pearce, G. 2007. Introduction to membranes: Filtration for water treatment, *Filtration and Separation*, V.44, pp. 24–27.

Petrova, V.A. & Pashkevich, M.A. 2011. The state of the water bodies in the zone of influence of mining enterprises on the example of "Kovdorskij GOK", *Almanac of the modern science and education*, №4, pp. 107–109.

Prakash, N.B. et al. 2014. Waste water treatment by coagulation and flocculation, *International Journal of Engineering Science and Innovative Technology*, Vol. 3.

Rumjanceva, V.A. et al. 2015. *Lakes of the European part of Russia*. St. Petersburg: LEMA.

Stepannikova A.S. & Artem'eva O.A. 2016. Improving the efficiency of the wastewater treatment of the mining enterprises from multicomponent impurities. *In the collection: Problems of the subsoil development in the XXI century through the eyes of the young. Materials of the 13th International Scentific School of Young Scientists and Specialists*, pp. 296–299.

The intensifying pulverized coal combustion in blast furnace

V.V. Kochura & A.S. Novikov
Donetsk National Technical University, Donetsk, Ukraine

ABSTRACT: Replacement of metallurgical coke by pulverized coal (PC) injected in blast furnace (BF) tuyeres is a major economical challenge, due to the high price of coke and unfavourable effect of its production for the environment. Theoretical and experimental research of PC burning process under conditions of raceway have been carried out. Among them there are enriching blast with oxygen and its rational use. Recommendations to coal grinding and method of complex compensation for the changes in technological regime have been developed.

1 INTRODUCTION

Pulverized coal injection into the blast furnace is an effective technique as a means of stable operation, cost reduction, productivity increase and environmental protection. The first in Europe commercial complex for PC preparation and injection was built in 1980 at Donetsk Metallurgical Woks (DMW) (Yaroshevskiy 2006). At present more than 100 complexes operate with PC injection more than 200 kg/tHM in countries of European Community, China, Japan, Korea, USA and other countries. The best results have been achieved at Kakogawa BF-1, Hoogovens BF-6, Bao Steel BF, Scunthorpe Victoria BF, and Gary Works BF-13, with PCI about 200 kg/tHM (Savchuk 2000, Kurunov 2010, Geerdes 2016, Bosenhofer 2019).

The effective use of expensive PCI complexes can be attained when consumption rate of coal gets to maximum.

The main factors which hinder the increase of PC quantity being injected into the hearth of furnaces is the maintenance of its complete gasification within the raceway and smooth operation of a furnace due to change of heat, slag-formation and gas distribution conditions.

The results of research of PC burning process in the raceway, investigations of different type of coals and changes in reduction and temperature processes are given in this paper.

2 INTENSIFICATION OF PC COMBUSTION IN THE RACEWAY

2.1 *Theoretical research*

The relationship between incompletely burnt fuel, q, and the flowing size of particles of different fractions in polydispersioned flame, is expressed in the following way (Pomerantsev 1986).

$$q = K^P \int_{(\delta_{0i})_\tau}^{\delta_{01}} \frac{|dR_{0i}|}{d\delta_{0i}} \left(\frac{\delta_i}{\delta_{0i}}\right)^3 d\delta_{0i}, \tag{1}$$

where q = incompletely burnt fuel (-); K^P = relative content of coke residue in working mass of fuel (kg/kg); δ_{0i}, δ_i = correspond to initial and flowing size of particle of i-fraction (m); δ_{0i} = initial size of the biggest particle (m); $(\delta_{0i})_\tau$ = initial size of particle completely burnt by the moment of time τ (m); R_{0i} = relative mass content (remain of sieve) of particles with size, equal to or bigger than δ_{0i} in the initial dust (-).

The character of burning process of PC flow is expressed by non dimension expression:

$$\frac{K_1 \cdot \delta_1}{Nu \cdot D} \tag{2}$$

where K_1 = constant of burning velocity (m/s); δ_1 = maximum size of particle (m); Nu = diffusional criterion of Nusselt (-); and D = coefficient of oxygen diffusion in gas into the zone of burning (m²/s).

2.2 Individual delivery of oxygen to BF tuyeres

Oxygen enrichment of blast under PC injection was proved by industrial experiments (Langer 2005, Ryzhenkov 2005, Murai 2016, Zhou 2017, Zhang 2017, Chai 2019). However these methods do not ensure maximum oxygen utilization efficiency because design of tuyere doesn't give even mixture of coal particles with blast flow.

Due to complication of complete PC mixing with blast is proposed to increase the local oxygen concentration in the place of moving PC jet. These proposals are known and put into practice at Thyssen Stahl and British Steel BFs. However they do not provide rational use of oxygen because of decreasing due to diffusion oxygen concentration by the beginning of burning coke residue of coal. Therefore it is necessary to determine place and time for adding oxygen to PC flow with purpose to provide maximum oxygen concentration in the moment of coke residue burning.

This is due to the fact that combustion of coal particles occurs in four stages (Babiy 1986).

$$\tau_1 = k_1 \cdot 5.3 \cdot 10^{14} \cdot T^{-4} \cdot d^{0.8} \tag{3}$$

$$\tau_2 = k_2 \cdot 0.5 \cdot 10^6 \cdot d^2 \tag{4}$$

$$\tau_3 = k_3 \cdot 5.36 \cdot 10^7 \cdot T^{-1.2} \cdot d^{1.5} \tag{5}$$

$$\tau_4 = k_4 \cdot 2.21 \cdot 10^8 \frac{100 - A}{100} \frac{\rho \cdot d^2}{T^{0.9} \cdot O_2} \tag{6}$$

where: τ_1, τ_2, τ_3, τ_4 = times of heating of particle before ignition of volatile matter, volatile matter burning, heating of coke residue and coke residue burning, respectively, (s); k_1, k_2, k_3, k_4 = coefficients which depend on the coal grade (-); T = the temperature of the surrounding atmosphere (K); d = the average initial particle size (m); A = ash content in fuel (%); O_2 = oxygen concentration in the blast (m³/m³); ρ = the apparent density of coke residue (kg/m³).

An increase in oxygen concentration affects only the time over which the last stage occurs. Therefore preliminary mixing of fuel with oxygen is inexpedient: when it enters the tuyere a considerable portion of the PC particles burn before they emerge into the oxidizing zone because of the high local oxygen concentration around the coal particles. However, combustion PC in the tuyere assembly is undesirable because of the increase in gas temperature and melting of ash, and on account of deviation of the stream of combustion products and unconsumed particles upwards along the furnace periphery. Furthermore, as usual, the fine particles burn first, and consequently their combustion occurs under more favourable conditions. Burning of coarse particles, which occurs later, already proceeds in a region of lower oxygen concentration.

Consequently it is most effective to increase the oxidising agent concentration after heating and emergence of volatile matter and heating of the coke residue, i.e. after time $\tau_1 + \tau_2 + \tau_3$. Calculations show that this time is 0.005 – 0.020 s, depending on PC characteristics and the temperature and blast conditions of the operation (Babich 1991, Babich 2014).

Table 1 gives the results of calculations of PC combustion time with individual delivery of process oxygen into the tuyeres for the conditions at DMW. The zone of active oxygen

Table 1. Effect of oxygen rate on PC combustion time.

Oxygen consumption, m^3/tHM	Blast oxygen content, %	Local oxygen concentration, %	Time τ_4, τ_4, s	Total combustion time, s
-	21	21	0.045	0.065
0.2	22	26	0.035	0.055
0.35	23	30	0.025	0.045
0.5	24	35	0.020	0.040
0.7	25	40	0.015	0.035

Table 2. The depth of jet penetration into the blast.

Diameter of oxygen line, mm	10	18	25	10	18	25	10	18	25
Oxygen flow rate per tuyere, m^3h^{-1}	200	200	200	300	300	300	400	400	400
Depth of jet penetration, mm	37	21	15	56	31	22	75	42	30

diffusion to PC particles was taken to be 25% of the tuyere cross-sectional area. With PC delivery into tuyere, this quantity can vary over the range 15-35%, depending on the diameters of the dust line and tuyere, and the design of the dust introduction unit.

Complete combustion of 100 kg of PC requires 35-50 m^3/t HM of oxygen spread among the tuyeres if oxygen enrichment of blast is excluded.

To ensure timely arrival of oxygen in the stream of preheating pulverised coal, the depth of its penetration and flow trajectory in the tuyere cavity must be known. For determination of these parameters there are analytical procedures in jet hydraulics and aerodynamics theory and experimental procedures for calculating the trajectory of a round jet in a carrier stream. The trajectory of the jet axis in a carrying stream is described by the equation (Pomerantsev 1986):

$$\frac{a \cdot x}{d} = 195 \left(\frac{\rho_1 \cdot v_1^2}{\rho_2 \cdot v_2^2} \right)^{1.3} \cdot \left(\frac{a \cdot y}{d} \right)^3 + \left(\frac{a \cdot y}{d} \right) \cdot \cot \alpha \tag{7}$$

where a = a jet structure coefficient taking account of its initial turbulence and the degree of non-uniformity of the velocity field at the outlet from the nozzle (-); d = the jet diameter at the mouth (mm); ρ_1, ρ_2 = the density of the carrying stream and the jet (kg/m^3); v_1, v_2 = the carrying stream and jet velocities (m/s); α = the jet angle of incidence (degree); x, y = the axis of the jet velocity and axis of the carrying stream velocity (mm).

The depth of jet penetration into the stream is calculated as

$$h = k \cdot d \frac{v_2}{v_1} \left(\frac{\rho_2}{\rho_1} \right)^{0.5} \cdot \sin \alpha \tag{8}$$

where h = the depth of jet penetration into the stream (mm); k = a dimensionless coefficient dependent on the angle of incidence and the structure coefficient (-).

Results calculated with this formula for blast furnace operation conditions at DMW are given below.

To ensure a maximum throughput capacity of oxygen of 600 m³/h to a tuyere, the connecting pipe must be half an inch in diameter. Data characterising the oxygen jet trajectory for this case are given in Table 3.

A tuyere assembly design with individual oxygen delivery has been developed at the DMW (Figure 1).

Table 3.   Characteristics of oxygen jet flow.

Oxygen flow rate, m³/h	Oxygen pressure, kPa	Depth of penetration, mm	Distance from delivery point, mm
200	800	34	118
200	1000	27	102
300	800	51	200
300	1000	46	172
400	800	68	290
400	1000	61	250

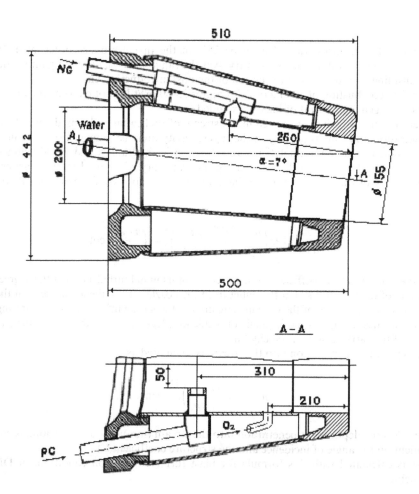

Figure 1.   Tuyere assembly design for co-injection of natural gas, pulverized coal and oxygen.

718

An experimental industrial installation was created to carry out investigations on a blast furnace. Oxygen delivery into the tuyere was effected from a manifold through an oxygen line provided with the necessary fitting and instruments. As for the natural gas delivery system, a return valve, which operates if the oxygen pressure falls, was provided.

During the investigations, the 1033 m^3 blast furnace produced steelmaking hot metal on an iron ore burden consisting of 94% sinter from the Southern Mining and Benefication Plant and pellets from the Central Mining and Benefication Plant in a ratio of 1.5:1 with the necessary additions of iron and manganese ores. The average consumptions were as follows: coke 487 kg/tHM, pulverised coal 76 kg/tHM, natural gas 84 m^3/tHM, oxygen 75 m^3/tHM. The furnace was equipped with a system for controlling the distribution of blast and PC between the tuyeres. The oxygen pressure in the oxygen line was 570-1000 kPa, the oxygen flow rate per tuyere was 320-350 m^3/h.

The effectiveness of the proposed method of delivering oxygen was assessed with the aid of a quantitative method for determining the completeness of PC combustion in the tuyere zones, which consists in probing of the hearth with sampling of materials from different points along the tuyere assembly and the oxidising zones with a water cooled tube (Yaroshevskiy 1988).

Investigations were carried out on blast oxygen enriched to 25% and atmospheric blast (oxygen delivery to the blower inlet was stopped). The amount of oxygen delivered through the tuyere corresponded to the oxygen consumption for blast enrichment of the furnace. Each trial included two experiments, carried out in sequence with an interval of several minutes: probing of the hearth while oxygen was delivered into the tuyere and with the oxygen turned off.

The results of mineralogical analysis of samples of materials taken from the hearth showed that, with and without oxygen delivery to the tuyere, fine, acute angle coal dust particles of 0.004-0.008 mm size, and more rarely up to 0.02 mm in size predominate at the oxygen delivery point. At the tuyere end there are residues of burnt PC particles in the form of thin rings, as well as unchanged PC particles. At a distance of 0.25 m from the tuyere nose, PC particles no longer exist. Only residues of dust particles in the form of 0.004-0.012 mm rings remain. At a distance of 0.25 mm and 0.5 m, rounded globules of acid and iron containing slag appear in samples. The particles are generally irregular in shape, up to 0.4 mm in size.

The graphs shown in Figure 2 were plotted from an estimate of the percentage content of components in the samples investigated.

In all experiments, the PC consumption was 8 t/h and the consumption of blast enrichment oxygen 6000 m^3/h, the flow rate of oxygen delivered to the tuyere being 320 m3/h (*a* and *b*) and 340 m^3/h (*c* and *f*). The natural gas flow rate was 6200 m^3/h (*a* and *b*), 5500 (*c* and *d*), and 4200 m^3/h (*e* and *f*).

Comparison of corresponding pairs of experiments shows that the content of PC particles at the end of the tuyere with individual delivery of oxygen into it is on average 15-20% lower than with the conventional method. At a distance of 0.5 m from the tuyere nose the amount of dust particle residue is also lower in experiments with local oxygen delivery. The results obtained point to acceleration of pulverised coal combustion with individual oxygen delivery.

The design of an industrial system for individual oxygen delivery to the tuyeres of a blast furnace has been proposed for DMW.

3   OPTIMASATION OF COAL GRINDING

Increase in the level of coal grinding leads to intensification of its combustion, but certainly increases the cost of PC complex. Overgrinding of coal dust does not only increase energy-waste on grind and reduce mill productivity, but at the same time worsens transportation of PC, since the smallest particles have the tendency to stick together. Therefore in order to increase productivity of mills it is profitable to produce PC of coarse grind. However with coarseness of grind combustion occurs with difficulties in the raceway that causes a whole complex of negative phenomena: reducing gaseous passage capacity through burden column, increasing viscosity of slag, reducing the coke/coal replacement ratio and others.

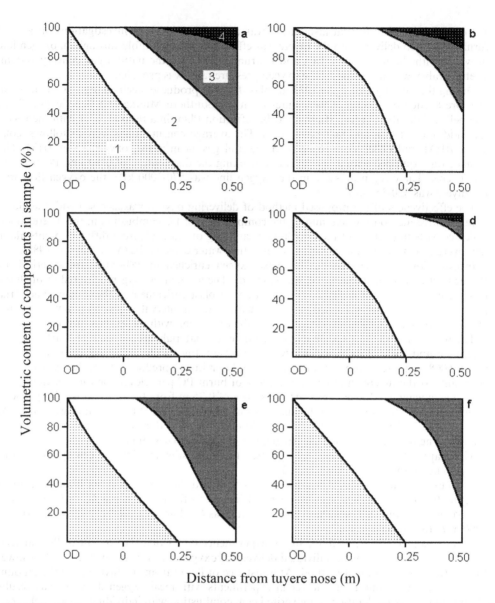

Figure 2. Content of component in materials taken along hearth radius in relation to flow rate of oxygen delivered into tuyere (indicated in text).

a, *c*, and *e* experiments with oxygen delivery; *b*, *d*, and *f* experiments without oxygen delivery; 1- PC (unchanged); 2 - PC residues (rings); 3 - slag; 4 - coke; OD - oxygen delivery point

Economy of PC injection is the highest when total consumption of energy for preparation and combustion of dust get to minimum.

To substantiate optimum fractional content of PC, research were carried in different ways with dust, prepared from concentrate of lean coal and used for injection into BFs in Ukraine.

The desire to reduce size of particles is explained with the hope of increasing the reaction surface area of PC, which at the determining level indicates its combustion degree in the furnace. At the same time unit surface of particles increases also in consequence of high temperature thermal hit at their arrival in the tuyere cavity.

Table 4 gives the results of gas chromatographical analysis which show that increase of a unit surface of coal dust as result of series of changing to smaller fractions isn't obvious compared with thermal shock.

Redistribution of chemical components in different fractions of PC happens which is confirmed by results of analysis of examined samples (Table 4). As seen, ash concentrates in fine fractions reducing content of fixed carbon and, as a result, effective use of PC.

The kinetic of coal burning phenomenon was examined using the Q 1500 D derivatograph with a constant quantity of 700 mg of PC, placed in a platinum crucible and heated in atmosphere of air from 25 to 1000°C at a rate of between 7.5 and 20 °C/min. Results of work up of the derivatograms received are given in Table 5. From analysis of available facts we can see that refining of PC grind changes physical-chemical processes of its burning in the following ways.

The surface area increases, but for small particles (less than 50 μm) this increment starts already to decrease. The optimum, from the viewpoint of unit surface value, is expressed as fractions 50-63 μm. The total time for process of burning of polyfractional dust equals to the time for burning of its smallest fractions < 50 μm, average geometrical size of which is 11 μm. The presence of fine particles in polyfractional dust is responsible for its ignition, accelerating combustion of the biggest particles of PC.

Therefore optimum coal grinding for injection into blast furnaces exists: main mass (60-80%), obviously, must consist of fraction 50-100 μm. Grinding of PC 100 μm below 50 μm is not proved technologically.

Table 4. Unit surface and analysis of different coal fractions.

	Unit surface, m^2/g		Proximate analysis, wt.%		
Fraction, μm	Initial samples	After thermal shock	Ash	VM	Fixed carbon
< 50	2.73	5.27	7.9	13	76.6
50-63	2.17	4.98	6.6	10.74	80.16
63-100	2,0	4.82	6.4	10.26	80.84
>100	1.34	4.51	5.4	9.70	82.40
Polyfractional	3.87	6.09	8.5	12.07	77.93

Table 5. The parameters of PC burning process.

	Fraction composition of samples, μm				
Parameters of process	<50	50-63	63-100	>100	Total
Increases in mass during chemisorption, μg	11	12	8	4	14
Total time, min	99.5	105.5	100	111	99.5
Relative time of stage, %:					
evolution of moisture	6	5	5	5	6
chemisorption	8	7.7	5.5	6.5	7
escape of volatile	26.8	21.3	20	16.5	27.5
burning of coke residue	60.2	66.7	69.5	72	59.5
Temperature, °C:					
initial stage of chemisorption	100	102	110	110	119
initial stage of volatile escape	310	347	353	363	326
ignition	551	563	568	565	557
maximum	875	927	932	934	898
end of burning	882	898	902	914	884

4 A COMPLEX COMPENSTING FOR THE NEGATIVE CHANGES IN BF OPERATION WITH PCI

Attaining maximum effectiveness of PC use is possible with the help of principle of complete and complex compensation for the negative changes of heat fluctuation, slag-formation and gas distribution processes determined by the combustion of PC and simultaneously reducing of coke rate in burden.

To evaluate the effect of PC on heat regime of hearth, the value of flame temperature was determined from equation of heat balance for lower zone of heat exchange. Viewing two technological regimes – with (index 1) and without using PC (index 0), equation is got for flame temperature, needed to save initial heat and composition of pig iron under new technological conditions (Yaroshevskiy 1988):

$$T_1 = T_n + (1 - A\frac{r_{d0} - r_{d1}}{r_{d0}}) \cdot \frac{Q_{k0} \cdot V_{g0}}{Q_{k1} \cdot V_{g1}} \cdot (T_0 - T_n) \tag{9}$$

where T = flame temperature (°C); T_n = temperature of burden and gases in the slowed down heat exchange zone (°C); r_d = direct reduction rate (-); Q_k = coke consumption (kg/tHM); V_g = output of hearth gases (m^3/t of coke); A = constant, determined by initial value of water equivalent of burden (-).

From Eq. (9) it follows, that technological conditions of BF operation functionally determine level of necessary flame temperature. E.g., reducing of coke consumption or raising the level of direct reduction by 1% leads to necessity of increase in temperature T_1 correspondingly by 20-30°C or 15-25°C.

Injection of PC into the hearth is accompanied by reducing in theoretical temperature of burning by 100-200°C for every 100 kg of fuel/t HM. Lowering the percentage of direct reduction by 2-4% for every 100 kg fuel per 1 t HM makes it possible to reduce necessary flame temperature by 30-100°C, which cuts down by the indicated value the temperature compensating calculated on the basis of maintaining initial temperature level of products of melting.

Table 6 shows the results of calculation of effectiveness of PCI compensating by blast temperature, NG consumption, oxygen and moisture content in blast.

Compensation with increase in the blast temperature (by 250 °C for every 100 kg of Pc/t HM) maintain the best result: total coke/coal replacement ratio 1.45-1.55, saving of 40-45 kg/t HM (8-10%) of additional fuel, increasing the productivity by 1.4-7.0%, improving of tuyere gas efficiency (2-6%), and reducing direct reduction rate by 3.6%. Using this method coke/coal replacement ratio does not practically reduce when increasing PC injection from 0 to 300 kg.t HM. When compensating with increasing oxygen content in the blast (by 53.5 m^3/t HM or 3% for every 100 kg of PC/t HM) productivity increases, but effectiveness of PC use constantly falls: thus, at PC consumption of about 300 kg/t HM, value of compensation increases more than 3 times, and the CO utilization rate decreases as PC consumption increases.

5 CONCLUSIONS

Theoretical and industrial research of PC burning showed that under conditions of BF operation process of PC burning exists in the diffusional regime. Complete gasification of high PC quantities can be provided by special measures on intensifying its combustion in the raceway: enriching blast with oxygen, complete mixing of PC and oxidizer.

To accelerate PC combustion it is proposed to increase the local oxygen concentration in the region where the jet of suspension is flowing in the tuyere assembly, rather than the total oxygen content of the blast. This should be done 0.005-0.020 s after the coal dust particles have emerged into the inner cavity of the tuyere assembly. Experimental industrial trials with probing of the hearth have confirmed the faster disappearance of PC particles on leaving the tuyere compared with the traditional method of blast oxygen enrichment.

Table 6. Effectiveness of compensating measures under PCI.

Parameters	Base period (P-C=0)	Change of parameter per every 100 kg PC/tHM compensating for flame temperature by			
		T_{blast}	NG	O_2	φ
Blast temperature,°C	1100	250	0	0	0
NG consumption, m³/tHM	80	0	-36.4	0	0
O_2 volume, m³/tHM	23	0	0	53.5	0
Moisture in blast (φ), %	1.1	0	0	0	-2.5
Productivity, %	100	10.8	-0.8	7.1	-4.5
Coke consumption, kg/tHM	440	-171	-92	-117	-125
Tuyere gas flow (m³/tHM)	2100	-167	-42	-203	-25
CO utilization rate, %	35.5	1.88	0.83	1.25	2.25
Direct reduction rate, %	30	-3.6	0	-3.6	0
Total fuel consumption, kg/tHM	610	-45.8	-13.3	-1.25	-75

Using the principle of complex compensating enables to get the value of total coke/coal replacement ratio of more than 1.0 kg/kg. As a result the output of blast furnace is increased and the utilization rate of gases is improved.

REFERENCES

Babich, A., Senk, D. & Born S. 2014. Interaction between co-injected substances with pulverized coal into the blast furnace. *ISIJ International*, 12: 2704–2712.

Babich, A.I., Kochura, V.V., Nozdrachev, V.A. et al. 1991. Individual delivery of process oxygen to blast furnace tuyeres. *Steel USSR-Eng. Tr.* 21 (12): 538–540.

Babiy, V.I. & Kuvaev, Yu.F. 1986. *Combustion of coal dust and calculations on coal dust flame.* Moscow: Energoatomizdat.

Bosenhofer, V., Wartha, E., Jordan, C. et al. 2019. Suitability of pulverised coal testing facilities for blast furnace applications. *Ironmaking & Steelmaking* 1: 1743–2812.

Chai, Y., Zhang, J., Shao, O. et al. 2019. Experiment research on pulverized coal combustion in the tuyere of oxygen blast furnace. *High Temp. Mater. Proc.* 38: 42–49.

Geerdes, M., Chen'o, R., Kurunov, I. et al. 2016. *Modern blast furnace process. Introduction.* Moscow: Metallurgizdat.

Kurunov, I.F. 2010. The blast furnace production of China, Japan, North America, Western Europe and Russia. *Proceedings of the International Domenshchik Congress.* Moscow: Kodeks: 6–17.

Langer, K. 2005. Injection of pulverized coal at Thyssen Krupp Steel. *Stahl and Eisen.* 11: 591–594.

Murai, R., Kashihara, Y. Murao, A. et al. 2016. Convergent-divergent injection lance for the enhancement of combustion efficiency of pulverized coal at blast furnace. *ISIJ International*, 5: 770–776.

Pomerantsev, V.V. 1986. *Principles of Practical Combustion Theory.* Leningrad: Energoatomizdat.

Ryzhenkov, A.N., Yaroshevskiy, S.L., Krikunov B.P. et al. 2005. Melting technologies using pulverized coal and natural gas on an oxygen-enriched blast. *Steel.* 12: 13–18.

Savchuk, N.A., & Kurunov, I.F. 2000. *Blast furnace production at the turn of the XXI century. News of ferrous metallurgy abroad.* Moscow: Chermetinformacija.

Yaroshevskiy, S.L. 1988. *Ironmaking with pulverised coal injection.* Moscow: Metallurgy.

Yaroshevskiy, S.L. 2006. Coal fuel - a real and effective alternative to natural gas in metallurgy. *Metal and casting of Ukraine.* 3:15–20.

Zhang, R., Cheng, S. & Guo, C. 2018. Detection method for pulverized coal injection and particles in the tuyere raceway using image processing. *ISIJ International*, 2: 244–252.

Zhou, Z., Liu, Y., Wang, G. et al. 2017. Effect of local oxygen-enrichment ways of oxygen-coal double lance on coal combustion. *ISIJ International.* 2: 279–285.

Topical Issues of Rational Use of Natural Resources 2019 – Litvinenko (Ed)
© 2020 Taylor & Francis Group, London, ISBN 978-0-367-85720-2

Matter of the environmental safety markers identification

O.G. Kuznetsova
Moscow State University of Civil Engineering, Moscow, Russia

ABSTRACT: The study of the conditionally undisturbed – reference landscapes (State Nature Reserve Prisursky) in the territory of the Sura-Sviyaga interfluve in the Volga Upland, and the landscapes, slightly disturbed hypsometrically below the cities Shumerlya and Vurnary are carried out. The conducted analytical research allows to conclude the following: 1) the accumulation of heavy metals and iron by the slightly disturbed landscapes in soils, waters and vegetation is registered; 2) the increased concentration of iron is connected with the fact that iron is a tipomorphic element for this territory; 3) heavy metals, such as copper, lead, zinc are offered as the markers of the initial degradation of the ecological structure of the landscapes.

1 INTRODUCTION

Nowadays, the main focus of the research is on the anthropogenic disturbed areas. The data, obtained within the chemical analysis result in the comparison with the MAC (maximum allowable concentration) of a definite substance. Undisturbed areas are studied in order to replenish the database, to study the mechanisms of accumulation, transformation of substances. The areas, slightly disturbed technologically, do not get enough attention. Meanwhile, these areas allow us to estimate the degree of the increase of the anthropogenic impact, comparing the data with the "reference" areas (special protected areas). In this respect, the matter of the environmental situation markers identification is particularly relevant. The environmental situation markers allow to determine the state of the environment quickly and reliably.

2 METHODOLOGY

Our study focuses on the landscapes, slightly disturbed technologically. The study of soil-forming rocks, vegetation, mud and water was carried out. The study was carried out in the territory of Sura-Sviyaga interfluve in the Volga Uplands (Gorshkova (2009)). The objects of the study are the slightly disturbed (Figure 1a) and technologically unchanged landscapes in the territory of the State Nature Reserve "Prisursky" (Figure 1b).

The soil sampling according to the genetic horizons in accordance with common techniques (Glazovskaya (1964), Evdokimova (1987), Guidelines for conducting field and laboratory.... (1981)) was made in all the key areas. The samples of forming rocks, soil, vegetation, water, and sediment were the subjects to analytical studies.

Most of the natural landscape occupies the State Reserve "Prisursky". The coniferous and deciduous forests dominate in the territory. Spruce, pine and less frequently birch dominate in the forests. The soil surface is covered by a carpet of green moss. The vegetation of the studied area is typical for the green moss spruce forest.

To assess the geochemical anomaly of the area it is necessary to have data on the background. The background areas are selected among those, located far from local industrial sources of pollutants, usually more than 30-50 km (Perelman & Kasimov (1999)). The distance from the independent landscape to the city Vurnary is 54 km.

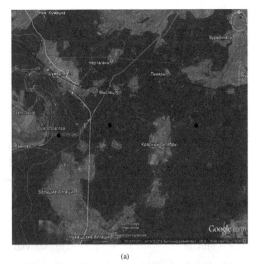

(a)

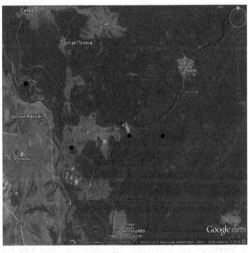

(b)

Figure 1. a. Disturbed landscapes. Photos from the satellite (Google Earth program.) The dots show the main places of sampling. Scale: 1 cm - 3.75 km. b. Natural or reference landscapes. Photos from the satellite (Google Earth program.) The dots show the main places of sampling. Scale: 1 cm - 3.75 km.

The anthropogenic impact is minimal and it is detected mainly in the influence of precipitation.

The main soil-forming rocks of the studied area are clay and heavy parent rocks, as well as ancient alluvial sands. The autonomous landscapes in the studied areas are mainly the sod-podzolic soils. The particle size distribution is non-uniform because of the heterogeneity of the parent material of the soils.

The landscapes with anthropogenic load, selected for the study, are disturbed. Superaquatic and aquatic landscapes, located below the hypsometrically transit, are influenced by the cities Shumerlya and Vurnary and operating enterprises, agricultural areas, they are exposed to the effects of enrichment arable fertilizers, herbicides and other chemical products. Meanwhile the lake Urgull is located at the distance of 4 km from the city Shumerlya and at the distance of 32 km from the city Vurnary, the distance from the Vurnary mountain to the autonomous landscape, we studied, is 14 km. Superaquatic landscape, being a natural monument, is still used for recreation.

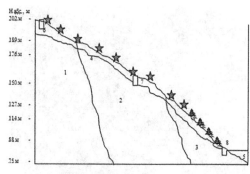

Nature management	Cordon	Cordon	Nature monument Recreation
Soils	Sandy, podzolic, gley-and-loam	Sandy, podzolic	Gley-and-loam, layered
Elementary geochemical landscapes	eluvial	Transeluvial	superscuba Scuba

1. Jurassic clay deposits; 2. Chalk sands; 3. Quaternary sand deposits; 4. Groundwater level; 5.Urgul lake.

Figure 2.   Schematic catena of the disturbed landscapes.

Eluvial autonomous disturbed landscapes is formed with sand, podzolic, gley-and-loam soils on the Jurassic clayey sediments. The vegetation of the autonomous landscape is presented by the birch forest with aspen, lime and other breeds. Transeluvial landscape is formed with sandy podzolic soils on the Cretaceous sands. Superaquatic landscape is formed with gley-and-sandy soils on the stratified Quaternary sandy parent rocks. The vegetation of the superaquatic Urgull lake landscape is represented by pine forests with of birch, aspen, and less often with spruce, linden (Figure 2).

The type of nature management in the autonomous and transit landscapes is the cordon. Lake Urgull is a natural monument, the protection regime is set. However, it suffers the anthropogenic impact in the form of recreation. Moreover, the impact comes from the cities Shumerlya and Vurnary, located hypsometrically higher. The chemical plant and other enterprises are located in the territory of Vurnary. In addition, the city has a dilapidated sewerage system.

The method of geochemical analysis of the conjugated landscapes is unique, as it is possible to identify the source of pollution even "after the fact". Due to the geochemical migration the pollutants are accumulated in the aquatic and superaquatic landscapes.

To evaluate the environmental hazard and variations from the natural background level the pollution indication, applicable thanks to the calculation of the coefficient of technogenic concentration or abnormality (Kc). The technogenic concentration factor indicates how many times the content of the element in the disturbed landscape exceeds its content in the background landscape (Perelman & Kasimov (1999)). Kc describes the intensity of the pollution, but does not show the level of its danger.

3   LABORATORY RESEARCH

The vegetation of the disturbed landscapes in all the samples contained greater amounts of heavy metals than the natural landscape plants (Table 1).

Thus, in the disturbed landscapes grains contain magnesium in 2.5 times the magnesium concentration in the reserve soils; copper - 4.5 times; Zinc - 6 times. Woody plants in the damaged landscapes contain magnesium in 1.5 times; copper - 2 times; Zinc - the same amount. Kc (Mg in the ash vegetation) = 2.5; Kc (Cu in the ash vegetation) = 4.5; Kc (Zn in the ash

Table 1. Comparative characteristics of heavy metals in natural vegetation and vegetation from the disturbed landscapes.

Heavy metals, mg/kg	Natural landscapes	Disturbed landscapes
Grass		
Mn	812.05	2869.68
Zn	67.54	395.29
Cu	38.58	171.71
Goat willow (green weight)		
Mn	8718.07	11305.26
Zn	747,50	767.66
Cu	127.44	346.07

vegetation) = 6. We calculated the total pollution index (Zc), offered by Y.E. Saet and characterizing the degree of pollution of the association members to the background. The total pollution index for heavy metals vegetation is: Zc = (2.5 + 4.5 + 6) - 2 = 11.

All water the samples did not exceed MPC according to the general chemical indicators. However in the damaged landscapes water contains a dry residue of 2.5 times the value of natural landscapes; chloride 1.5. Sulfates are over 2.6 times in the lakes of the natural landscape. This is probably connected with the shallow gypsum deposits (tabl.4.2.2.). Active reaction of the environment in the lakes in the damaged landscapes is slightly acid (5.6-6.7); Lake Urgull is close to neutral (7.3). Nitrates in waters exceed the amount in the lakes of the nature reserve (1.92-1.68 mg/l). These indicators are 1.5 times higher than those for Urgull lake water (1.06 mg/l) experiencing anthropogenic interference. Kc (dry residue) = 2.5; Kc (Cl-) = 1.5. Total pollution indicator for general water chemical parameters is: Zc = (2.5 + 1.5) - 1 = 3.

Thus, in the damaged landscapes water contains greater amount of dry residue and chlorides than water in the reserve lakes. However, they do not exceed the maximum permissible concentration. Accordingly, the damaged landscapes can be considered to be slightly damaged.

Heavy metals and their salts are primarily industrial pollutants. Lake water Urgull exceeded MPC for fishery for copper, zinc and lead (Table 1; Figure 3).

The excess of the MPC is because of the city Vurnary, located hypsometrically higher and its operating chemical plant "Mixed preparations", which produces chemical pesticides, fertilizers and household chemical products. Metals are included in the fertilizer "superphosphate" and fungicides which are put into the soil, located hypsometrically above fields. They arrive into the water of the hydromorphous landscapes of Vurnarka river and Apnerka river, located hypsometrically higher, into which the insufficiently treated waste water capital residential and public buildings, "Plant mixed preparations "and" Factory of skimmed milk powder "and other enterprises of the city Vurnary drain. In addition, on the 10 of October, 2012 the prosecutor's office of the Vurnarsky district revealed some violations in the supports of the construction sewage pipes at "Vurnary. Skimmed milk powder" Ltd.

According to the official statements the standard-treated wastewater from the city Vurnary comprise 95%. Despite this, the degree of pollution of the Vurnarka river is estimated from mild (0.1-3) to high (3-10) at its tributaries from high (10-100), and even extreme settlement (more than 100). It has the greatest concentration of iron, manganese, sodium and cadmium.

The uninterrupted reception of sewage is mainly effected by collectors wear, depreciation of the equipment and buildings of the main biological treatment facilities and sewage pumping stations. In general, the urban areas had the tense situation with the system of household sewage. Most of the networks are worn more than 70%.

Figure 3 is a diagram of the heavy metal content of natural waters and lakes in the technogenically -damaged landscapes. The diagram shows that the maximum values are 6-13 times

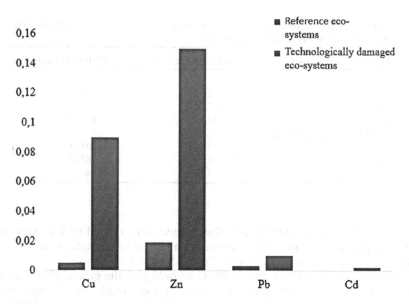

Figure 3. Comparative characteristics of heavy metals in water of "standard" and technogenically damaged ecosystems.

greater than the minimum values of the concentration of the natural landscape. Comparison charts of heavy metals and natural slightly damaged landscapes showed that:

- lead exceeds 25 times; Kc (Pb in water) = 25;
- zinc exceeds 50 times; Kc (Zn in water) = 50;
- copper is 128 times more in the slightly damaged landscapes than in nature. Kc (Cu in water) = 128.

Summary indicator for heavy metal contamination is water: Zc = (25 + 50 + 128) - 2 = 201. This shows the intracellular accumulation of heavy metals in the waters of the damaged landscapes lakes.

4 CONCLUSIONS

Thus, in the studied area Sura-Sviyaga interfluve in the Volga Uplands heavy metals may be indicative of initial degradation of soils, water and vegetation. Accumulation of metals in the superaquatic landscapes is the most expressive indicator, such data can be used to identify the source of pollution "after the fact". So, to determine the quality of the entire catchment detection can be used metals in soils of superaquatic landscapes. Heavy metals in water samples are also a good indicator, but for a limited time - one season, i.e., to settling metals in sediment and with an organic binding agent. The vegetation is species-specific in the relation to the accumulation of heavy metals, i.e. different types of vegetation tend to accumulate different kinds of heavy metals or on the contrary, rejection, despite the metal pollution of the soil. Moreover, a barrier to the accumulation of metals from the soil is the root of the plant. However, as pointed out by the researchers recorded a tendency to increase the concentration of heavy metals in grains even three years after the registration of the maximum permissible concentration of the metal in the soil.

Thus, heavy metals: lead, copper, cadmium and others can be used as the markers of the environmental situation.

Geochemical methods allow makes reliable and complete characterization of the environmental pollution caused the landscape in the past, to show the current status and make the forecast development of the territory. Comparative analysis of natural and slightly damaged landscape allows us to trace the degree of increase of anthropogenic pressure area, as well as to make a prediction of the landscape development.

Thus, all the samples of water and vegetation keep the tendency of accumulation of heavy metals in the technogenically damaged landscapes. Slightly damaged landscapes can be used as a medium quality indicators territory as on their territory can be traced to the degree of accumulation of man-made load. Environmental markers of the degree of increase of anthropogenic impact area can be heavy metals.

5 RECOMMENDATIONS

The following activities can be used to optimize the use of land:

- Using "zero tillage", no-till technology (a technology of conservation agriculture, in which there is no tillage, and crop residues remain on the soil surface. Ideally, the seeds are introduced into the soil without damaging it).
- The breakdown of agricultural land plots 1-1.5 ha.
- Planting of conifers and deciduous trees around the perimeter.
- The use of crop rotation and green manure.
- Using hydromorphic landscapes as agricultural land.
- Refusal to make the soil herbicides and fertilizers.

REFERENCES

Glazovskaya M.A. Geochemical basics of typology and methodology of the natural landscape studies. Moscow, Moscow State University, 1964. -230 p.

Gorshkova O.G. On the landscape-geochemical characteristics of hydromorphic landscapes in the middle basin of the river Sura // Bulletin of Moscow State Open University. "Science" series. - 2009. -2. – pp. 39–43.

Evdokimova T.I. Soil mapping. Moscow, Moscow State University, 1987. -269 p.

Kostovski S.K., Kuznetsova O. Identification of anthropogenic landscapes history by the hydrochemical methods // Nature and society. Crisis dynamics. Series "Social –and-natural history. Genesis of the crisis of nature and society in Russia. "Edired by Borisova E.A. Volume HHHIH. Moscow, MBA, 2015. pp. 148–152.

Guidelines for conducting field and laboratory studies of soils and plants, within the control of the environment pollution by metals / Edited by N.G. Zyrin, S.G. Malakhov. Moscow, Gidrometeoizdat, 1981. -109 p.

Perelman A.I., Kasimov N.M. Geochemistry landscape: Textbook. 3rd edition, revised and enlarged. - Moscow: Astrea 2000, 1999. - 768 p.

Topical Issues of Rational Use of Natural Resources 2019 – Litvinenko (Ed)
© 2020 Taylor & Francis Group, London, ISBN 978-0-367-85720-2

Double-column double-filter water well for groundwater abstraction

J.A. Medvedeva
BNTU, Minsk, Belarus

ABSTRACT: In this article, the author proposes a new design of a two-column water well, which provides uninterrupted water supply to consumers, and has an increased service life compared to standard designs. As a research task, the author has identified an attempt to assess the impact of two filter columns of the well on the work of each other in the laboratory, as well as to compare the performance of a two-column well with a typical in situ, drilled in the same geological conditions.

1 INTRODUCTION

Currently, groundwater is extracted in most European countries through water wells, which are the most common type of water intake facilities. To ensure the uninterrupted supply of water to the consumer, it is often provided for the device of two water wells: the main and the reserve, with individual buildings of pumping stations, as well as a zone of sanitary protection of strict regime with an external fence (TKP 45-4.01-199-2010, 2011). Such a scheme of placement of working and reserve wells involves significant material costs for the construction of individual wells with pavilions and networks and requires the alienation of significant areas under the territory of sanitary protection zones.

2 METHODOLOGY

To reduce capital costs for the construction of reserve wells in BNTU, the design of a double-column double-filter water well located in one common conductor (Figure 1), which performs the functions of both working and reserve wells (Ivashechkin et al. 2017), is proposed.

Each of the two well columns have its own head, a filter pipe to accommodate the pump, a filter and a sump. The columns are hydraulically connected to each other by means of upper and lower connecting bridges placed above and below the filters. In a well in each column on water-lifting pipes submersible pumps which can work both together, and alternately are mounted. Thus, we have a pumping station of the first lifting with the general pavilion having one water intake structure representing the two-column well equipped with two identical pumps.

When working with one pump, two filters connected by jumpers work in parallel. This reduces operating costs for lifting water, because the specific yield of the two-column well higher than that of a drilled one-filter one-column water well. In case of failure of one of the pumps, immediately turns on the other, which ensures uninterrupted water supply to the consumer. In case of failure of one filter or its current or major repairs, another filter can work. This significantly improves the reliability and performance of the proposed well in comparison with the well-known design of the two-column well developed by Tkachenko V. P., in which the filter is installed only in one column, and both columns are connected below the bypass pipe (Tkachenko, 1988), when the filter fails, such a well is decommissioned.

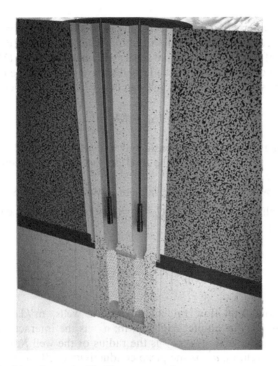

Figure 1.   Design of double-column double-filter water wells.

Consider the case when two identical pumps in two columns operate simultaneously. Then the work of the proposed two-column water well can be considered as the work of two interacting wells with the same debit.

The interaction of a group of water wells to work each other in the steady-state regime of groundwater movement is shown in Figure 2 (Maksimov et al. 1979).

As a result of pumping from the well №0 at the flow rate Q_0, the position of the water level in it was determined by the value h and a decrease of S_0. If the distance between the wells is less than the radius of influence R of the well №0, then in the well №3 the pressure decrease will be set at a level corresponding to the decrease S_3. Well flow rates, respectively, Q_0; Q_1; Q_2; Q_3; Q_4.

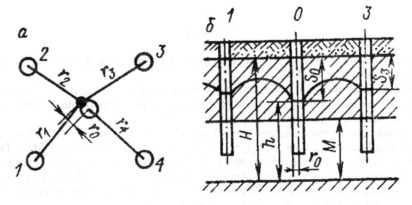

Figure 2.   Scheme of wells interaction.
a – plan of location of wells; b – section of the pressure horizon

The calculation of interacting wells is generally carried out based on the method of super-position of flows, according to which the resulting field is determined by a simple algebraic addition, regardless of the fields considered separately operating wells (Maksimov et al. 1979).

With regard to the determination of the level decrease in one of the n interacting wells, the solution based on the superposition principle can be written in the general form as follows:

$$S_c = S_0 + (\Delta S_1 + \Delta S_2 + \ldots + \Delta S_n), \tag{1}$$

where S_0 - lowering the level in the considered well without considering the interaction (as a single); ΔS_1, ΔS_2, ..., ΔS_n - lowering the level on the wall of the considered well from the action of all other interacting wells.

Based on the general expression (1) the value of lowering the level S at the considered point on the wall of the well №0 (Figure 2) at $Q_0 = Q_1 = Q_2 = Q_3 = Q_4$ and $\frac{r_i^2}{at} \leq 0.05 \div 0.1$ can be determined by the formula

$$S = \frac{Q_{sum}}{4\pi km} \alpha \left[\ln \frac{2,25at}{r_0^2} + \sum_{i=1}^{n} \ln \frac{2,25at}{r_i^2} \right], \tag{2}$$

where Q_{sum} – is the total flow rate of interacting wells, m^3/day; k – is the filtration coefficient, m/day; m – is the aquifer capacity, m; α – is the interaction coefficient (interference) of wells in the debit ($\alpha = \frac{Q_i}{Q_{sum}}$); r_0 – is the radius of the well №0, m; r_i – is the distance from the point to the well, m; a – is the piezo conductivity coefficient, m^2/day; t – is the duration of pumping, day.

Applying the solution of equation (2) to the scheme of a two-column well, where the system consists of two interacting perfect artesian wells with radii r_1 and r_2, respectively, located at a distance r_i from each other and working with debits $Q_1 = Q_2 = Q$ (Figure 3), it is possible to determine the decrease in groundwater level on the wall of the wellbore №1 at $\alpha = 0.5$; $Q_{sum} = 2Q$:

$$S = \frac{Q_{sum}}{4\pi km} \alpha \left[\ln \frac{2,25at}{r_1^2} + \ln \frac{2,25at}{r_i^2} \right] = \frac{Q}{4\pi km} \left[\ln \frac{2,25at}{r_1^2} + \ln \frac{2,25at}{r_i^2} \right], \tag{3}$$

When constructing a piezometric curve around a two-column well, in order to determine the level decrease at any point at a distance from it, you can use an approximate technique based on the replacement of a group of interacting wells with a single well (Artsev et al, 1976). Lowering the level at a point remote from the "center of gravity" of such a well at a distance r is determined by the formula:

$$S = \frac{Q_{sum}}{4\pi km} R_c, \tag{4}$$

where R_c – hydraulic resistance caused by pumping from the well, $R_c = 2\ln \frac{r_{in}}{r}$; r_{in} – time-dependent parameter, $r_{in} = 1.5\sqrt{at}$,

Considering a two-column well running two pumps with a total supply of Q_{sum}, as a single well with a "center of gravity", located in the middle of the bridge connecting the filters, you by formula (4) can get a decrease in any point at a distance r.

An example. Let the filter columns №1 and №2 of the two-column water well have radii $r_1 = r_2 = 0,04$ m and work with flow rates $Q_1 = Q_2 = 5$ m^3/h = 120 m^3/day for $t = 90$ days. Determine the decrease on the wall of the column №1, if the distance from the center of the column №2 to the point of the column №1, which is determined by the decrease $r_i = 0,055$ m (Figure 3). For medium-grained sand, we take the piezo conductivity coefficient $a = 0.5 \cdot 10^5$ m^2/day, the filtration coefficient $k = 10$ m/day. The aquifer capacity $m = 10$ m.

Then the decrease in the two-column well according to (3) will be

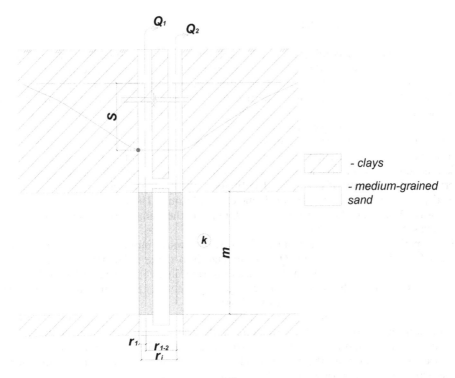

Figure 3.   Design scheme of a two-column water well.

$$S = \frac{120}{4 \cdot 3.14 \cdot 10 \cdot 10} \left[\ln \frac{2.25 \cdot 0.5 \cdot 10^5 \cdot 90}{0.04^2} + \ln \frac{2.25 \cdot 0.5 \cdot 10^5 \cdot 90}{0.055^2} \right] = 4.25 m.$$

We define depressions at points at distances from the "center of gravity" of a two-column well $r = (1, 5, 10, 100, 200)$ m in expression (4).
Determine the parameter $r_{вл}$:

$$r_{in} = 1.5\sqrt{0.5 \cdot 10^5 \cdot 90} = 3182 m.$$

The calculation of the position of the piezometric curve in the reservoir is summarized in Table 1.
To study the hydraulics of a two-column two-filter water well, several laboratory and field studies were carried out (Medvedeva et al. 2018).

Table 1.   Determination of level reduction in the reservoir.

The distance from the well to the point where the decrease is determined r, m	Hydraulic resistance, R_c, m	Lowering the level in the reservoir, S, m
1	16,1	2,26
5	12,9	1,81
10	11,5	1,6
100	6,9	0,97
200	5,5	0,77

3 LABORATORY RESEARCH

The BNTU laboratory conducted several experiments on the model of a two-column two-filter well to determine the effect of jumpers on the well hydraulics (Figure 4).

The installation consisted of a radial filtration tray with a diameter of 1,22 m and a height of 1,0 m with a round pond. Inside the tray, a model of the reserve (C2) and the main (C1) water wells connected to each other by means of the lower and upper jumpers were installed. Four small-tube wells and piezometers have been installed in the near-filter zone of the wells. The tank is filled to a height of 0,7 m quartz sand. To ensure the conditions of pressure filtration in the soil during circulation, a polyethylene film with a thickness of 2 mm with a clay lock was laid on its surface, which was additionally loaded. Thus, the model of the pressure reservoir had a capacity of m=0.3 m. The filters of the main and reserve wells were a tubular polyethylene frame with an internal diameter of 125 mm. The diameter of the holes in the frame was 12 mm. The holes were staggered. Outside the frame was wrapped with polyethylene canvas thickness δ=7,5 mm.

The experiments were carried out as follows. The water pump pumped from a barrel C1 model of a well with a flow rate Q_c and directed into the annular downstream to maintain levels in the tank. The pressure in the reservoir during filtration was recorded in the wells and piezometers with the help of an electric level gauge. The water flow rate was measured using a flow meter mounted on the pump pressure hose.

Study of parameters of the pressurized flow in near-filter zone area model two-column well equipped small-tube washing wells was performed by determination of the pressure h_n in the aquifer. With a known h_n pressure, it is possible to determine the decrease and increase in the water level S in wells and piezometers during pumping.

The value of S is determined by the formula: $S = h_n - h_c$,

where h_c – is the static head in the reservoir.

The experiments were carried out at different combinations of connection of filters of the two-column well model to determine the effect of the jumpers on the well hydraulics:

Scheme 1. Water is pumped out from column C1, the upper and lower jumpers are closed (Figure 5, a);

Scheme 2. Water is pumped out from the column C1, the upper jumper is open, the lower – closed (Figure 5, b);

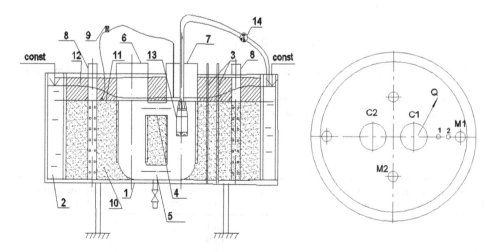

Figure 4.   Scheme of the experimental setup.

1 - filtration tray; 2 - round pond; 3 - piezometers; 4 - upper jumper; 5 - lower jumper; 6 – the second column (C2); 7 – the first column (C1); 8 - small-tube washing wells (M1,M2); 9 - level gauge; 10 - water-containing soil; 11 - polyethylene film; 12 - clay lock; 13 – pump; 14 – flowmeter.

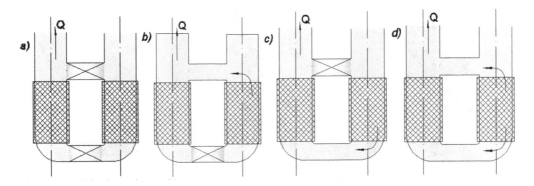

Figure 5. Combinations of work filters.

Table 2. Results of measurements in a two-column well.

	Consumption Q, l/s	Decrease S, cm	Specific flow q, cm^2/s
Scheme 1	0,14	5,7	24,5
	0,135	5,2	25,9
Scheme 2	0,17	5,2	**32,7**
	0,135	4,5	**30**
Scheme 3	0,19	7,5	25,3
	0,12	5	24
Scheme 4	0,19	6,3	30,2
	0,11	4	27,5

Scheme 3. Water is pumped out from the column C1, the upper jumper is closed, the lower one is open (Figure 5, c);

Scheme 4. Water is pumped out from column C1, the upper and lower jumpers are open (Figure 5, d).

The results of the experiments are shown in Table 2.

As a result of processing of measurement materials, it is obtained that with simultaneous operation of filters of two wells, the water intake well, with the blocked lower jumper (scheme 2), has the maximum specific flow rate.

4 EXPERIMENTAL RESEARCH

Two wells of the following structures were constructed to conduct full-scale studies at the site in Primor'e, Minsk region: one-column (typical) and two-column (Figure 6).

The typical design of the well drilled depth of 16 meters and has a filter length $l = 2$ m with a diameter $d = 127$ mm. The interval of the filter 12 – 14 m. The two-column well is also built depth of 16 meters and has a filter length $l = 2$ m each with a diameter $d = 76$ mm.

Preliminary measurements of the specific flow rate of wells were carried out. Pumping was performed by a screw pump. The flow rate was determined by a volumetric method using a 200 l barrel and a stopwatch. The measurement results are summarized in Table 3.

As a result of the processing of measurement materials, it was found that when water is taken from a two-column water well, the specific flow rate exceeds the value obtained when pumping from a one-column well.

Figure 6. One-column and two-column wells for field studies.

Table 3. Full-scale measurements of specific well flow rate.

Well construction	Volume V, l	Time t, sec	Consumption Q, m³/h	Decrease S, m	Specific flow q, m²/h
One-column		573	1,26	0,40	3,15
Two-column	200	505	1,43	0,42	**3,40**

5 RESULTS

According to the research results, it can be concluded that the application of the proposed design of a two-column two-filter water well will allow:

✓ to increase the specific flow and lifetime in comparison with one-column well;
✓ provide uninterrupted water supply to the consumer, having a separate pump unit in each barrel;
✓ significantly reduce the area of land alienated for construction and the length of communications;
✓ to reduce capex compared to building two separate wells in one drilling platform and one pavilion instead of two.

REFERENCES

Abramov S.K. & Alekseev V.S. 1980. *Water intake from underground sources.* Moscow: Kolos.
Abramov S.K. & Semenov M.P. & Chalishev A. M. 1956. *The underground water.* Moscow: Gosstroyizdat.
Artsev, A.I. & Bochever F.M. 1976. *Groundwater intake design.* Moscow: Stroyizdat.
Borevskiy B.V. & Drobnokhod N.I. & Yazvin L.S. 1989. *Assessment of groundwater.* Kiev: Higher school.

Ivashechkin, V.V. & Medvedeva Yu. A. & Kurch A.N. 2017. Double-column filter water well for the operation of a single aquifer. *Melioration*. № 3 (81). 36–41. Minsk.

Maksimov, V.M. & Grandma V.D. 1979. *Reference guide hydrogeologist*. Moscow: Nedra.

Maksimov, V.M. et al. 1979. *Hydrogeologist reference guide*. Leningrad: Needra.

Medvedeva, Yu. A. & Ivashechkin V.V. & Kochergin A.Yu. 2018. Laboratory and field studies of a two-column water well. *Geotechnics of Belarus: science and practice: materials of the International Conference*. Minsk: BNTU. 132–138.

Tkachenko V.P. 1988. Water well: a.s. 1448002SU, MKI E O3V 3/18 *Hydrological expedition of the Ministry of Land Reclamation and Water Management of the Ukrainian SSR*. - № 4235664. 29–33.

TKP 45-4.01-199-2010, 2011. Borehole intakes. Design rules. *Ministry of architecture and construction of the Republic of Belarus*. Minsk.

Topical Issues of Rational Use of Natural Resources 2019 – Litvinenko (Ed)
© 2020 Taylor & Francis Group, London, ISBN 978-0-367-85720-2

Kholodninskoe deposit as a threat to the water quality of Lake Baikal

T.S Koshovskiy, T.A. Puzanova & O.V. Tkachenko
Faculty of geography, Lomonosov Moscow State University, Moscow, Russia

A.Y. Sanin
Zubov State Oceanographic Institute, Moscow, Russia

ABSTRACT: The features of pollution of aquatic environments in the area of impact of Kholodnenskoe polymetallic deposit, located in the water basin of the Lake Baikal is considered. The main existing sources of pollution and the scale of the impact on the drainage watercourses were determined. Today, it is part of the basin of the River Kholodnaya, which flows near the mine and takes water streams flowing from abandoned tunnels. In the waters of these streams high concentrations of pollutants for a number of heavy metals have been revealed. They in the hundreds and thousands of times higher as compared with the maximum allowable concentration of pollutants. Directly in waters of Lake Baikal, where the Kholodnay flows, exceeding the permissible values are not identified. Based on the analysis of the selected samples, conclusions are made about the peculiarities of migration of chemical elements in the "Kholodnaya River–Lake Baikal" system.

1 INTRODUCTION

The water area of Lake Baikal was included in the UNESCO World Heritage Site in December 1996. For many foreigners it is the most famous natural object of Russia, in some way its symbol.The Russian government is aware of the huge role of the Lake and its waterbasin area for the nature of the Earth. This is evidenced by the adoption of a separate law "On protection of Lake Baikal" in 1999 (Verhozina E.V et al., 2011), as well as the development of the federal target program "Protection of Lake Baikal...". This law regulates economic activity within water basin of Baikal imposing serious restrictions on it. It is in accordance with this law that «the Power of Siberia» pipeline, which has a strategic importance for Russia, was built at some distance from the Lake outside its water protection zone, although its first route had laid within it.

The special status of the Lake and its exceptional importance for humanity and the biosphere as a whole forced us to predict the consequence of anthropogenic impact not only on the Lake, but on its drainage basin too. It is necessary to identify the sources of pollutants in the present and potential sources of pollution in the future for the entire drainage basin, as any of them can impactthe quality of the Lake waters. Sources of pollutants can be both natural and anthropogenic, which are associated with various types of economic activity. These include industrial enterprises, for example, pulp and paper plant and mining, municipal and agricultural wasteinputetc.

83% of pollutants come to Baikal with the waters of tributaries, by comparison, as a result of the activities of the Baikal pulp and paper mill – approximately 0.5% (Zubov & Mikhaylenko, 2011). It is necessary to take into account all anthropogenic impacts within the Baikal drainage area, which can significantly affect the ecological condition of the Lake. Some of them, which are located in the immediate vicinity of the Lake, can be especially dangerous because in this case contaminated water may not have time to be diluted by clean water in sufficient quantities before entering the Lake. An example of such an impact is the production of non-ferrous metal ores in the territory of Mongolia (for example, in the Selenga River basin), and Russia, in the Irkutsk region and the Republic of Buryatiya. The influence of this type of economic activity is considered by the example of the Kholodninsky deposit, located in the Republic of Buryatiya to the North-West of Lake Baikal within its water protection zone.

Currently, the deposit is a virgin field. However, the Company Metals Of Eastern Siberia, a subsidiary of Russia's privately owned Metropol group, has a license to develop it.

However, due to current restrictions on economic activity within the drainage area of the Lake, it is difficult for the company to avail of the license and to start developing the deposit. Of all its many assets, this is not only one of the most promising ones due to the huge reserves of the deposit, but also the most problematic one.

The deposit is the largest by reserves of lead and zinc in Russia: it contains 34.1% of all Russian zinc and 11.2% of lead reserves (according to other data: up to 30% of zinc and 15% of lead). The content of elements in the ore is quite high: zinc – 3.5-6.5%, lead – 0.4-1.2% (Zubov & Mikhaylenko, 2011, Kislov & Plyusnin, 2009). There are such elements-impurities as silver, gold, copper, cadmium, arsenic, antimony, thallium and others. Experts say, that the Kholodninskoye is the planet's third largest lead and zinc deposit. It has 13.3 million tons of zinc and 2 million tons of lead.

Currently the deposit is not developed but water enriched in heavy metals continues to flow from the abandoned exploration tunnels. It enters to the Kholodnaya River, which falls into the Angarsky Sor Bay of Lake Baikal.

Detailed exploration with construction of two mine galleries was conducted in the period from 1969 to 1984. The deposit is located within the so-called central ecological zone of the Baikal natural area, recognized as a UNESCO world heritage site.

According to public data, the following amount of substances is washed out from the mine gallery №2 by groundwater: zinc – 1480 kg, manganese – 136 kg. The content of zinc is 424 times more than the maximum permissible concentration that of manganese – 49 times more. Monthly a ground stream brings about 440 kg of zinc to the Kholodnaya River from the mine gallery №.2. Several times less of metals is delivered to the River from the mine gallery №.1.

2   EXPERIMENT

Area of current research includes:

1. Streams, abandoned draining tunnels of the field.
2. The River Kholodnaya, receiving these streams.
3. The River Tyya, flowing in the immediate vicinity of the Deposit, draining the ore body but not flowing through the tunnels.
4. Bay of Baikal Angarsk Sor, receiving the rivers of Kholodnaya and Tyya.
5. Streams outside the drainage basin of the rivers Tyya and Kholodnaya, where background samples were taken.

During the research water samples of all listed rivers and streams were taken.

3   METHODS

A gross content of more than 70 chemical elements was determined in the samples of water, bottom sediments and suspension by mass-spectral methods with inductively coupled plasma (by the method of MP-2.4) and atomic emission with inductively coupled plasma (by the method of NSAM 487-XC). The suspension was studied by separating it on a membrane filter from a sample of a known volume;concentration of the main ions was determined in the water samples. The results were compared with the selected background samples and with the average content of elements in river waters (Dobrovolsky, 2011). To assess the risk of contamination the results were compared with the maximum permissible concentration (MPC) of the fisheries regulations for water and MPC of soils for bottom sediments. The integrated assessment was carried out on the basis of the total pollution coefficient Zc (Alekin, 1970).

4 RESULTS

The main existing sources of pollution, the degree of toxicity and the scale of the impact on the drainage watercourses, as well as the peculiarities of transportation of pollutants have been determined as a result of research. Existing spatial zones of ecological risk for aquatic landscapes associated with modern geochemical impact of Kholodnenskoe deposit despite the fact that it is not developed now. Today, it is the part of the basin of the River Kholodnaya, which flows near the mine and takes water streams flowing from abandoned tunnels.

In the waters of these streams, there have been revealed very high concentrations of pollutants for a number of heavy metals. Their amount exceeds the maximum allowable concentration in the hundreds and thousands of times. After the confluence of streams in Kholodnaya River pollution zone have been formed, within which the concentration of zinc in the waters of the River exceeds the normative values. For the Tyya River that drains the ore body but does not receive streams from the mine galleries, significant excess standards for the content of heavy metals have not been found. Directly in the water area of Lake Baikal, including the Gulf of Angarskiy Sor, where the Holodnaya River flows, exceeding the permissible values are also not identified.

Studies have shown that in general, the concentration of heavy metals in the region does not exceed the permissible and is close to the background. However, in case of exceeding the maximum permissible concentrations, they can be significant. In some cases waters contains Zn – 17000 times; Cd, Mn and Ce – more than 1000 times; Al, Co, rare earth metals (La, Nd, Pb, Tl, Pr, Dy, Gd, Sm, Ho) – from 100 to 1000 times; Y, Li, Rb, Eu, Er, Ga, Yb, Tb, Cu, Lu, Sb, Th, Tm – from 10 to 100 times more than it should be according to maximum permissible concentrations. The concentrations of Ge, Zr, W, U, Hf, Sn are less than 10 times more than maximum permissible concentrations.

For example, Figure 1 shows the concentrations for zinc with the highest exceedances.

For other heavy metals, there is a similar distribution. Abnormally high concentrations are recorded for streams draining deposits. After their confluence into the River Kholodnaya there has also been recorded significant excess concentrations of pollutants for its waters, but

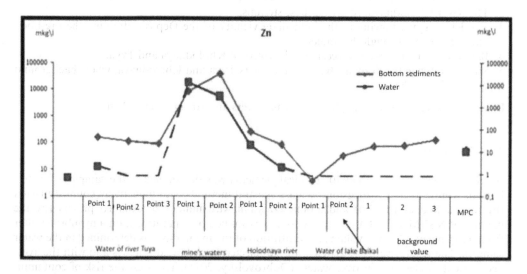

Figure 1. Zinc concentrations at different water and sediment sampling points.

closer to the mouth, they are reduced and generally no longer exceed the maximum permissible concentrations.

The excess was noted for bottom sediments as well: Zn – 390 times, As 180 times, Sb, Mn – 15-30 times, Pb, Cu – 3-6 times more than maximum permissible concentrations. For bottom sediments, it is more complicated to estimate concentrations than for water because, unlike water, the maximum permissible concentrations are not approved for them. However, there are common requirements for the quality of soils that have been used. The approved maximum allowable concentrations specially for bottom sediments would let, apparently, to evaluate the level of pollution and the danger of such pollution to aquatic landscapes, but currently they do not exist.

According to the criteria of sanitary and epidemiological requirements for soil quality, bottom sediments of mine streams belong to an extremely dangerous category, most samples taken from rivers belong to dangerous or moderately dangerous category, the permissible level corresponds only to samples from two points.

5 DISCUSSION

Geochemical effects of Kholodninskoe lead-zinc deposit exists and can be traced down the River Kholodnaya. Due to the large dilution, when moving from the field downstream, it significantly decreases. In the lower of River and when it reaches the Lake Baikal, concentrations of pollutants currently does not exceed the maximum permissible concentrations. However for achievement of the sustainable state of unique aquatic ecosystems of Lake Baikal in the long term, it is necessary to limit any significant anthropogenic impact. Further accumulation of heavy metals may result in exceeding the accepted standards. For a more accurate assessment of the impact of the Kholodninskoe deposit on the environment, regime observations should be carried out as well as hydrochemical testing. Moreover it is necessary to assess possible human impact on other components of local landscapes.

Nevertheless, for the sustainable state of unique aquatic ecosystems of Lake Baikal in the long term, it is necessary to prevent any significant anthropogenic impact (Koloboy, 2018, Ryzhenkov, 2018). Further accumulation of heavy metals may lead to exceeding the accepted standards, the need to comply with which is recognized by the supporters of the field development (Zubov & Mikhaylenko, 2011).

For a more accurate assessment of the impact of the Kholodninskoe deposit on the environment, it is necessary to apply a systematic approach that has already successfully proven itself to solve the problems in other similar researches and in land-using managements for other territories (Roger H. Charlier, 1989, Tulohonov et al, 2010, Kislov & Plyusnin, 2009).

Despite the formally (in comparison with maximum permissible concentrations) permissible concentrations of heavy metals in the Angarskij Sor Bay, their income into the Lake in the volumes recorded by current research seems to be an alarming trend. They inevitably accumulate in living organisms, which are used for food. Strong pollution can lead to the disappearance of certain species or a strong reduction in their number.

In order to organize the environment management of the coastal area of Lake Baikal, the most important goal is to preserve the quality of its waters and minimize their pollution. The arrival of pollutants into the waters of the Lake takes place both due to activity of natural factors and from anthropogenic sources. The most important natural factor is the flow of rivers. In some cases, however, anthropogenic forcing can significantly increase the amount of pollutant transferred in this way. An example of such situation is the increase in the volume of incoming heavy metals due to the exploration of reserves of the Kholodnenskoye deposit.

In the case of the start of mining lead and zinc in this deposit is expected to be further significant growth in the coming contaminants. As a possible action to prevent this, the diversion of water from the mine tunnels beyond the Lake Baikal drainage area (approximately 20-25 kilometers) may be considered. The introduction of a closed cycle of use for the processes of enrichment of underground mine waters can significantly reduce its negative impact in this case (Distanov, 1982, Zubov & Mikhaylenko, 2011). The minimum necessary measure, even if the development of deposit does not begin, is the constant monitoring of the quality of the waters of the rivers Kholodnaya, Tyya, streams that drain the ore body and flow into the Kholodnaya and waters of the Bay of Angarsk Sor of Baikal, receiving these rivers. Such monitoring will record changes in the content of pollutants over time. Especially such monitoring is necessary in the case of any, including exploration, works at the deposit.

The less difficult solution is simply to prevent region from the deposit's development, but in this case the country's industry will not receive valuable raw materials.

In the case of the development of the deposit by Metropol, its experts suggest some measures aimed at reducing the anthropogenic impact on Lake Baikal:

1. The location of the enrichment plant outside the drainage area of Lake Baikal (outside of its protected area).
2. Use of underground space for ore separation and crushing.
3. Transportation of ore to the concentrator on the conveyor, hidden in the main gallery.
4. Capture, treatment and transfer of mine water to a concentrator for use in a closed water cycle system.
5. The waste resulting from ore processing is planned to be laid with the use of a binder back into the mine workings, which will reduce the scale of possible pollution and preserve the natural relief of territory around deposit.

However, the plans of Metropol Company to develop the deposit, although accompanied by the proposed measures to prevent environmental pollution of the Lake, raise concerns of scientists and environmentalists both from regional and federal levels. Accordingly, further research in this region is necessary. As a possible solution geological exploration and development of other lead-zinc deposits available in Russia can be considered, but this requires a lot of time and huge material costs.

The example of the Kholodninsky deposit shows that it is not enough just to stop the activity of economic facilities or to avoid starting it in order to avoid environmental pollution. Currently the deposit is not developed, but it largely pollutes the environment. Fortunately, it is not located directly on the coast of Lake Baikal, but in a distance of a few dozen kilometers from the Lake. This allows the waters of the Kholodnaya River that receives polluted streams from the tunnels to dilute the polluted water, which reduces the concentration of pollutants to acceptable values before the River falls into the Lake. However, in case of development of the deposit it is possible that the volume of incoming pollutants will increase significantly, and contaminated water will not be diluted in time and will fall into the Lake.

6 CONCLUSION

In case the exploitation of deposit is resumed, strict measures should be taken to prevent pollutants' output especially heavy metals into the waters of Lake Baikal. In this regard proposed measures are to be thoroughly examined. Moreover, if necessary, the new ones are to be developed.

Even if the exploitation of the deposit will be not resumed, continuous monitoring of the quality of the waters of the streams and rivers in the area of Lake Baikal and the waters of the Angarsk Sor Bay into which they flow is vital.

REFERENCES

Aibulatov N.A. 2007. Russia's activities in the coastal zone of the sea and environmental problems. Shirshov Institute of Oceanology. Moscow.: Nauka, 364 p.
Alekin O.A. 1970. Basics of hydrochemistry. - Leningrad.: Hydrometizdat, 444 p.
Avdonin V.V., Bojtsov V.E., Grigoriev V.M. 2005. Deposits of metal minerals. Moscow: Academic project, Triksta 2005, 720 p.

Chief state sanitary doctor of the Russian Federation. 2003. GN 2.1.5.1315-03 "Maximum permissible concentrations (MPC) of chemicals in water of water objects of economic and drinking cultural and household and water use". Moscow.: Ministry of Health of Russia, 213p.

Distanov E. G. 1982. Kholodninskoe pyrite-polymetallic Deposit in Precambrian of the Baikal region. Novosibirsk: 1982, 208 p.

Dobrovolsky V.V. 2003. Basics of biogeochemistry. Moscow: Academy, 397 p.

Federal law of the Federation «On sanitary and epidemiological welfare of the Russian population "№ 52-FZ. 1999. Sanitary and epidemiological rules and regulations 2.1.7.1287-03. Sanitary and epidemiological requirements for soil quality». [On the Internet.] Available at: http://www.consultant.ru/document/cons_doc_LAW_22481/. [Date of application: 14.11.2018].

Kislov E. V., Plyusnin A. M. 2009. Problems of development of Kholodninskoe lead-zinc Deposit (the North-Baikal region). Geography and natural resources, V. 30, Issue 4: 340–344.

Kolobov, R.Y. 2018. Baikal as an object of UNESCO world importance: history and modernity. Prologue: journal of law, 2018: 33–39.

Korotaev. V.I. 2012. Essays on the geomorphology of estuaries and coastal systems. Moscow: Publishing house of the faculty of Geography of MSU, 2012, 540p.

Limarev V.I. 1986. The seacoasts and the people.Moscow: Science, 160 p.

Perelman A.I., Kasimov N. S. 2000. Landscape Geochemistry. Moscow: Astreya, 768 p.

Resolution of the Government of the Russian Federation. 2018. About the Federal target program" Protection of Lake Baikal and social and economic development of the Baikal natural territory for 2012-2020" (with changes for March 17, 2018)]" [online]. Available at: http://docs.cntd.ru/document/902365895 [date of access: 18.11.2018].

Roger H. Charlier 1989. Coastal zone: employment, management and economic competitiveness. Ocean and coastline management, V. 12, issues 5-6, p 383–402.

Ryzhenkov, A. Ya. 2018. On the principles of Lake Baikal protection. Bulletin of Omsk law Academy: 137–141.

Solovov, A. P. 1985. Geochemical methods of mineral deposits prospecting. Moscow: Nedra, 394 p.

Tulokhonov A. K., Slipenchuk M. V., N. G. Dmitrieva. 2010. The development of the mining cluster in the Transbaikalia: problems and solutions. Geography and natural resources, V. 31, issue 1: 53–57.

Guide to inductively coupled plasma mass spectrometry. 2005. Ed. M. S. Nelms. CRC Press: Boca Raton, 244p.

Verkhozina E.V., Verkhozina V.A., Protasova L.E. 2011. Ecological problems in the sphere of water use in the Baikal region. Bulletin of Irkutsk State Technical University, Vol. 48, № 1, 237–241.

Zubov V.P., Mikhaylenko O.V. 2011. Organizational-technical problems of development Kholodnenskoe polymetallic deposits. Notes Mining Institute,-Vol. 190: 318–322.

Topical Issues of Rational Use of Natural Resources 2019 – Litvinenko (Ed)
© 2020 Taylor & Francis Group, London, ISBN 978-0-367-85720-2

Alternative ways to reduce carbon dioxide emissions - the main objectives of the Greenpower Poland charity

M.A. Sobek, Ł.M. Grabowski & A. Werner
Silesian University of Technology, Poland

ABSTRACT: Despite the increasingly stringent emission regulations, the purity of the ecological environment is deteriorating day by day. The situation is much worse in the case of large cities or urban agglomerations. The increasing number of vehicles powered by internal combustion engines causes an increase in carbon dioxide emissions to the atmosphere. The situation is even worse when factors related to heating of residential buildings, which in Poland are still mostly heated with boilers powered by hard coal are added to the increasing number of vehicles. The combination of these factors causes that the condition of air in Polish cities is the worst since the records of its state began to be maintained. It is not easy to solve such a difficult situation. This problem should be solved on many levels. One of them is to educate young people according to ecological guidelines. The task of the Greenpower Poland Charity is to conduct workshops and lectures for young people. These lectures are related to the direct environmental and educational activities of the charity. The workshops are aimed at the construction of electric vehicles, which, thanks to common constructional features and common regulations, can compete in electric car races organized by the global organization which is Greenpower. The paper titled "Alternative ways to reduce carbon dioxide emissions - the main objectives of the Greenpower Polska charity" raises problems related to the environmental problems that in the long run can be reduced by proper preventive action consisting in raising awareness of young people and reaffirming them in ecological ideas

1 INTRODUCTION

For over a dozen years there has been an opinion in society that for various climatic changes we are witnessing right now are the effect of emitting a number of chemical compounds, i.e. methane, ozone, freon or even water vapor. The most responsible for this process is the drastically increasing content of carbon dioxide in the Earth's atmosphere (approximately 50% of green gas emission). There are numerous studies to prove the effect of this compound on the ecosphere of our planet. A large part of the research is also devoted to the detection of the main sources of the most destructive greenhouse gas, which is carbon dioxide.

Observing the increase in the number of vehicles on the roads of Europe, it is not difficult to conclude that the main source of CO_2 emission to the atmosphere is the burning of fossil fuels. Analyzing the situation in Poland, it turns out that a very big problem, maybe even the biggest one, is running the entire energy industry on fossil fuels such as black coal or lignite.

The European Union requires its countries to strictly comply with greenhouse gas emission allowances. This is directly related to the ecological and economic approach, so it is important to take care of the number of harmful gasses emitted.

2 MAIN SOURCES OF CARBON DIOXIDE

Before examining the main sources of carbon dioxide emitted into the atmosphere, it should first be outlined why this compound is the main issue. It is estimated that it is in

DISTRIBUTION OF ANNUAL GREENHOUSE GAS EMISSIONS

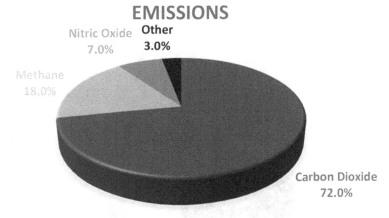

Figure 1.   Distribution of annual greenhouse gas emissions (https://ziemianarozdrozu.pl/encyklopedia/38/energia-i-emisje-co2).

the case of greenhouse gases, the CO_2 share is somewhere around 72% of emissions and it is constantly growing in comparison with other greenhouse gases, i.e. methane or nitrous oxide (Figure 1).

Daily car driving, every household item purchased, or kilowatt-hour of electricity accidentally used by us does not seem to generate a big problem. However, it should be remembered that those factors multiplied by billions of people repeating these mistakes can mean a disaster in the long run. Analyzing the graphs of the annual share of particular greenhouse gas emission sources, the general impact of the agricultural industry on the emission of methane and nitrous oxide can be easily noticed. However, this branch of the industry has no effect or is negligible on carbon dioxide emissions. However, considering the main shareholder in the emission of greenhouse gases such as CO2, it can not be missed that three industries are responsible for more than 50% of its emissions: energy industry (29.5%), manufacturing industry (20.6%) and fossil fuels used in transport (19.2%) (Figure 2). (Alami, Hawili, Hassan 2019, Wenli, Yang, Xianneng, Sun, Wang 2019, https://ziemianarozdrozu.pl/encyklopedia/38/energia-i-emisje-co2).

In the natural circulation of carbon, all emissions and absorption of carbon fraction are balanced. A small fraction of the carbon absorbed by nature turns into fossil fuels which are then deposited in the rocks. However, this is a long-term process. If it comes to human CO_2 emission, as an inherent element of the biosphere, his "biological" emissions are part of the carbon cycle (Figure 3). Plants collect carbon in the form of atmospheric CO_2. By eating plants human acquires coal in the form of carbohydrates. Then he exhales coal in the form of CO2. In the final analysis, the balance of the entire operation called the carbon cycle is equal to 0.

Combustion of fossil fuels that industries are feeding on is no longer a natural element of the carbon cycle. They have been accumulated by nature underground for tens or hundreds of millions of years. Running a worldwide industry based on fossil fuels instantly (in the perspective of millions of years) releases the carbon contained in them into the atmosphere. Therefore, it is necessary to consider the use of alternative methods of reducing carbon dioxide emissions to the atmosphere of the Earth. (Wenli, Yang, Xianneng, Sun, Wang 2019, Mukherjee, Okolie, Abdelrasoul, Niu, Dalai 2019, Shijie, Chunshan, Shaojian 2019, Rzepiela., Sobek, Grabowski 2017, https://ziemianarozdrozu.pl/encyklopedia/38/energia-i-emisje-co2).

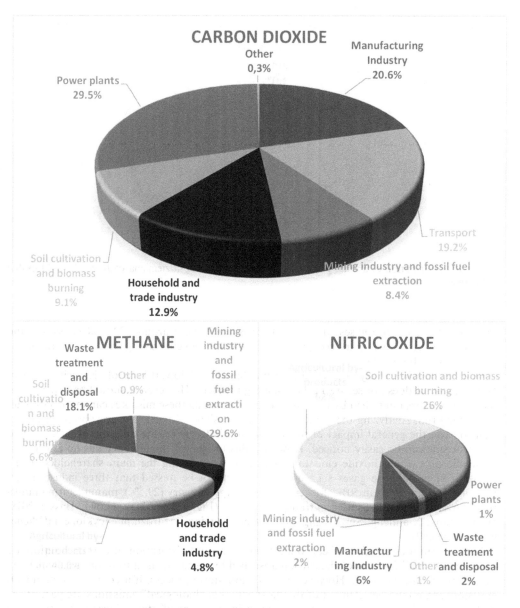

Figure 2.   Main sources of greenhouse gases (https://ziemianarozdrozu.pl/encyklopedia/38/energia-i-e misje-co2).

3   ALTERNATIVE METHODS FOR THE REDUCTION OF CARBON DIOXIDE EMISSIONS

Considering the aspects cited in the previous chapter, the question arises - how to implement a plan to reduce the emission of carbon dioxide to the atmosphere? First of all, three areas, which from the CO_2 emissions point of view are the most threatening to the natural environment should be considered.

In Poland, coal is still the main source of energy for power plants. According to the Central Statistical Office, in 2016 black coal consumption in Poland reached 74.17 million tonnes. About 59% of this amount (43.78 million tonnes) was used in power plants, heating boilers and heating plants to generate electricity and heat. It makes Poland the second largest coal market in Europe.

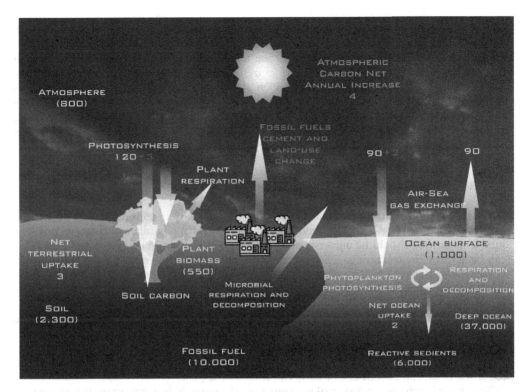

Figure 3. Carbon cycle in nature. Red numbers indicate human contribution in CO2. Yellow numbers are natural fluxes. Numbers are presented in gigatons per year. (https://genomicscience.energy.gov/11).

The first solution in the case of a reduction of carbon dioxide emissions in the production of electricity is the use of renewable energy sources. In this case, we can consider energy from four sources: water, air, sun or geothermal energy. Despite the very good effects and increasing efficiency of the devices used to collect each of these energies, they carry many negative effects.

In the case of capturing solar energy, we are dealing with photovoltaic panels, which require a large space to capture a small amount of energy. Despite the high price of the investment, it is currently the most available form of renewable energy in Poland.

Wind energy is mostly reserved for energy concerns or large investors. It is associated with the huge cost of the device itself, which in this case is a windmill. This device is huge in size, which also means the need to guarantee a suitable location. The location of a wind farm is also very important in terms of its efficiency. Not every location is abundant in the winds that feed the power plant. An unfavourable fact is also the high noise level reported by many people living next to wind farms. The high noise level accompanies the work of wind turbines themselves. Both in the case of solar power plants and in the case of wind turbines, it is possible to make such a device on its own and use it to support the home electrical installation. Such a solution does not involve large financial outlays but also the efficiency of those solutions in rather low. Also, the aesthetics of vertical turbines remains a matter of dispute.

Considering the acquisition of geothermal energy, at the very beginning we crash on to two serious problems associated with it. The first problem is the location of the hot springs. They usually lie at great depths, which makes it necessary to make deep and expensive wells. Not every well is guaranteed to encounter such a hot spring. The second problem is the low occurrence of geothermal sources. Unfortunately, they are not as accessible as the previously described methods. The effect of this unavailability is a very narrow number of facilities that can use such a source of energy. In practice in Poland, the group of recipients narrows to private investors and developers.

The last source of renewable energy is the energy that comes from the gravity-driven movement of watercourses, mainly the rivers and dam reservoirs. This solution is the least popular due to the need to interfere in the natural element of the river. The introduction of such a solution on a massive scale is impossible mainly because of the need for direct access to rivers.

The above-mentioned ways to reduce carbon dioxide emissions to the atmosphere seem to be the most reasonable solution for every citizen seeking a cleaner environment. However, each of them has various complications, which in many cases do not allow for their application in a wider perspective. (Alami, Hawili, Hassan, Al-Hemyari 2019, Wenli, Yang, Xianneng, Sun, Wang 2019, Mukherjee, Okolie, Abdelrasoul, Niu, Dalai 2019, Shijie, Chunshan, Shaojian 2019, Rzepiela., Sobek, Grabowski 2017, https://ziemianarozdrozu.pl/encyklopedia/38/energia-i-emisje-co2, https://earthobservatory.nasa.gov/features/CarbonCycle).

4 GREENPOWER POLAND CHARITY OBJECTIVES

In order to meet the current ecological problems, it is necessary to consider how to reduce the carbon dioxide emission with a method available to everyone, regardless of financial possibilities. This method should be universal and possible to implement in any place and time. This can be achieved by increasing environmental awareness among the public. Such activities are possible and should be implemented from an early age among children and adolescents. Such and other activities are carried out by the Greenpower Poland Charity (pol. Fundacja Greenpower Polska).

The Greenpower Poland Charity is an organization founded in Poland by a group of three students who want to bring to Poland the idea of clean electric vehicles racing known then only in Great Britain. In February 2019, the Charity celebrated its five-year anniversary. As part of the charity activities, many elements can be distinguished that indirectly affect the reduction of carbon dioxide emissions in the atmosphere.

The charity's activities include making young people aware of how important it is to care for the environment in every aspect of everyday life. Increasing the awareness of youth in the field of environmental protection in the perspective of time gives hope for an indirect reduction of greenhouse gases emitted into the atmosphere of the Earth. These activities are carried out in many areas, especially in areas where CO_2 emissions are the highest. (Rzepiela, Sobek, Grabowski 2017, Machoczek, Rzepiela, Sobek, Grabowski 2017, Werner, Rzepiela, Sobek, Grabowski 2017).

4.1 *Transport and electromobility*

The crowning of youth work is a taking part with a Greenpower electric car races in Poland (Figure 6). The rules of racing electric cars, which are organized by the members of the Greenpower Poland charity, consist of overcoming as much distance as possible during a limited time. An additional element to be taken into account is the fact of limited energy supplying the vehicle. Therefore, during the workshops conducted with the youth, a lot of attention is paid to the optimal energy consumption process. The introduction of energy optimization should translate directly into reducing the demand for it. The issue of energy saving does not have to be carried out only in the case of racing electric vehicles. The idea of energy saving works well in everyday life. Energy saving is not achieved only in the area of optimizing its consumption. This problem can be resolved at the beginning, during the process of designing goods – in this particular case by a properly designed electric race car.

Therefore, it should be considered what can affect the increased energy consumption. On the example of an electric vehicle, three main factors of the energy consumption rise in an electric vehicle can be pointed out. Factors such as:

a) High rolling resistance
b) Aerodynamic resistance
c) Large mass

The aforementioned factors should be reduced accordingly in order to obtain the best possible distance covered by the vehicle within only one battery. These issues are also addressed by the members of the Greenpower Poland Charity.

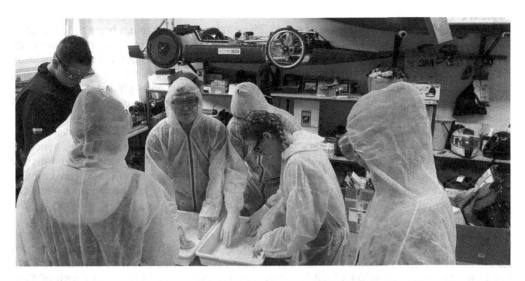

Figure 4.   Workshops on modern engineering materials run by the foundation's members.

During the workshop classes and lectures, young people find out what are the causes of the rolling resistance and how to safely reduce them. During the classes, young people also learn the secrets of vehicle design their aerodynamics taken into account. The fluid mechanics of air is a very important aspect of the engineer's work, but not only. Contrary to appearances, it is also reflected in everyday life, allowing young people to understand how a solid body moves in a fluid and what determines how much energy it will have to use to make this move.

The last element, which undoubtedly affects the amount of energy consumed during the movement of the electric vehicle is its mass. Unfortunately, electric vehicles are doomed to a large mass due to the large mass of the electric cells themselves. It is, therefore, necessary to look for weight reduction elsewhere. During the classes conducted by the members of the charity, a lot of attention is paid to the correct construction of the electric vehicle. This is due to the uniform technical regulations binding in each member country of the Greenpower family. This is, of course, related to the safety of the participants. Therefore, vehicles designed by young engineers as part of workshop classes are characterized by a large number of new technologies that allow maintaining an appropriate level of safety with reduced weight. Young people learn about ultralight construction materials as well as composite materials (Figure 4). Thanks to workshop classes, the young people can produce laminates by themselves, learning the principle of their operation and on their own skin, convincing of their advantages in the context of conventional materials. Thanks to the use of ultra-light construction materials in the construction of an electric vehicle it is possible to reduce the mass and thus the energy necessary for the vehicle to move. This rule applies to both electric vehicles and those powered by combustion engines.

Such final effects should translate into reducing the amount of energy needed to produce. Bearing in mind that the energy industry is largely dependent on coal in Poland, increasing public awareness in the area of energy management should directly translate into a reduction in CO_2 emissions by power plants in Poland. (Rzepiela, Sobek, Grabowski 2017, Machoczek, Rzepiela, Sobek, Grabowski 2017, Werner, Rzepiela, Sobek, Grabowski 2017, https://www.pgi.gov.pl/psg-1/psg-2/informacja-i-szkolenia/wiadomosci-surowcowe/10414-zapotrzebowanie-na-wegiel-kamienny-w-polsce.html).

4.2   CO2 Savings in industry

The activities of the Greenpower Poland charity may also bring positive effects in terms of introducing savings in the broadly understood industry. Considering the impact of this area on carbon dioxide emissions, it seems reasonable to look for a possible reduction there. Nowadays, humanity suffers from the satiation of goods. There are many more things being produced than we can manage as a community. This translates into a much higher production rate and over-stepping of the quantitative norms that factories must meet. Such an approach lets the industry to grow faster, which is obviously a positive phenomenon from the point of view of economics. Unfortunately, the industry is powered by energy and each of its areas generates a lot of green-house gases and waste. As a result of very strict regulation, corporations are forced to process their waste or make someone else do it on their behalf. Although the idea is very noble, it usu-ally takes a lot of energy to process every waste produced. It seems that it is a vicious circle.

The solution to this problem may be to raise public awareness about today's tendency towards unjustified consumerism. Of course, this solution will reduce demand and, at the same time, supply on the market. This would translate into a reduction in energy consumption and a reduction in greenhouse gas emissions, including CO_2. However, the approach proposed by the Greenpower Poland Charity is slightly different and consists rather in proposing the use of recycled materials or the use of modern technologies that do not generate CO_2 emis-sions as much as conventional engineering materials.

During the workshops conducted as part of Charity's activities, young adepts of engineering art learn about modern composite materials and 3D printing technologies. Composite materials allow reducing the weight of vehicles built by young people without sacrificing the strength of these

Figure 5.   CAD/CAE engineering workshops.

Figure 6.   Greenpower Poland race in Bydgoszcz, 5.06.2018.

structures. The use of 3D printing during the workshop allows expanding the imagination of young people. Thanks to the combination of CAD modelling and Rapid Prototyping technology.

Young people are able to design and manufacture structures that have no waste material or the fraction of waste is very small (Figure 5). Orienting awareness in this direction will undoubtedly allow indirectly reduce carbon dioxide emissions into the atmosphere in the future. (Rzepiela, Sobek, Grabowski 2017, Machoczek, Rzepiela, Sobek, Grabowski 2017, Werner, Rzepiela, Sobek, Grabowski 2017).

5 CONCLUSION

The fight against excessive emission of carbon dioxide into the atmosphere is a very important problem nowadays. Despite the actions being carried out, it is difficult to control the increased greenhouse gas emissions. All the methods of combating excessive CO_2 emissions quoted in this paper are undoubtedly very good solutions, but they are not addressed to each and every social group. What's more, they are usually not available due to many factors. The main reason is too high prices of ecological solutions, ie solar panels or wind farms. Otherwise, these are factors related to unavailability resulting from the necessary location requirements. Water, river and geothermal power plants are affected by such requirements.

Solutions proposed by the Greenpower Poland Charity are characterized by versatility. The workshops lead to the awareness and shape ecological ideas that are not even comparable to the cost of investments mentioned in this article. Changing the way of thinking and making the society aware of the effects of excessive greenhouse gas emissions will undoubtedly indirectly affect the size of emission. Conducting workshops indicating ecological ways of energy management contributes to its efficient use. In the long run, this approach should result in a cleaner atmosphere.

REFERENCES

Alami A. H., Hawili A., Hassan R., Al-Hemyari M. 2019. Experimental study of carbon dioxide as working fluid in a closed-loop compressed gas energy storage system. *Renewable Energy* 134 (2019) pp. 603–6611.

Machoczek S. Rzepiela K., Sobek M., Grabowski Ł. The possibility of using alternative energy sources in motorsport in the context of the Foundation Greenpower Poland. *Problemy nedropol'zovaniâ. Meždunarodnyj forumkonkurs molodyh učenyh*, Sankt-Peterburg, 19-21 aprelâ 2017 g. Sbornik naučnyh trudov. Čast' 2. Sankt- Petersburg: Sankt- Peterburgskij Gornyj Universitet, 2017, pp. 51–54.

Mukherjee A., Okolie J. A., Abdelrasoul A., Niu C., Dalai A. K. 2019. Review of post-combustion carbon dioxide capture technologies using activated carbon *Journal of Environmental Sciences* 8 3 (2019) pp. 46–63.

Rzepiela K., Sobek M., Grabowski Ł. 2017. The Greenpower Poland charity as an opportunity to promote and develop renewable sources of energy. *Problemy nedropol'zovaniâ. Meždunarodnyj forumkonkurs molodyh učenyh*, Sankt-Peterburg, 19-21 aprelâ 2017 g. Sbornik naučnyh trudov. Čast' 2. Sankt- Petersburg: Sankt- Peterburgskij Gornyj Universitet, 2017, pp. 98–100.

Shijie L., Chunshan Z., Shaojian W. 2019 Does modernization affect carbon dioxide emissions? A panel data analysis. *Science of the Total Environment* 663 (2019) pp. 426–435.

Wenli L., Yang G., Xianneng L., Sun T., Wang J. 2019. Cluster analysis of the relationship between carbon dioxide emissions and economic growth. *Journal of Cleaner Production 225* (2019) pp. 459–471.

Werner A. Rzepiela K., Sobek M., Grabowski Ł. (2017) The possibility of using alternative energy sources in the education process - the Foundation Greenpower Poland. *Problemy nedropol'zovaniâ. Meždunarodnyj forumkonkurs molodyh učenyh*, Sankt-Peterburg, 19-21 aprelâ 2017 g. Sbornik naučnyh trudov. Čast' 2. Sankt- Petersburg: Sankt- Peterburgskij Gornyj Universitet, 2017, pp.107–108.

https://earthobservatory.nasa.gov/features/CarbonCycle.

https://genomicscience.energy.gov/ US Department of Energy.

https://www.pgi.gov.pl/psg-1/psg-2/informacja-i-szkolenia/wiadomosci-surowcowe/10414-zapotrzebow anie-na-wegiel-kamienny-w-polsce.html.

https://ziemianarozdrozu.pl/encyklopedia/38/energia-i-emisje-co2.

Topical Issues of Rational Use of Natural Resources 2019 – Litvinenko (Ed)
© 2020 Taylor & Francis Group, London, ISBN 978-0-367-85720-2

Current state of the water and sanitation services of Russia in the regions

M.V. Suchkova
Saint Petersburg Mining University, Saint-Petersburg, Russia

ABSTRACT: The study deals with the state of water and sanitation systems in Russia in conditions of small and medium-sized cities. The characteristic features of water supply and sanitation are analyzed in detail on the example of one of the middle cities of the Volga region. According to the results of the technical survey and analysis of production activities, a list of the main required measures for the reconstruction of wastewater treatment facilities was compiled. The estimated cost of capital expenditures was taken into account. The study was supported by the Common Use Center of St. Petersburg Mining University in the framework of the State Order №5.12850.2018/8.9 of the Ministry of Education and Science of the Russian Federation.

1 INTRODUCTION

The problem of providing the population with drinking water of established quality as well as the environmental safety of water use in Russia are topical issues that largely determine the well-being of the state's inhabitants. That is why water and sanitation services (WSS) are one of the main directions of social policy of the Russia.

The necessary regulatory and legal framework in the field of WSS in Russia is represented by the Federal Law dated December 7, 2011 №416-FZ "On Water Supply and Wastewater Disposal" and the regulatory legal acts approved pursuant to this Law. However, despite the extensive legislative base, at present there is a gap between the developed regulatory acts and their implementation in practice (Danilovich et al. 2015).

In Russia, direct responsibility for the choice of wastewater treatment technology rests with the operating organizations, which are called "water utilities" ("vodokanal" in Russian). It should be noted that the requirements for wastewater treatment in Russia are more stringent than in Europe. Nevertheless, at the same time, we should not forget about the enormous difference in the equipment and condition of the wastewater treatment plants (WWTP) of Russia and foreign countries (Ngoc et al. 2018). The vast majority of modern Russian WWTP, especially those located in small towns, are in poor condition. There is no doubt that the problem of outdated equipment is one of the main factors for the deterioration of the quality of water purification at enterprises and cities.

WSS of Russia (despite all the existing objective difficulties and problems including unprofitableness) is one of the most ambitious and significant sectors of the state's economy. Meanwhile, a situation with water supply and sanitation in the country cannot be named as uniquely negative. It is necessary to take into account the fact that it is different for various regions and groups of the population. If the metropolises and cities in Russia look up to best practice and international standards in the field of WSS, the scale of problems related to water supply and sanitation in the regions should not be underestimated. Particularly vulnerable in these conditions are small and medium-sized towns, whose share is 84% of the total number of cities in Russia (as of 26/04/2019).

2 GENERAL INFORMATION ABOUT THE OBJECT OF STUDY

The state of WSS in regions in the framework of this study is considered on the example of one of the medium-sized towns (the population is about 60,000 people as of 2019) located in the region of the Middle Volga.

The climate of region is temperate continental with warm summers, moderately cold winters and well-defined seasons. The town is the capital of two specially protected natural areas (SPNA). This is a national park and state nature reserve. However, today's development of socio-economic initiatives using the potential of SPNA is very slow and has a number of negative trends.

The leading industries of the town are hydropower, mechanical engineering, food and pharmaceutical industry, mining of non-metallic materials and the production of cement. There are a cumulative effect of emissions from a cement plant, transport and other sources of pollution on the ecological condition of the town. It is reflected in changes in the concentrations of chemical elements in surface waters and the accumulation of pollutants in the soil and plants (Pashkevich et al. 2015, Smirnov & Suchkova 2017). In addition, the activities of enterprises of the mineral complex are associated with the problem of air pollution. In the case of cement production, the problem for the city is massive dust emission (Smirnov & Ivanov 2018).

In the last decade, a steady decline in population has been observed in the town. This is due to a decrease in migration growth, which, in turn, is the main source of compensation for natural population loss. Due to the special geographical position, the population increases at the expense of tourists in the holiday season. Although their stay is not long, however, it causes a certain increase in water consumption, the exact value of which is difficult to assess.

A characteristic feature of the town is the predominance of low-rise buildings with adjacent land. This led to a large occupied area (60.7 km^2) and a low population density (992 people/km^2), and as a result, a considerable length of water supply network, the density of which is 1.98 km/km^2.

3 FEATURES OF WATER SUPPLY AND SANITATION

Further development of the town is largely determined by the state of all elements of the industrial and social infrastructure, including WSS. Currently, the presence of dilapidated and emergency housing is 37% of the total housing stock, with an all-Russian value of indicator of 30%. Consequently, the priority development of WSS in the town should not be aimed at increasing the capacity for water production, but at its quality and reducing losses in the process of production and transporting (Rathnayaka et al. 2016).

It should be noted that the above industries are characterized by a high content of heavy metals in the wastewater of enterprises that also adversely affects the state of urban wastewater (Isakov & Matveeva 2016, Matveeva et al. 2018).

In general, it is possible to determine the following characteristic features of water supply and sanitation in the town:

- Low-rise buildings, large occupied territory and low population density are causing a high length of water supply network and increased energy consumption for water transportation;
- Presence of land plots adjacent to the houses, primarily in the villages, is a potential factor for increased water consumption;
- High volume of water consumption is combined with a low water tariff, which does not contribute to its economical spending;
- Low volume of work on the replacement of water supply network (1.2% of the total length with a normative indicator 4-5%) due to insufficient financial capabilities of the water utility. At the same time, sanitary sewer practically does not replace, the total length of sanitary sewer that require replacement is 84.7 km (90% of the total length);
- High level of depreciation of fixed assets of more than 65% and water supply network of more than 80%. At the same time, the wear of water supply network is 100% in many

areas. The length of water supply network that require urgent replacement is 36.32 km or 20% of the total length;

- High level of losses and unaccounted expenses of water supplied to the network (16%);
- Increase in population in the holiday season causes certain fluctuations in water consumption;
- Increase in the number of accidents and defects on the water supply network (1.93 accidents per 1 km of the network per year), which is more than 2 times higher than the average indicators for water utilities in Russia. This situation also indicates the critical condition of water supply network.

In other water utilities of Russia, the number of accidents and defects in the water supply network has also increased since the beginning of the 90s by more than 2.5-3 times, but is 0.8-0.9 defects per 1 km of network per year. It should be noted that in Western European countries this indicator is no more than 0.1 accidents per 1 km of network per year. In the 80s in Russia, this indicator also did not exceed 0.15-0.20 accidents.

All these and other problems in the activities of town's water utility are reflected in the results of both the quality of the provision of water supply and sanitation services, and the financial and economic activities of the utility.

Along with centralized sanitation in the town, part of wastewaters is discharged into the cesspools, from where it is moved by special transport to the treatment plant, where it is cleaned. The town is divided into five technological zones of sanitation.

A distinctive feature of the town's WWTP is that they do not constitute a single complex. WWTP are scattered in different localities, since a part of the territories were designed, built, and equipped with infrastructure as settlements isolated from the town.

In the town, household and industrial wastewaters through self-flowing sanitary sewer and pumping stations enter the WWTP with a capacity of 16,200 m³/day, the first line of which is designed for 12,000 m³/day in 1973.

The structure of the first stage of WWTP includes:

- Crusher grates - 2 pcs;
- Sand traps with circular rotation of water - 2 pcs;
- Aeration tanks - 2 pcs;
- Primary clarifiers - 6 pcs;
- Secondary clarifiers - 6 pcs;
- Contact tanks - 2 pcs;
- Blower station;
- Chlorinator;
- Sludge drying beds.

The first start-up complex of the second construction line (with a design capacity of 18,000 m³/day) was completed in 2003-2005 and included:

- Blower station;
- Mechanical dehydration workshop (engine and pump rooms, domestic premises);
- Aerobic mineralizer;
- Grit dewatering bays.

Wastewater from the town enters the reception chamber of WWTP. Mechanical cleaning facilities include: crusher grids, horizontal sand traps, vertical primary clarifiers. The sand retained in the sand catchers is removed with a hydraulic elevator and in the form of sand pulp is pumped to grid dewatering bays. After the sedimentation tanks, the clarified water is sent by gravity to the biological cleaning in aeration tanks. Then the water enters the secondary vertical clarifiers. To destroy pathogenic microbes, wastewater is disinfected by chlorination in contact tanks before descent. Excess activated sludge from secondary clarifiers enters sludge drying beds with a working depth of up to 2 m. The beds are organized in the form of rectangular reservoirs with a waterproof bottom and walls (Kulikov et al. 2014).

In fact, all installed equipment requires upgrading and replacement:

- Large physical deterioration of metal structures (gratings, drainage trays, metal pipes, stop valves, etc.);
- Destruction of concrete structures and panels of process tanks (primary and secondary clarifiers, aeration tanks, receiving and distribution chambers);
- Full wear of chlorination equipment;
- Large amount of raw sludge on sludge drying beds, reducing the time of sedimentation and drying of the sludge, the lack of sludge consolidation tanks.

There are no metering devices; water disposal is recorded by an indirect method, according to the duration and performance of the pumps. Discharge of treated wastewater is carried out in the northeastern part of the town below the hydroelectric station by 4 km into the Volga River. It is a reservoir of fishery value of the highest category. The discharge is carried out by two dissipating outfalls: №1 (from urban WWTP) and №2 (from sewage pumping station without cleaning). The wastewater for the outfall №1 is not enough cleaned and exceed all indicators of maximum permissible concentrations of pollutants, which adversely affects the Volga' waters.

Street sanitary sewer is built of asbestos-cement pipes of various diameters. These underground pipes are subject to many influences that it is difficult, if not impossible to predict their combined effect in advance. These influences include the corrosive action of the soil and sewage, ground movements, etc. Most of these factors are random, and are characterized by significant variability (Ermolin & Alexeev 2018). As mentioned earlier, the depreciation of fixed assets in the wastewater industry is more than 85% and sanitary sewer is more than 90%. Thus, the total length of network that require replacement is 84.7 km or 90% of the total length. In addition, the town requires the replacement of the main collectors with a length of 15.1 km, the street sanitary sewer (10.8 km) and the intra-quarter and in-yard networks (49.2 km). The wear of these networks is also 100%. Work on the renewal of fixed assets, including the replacement of sanitary sewer is practically not conducted.

All sanitation systems are not provided with automated systems for dispatch control, management, technological and commercial accounting.

Thus, as well as in the water supply industry, the problem of the physical and moral state of the fixed assets of sanitation in the town will intensify every year.

In the process of mechanical and biological wastewater treatment, various types of sludge are formed; those are containing organic and mineral components. The sludge has a high humidity of 99.2-99.7% (Barkan et al. 2007). The sludge dewatering is produced by the natural dewatering method on sludge drying beds. The total area of beds is 11,700 m^2. Further, the dehydrated sludge with a humidity of 77-78% is transported to the municipal solid waste landfill (Federal Agency on Technical Regulation and Metrology, 2011). At the same time, there are many ways to reclaim disturbed land using local waste and soil-forming substances (Danilov et al. 2017, Ivanov et al. 2018). Additional treatment and neutralization of sewage sludge would allow turning it into a promising product. However, currently with this method of dehydration, the sludge cannot be reused for the reclamation of disturbed lands due to its high contamination by pathogenic organisms and heavy metals (Antonova et al. 2014).

In most Russian towns, water utilities have no other systems for the treatment of sewage sludge, except sludge drying beds. At the same time, over the decades of operation, the beds have lost their original function of sludge dewatering. However, the sludge continues to be placed on the sludge beds, thereby turning them into disposal facilities. Due to the unsettledness of the legislative base, this leads to the fact that sludge drying beds, which belong to WWTP, are transformed under a different function as a "landfill". All this leads to legal difficulties with their further exploitation (Dregulo & Kudryavtsev 2018). At the same time, sludge beds and sand grounds are unorganized sources of air pollution in the urban area. They require mandatory environmental monitoring. In particular, there is a promising method of remote monitoring using small-sized unmanned aerial vehicles (Danilov et al. 2016, Ivanov & Smirnov 2016).

Due to overloading of WWTP, the numerical values of the actual quality of wastewater treatment are constantly observed above the normative values, which do not ensure compliance with the requirements of protecting surface water from pollution by wastewater. Currently, the town has two sanitation system: a centralized system for domestic and storm wastewater and a centralized system for wastewater without a cleaning element. Thus, the average annual difference between the volume of treated wastewater and the volume of water supplied to consumers is more than 2,000,000 m^3 (Sapkota et al. 2015).

Serious problems arise during the operation of sanitation systems, including the impact of wastewater itself on the sanitary sewer. The corrosive action of water additionally damages already worn-out pipes and causes significant leakages in the distribution network. A lack of water meters in the main part of consumers further aggravates the production and technical situation. All of this leads to an increase in the number of accidents and defects, the emergence of uncontrolled water losses and a number of problems on the maintenance and management of sanitary sewer (Capodaglio 2017).

4 SANITATION SYSTEM DEVELOPMENT PROJECT

Based on the results of a technical survey of WWTP, analysis of production activities, management structure and relations with consumers of the water utility, a list of the main required measures has been developed:

1 Reconstruction of existing WWTP in the town, including:

- Overhaul of the building of the grids with the replacement of the process equipment and the building of the blower station, as well as the reconstruction of the main sludge pumping station and the technological dewatering workshop.
- Replacement of a pressure collector.
- Reconstruction and replacement of drainage networks.

High wear and low power of WWTP, especially during the flood period, require the need to expand and reconstruct facilities. Replacement of pressure collectors with an increase in their diameter is also required. With an insufficient diameter and clogging of pipelines during the flood period, overflowed sewage wells lead that unpurified wastewater falls into the Volga River. Currently, the increase in depreciation (over 80%) and damage to fixed assets (primarily water supply networks – 1.93 defects per km), has led to a high probability of accidents affecting all town consumers. In this situation, increasing the reliability and sustainability of sanitation systems can only be achieved by one-time replacement of worn-out networks, primarily of at least 10% of the total length.

In the conditions of dense urban development of the town, there is a need to apply trenchless methods of repair and restoration of pipelines of sanitary sewer, which increases the cost of reconstruction and replacement of networks. A new method of repairing large-diameter pipelines "pipe in pipe" makes it possible to return to operation the pipelines that have lost their serviceability, to ensure their stable throughput for a long time (50 years or more). For newly laid sections of sewer pipelines, the most reliable and durable material is polyethylene. It withstands shock loads with a sharp change in pressure in the pipeline and has resistant to electrochemical corrosion.

2 Installation of dispatching, teleautomation and control systems at WWTP.

Installation of automation systems for sewage treatment processes will increase the energy efficiency of facilities and improve the quality of service provision.

3 Installation of water meters.

The absence of water metering devices in the places of supply and dictating points of consumption, as well as the requirements of the Federal Law of 23 November 2009 №261-FZ "On Energy Saving and Improving Energy Efficiency and Amending Certain Legislative Acts of the Russian Federation" and Decisions of the Government of the Russian Federation of 6 May 2011 №354 "On the provision of utilities to owners and users of premises in apartment buildings and residential buildings" require the installation of metering devices.

4 Reuse of wash waters.

It is known that one of the permanent sources of concentrated pollution of surface water bodies are resulting from the washing of filtering facilities of WWTP. Wash waters are often discharged without treatment. Suspended substances and components of technological materials in their composition, as well as bacterial contamination, when entering the water, increase the turbidity of the water, reduce the access of light to the depth, and, as a result, decrease the intensity of photosynthesis. This in turn leads to a decrease a processes of self-cleaning.

It is necessary to use resource-saving, environmental protection technology of reuse of filters' wash waters to prevent adverse effects on the reservoir in the process of water treatment. This technology allows improving the environmental safety of the Volga River, eliminating the discharge of wash water into the water body.

5 Improvement of wastewater disinfection technologies.

The introduction of automated complex disinfection of raw and purified water with sodium hypochlorite will eliminate the use of liquid chlorine. The transition to the use of sodium hypochlorite allows eliminating chlorine farming in the town, to ensure environmental and technological safety in the production of drinking water and the wastewater treatment.

The introduction of ultraviolet disinfection technology, successfully used in other cities of Russia and foreign countries, will also make it possible to abandon the use of liquid chlorine. It is recommended to use UV emitting with a flux density of at least 40 mJ/cm^2 and a wavelength of 253.7 nm and 185 nm (Mecha et al. 2017).

6 Modernization of dewatering sewage sludge.

In parallel with the solution of wastewater disinfection tasks, it is proposed to mechanically dewater sludge using a belt press filter, which will make it possible to reduce the consumption of electrical energy in relation to the existing technology more than 3 times, as well as not to expand the sludge drying beds.

It is important to note that the estimated capital costs for the reconstruction of the WWTP with a capacity of 15,000-20,000 m^3/day (including the overhaul of buildings, the construction of a UV-disinfection station and the introduction of a mechanized sludge dewatering system) will amount to about 20-25 million rubles (taking into account the Deflator Index by industry; for 2019, Index for Water Supply and Sanitation is 104.9% of the state for 1999) (Ministry of Construction and Housing and Communal Services of the Russian Federation, 2015; Ministry of Regional Development of Russia, 2012).

5 CONCLUSION

While in the major Russian cities, the problem of water disposal is being systematically solved, in medium-sized and small towns the WWTP and sanitary sewer are in a state of decline. The main reasons for the low efficiency of town's WWTP:

- Lack of budget funds for the reconstruction and modernization of facilities;
- Non-observance of the technological mode of their operation;
- Inconsistency of the incoming wastewater composition to treatment technologies;
- Significant physical deterioration of existing treatment facilities and sanitary sewer.

One of the National Projects of Russia (for the period from 2019 to 2024) is the National Project "Ecology" (Presidential Decree of 7 May, 2018). It includes, in particular, the Federal Project "Improvement of the Volga River", which is aimed at the environmental rehabilitation of water bodies, also at reduction by three times the proportion of polluted wastewater discharged into the Volga River as a result of construction or modernization of treatment facilities.

The solution of such local issues is a small but tangible step towards solving this problem as a whole.

REFERENCES

Antonova, I.A., Gryaznov, O.N., Guman, O.M., Makarov, A.B., Kolosnitsina, O.V. 2014. Geological conditions for allocation of solid municipal and industrial waste disposal sites in the Middle Urals. *Water Resources*, № 41(7): 896–903.

Barkan, M.Sh., Kuznetsov, V.S., Fedoseev I.V. 2007. Research of physical and chemical parameters of deposits of city sewage. *Journal of Mining Institute*, V. 172: 214–216.

Capodaglio, A.G. 2017. Integrated, Decentralized Wastewater Management for Resource Recovery in Rural and Peri-Urban Areas. *Resources*, 6(2): 22 p.

Danilov, A.S., Smirnov, Y.D., Korelskiy, D.S. 2016. Promising method of remote environmental monitoring of Russian oil and gas industry facilities. *Oil industry*, №2:121–122.

Danilov, A.S., Smirnov, Y.D., Korelskiy, D.S. 2017. Effective methods for reclamation of area sources of dust emission. *Journal of Ecological Engineering*, V. 18 (5): 1–7.

Danilovich, D.A., Epov, A.N., Kanunnikova, M.A. 2015. Data analysis of the treatment facilities of Russian cities is the basis for technological standardization. *Best Available Technologies*, № 3: 18–28.

Dregulo, A. M., Kudryavtsev, A. V. 2018. Transformation of techno-natural systems of water treatment to objects of past environmental damage: peculiarities of the legal and regulatory framework. *Water and Ecology*, №3 (75): 54–62.

Ermolin, Y. A., Alexeev, M. I. 2018. Reliability measure of a sewer network. *Water and Ecology*, №2 (74):51–58.

Federal Agency on Technical Regulation and Metrology. 2011. State standard R 54535-2011 Resources saving. Sewage sludge. Requirements for waste dispose and use at landfills. Moscow: Standards Publishing House.

Isakov, A.E., Matveeva, V.A. 2016. OAO "Kovdorsky MCC" manganese-containing waste water purification study. *Obogashchenie Rud*, 2016 (2): 44–48.

Ivanov, A.V., Smirnov, Y.D. 2016. The impact of anthropogenic alluvial arrays on areas settlements depending on the particle size distribution of stored taillings. *Journal of Ecological Engineering*, V. 17 (2): 59–63.

Ivanov, A.V., Smirnov, Y.D., Petrov, G.I. 2018. Investigation of waste properties of subway construction as a potential component of soil layer. *Journal of Ecological Engineering*, V. 19 (5): 59–69.

Kulikov, N.I., Nozhevnikova, A.N., Zubov, G.M. et al. 2014. *Municipal wastewater treatment with reuse of water and treated sludge: theory and practice*. Moscow: Logos, 400 p.

Matveeva, V.A., Petrova, T.A., Chukaeva, M.A. 2018. Molybdenum removal from drainage waters of tailing dumps of Apatit JSC. *Obogashchenie Rud*, 2018 (2): 42–47.

Mecha, A.C., Onyango, M.S., Ochieng, A., Momba, M.N.B. 2017. Ultraviolet and solar photocatalytic ozonation of municipal wastewater: Catalyst reuse, energy requirements and toxicity assessment. *Chemosphere*, V. 186: 669–676.

Ministry of Construction and Housing and Communal Services of the Russian Federation. 2015. State estimated standard SBCP 81-02-17-2001 Reference book of basic prices for project works in construction SBCP 81-2001-17 Water supply and sanitation facilities. Moscow: Ministry of Construction of Russia.

Ministry of Regional Development of Russia. 2012. Code of Practice 32.13330.2012. Sewerage. Pipelines and wastewater treatment plants. Building Codes and Regulations 2.04.03–85. Moscow: Ministry of Regional Development of Russia.

Ngoc, H.T., Reinhard, M., Yew-Hoong, G. K. 2018. Occurrence and fate of emerging contaminants in municipal wastewater treatment plants from different geographical regions-a review. *Water Research*, V. 133: 182–207.

Pashkevich, M.A., Alekseenko, A.V., Vlasova, E.V. 2015. Biogeochemical and geobotanical assessment of marine ecosystems conditions (Novorossiysk city). *Water and Ecology*, 2015 (3): 67–80.

Presidential Decree of 7 May 2018 №204 On the national goals and strategic objectives of the development of the Russian Federation for the period until 2024. *The collection of legislation*, №20. – Art. 2817.

Rathnayaka, K., Malano, H., Arora, M. 2016. Assessment of Sustainability of Urban Water Supply and Demand Management Options: A Comprehensive Approach. *Water*, 8(12): 595.

Sapkota, M., Arora, M., Malano, H., Moglia, M., Sharma, A., George, B., Pamminger, F. 2015. An Overview of Hybrid Water Supply Systems in the Context of Urban Water Management: Challenges and Opportunities. *Water*, 7(1): 153–174.

Smirnov, Y.D., Ivanov, A.V. 2018. Investigation of dust transfer processes during loading and unloading operations using software simulation. *Journal of Ecological Engineering*, V. 19 (4): 29–33.

Smirnov, Y.D., Suchkova, M.V. 2017. A complex assessment of the ecological condition of waters Murinsky creek. *Water end Ecology*, 2017 (3): 35–48.

Topical Issues of Rational Use of Natural Resources 2019 – Litvinenko (Ed)
© 2020 Taylor & Francis Group, London, ISBN 978-0-367-85720-2

Liquidation of the uranium mine Rožná I in Dolní Rožínka

M. Vokurka

VSB – Technical University of Ostrava, Ostrava, Czech Republic

ABSTRACT: The liquidation of the Rožná I uranium mine in Dolní Rožínka is a current problem, as the uranium ore mining was terminated in 2017 and the deposit will have to be liquidated after all mining activities in this area are finished.

As soon as the mine areas cease to be used, the mine will be liquidated. The liquidation will start with the stopping of the mine water pumping, which will cause gradual flooding of the mine. The uranium mine, as a non-gassing mine, will be liquidated by a complete, non-strengthened backfill of the initial mine workings opening to the surface. After filling up the initial mine workings, a closure sinking platform will be built at their mouth.

The next part is devoted to the problems of mine water during gradual flooding. During the flooding process, the geochemistry of the waters changes and their stratification occurs as a result of the dissolution of the secondary minerals from the old mine workings.

Once the deposit is flooded, the mine water will be pumped and cleaned for a certain time horizon. As soon as the water contamination no longer exceeds the permitted limits, the water will be discharged freely to the local watercourse without prior cleaning.

1   INTRODUCTION

Upon completion of the uranium ore mining from the Rožná deposit, it is the duty of the organization performing the mining activity to carry out a complete liquidation of the mine to ensure the safety of the area and to terminate the mining activity on the site. At present, underground areas are used for non-mining activities, mostly of a scientific nature, and thus extend the life of mining activities in the given area. As soon as the premises are not used, the deposit will be liquidated by stopping the pumping of mine water and subsequent backfilling of mine workings to the surface with non-strengthened backfill material. When the liquidation work is completed, the controlled discharge of mine water will follow after the water is cleaned.

2   GENERAL INFORMATION ABOUT THE ROŽNÁ I URANIUM MINE IN DOLNÍ ROŽÍNKA

The uranium mine Rožná I is located on the edge of the Bohemian-Moravian Highlands (approx. 550 m above sea level) in the district of Žďár and Sázavou, in the Vysočina Region, in the central part of the Czech Republic. In the mining area of 8.76 km², the spin-off plant GEAM Dolní Rožínka, which is part of the state enterprise DIAMO in Stráž pod Ralskem, carried out a permitted mining activity. Until 2017, the main activity was deep mining of uranium ore from the Rožná deposit and its subsequent treatment into the uranium concentrate - ammonium diuranate $(NH_4)_2U_2O_7$. At present, deep mining of uranium ore and its subsequent treatment is terminated and the underground areas at the Rožná deposit are used for non-mining activities, mostly of a scientific nature.

The Rožná deposit was discovered as part of the nationwide uranium exploration in 1957. Uranium ore mining was terminated in 2017 after 60 years. A total of 16,927,897 tons of broken ore were extracted, which represent a yield of 20,220 tonnes of uranium concentrate at an average content of 0.119% of U. (Vokurka, 2018)

3 LIQUIDATION WORK

In the course of mining activities at the Rožná deposit, a total of 55 main mine workings opening to the surface were excavated in the specified mining area. Of this number, 43 main mine workings have been liquidated so far, mainly by backfilling. Except for the liquidation of the 12 main mine workings, see Table 1, the liquidation of the already liquidated mine workings will have to be completed by constructing sinking platforms. (Vokurka, 2018; DIAMO, 2019)

3.1 Planning liquidation works

The main mining works will be liquidated according to the principles set out in the Decree of the Czech Mining Authority No. 52/1997 Coll., as amended, No. 32/2000 Coll., No. 592/2004 Coll., No. 176/2011 Coll., No. 237/2015 Coll. (hereinafter referred to as the Decree) laying down requirements to ensure occupational safety and health and safety at the liquidation of main mine workings by full backfilling with non-strengthened waste rock material and constructing a closure sinking platform at their mouth.

Before proceeding to the planned liquidation of pits from the surface, dam structures on designated levels, the so-called supporting dams must be constructed and shaft bottoms of the levels and manways of the pits must be dismantled underground, and shaft bottom must be prepared for backfilling with non-strengthened backfill material. (Decree, 1997)

3.1.1 Preparation of mine workings for liquidation

The supporting dams are built to prevent the displacement of non-strengthened backfill material in horizontal mine workings due to the horizontal stress exerted by the backfill material column in the pit shaft of the mine workings. Otherwise, the backfill material could

Table 1. Parameters of the main mine workings opening to the surface not yet liquidated and requirements for backfill material.

Pit	Rough profile m^2	Clearance m^2	Depth m	Number of contacts pc	Shaft bottom m^3	Pit shaft m^3	Fill-up m^3	Backfills m^3
R1	10.95	6.83	667.87	14	3 981	3 683	383	8 047
R2	10.95	6.83	541.70	9	2 559	2 988	277	5 824
R3*	20.43	14.52	1 200.40	20	-	-	-	-
R4	10.95	6.83	313.00	5	1 280	1 726	150	3 156
R6	25.52	20.36	784.20	9	2 559	13 571	807	16 934
B1	10.95	6.83	651.2 0	9	2 559	3 592	308	6 459
B2	20.43	15.90	541.80	2	368	7 323	385	8 076
Š11	7.68	4.56	134.46	3	620	495	56	1 171
Š13	6.68	4.36	59.60	1	207	216	21	444
Š37	7.06	4.27	351.10	3	414	1 231	82	1 727
VK-4/0-1	7.95	7.95	200.30	1	142	1 327	74	1 543
VKS-7/0	13.20	9.62	305.50	1	362	2 498	143	3 003
In total				77	15 051	38 650	2 686	56 387

*The pit R3 will not be liquidated by backfilling, as it will serve as a water pit of the Rožná deposit

move over horizontal workings by several tens of meters and stabilization of the backfill would become an economically demanding and long-term matter.

In order to build the supporting dams, it is necessary to select such places at the shaft bottoms of the pits where:

- there is a change in the profile of the mine working – i.e. a change to a smaller profile, change of shape,
- there is a change in the direction of the mine working,
- the design of the existing reinforcement allows its incorporation into the retaining dam itself.

The supporting dams have the character of a „cage-like welded steel structure", the authors of which proceed from the possibility of using materials that can be found on the corresponding levels. A pair of I 240 steel profiles are placed on the roof, floor and sides of the dam's front and are fixed to the rock by means of shackles by a couple of glued bolts, which are then placed on each meter of the I 240 profile. The profile frame is welded at all the contact points and the cross of the I 240 profile is welded into the frame. Then, a cage steel structure of up to 5 m is welded from the K 93 rails to support the dam face. The dam is closed with the rear frame of the I 240 profiles. The dam body is then filled with the available metal material of up to 1 m in size, possibly with aggregate. Finally, a grid of K 93 rails is welded to the face of the dam at the distance of 0.25 m, see Figure 1. (Vokurka, 2018; Grygarek, 2002).

Dismantling on horizontal mine workings will be carried out in the area of the contact of the pit with the accessible floor, i.e. at the shaft bottom. Steel structures and equipment that could prevent filling the shaft bottom with unpaved backfill material will be removed. Before leaving the levels, the shaft bottom frames, i.e. the barriers and the shaft way door will be stripped down, the chambers and easily removable parts of the hinged bridges, i.e. the landing chairs and the striking device will be covered with a metal sheet. The dismantled structures and equipment will be left on the level. Dismantling will not be carried out on inaccessible levels, but it will be assumed that there are no structures on the levels to prevent filling the contact with the floors with the backfill material.

Dismantling the equipment of vertical mine workings, i.e. pits will be carried out on the basis of occupational safety assessment and those parts that could prevent or limit the penetration of backfill material, i.e. wooden landings in the manway, will be dismantled. Wooden shaft guides, pipelines, cable ducts, or ventilating pipes will be left in the pit. (Vokurka, 2018; Grygarek, 2002, Wojtkowiak, 1999)

3.1.2 *Filling the mine working with the non-strengthened rock material*
The waste rock which was created in the past by long-term mining activity and was deposited on the Rožná I mine slag heap will be used as the backfill material. It is a material that is non-

Figure 1. Welded steel construction of the supporting dam at the shaft bottom of the B1 pit 5th level.

flammable, insoluble, non-poachable and non-swellable, which does not endanger the environment and does not contain metal objects. Above all, fine- to medium-grained biotite locally migmatitized paragneiss and amphibolites were deposited on the slag heap. Only a small proportion is formed of crushed rock from the zone and vein filling and of clay components. The fragmentary character of the deposited rock was evaluated as favourable for backfilling the mine workings, see Table 2.

A small amount of the fraction above 250 mm will facilitate the compliance with the Decree, according to which the grain size used for filling a mine working must be ensured by passing through a grate of 250 x 250 mm mesh size. The backfill material of less than 30 mm will contribute to filling the areas between pieces of larger fractions. The proportion of clay component can be neglected in such a small proportion.

The weight of 1 m^3 of the backfill material in this grain size composition was set at 2,000 kg with an average rock density of 2,700 kg.m^{-3}, with the swell factor of 1.35. After compaction of this material due to its fall into the mine working, the immediate swell factor is changed to the permanent swell factor, which reaches the value of 1.2. The weight of 1 m^3 will then be 2 250 kg. The gradual flooding of the deposit will cause the grains to float into gaps that have not been filled yet, and thus increase the weight of 1 m^3 up to 2,300 kg. The waste rock material must be particularly stable for the given conditions. Therefore, it is necessary to place demands mainly on its compressibility. For this backfill material, the average compressibility set in the laboratory conditions was 15%. For loose aggregates, there is a so-called natural bulk angle of the slope of a loosely heaped rock with a horizontal plane. Its size depends on the shape and character of the grains, on their moisture, density, granulometric composition and other factors. For the given material of the above-mentioned grain size composition, the natural bulk angle was almost identical to the internal friction angle of the rock, which only differs in the fact whether the backfill will be carried out to a dry pit or to a pit that has already been flooded. In the case of backfilling into a dry pit, the angle of internal friction of the rocks will be 35°, otherwise 25°. (Vokurka, 2018; Decree, 1997; DIAMO, 2019; Bukowski, 2013)

The principle of backfilling mine workings is given by the Decree, so in all cases, it will be subject to the same procedures. Before carrying out the mine work itself, it is necessary to prepare a shaft bottom for this purpose. At the mine workings which were in operation until the end of mining, the shaft building and the mining tower will be left during backfilling; therefore the immediate surroundings of the pit must be secured to make it possible for the means of transport to access the pit. In other cases, the surface infrastructure is already removed, so the liquidation will be simpler, but even in these cases, a truck driveway will have to be established. A backfill grate with a mesh size of 250 x 250 mm will be placed on the pit-bank with a directional hopper and a stop against the crossing of trucks into the pit profile. The backfill material will thus be cyclically heaped into the pit directly from the truck body or by loaders, see Figure 2. The backfill inspection will be carried out after filling in the converted height of 50 m of the mine working. If the backfill difference is greater than 20% of the calculated value, the work will be interrupted and the next procedure determined. In accordance with the Decree, graded aggregate of a piece size of 63 - 125 mm, possibly supplemented with a sharp ballast stone not

Table 2. Fragmentary character of the deposited material at R I slag heap.

Grain size mm	Proportion %
above 250	1 – 2
150 to 250	5 – 10
30 to 150	45
below 30	40 – 45
clay component	3

Figure 2.   An example of the cyclic backfill of pit no. 13 from the Pucov uranium deposit.

exceeding the size of 250 mm, will be used for backfilling the mine working and its contact with horizontal mine workings where it was not possible to build the supporting dams for technical or safety reasons. This fraction will be obtained from the Rožná I Mine slag heap, where a mobile sorter will be used. Since the Rožná I Mine slag heap contains 508,760 m³ (as of 1 January 2019) of rock material, which is about nine times more than is required for the liquidation of the main mine workings, a safer possibility of securing the contact between the vertical and horizontal mine workings will be selected. The sorted aggregate of a piece size of 63 - 125 mm sorted by the mobile sorter from the R I slag heap will serve to backfill all level contacts, whether or not the supporting dam will be built on it. (Vokurka, 2018; Decree, 1997; DIAMO, 2019; Bukowski, 2013).

3.1.3 *The mine working surface protection*

As soon as the mine working is backfilled, a sinking platform will be built at its mouth according to the Decree. The platform will be designed for a uniform load of 33 $kN.m^{-2}$, and with regard to the non-strengthened backfill, the sinking platform will also be dimensioned for an additional unpredictable load from the suction forces of 80 $kN.m^{-2}$. The closure sinking platform will be designed as a reinforced concrete square slab made of C30/37 concrete, reinforced with continuous two-way reinforcement Ø 22 B500A. The closure sinking platform will be installed on the C30/37 concrete foundation strip. The foundation strip is designed in the dimensions of 1.0 x 1.0 m. The foundation strip will be mounted on micro-piles, which will be realized in the borehole diameter min. Ø 156 - 200 mm. The micro-piles will consist of reinforcement tube min. Ø 102/10 mm and assumed length 9.5 m. When realizing the micro-piles, the length of the fixed part is min. 1.5 m into the rock massif. The grouting will be carried out with the cement compound in the C: V = 2.5:1 ratio with an assumed consumption of 20 - 30 l of the compound for 0.5 m of grouting. The connection of the micro-piles to the foundation strips will be realized by leading the micro-pile reinforcement tubes to the lower level of the foundation strip, on which the 0.2 x 0.2 m steel plates, thickness 20 mm, will be welded, thus creating a foundation for the sinking platform steel structure. A closable opening of min. 0.6 x 0.6 m for checking the filling condition and for the possibility of filling the pit will be left in the closure sinking platform.

According to the Decree, the safety zone will be set for the liquidated pits, and it will be indicated in the relevant map documentation. The safety area will be the same as the safety zone, as there is no gaseous mine or mine with a coal dust explosion hazard at the Rožná deposit.

Figure 3. The liquidated R2 pit of the Rožná deposit with the sinking platform.

The fencing of the closure sinking platform will be realized on its perimeter by means of supporting posts Ø 51/4 mm in combination with reinforced corner and gate posts Ø 89/5 mm. All posts will be embedded in the concrete lining to a depth of 0.9 m and will be fitted with a welded protective cap to prevent rainwater from entering. Corner and gate posts will be further stabilized by angled struts Ø 51/4 mm. Galvanized wire mesh Ø 4 mm, as shown in Figure 3, will be bound between the supporting posts; it will be bound in three rows to a tensioning wire provided with M12 tensioners. Barbed wire will be pulled over the wire mesh in three rows on the higher posts. The fence will reach a height of 2.4 m and will be provided with the warning signs "No entry – an undermined area". The sinking platform will be permanently marked. (Vokurka, 2018; Decree, 1997).

4 FLOODING THE MINE AND MINE WATER MANAGEMENT AFTER THE MINING ACTIVITY TERMINATION

The termination of mining and treatment of the Rožná uranium ore deposit does not mean the end of the working activities at the deposit. The organization is under an obligation to drain mine water permanently, subject it to systemic monitoring for contaminants, and subsequently, to purify it until the contaminant contents fall to such values that this water can be discharged spontaneously, without pumping and cleaning, the local watercourse.

The first step in mine water management after the termination of mining is to flood the extracted areas to a predetermined level where the subsequent pumping will prevent possible uncontrolled discharges of contaminated mine water to the surface. The R3 pit, which will become a water pit, will serve to drain the Rožná deposit permanently. From this pit, a drainage tunnel will be led through which the water will be pumped to the cleaning technology.

As soon as systematic monitoring records the decrease of contaminants in the pumped mine water to the permitted limit, pumping of water will stop and the mine water rises to the level of the drainage tunnel floor and start flowing spontaneously through the drainage tunnel to the local watercourse. (Vokurka, 2018).

4.1 Variants of flooding the mine space in the Rožná I Mine

The full flooding of the Rožná I Mine depends on the concept of exploiting mine areas for non-mining activities, i.e. the alternative use of mine workings. At present, the mine areas are being hired by the Radioactive Waste Repository Authority for research and experimental work at a depth of about 550 m where a generic laboratory has been built.

The use of the mine areas for non-mining activities places high demands on the financial resources necessary to ensure not only the pumping and cleaning of mine water but also the ventilation and maintenance of the mine workings. Therefore, the following variants of abandonment of mine areas and their subsequent flooding were proposed to make the best use of the mine areas, but in a limited time period:

- Dry variant (the mine is in operation up to the 24th level)
- Wet variant (the mine is in operation up to the 12th level).

A few default parameters are needed to determine the time of flooding:

- Total excavated volumes: 12 024 356 m^3,
- Average groundwater inflow (for the period from 2000 to 2017): 1 473 thousand m^3.year^{-1},
- Average surface water inflow (for the period 2000 - 2017): 101 thousand m^3.year^{-1}.

In the first three years, an expected average inflow to the flooded mine is 49.9 l.s^{-1}, i.e. 1,574 thousand m^3.year^{-1} (including rinse water from the premises), i.e. within three years approx. 4 700 thousand m^3, and the mine will be flooded between the 14th and 15th levels.

Available data from flooding other deposits, predominantly the nearby uranium deposit Olší-Drahonín, which was liquidated in the 1990s, can indicate the likely course of the Rožná deposit flooding. It is expected that after three years, the inflow on the lower levels will decrease and the total inflow into the mine can be expected to decrease by about 15%, i.e. 42.4 l.s^{-1}. The annual inflow will be 1 337 thousand m^3.year^{-1}, i.e. 4 000 thousand m^3.

After six years, the volume of 8 700 thousand m^3 will be flooded and the mine will be flooded below the 9th level. Similarly, further decrease of the inflows into the mine by approximately 20%, i.e. 33.9 l.s^{-1}, can be expected. The annual inflow will be 1 069 thousand m^3.

The volume of approximately 2 600 thousand m^3 still remains to be flooded to the level of the second floor; it will be flooded in about two and a half years. Subsequently, it will be necessary to start pumping mine water and maintain the level in the retention area below the level of the drainage tunnel floor. When pumping, it is necessary to take into account the average inflow of mine and rinse water of approx. 25 l.s^{-1}, i.e. the average pumping of mine and rinse water amounting to 25 l.s^{-1}, i.e. 788 thousand m^3.year^{-1}. (Vokurka, 2018; DIAMO, 2019)

4.1.1 Dry variant

The dry variant, as shown in Figure 4, follows the termination of uranium ore mining when the mining activities stop, but the areas from 12th to 24th level (550 - 1200m depth) are still used for scientific purposes. In the case of this variant, it is necessary to ensure complete pumping and cleaning of mine water at the level of approximately 50 l.s^{-1}, i.e. 1577 thousand m^3.year^{-1} and rinse water in the volume of about 1.7 l.s^{-1}, i.e. 53 thousand m^3.year^{-1}. It will be necessary to clean 1 630 thousand m^3 of water. (Vokurka, 2018; DIAMO, 2019).

4.1.2 Wet variant

The wet variant will be realized when the dry variant is no longer needed. This variant means that the mine areas below the 12th level will be flooded with water. This is preceded by the liquidation of the mine below the 12th floor according to the approved technical project of the mine liquidation. Prior to flooding, environmental remediation will be carried out, i.e. the removal of lubricants, petroleum substances, etc., including partial pillage of mining equipment and dismantling of shaft bottoms. Pits below the 12th level must be adapted to a change in the ventilation and drainage modes of the mine. The adaptation of the extend of the maintained mine workings necessary for the operation of the generic laboratory on the 12th level

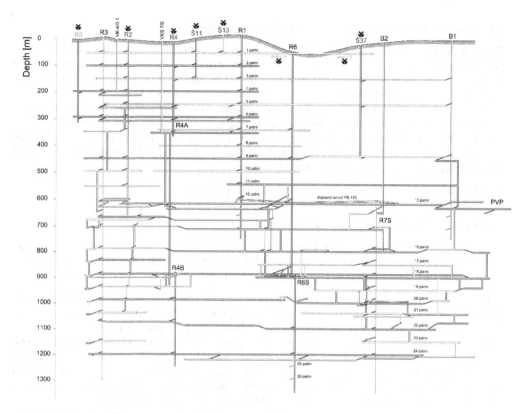

Figure 4.   Current Rožná deposit development – the dry variant.

(depth of 550 m) is also related to the liquidation. Surface infrastructure will be gradually removed, but continuous maintenance and operation of the machinery, including the equipment necessary for the operation of mine workings, i.e. mining machines, transformer stations, decontamination stations, etc., must be ensured.

Partial flooding will be realized by stopping the pumping of mine water until the mine water rises to space between the 12th and 13th levels, which will serve as a mine water retention space for about 84 days without pumping water. (Vokurka, 2018; DIAMO, 2019).

4.1.3 Full flooding of the deposit

As soon as the mines are not used, they will be abandoned, the pumping system will be shut down and the mine will be gradually flooded up to below the 2nd level (451.0 m above sea level, 96.19 m below the surface), see Figure 5. The level of the mine water will be maintained in the retention area below the drainage tunnel (485.0 m, 62.19 m below ground level). The mine water level will be maintained by pumping through this drainage tunnel to the cleaning technology and from there to the local watercourse. As soon as the contaminants in the water meet the permissible limit, pumping will be stopped and the water will spontaneously flow into the local watercourse. The total duration of the Rožná deposit flooding is set at eight and a half years after the pumping of mine water from the underground and the transfer of rinse water from the surface areas to the underground. (Vokurka, 2018; DIAMO, 2019).

4.2 Assumption of geochemistry and stratification of mine water during the mine flooding

During the mine flooding, there will be fundamental changes in mine water chemistry, which will be caused by a change in oxidation-reduction conditions in the mine space. These changes are caused by long-abandoned mine workings on the upper levels of the deposit, which have

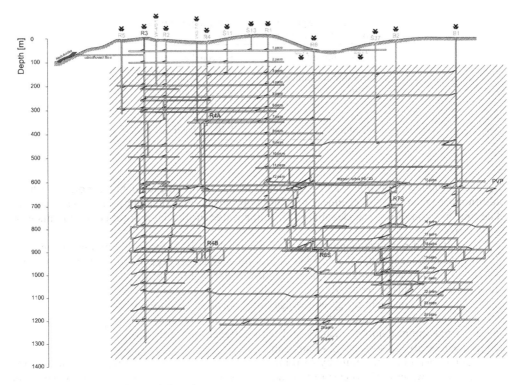

Figure 5.   Full flood of Rožná deposit.

been exposed to atmospheric oxygen and humid environments for a long time. As a result of the mine flooding, these areas will once again come under groundwater to reducing conditions. In the original oxidation zone, reductive dissolution of secondary minerals, both uranium minerals, and especially oxo-hydroxides of trivalent iron, four-valent manganese, and sulphide compounds begins to occur. Thus, the mine water becomes highly mineralized and gains a rather reductive character. Another process that can be expected is the reduction of nitrate ions down to NH_4^+ with the aid of microbial processes. Thus, after the mine is flooded, an increase in the U, ^{226}Ra, Fe, Mn, SO_4^{2-} and, in lesser amounts, other components may be expected. (Paul, 2016; Rapantova, 2013; Geller, 1998).

Apart from the fact that the mine water will be highly concentrated, there will be a significant stratification. When the mine is flooded, low-mineralized water will flow into the deposit to dilute the upper surface of the mine water. As soon as the mine water enters the drainage tunnel level, the concentration of minerals in the mine water will be low, but after some time a so-called First flush will occur when the concentration of minerals will increase extremely. This is due to the removal of the diluted mine water layer. Concentrations will slowly decrease as a result of continuous dilution with low mineral water inflows. It is assumed that the concentration decrease will be influenced by the deposit development when the concentration will rise step by step due to the level of the individual levels. After some time, the concentration of minerals in the water will be at the natural level of the low-mineralized surrounding waters. (Geller, 1998; Rapantova, 2013; Zeman, 2008).

4.3   Mine water treatment technology

Mine water will be pumped to the mine and wastewater treatment plant, which will be used to remove contaminants from mine water so that it can be discharged to the local watercourse. Due to the anticipated chemistry of mine water after the flooding of the deposit, see Table 3,

Table 3. Expected content of contaminants in pumped mine water.

Contaminant	Expected content
Soluble substances	2 500 – 4000 mg/l
Iron	30 – 40 mg/l
Manganese	7 – 10 mg/l
Uranium	15 – 20 mg/l
Radium	1 500 – 2 500 mBq/l

the decontamination of mine water is proposed by means of ion exchange technology with the additional recovery of the chemical uranium concentrate and precipitation of heavy metals. The time horizon of natural cleaning of the mine water surface layer from contaminants is 30 to 35 years.

The controlled mine water pumping will be carried out by submersible pumps of the UPA 250 B type with a capacity of 2x 50 l.s^{-1} into a mine water storage tank of 190 m^3. The mine water will be pumped through the induction flow meter and the control valve to the bottom of the sorption columns as required.

Uranium sorption technology will consist of 3 sorption columns of 2x 26 m^3 and 7.5 m^3. Mine water will be fed to the bottom of the columns and filtered up through the ion exchanger layer. At maximum column capacity and with sufficient ion exchanger and mine water, the column will be able to sorb at 1.2 - 2 m^3.min^{-1}.

After removal of the uranium by sorption on the ion exchanger, water will flow through the ion exchanger water curb to ensure that no ion exchanger remains in the water. Mine water will continue to flow into the distributor where the barium chloride solution ($BaCl_2$) will be dispensed to remove the radium and the hydrate lime suspension to remove heavy metal ions.

The water will flow from the distributor into parallel aeration reactors where the Fe and Mn ions present in the mine water will be oxidized and the free carbon dioxide displaced by pressurized air. The aeration system will consist of a tubular distribution on which aeration elements with a perforated membrane will be placed. The method of perforating the membrane will ensure the constant formation of fine bubbles and thus the maximum use of oxygen from the pressurized air.

The mine water treated in this way will flow from the aeration reactors to the distributor where the flocculant will be dispensed and the water will be discharged to the thickeners. Concentrated sludge is automatically pumped out at set intervals and deposited in the sludge bed. Cleared water will fall over the thickeners edge into the tank with pumps, from where it will be pumped for pressure filtration to sand filters. After passing through the sand filters, the filtered water will flow into the retention tank, from where it will be discharged into the watercourse. (Vokurka, 2018; DIAMO, 2019).

5 CONCLUSIONS

The liquidation of each mine with such a long history brings about negative effects on the entire region from employment to the mitigation of the effects of mining activity and its impact on the environment around the mine. Proper distribution of the liquidation operations over time and possible alternative use of mine workings, such as the generic laboratory on the 12th level, can partially reduce or even eliminate these negative effects.

The liquidation of the main mine workings opening to the surface will be carried out by non-strengthened backfill material from their own slag heap. Subsequently, these mine workings will be secured on the surface by a closure sinking platform at their mouth.

In addition to the liquidation works, the mine water management is described here after the pumping is stopped when the mine is flooded. Depending on the monitored inflows and experience of flooding other deposits, it is possible to assume the course of flooding including time. It would take about 8.5 years for the mine water to reach a set level below the 2nd level, where water would be retained in the retention space. From there, water will be drained through the drainage tunnel leading from the R3 pit to the mine and wastewater treatment plant. The mine water geochemistry will be adjusted to the permitted limit while uranium will be separated. As soon as these contaminants do not exceed the set limits, the pumping from the R3 pit will be stopped and the mine water will flow out spontaneously to the local watercourse.

REFERENCES

Bukowski, P., Niedbalska, K. (2013). The analysis of selected properties of solid rock materials designed for shafts liquidation. Geoconference on Science and Technologies in Geology, Exploration and Mining, SGEM 2013, VOL II. 467-474.

Decree of the Czech Mining Authority No. 52/1997 Sb., as amended, No. 32/2000 Sb., č. 592/2004 Sb., č. 176/2011 Sb., č. 237/2015 Sb., laying down requirements to ensure occupational safety and health and safety at the liquidation of main mine workings.

DIAMO Stráž pod Ralskem, the state enterprise. (2019). Internal materials.

Geller, W., Klapper, H., Salomons, W. (1998). Acidic mining lakes. Acid mine drainage, limnology and reclamation. Springer, Heidelberg. ISBN 978-3-642-71954-7.

Grygarek, J. (2002). Proposal of the construction and designing of the supporting dams at expected liquidation R2 and R3 pits for GEAM Dolní Rožínka by non-strengthened material.

Paul, M., Rapantova, N., Wlosok, J., Licbinska, M., Jenk, U., Meyer, J. (2016) Experience of Mine Water Quality Evolution at Abandoned Uranium Mines in Germany and the Czech Republic. Proceedings INWA 2016. p. 636.

Rapantova, N., Licbinska, M., Babka, O., Grmela, A., Pospísil, P. (2013). Impact of uranium mines closure and abandonment on groundwater quality. Environmental Science and Pollution Research. 7590-7602. Doi: 10.1007/s11356-012-1340-z.

Vokurka, M. (2018). Liquidation of the mine Rožná I, history and the possibility of using mining areas after termination of uranium ore extraction. Ostrava: VSB - Technical University of Ostrava.

Wojtkowiak, F. Didier, C. (1999). Principles for a safe closure of old mine shafts and adits. Ninth International Congress on Rock Mechanics, Vols 1&2. 25-30.

Zeman, J., Cernik, M., Supikova, I. 2008. Mining and its environmental impact II - Stratification of mine water after flooding, its causes and consequences. Chrudim: Ekomonitor. ISBN 978-80-86832-36-4.

Project work with pupils aimed at increasing interest in environmental protection measures

S. Walter

Technical University Bergakademie Freiberg, Institute for Mining and Special Civil Engineering, Germany

ABSTRACT: As the "decision-makers of tomorrow", today's children and adolescents will decide how valuable nature and the environment are to society, and how future environmental policy and, consequently, environmental protection measures will be shaped. However, according to teachers and parents, adolescents are increasingly alienating themselves from their natural environment and spending more of their time in social media. With the help of project work, which addresses their interests such as media and technology, their interest in environment and environmental protection can be developed and at the same time their knowledge in this field can be increased. This article outlines exemplarily how such project work should be designed and demonstrates that nature conservation and environmental protection can also be practically experienced by children and young people at school and have a lasting impact on them. After all, environmental protection is of interest to all of us.

1   THEMATIC INTRODUCTION AND PROBLEM STATEMENT

The basis of human life is the nature (Possit & Zander 2017): Plants supply the indispensable oxygen, they and animals serve as food sources, relaxation takes place around clean stretches of water and in forests, etc.

However, in order to maintain their partly exalted lifestyle, humans had to and still have to homogenise surfaces, e.g. in order to be able to produce a sufficient amount of food. To some extent, vegetated areas are deforested and converted into mining sites in order to extract various raw materials – as well as for the construction of renewable energy facilities. Extraction sites in particular, however, do not harm nature as much as commonly suspected and reported in many media reports. During mining but especially after recultivation and renaturation, unique areas develop which become unique ecosystems due to very specific factors Various european studies demonstrate that the resulting specific flora and fauna provide an exceptional opportunity for environmental protection and species conservation (Wittig & Niekisch 2014; Wasserbach 2008). On the other hand, it is scarcely known among general public that these measures create habitats that are of greater importance than the same areas were prior to extraction.

School children in particular should have an interest in these topics (environmental protection, raw materials policies) as they are the "decision-makers of tomorrow" and should therefore have sufficient knowledge about these topics. This is the only way they can shape their own views on environmental policies and development policy so they can make decisions that are in the interests of the environment in the future (Krombaß 2007). This statement is backed by scientific evidence indicating that by experiencing nature directly in the early stages of their lives, children become more aware of it and are more likely to be more willing to engage themselves for environmental protection when they are adults (Gerl & Almer & Gerl 2018).

For example, pupils aged 15 to 16 at a German grammar school considered their knowledge of nature and the environment to be on or below average. When asked about their knowledge about the raw materials sector, it became clear that they are aware that their knowledge is very limited (Figure 1).

2 METHODOLOGICAL APPROACH

Children and adolescents all over the world are interested in social and digital media. Facebook, ВКонтакте (VKontakte), Whatsapp, Youtube and Instagram have established themselves as elementary components of the lives of adolescents worldwide (Meckel 2008; Fieseler & Hoffmann & Meckel 2010). Russian residents for example are among the most active social media users in a worldwide comparison (Parkhimovich n.d.). In Germany the situation is quite similar: 97.6 percent of 14 to 19-year-olds are online each day (Koch & Frees 2017), and the average duration of time spent on social media is over 172 minutes per day (DAK-Gesundheit 2018). Interest in social media is very high among the school class surveyed (Figure 2). Comparing the data on students' interest in raw materials, nature/environment and social media, it becomes clear which are the areas that adolescents prefer to engage with and what they would like to learn more about (Figure 2).

In order to introduce children and adolescents to raw materials and nature/environmental protection while they are still at school, their interest in social media should be used.

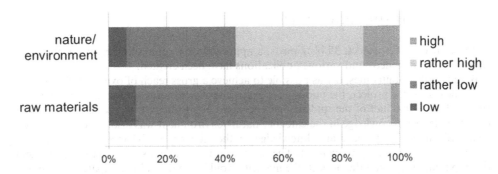

Figure 1. Level of knowledge in the fields of nature/environment and raw materials among the project's participants.

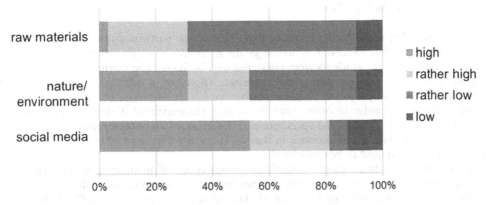

Figure 2. Interest in raw materials, nature/environment and social media among the project's participants.

There are different ways of doing this. All of them have in common that they are most easily achievable with the support of the respective school administration and teachers of biology and other natural sciences. At German schools there are a number of unrestricted content areas in the curricula that teachers can shape themselves, as well as project days or project weeks that usually take place at regular intervals of a school year. During these project days or weeks, the teaching staff can also work with the pupils on specific topics according to their own interests. For the project work, natural or near-natural places such as forests or bodies of water are ideal (Krombaß 2007).

The leeway in the curricula was used to carry out the project SOCIAL NATURE in cooperation with a company involved in quarrying and a grammar school in Saxony (Germany). In the course of the regular natural science lessons, the pupils witnessed the life of a bird and prepared their findings for publication in social media. The chosen bird was the whitethroat which is a rare bird in Germany nowadays and which can be found mainly in former quarries after the extraction of raw materials due to its special characteristics. In the two weekly school lessons the pupils were informed about the necessity to extract raw materials, they visited the quarry, researched the animals on location, interviewed a biologist and a social media professional, created posters, gave presentations, organised workshops etc. As a result, a dedicated project blog containing all relevant information, a Facebook page featuring various posts including photos and 360° shots as well as a Youtube channel were created throughout the project's period of ten months.

3   RESULTS OF THE PROJECT WORK

In the course of the project, 35 blog posts, a approximately equal number of Facebook posts (Figure 3) eleven videos and various publications have been created.

Based on these outcomes, it was possible to achieve a gross reach of over 10,000 people.

At the end of the project, the pupils were again interviewed using written questionnaires. The survey showed that half of the students had increased or significantly increased their knowledge of raw materials. Knowledge in the category nature/environment was also increased or strongly increased among 44 percent and knowledge in the category social media was increased or strongly increased among 61 percent. The interest in raw materials grew among 11 percent of the pupils and they liked the project: the majority would recommend participation in the project to friends and more than three quarters of the participants rated the project as very good or good.

4   CRITICAL DISCUSSION

It turned out that social media are very important to the participating students, but it also became apparent that Facebook is no longer the most important platform in this age group: Only 25 percent of the students were registered on Facebook. The German participants preferred to be active on Instagram and Snapchat. Therefore, a different channel should be chosen for a renewed project work in Germany.

In addition, a transfer to other countries is only recommended if the concept is adapted appropriately. For instance, the social network ВКонтакте (VKontakte) must be regarded as a far more important social network in Russia. It has a different structure and has different characteristics that need to be considered.

Furthermore, it is interesting to note that although cooperation agreements between schools and companies in the extraction industry can be strongly recommended, they often do not (yet) exist or are not sufficiently exploited. The school involved in the SOCIAL NATURE project already had a long-standing cooperation with the company involved in the raw materials extraction, but only about half of the pupils had already visited the extraction site before the project began. Such cooperations are still to be recommended, but should be designed and maintained appropriately.

Figure 3. Detail of the created Facebook page.

5 CONCLUSION

The extraction of raw materials is necessary and environmental compatibility is taken into account as far as possible while the areas are restored or redesigned after dismantling activities have been completed (Wasserbach 2008). Various scientific studies confirm that thereby habitats are created for animals and plants that did not previously occur in these areas and that these are thus of particular importance for biodiversity (Wittig & Niekisch 2014). This results in additional areas for nature conservation that offer unique conditions without which certain species would no longer exist (Wasserbach 2008).

The request of various researchers to integrate the topics environmental protection and sustainability in school education in a consistent and interdisciplinary manner has to be fully supported (Menzel & Bögeholz 2006; de Haan & Harenberg 1999; van Weelie & Wals 2002). It is essential that pupils are made aware of their own role in dealing with natural resources and the environment. Project work may already begin in kindergartens and continue in schools and universities culminating in adult education (Gross & Behme 2017). Each individual educational measure can contribute to improving the level of knowledge, the range of possibilities is very large (Röser & Behme 2016).

Up to now, the subjects of nature conservation and environmental protection have often not been taught in a manner adapted to the respective target group. These topics are essentially not uninteresting subjects for children and adolescents, but they should be communicated using specific examples, communicated in a mass-compatible style and in an age-appropriate manner (Hey & Obermayr 2008). In particular, the prevailing interest in social media and the freedom in curricula that is given in Germany and elsewhere should be used to motivate young people to educate themselves in these fields.

ACKNOWLEDGEMENT

The project work involving the pupils was enabled by representatives of the Freies Gymnasium Penig, Sandwerke Biesern GmbH and HeidelbergCement AG. The superordinated promotion project is financially supported by the Free State of Saxony and the European Social Fund. Its cooperation partners are the Technical University Bergakademie Freiberg, Mittweida University of Applied Sciences and MIBRAG Neue Energie GmbH.

REFERENCES

DAK-Gesundheit 2018. WhatsApp, Instagram und Co. – so süchtig macht Social Media. Viewed 28 April 2019, URL: https://www.dak.de/dak/download/dak-studie-social-media-nutzung-1968596.pdf

de Haan, G. & Harenberg, D. 1999. Bildung für eine nachhaltige Entwicklung, Gutachten für das BLK-Programm (Materialien zur Bildungsplanung und Forschungsförderung Nr. 72), Bund-Länder-Kommission, Bonn.

Fieseler, C. & Hoffmann C.P. & Meckel M. 2010. CSR 2.0 – Die Kommunikation von Nachhaltigkeit in Sozialen Medien. *Marketing Review St. Gallen*, 27(5),22–27.

Gerl, T. & Almer, J. & Gerl, A. 2018. Das BISA-Projekt – Biodiversität im Schulalltag. *Anliegen Natur*, 40 (1): 95–101.

Gross, J. & Behme, D. 2017. Jeder kann mehr für Insekten tun. *agrarzeitung*, 43, 3.

Hey, H. & Obermayr, H. 2008. Was mich fasziniert, will ich auch erhalten. *Politische Ökologie*, 109: 30–31.

Koch, W. & Frees, S. 2017. ARD/ZDF-Onlinestudie 2017: Neun von zehn Deutschen online. *Media Perspektiven*, 9: 434–446.

Krombaß, AS. 2007. Lernen über das Thema Biodiversität im Naturkundemuseum. Empirische Untersuchungen zu kognitiven und motivationalen Wirkungen eines computergestützten Informationssystems, Munich: Ludwig-Maximilians-Universität München, viewed 28 April 2019, https://edoc.ub.uni-muenchen.de/11587/1/Krombass_Angela.pdf

Meckel, M. 2008. Reputationsevangelisten und Reputationsterroristen – Unternehmenskommunikation 2.0. In Miriam Meckel & Katarina Stanoevska (ed.), *Web 2.0: Die nächste Generation Internet*, Baden-Baden: Noms: 109–128.

Menzel, S. & Bögeholz, S. 2006. Vorstellungen und Argumentationsstrukturen von Schüler(innen) in der elften Jahrgangsstufe zur Biodiversität, deren Gefährdung und Erhaltung. *Zeitschrift für Didaktik der Naturwissenschaften*, 12: 199–217.

Parkhimovich, A. n.d. Internet und Social Media in Russland. Viewed 28 April 2019, URL: www.dt-trend.com/social-media-in-russland.html

Possit, K. & Zander, K. 2017. Kommunikation von Naturschutzmaßnahmen in der ökologischen Direktvermarktung, *Tagungsband WITA 2017*, Munich: Technische Universität München – Lehrstuhl für Ökologischen Landbau und Pflanzenbausysteme: 782–785.

Röser, M. & Behme, D. 2016. Wissenslücken sind kein Argument für Untätigkeit. *agrarzeitung*, 28: 18.

van Weelie, D. & Wals, A.E.J. 2002. Making biodiversity meaningful through environmental education. *International Journal of Science Education*, 24: 1143–1156.

Wasserbacher, R. 2008. Rohstoffgewinnung und Biodiversität. *BHM Berg- und Hüttenmännische Monatshefte*, 153 (4): 143–145.

Wittig, R. & Niekisch, M. 2014. Veränderungen der Biodiversität durch den Menschen. In Rüdiger Wittig & Manfred Niekisch (ed.), *Biodiversität: Grundlagen, Gefährdung, Schutz*. Berlin / Heidelberg: Springer-Verlag: 47–64.

New approaches to resolving hydrocarbon sector-specific issues

Simulation of organic solids formation process in high-wax formation oil

A.N. Aleksandrov, M.K. Rogachev, Thang Nguyen Van & M.A. Kishchenko
Saint Petersburg Mining University, Saint Petersburg, Russia

E.A. Kibirev
Gazpromneft-NTC LLC, Saint Petersburg, Russia

ABSTRACT: The novelty of this paper is to provide a detailed study of the effect of oil blend composition on conditions and nature of wax deposition based on a developed formation oil model. In addition, this work also proposes a phase equilibrium diagram for given conditions, which play a pivotal role in predicting and measuring wax formation. The paper presents the results from the modeling of the formation of organic deposits, through the processes of simulation using the PIPE SIM software together with Multiflash Wax module. The work studies the effect of high-wax oil components on conditions and nature of wax solidification based on the developed formation oil model. Plotting of the phase diagram of the hydrocarbon system allowed us to determine the value of the wax appearance temperature (WAT), which was confirmed by laboratory studies. Changes in the molecular mass of organic deposits during wax crystallization were established.

1 INTRODUCTION

Nowadays, the international petroleum industry is characterized by increasing in wax, asphaltenes, resins content of oil and other problematic factors, which have been causing a considerable amount of obstacles for petroleum engineers and operators working in oil fields. During oil recovery, physical and chemical properties, as well as the composition of the fluid changes. As a result, oil production is continually accompanied by organic precipitation on the surface of the down hole apparatus, and on the sidewalls of wells, and processing system (Goual 2012). Wax deposition and its crystallization are considered serious problems during production, exploitation, and transportation, resulting in obstruction of flow strings, formation damage as well as loss of hydrocarbons. Eventually, they lead to dramatic reduction in production and demand a great deal of workforce and money to handle the problems.

Wax is a complicated blend of n-paraffin, i-paraffin, and c-paraffin with carbon numbers from 18 to 65 (Lake 2006). Waxes are precipitated as solids with the changing thermodynamic equilibrium when the oil temperature cools down lower than the wax appearance temperature (WAT) (Fadairo et al. 2010 & Kasumu et al. 2013). In other words, the temperature at which the first organic wax crystalizes in oil as a deposit is referred to as the WAT. The WAT is one of the main factors, which needs to be taken into account when developing preventative strategies for handling wax deposition in oil wells (Berne-Allen et al. 1938 & Paso et al. 2009).

For the most part, the mechanism of wax formation is associated with molecular diffusion and shear dispersion. Molecular diffusion is acknowledged as the main mechanism behind wax deposition. During production, pipeline temperature drops gradually. Molecular diffusion influences on wax deposition until the temperature reaches the WAT. On the pipeline wall, there is a concentration gradient between the dissolved wax and remaining wax in the oil owing to wax precipitation. Wax will be precipitated as a result of the diffusion of dissolved wax onto the pipeline wall (Jiang et al. 2014 & Jung et al. 2014).

An additional mechanism to take into consideration is shear dispersion, which is considered to be likely important only in situations where the concentration of deposited wax in a fixed volume of oil is high, this happens when the temperature of oil is much lower than the WAT (Bern et al. 1980 & Burger et al. 1981). Wax precipitation caused by shear dispersion does not appear when other mechanisms are not present (Brown et al. 1994). Another factor influencing wax crystallization to be considered is surface roughness of the pipeline wall, which provides an excellent area for crystallization (Creek et al. 1999).

In the majority of cases, oil is a complicated dispersion system, which contains a variety of components. In addition, the WAT and other parameters of oil will relatively be affected by changes in oil composition. Thus, only by elaborating and studying all information about the effect of external factors on changes in the WAT can engineers accurately predict the potential wax formation zone on the wall of a pipeline as well as the timing of its occurrence (Guozhong et al. 2010). One of the ubiquitous factors influencing wax precipitation is temperature. During production, the temperature drops and as a result, the wax solubility (the ability of wax to dissolve in oil) decreases. There are some circumstances and factors resulting in wax deposition such as pressure changes, temperature cooling rate, changes in oil composition, surface roughness, the concentration of paraffin, the molecular mass of paraffin, water cut and water-oil ratio (Kohse et al. 2006; Mordvinov et al. 2010 & Huang et al. 2016).

One of the best ways to cope with wax deposition would be to predict its occurrence and propose effective approaches to eliminate it. A number of models have been presented in order to illustrate wax precipitation appearance as well as to predict wax formation in pipelines depending on known parameters of oil wells.

Despite the fact that wax formation stimulation is one of the more challenging questions facing engineers owing to the extremely complicated and limited data, there are a number of wax stimulation programs developed to predict and control wax formation, the amount of precipitated wax, and wax deposition rates.

Detailed design of production systems is required to improve the efficiency of oil production and exploitation. There are a tremendous number of software packages used to predict the crystallization process of high-molecular components of oil such us the PVT sim software package (developed by the Calsep company), the PVTP software package (developed by the Petroleum Experts company), and the OLGA and PIPESIM software packages (developed by the Schlumberger company).

One of the leading software systems in the field of analysis and optimization of oil production systems is the product of the Schlumberger Company - PIPESIM. For the most part, this software package is a wax simulation used for building and calculation of a static model of steady-state multiphase flow moving from the reservoir to the objects of the collection and preparation system. PIPESIM allows for the resolution of many complicated problems with the help of numerous modules included in this package.

The novelty of this paper is to provide a detailed study of the effect of oil blend composition on conditions and nature of the wax deposition based on the developed formation oil model. In addition, this work also proposes a phase equilibrium diagram for given conditions, which play a pivotal role in predicting and measuring wax formation.

2 SIMULATION SOFTWARE

Currently, petroleum companies are becoming increasingly interested in computer generated simulations of formations, wells and oil-gathering systems. With multifunctional software systems the companies are able to optimize equipment and staff operation while minimizing the costs and reaching a new level of strategic decisions.

Changes in blend composition and properties of oil, as well as increased water cut have a significant effect on the composition and properties of organic deposits that form on the surface of downhole equipment. The objective of this work is to study the effect of oil blend composition on conditions and nature of wax solidification based on the developed model of high-wax oil.

Detailed design of the production system is required in order to enhance the efficiency of wells operation. One of the leading software packages for analysis and optimization of production systems is the Schlumberger's PIPESIM. This software is primarily intended for building and calculating a static model for an established multiphase flow from a formation to a gathering and processing system. The PIPESIM can solve many applied tasks using different modules built-in in the system.

In this study, we use Multiflash Wax module. Numerical simulation is accomplished simultaneously with consideration of 2- or 3-phase model of the fluid condition. This module provides a detailed description of wax formation process and behavior of the liquid phase in the petroleum field equipment based on a model of thermodynamic equilibrium.

The accumulated experience of laboratory studies of wax shows that wax melting temperature is influenced by solid compounds which are a mixture of waxy crystalline hydrocarbons and ceresins. Waxes and ceresins are similar in formula C_nH_{2n+2}, but belong to different homologous series. The most frequent ratio of waxes to ceresins in oil is 4:1.

The paraffin molecules are mostly normal structure hydrocarbon radicals with carbon atoms number from 16 to 35. In the solid state, they form a crystalline structure with a molecular mass of 300 – 450 g/mol. The most heat-resisting components from the homologous series of alkanes are low-molecular waxes.

The ceresin molecules are radicals of naphthene, naphthene-aromatic and aromatic bases with isoalkane. The number of carbon atoms in a ceresin molecule ranges from 36 to 55, and the molecular mass – from 500 to 750 g/mol.

The naphthene hydrocarbons in the composition of ceresins contain side chains in their molecules of both normal and isomeric structure. Thus, with the same melting temperature as waxes, they have higher density, viscosity, and molecular mass, together with lower solubility in identical solvents. Lower chemical resistance of ceresins is due to their microcrystalline structures.

Both are not soluble in water, alcohol and aqueous solutions of mineral bases and acids, but are well soluble in hydrocarbons of all classes and polar non-electrolytes.

In order to determine the convergence of experimental data given in works (Kamenshchikov et al. 2005 & Glushchenko et al. 2009) and the data simulated with Multiflash Wax, and wax melting temperature was plotted against the number of carbon atoms (Figure 1).

Figure 1 features a curve that shows the entire range of changes in the melting temperature of waxes included in asphaltene-resin-paraffin depositions (ARPD) composition. It is obvious that higher content of high-molecular hydrocarbon causes melting temperature of wax to be higher, so that wax appears earlier in the well and deeper in the tubing.

High convergence of the dependences shown in Figure 1 makes it possible to create a model of high-wax oil with a precise description of physical and chemical properties of reservoir oils. In addition, it helps to conduct various studies based on this model in the future to have reliable results.

Initial reservoir temperature for the D_{2ef} pool is 62°C, reservoir pressure – 29.1 MPa. Degassed oil belongs to extremely light (density 802 kg/m^3). The pour point is 39°C. In terms of group analysis, the oil is high-wax (wax content is 27.12 % by mass), resinous (resin content is 8.25% and asphaltene content is 3.01% by mass). Physical and chemical properties of degassed oil from the D_{2ef} pool are given in Table 1.

The physical and chemical characteristics and the blend composition of reservoir oils were used to develop a model of high-wax oil in PIPESIM software package using the Multiflash Wax module. Numerical values of the components shown in Table 2 are input to the program.

Under normal conditions, different hydrocarbon components contained in the oil form a three-phase system. At reservoir conditions (high temperature and pressure), all the components form one phase which is a liquid mixture. Oil is a multicomponent system, so the phase equilibrium diagram in the "temperature-pressure" coordinates, unlike individual substances, is not a single curve, but an area limited by condensation line and evaporation line.

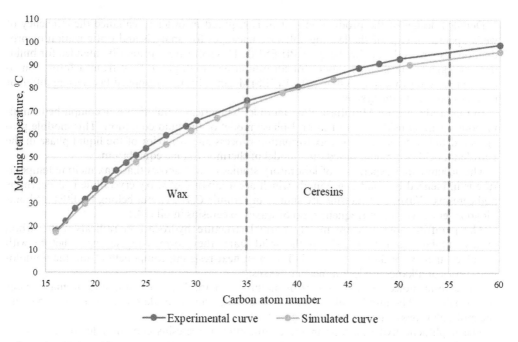

Figure 1. Wax melting temperature against carbon atomic number defining the type of n-alkanes.

Table 1. Physical and chemical properties of degassed oil from the D_{2ef} pool.

Main properties	Value
Density at 20°C, kg/m³	802
The pour point, °C	39
WAT, °C	43.5
Hydrocarbon group composition, % by mass:	
Wax content (wt %)	27.12
Asphaltene content (wt %)	3.01
Resin content (wt %)	8.25
Reservoir temperature (°C)	62
Reservoir pressure (MPa)	29.1

Table 2. Blend composition of reservoir oil of the D_{2ef} pool.

Component name	Chemical formula	Mole fraction, %
Carbon dioxide	CO_2	0.13
Nitrogen	N_2	1.03
Helium	He	0.01
Methane	CH_4	24.4
Ethane	C_2H_6	8.71
Propane	C_3H_8	10.3
Butane	C_4H_{10}	6.7
Pentane	C_5H_{12}	4.26
Hexane	C_6H_{14}	6.6
Heptane	C_7H_{16}	6.12
Octane	C_8H_{18}	7.43
C_{9+}	C_9H_{20} +	24.31

The first stage of the simulation is to plot a phase diagram of the hydrocarbon system for the D_{2ef} pool (Figure 2). To the left of the critical point (C) and above the evaporation line (area 1) hydrocarbons are in a liquid state; to the right and below the condensation line (area 2) hydrocarbons are in a gaseous state. Inside the two-phase boundary between the evaporation and condensation lines (area 3) gas and liquid can coexist. When the temperature drops below the wax formation line, the system turns into a three-phase state (area 4) with an additional solid phase – wax.

By setting blend composition in mole fractions, it is possible to determine wax formation conditions in the gas-liquid mixture. The wax formation line shown on the phase diagram corresponds to the temperature of the studied wax appearance temperature – 44°C.

Furthermore, the use of the Multiflash Wax module allows us to determine the dynamics of changes in the mass content of wax in the liquid when oil cooling below the WAT (Figure 3).

According to the graph, at the temperature of 20°C, the wax content is 27% by mass. With a further decrease in temperature, its content continues to grow, reaching up to 52% by mass at 0°C. In the oil and gas industry, the temperature at the wellhead is usually higher than 20°C, temperatures below this point do not occur in a well, so they are not covered in this work.

The values of WAT and wax mass content in the presented model are quite close to the experimentally obtained data at 20°C and the atmospheric pressure, given in Table 1.

In addition, there is a decreasing function of the changes in the molecular mass of organic deposits, which was obtained during the phase transition of wax to the crystalline state (Figure 4). At the WAT (44°C), its molecular mass is 522 g/mol, at 20°C it reaches 398 g/mol.

Thus, when the flow is cooling down, the components with the highest molecular mass are the first to crystallize. With further cooling, low-molecular hydrocarbons will predominate in the deposits.

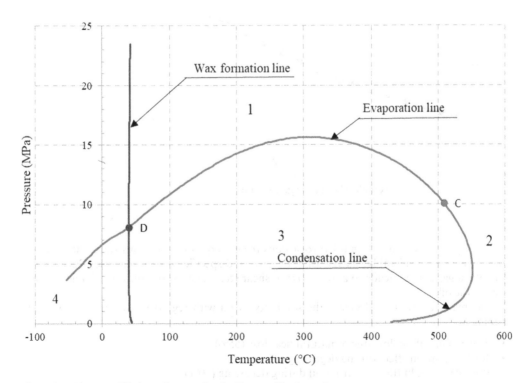

Figure 2.   Phase equilibrium diagram for the D_{2ef} pool hydrocarbon system.

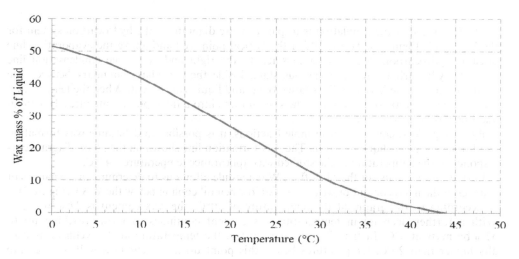

Figure 3.   Mass content of wax in the liquid plotted against temperature in Multiflash Wax.

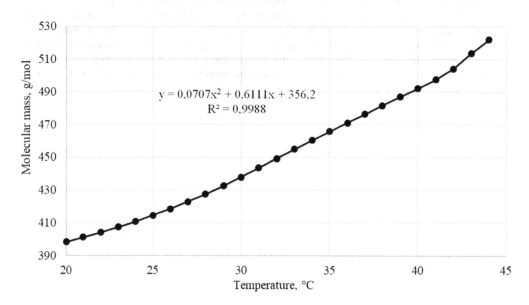

Figure 4.   Changes in wax molecular mass against temperature.

During the transition of light hydrocarbons from a liquid to a gaseous state, the properties of oil change considerably: its viscosity, density, and specific volume increase. Decreasing of dissolved gas content leads to a quick rise of shear stress due to structural-mechanical properties of the oil.

The main factors that decrease the solubility of oil with regard to wax moving along the tubing are:

- The cooling of oil flow due to natural heat loss (55%);
- Joule-Thomson effect during degassing (25%);
- Removal of light fractions from oil during degassing (20%).

Since the process of oil degassing has a significant effect on the reduction of oil solubility, the effect of light hydrocarbons on wax formation has to be thoroughly studied.

Figure 5 shows the effect of the mole fraction of heptane on the change in paraffin deposition parameters (WAT and mass content of wax).

This hydrocarbon is one of the outstanding representatives of the methane series to demonstrate the dependencies studied in this paper. The initial mole fraction in the model is 6.12%. The study of the effect of the system on the change in the mole fraction of heptane was carried out at an interval of 2%.

The resulting dependency allows us to say that an increase in the proportion of heptane in the developed model contributes to an increase in oil dissolving capacity and slows down the process of wax formation. To evaluate the changes of parameters, their precise values are given in Table 3.

When mole fraction of heptane in oil changes from 0 to 10%, WAT decreases by 1.41°C and the mass content of waxes reaches 5.97%. Such significant changes in wax formation parameters emphasize the importance of the study of the mole fraction of components, most frequently found in oil.

The influence of the composition and the amount of dissolved gas on the structural-mechanical properties of oils are studied at temperatures higher than WAT, when the structure in oil is formed by asphaltene (Devlikamov et al. 1975). Increasing the content

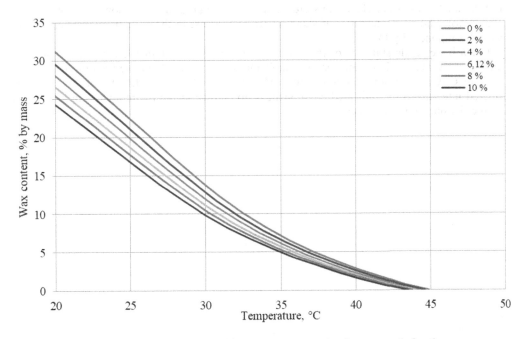

Figure 5.   Wax appearance temperature and its mass content against heptane mole fraction.

Table 3.   Wax appearance temperature and its mass content against heptane mole fraction.

Parameter name	Heptane mole fraction in oil, %					
	0	2	4	6.12	8	10
WAT, °C	44.89	44.59	44.30	44.00	43.75	43.48
Wax content in oil at 20°C, % by mass	30.95	29.54	28.25	27.00	25.98	24.98

of nitrogen, methane, and ethane in reservoir oil results in increasing critical dynamic shear stress.

The authors of this work have conducted researches in order to study the conditions of wax formation depending on the composition and content of gaseous and liquid components in high-wax oil. The simulation made it possible to evaluate the changes of WAT and wax mass content with the mole fraction of the components changing by 1% (Figures 6 and 7).

In both figures, graphs converge at the same point corresponding to the parameters of the initial model. The increase in contents of methane, ethane, and propane cause waxes to form earlier. On the contrary, the remaining components (butane, pentane, etc.) slow down the wax formation process.

The increase in the number of carbon atoms up to octane (C_8) leads to the strengthening of their effect on wax formation parameters. The proximity of the dependencies for nonane (C_9) and octane indicates a drop in the growth of their influence on WAT (Figure 6).

Different picture when observing changes in the wax content (Figure 7) was noticed. After the component C_8, the increase in the number of carbon atoms does not result in attenuation of the effect and the graphs tend to overlap. A further increase in the molecular mass of the components leads to an increase in the value by which the wax content changes in the mixture. This growth is unlimited, since the same effect is observed when the molecular mass of wax increases, which is already covered in this study.

The simulation gave us a chance not only to trace general trends in system changes but also to evaluate the values of change in system parameters. Figures 8 and 9 show how these values change for a homologous series of alkanes with an increase in the mole fraction of each individual component by 1%.

For a better understanding of the nature of changes in the simulation system, Figures 10 and 11 show changes in parameters of wax formation by setting the mole fraction of the components C_2-C_8 in the range from 0 to 10%. Methane mole fraction (24.4%) is beyond the considered interval, so the indicated component is not shown in the figure and its influence will further be shown in this paper.

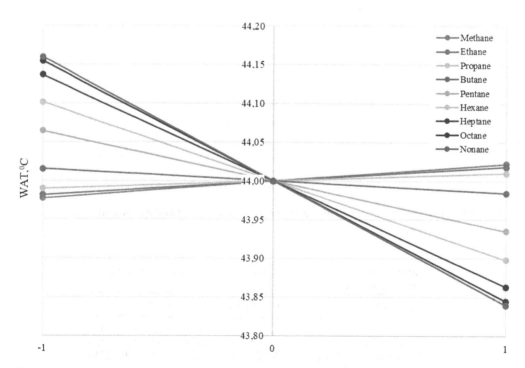

Figure 6. Wax appearance temperature against changes in components mole fraction.

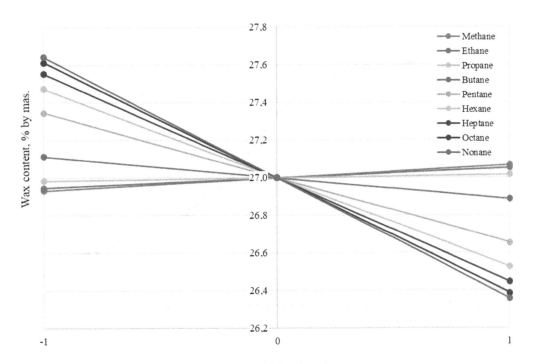

Figure 7. Wax content in oil against component mole fraction changes.

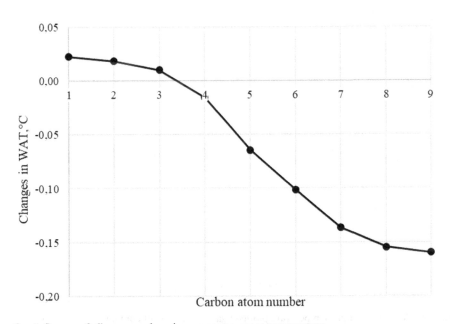

Figure 8. Influence of alkanes on changing wax appearance temperature.

The final part of the paper also studied the influence of non-hydrocarbon components that most frequently occur in oil production such as carbon dioxide, nitrogen and hydrogen sulfide on wax formation process. The mole fraction for these gases was limited to 2% since this content is the most frequent in oil composition (Kamenshchikov 2005 & Turbakov et al. 2014).

787

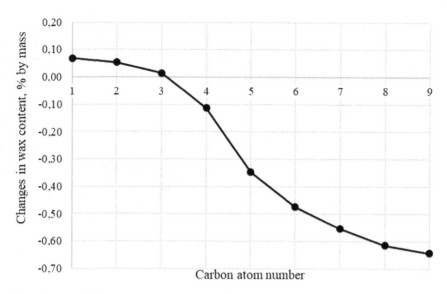

Figure 9. Influence of alkanes on changing mass content of waxes in oil.

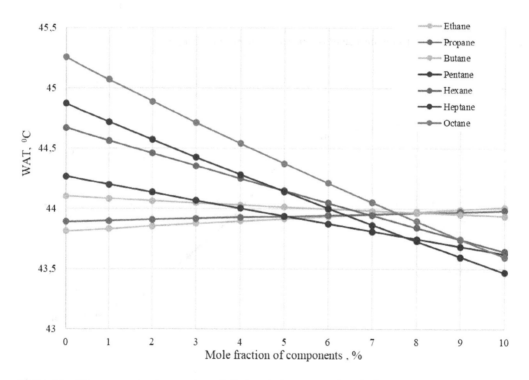

Figure 10. Wax appearance temperature plotted against component contents.

With an increase in the content of non-hydrocarbon gases in the oil (nitrogen, hydrogen sulfide and carbon dioxide), a slight increase in the WAT is perceived, which can be compared with the effect of ethane (Figure 12).

This example explicitly shows that all basic dependencies for the developed model of high-wax oil can be used for real life objects.

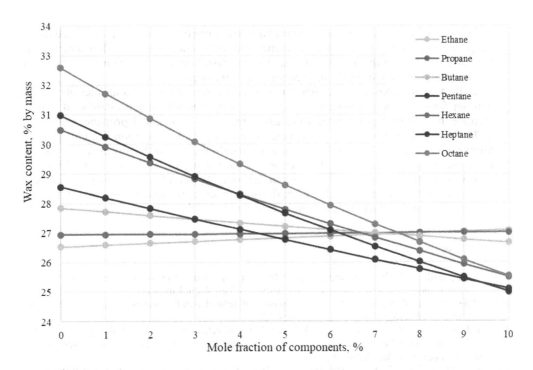

Figure 11. Wax mass content plotted against component contents.

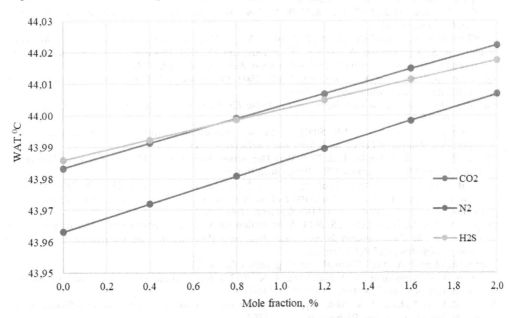

Figure 12.   Influence of non-hydrocarbon gases on wax appearance temperature.

3 CONCLUSIONS

1. The developed model of high-wax oil allowed us to build a phase equilibrium diagram of the hydrocarbon system for the considered pool and determine the wax appearance temperature – 44°C. The dynamics of wax formation was studied, the content of which reaches up to 27% by mass at 20°C.

2. A decay function of the changes in molecular mass of organic deposits was obtained during the phase transition of wax to the crystalline state. With a decrease in the flow temperature, the components with the highest molecular mass crystallize first, and with further cooling, low molecular mass hydrocarbons predominate in the organic deposits.

3. The effect of the component composition of oil on the conditions and intensity of wax deposits formation was studied based on the developed model. It was established that the increase in the content of hydrocarbon gases (methane, ethane, and propane) leads to an earlier transition of wax to the solid phase. The remaining hydrocarbon components from butane to nonane, on the contrary, act as solvents, slowing down the process of wax formation, and with an increase in the number of carbon atoms, their effect on the system increases. Furthermore, with an increase in the content of non-hydrocarbon gases in the oil (nitrogen, hydrogen sulfide and carbon dioxide), a slight increase in the wax appearance temperature can be observed, that can be compared with the effect of ethane.

REFERENCES

Bern, P. A., Withers, V. R., & Cairns, R. J. 1980. Wax deposition in crude oil pipelines. In *European offshore technology conference and exhibition*. Society of Petroleum Engineers.

Berne-Allen, Jr., A., & Work, L. T. 1938. Solubility of refined paraffin waxes in petroleum fractions. *Industrial & Engineering Chemistry*, *30*(7): 806-812.

Brown, T. S., Niesen, V. G., & Erickson, D. D. 1994. The effects of light ends and high pressure on paraffin formation. In: *SPE Annual Technical Conference and Exhibition*, 25-28 September, New Orleans, Louisiana. Society of Petroleum Engineers.

Burger, E. D., Perkins, T. K., & Striegler, J. H. 1981. Studies of wax deposition in the trans Alaska pipeline. *Journal of Petroleum Technology*, *33*(06): 1-075.

Creek, J. L., Lund, H. J., Brill, J. P., & Volk, M. 1999. Wax deposition in single phase flow. *Fluid Phase Equilibria*, *158*, 801-811.

Devlikamov, V. V., Khabibullin, Z. A., & Kabirov, M. M. 1975. Anomalous crude oils. *Nedra*, Moscow.

Fadairo, A. S. A., Ameloko, A., Ako, C. T., & Duyilemi, O. 2010. Modeling of wax deposition during oil production using a two-phase flash calculation. *Petroleum & Coal*, *52*(3): 193-202.

Glushchenko, V.N., Silin, V.N., & Gerin, Yu.G. 2009. Prevention and elimination of asphaltene-resin-paraffin deposits. Oilfield chemistry: 480 p.

Goual, L. 2012. Petroleum asphaltenes. *In Crude Oil Emulsions-Composition Stability and Characterization*. IntechOpen.

Guozhong, Z., & Gang, L. 2010. Study on the wax deposition of waxy crude in pipelines and its application. *Journal of Petroleum Science and Engineering*, *70*(1-2), 1-9.

Huang, Z, Zheng, S., & Fogler, H.S. 2016. Wax Deposition: Experimental Characterizations, Theoretical Modeling, and Field Practices. CRC Press, Boca Raton.

Jiang, B., Ling, Q. I. U., Xue, L. I., Shenglai, Y. A. N. G., Ke, L. I., & Han, C. H. E. N. 2014. Measurement of the wax appearance temperature of waxy oil under the reservoir condition with ultrasonic method. *Petroleum Exploration and Development*, *41*(4), 509–512.

Jung, S. Y., Lee, D. G., & Lim, J. S. 2014. A simulation study of wax deposition in subsea oil production system. In *The Twenty-fourth International Ocean and Polar Engineering Conference*. International Society of Offshore and Polar Engineers.

Kamenshchikov, F.A. 2005. Thermal dewaxing wells. Izhevsk: 254 p.

Kasumu, A. S., Arumugam, S., & Mehrotra, A. K. 2013. Effect of cooling rate on the wax precipitation temperature of "waxy" mixtures. *Fuel*, 103, 1144-1147.

Kohse, B. F., & Nghiem, L.X. 2006. Asphaltene and waxes. Petroleum. Engineering Handbook. Society of Petroleum Engineering, Richardson Texas, USA.

Lake, L. W., Kohse, B. F., Fanchi, J. R., & Nghiem, L. X. 2006. Petroleum Engineering Handbook: Volume I General Engineering. *SPE, Texas, USA*: 397-453.

Mordvinov, V.A., Turbakov, M.S., & Erofeev, A.A. 2010. The estimation technique of depth of intensive formation of paraffin sediments on downhole equipment. *Oil Industry* (7): 112-115.

Paso, K., Kallevik, H., & Sjoblom, J. 2009. Measurement of wax appearance temperature using near-infrared (NIR) scatter-ing. *Energy & Fuels*, *23*(10), 4988–4994.

Turbakov, M. S., & Riabokon, E. P. 2014. Improving of cleaning efficiency of oil pipeline from paraffin. In *SPE Russian Oil and Gas Exploration & Production Technical Conference and Exhibition*. Society of Petroleum Engineers.

Topical Issues of Rational Use of Natural Resources 2019 – Litvinenko (Ed)
© *2020 Taylor & Francis Group, London, ISBN 978-0-367-85720-2*

Method of calculating the wear and PDC bit operating time

K.A. Borisov & A.A. Tretyak
Platov South-Russian state polytechnic university (NPI), Rostov region, Novocherkassk, Russia

ABSTRACT: The experiment was carried out by the authors to research a process of PDC plates wear and evaluate its intensity. Time-varying wear of cutting elements was investigated along the experiment, as well as a connection was determined between the abrasiveness and PDC bit operating time. Wear resistance research was performed at the accelerated material testing facility (AMT-2) with the horizontal axis of rotation.

1 INTRODUCTION

Synthetic diamond and a carbide base are component parts of PDC cutting element. Their relative wear resistances are distinct from each other in 95-100 times, that leads to earlier carbide base wearing out; therefore, the PDC self-sharpens, forming the back angle α. varying in sizes of the area of bluntness depends on the width of the cutting part F, the length of the cutting part B and the back angle a (Dvoynikov & Blinov 2016).

$$F = F_i \cdot B \text{ mm}^2 \qquad (1)$$

The experiment was carried out by the authors to research a process of PDC plates wear and evaluate its intensity. Time-varying wear of cutting elements was investigated along the experiment, as well as a connection was determined between the abrasiveness and PDC bit operating time.

We apply the diagram (Figure 1) valid for the cylindrical PDC to define a growth in height of area of cutting and wear. We call diamond layer thickness as F, and the amount of B should be calculated by reference to height wear, using the formula:

$$B = 2\sqrt{R(R-h)^2} \text{ mm}^2 \qquad (2)$$

where R = 6,75 mm is PDC radius.

2 MATERIALS AND METHODS

Wear resistance research was performed at the accelerated material testing facility (AMT-2) with the horizontal axis of rotation. The facility is based on has in its basis a table-type horizontal milling machine with an integrated loading jig. The scheme of the laboratory facility AMT-2 is represented in Figure 2 (Tretyak 2017; Tretyak 2010; Krapivin 1969; Arcimovich 1969).

Inside the loading jig body, there is a spindle with a barrel and a feeding mechanism, in the form of a rack pair. The test sample 4 is fixed in the holder 3 of the loading jig. The samples are abraded along the surface generator of the grinding wheel 5. (researches were performed on "PP" wheels) 300x35x127 6377K3 40CM1 produced by the Volzhsky Abrasive Plant, with categorized penetration index). The grinding wheel drive is rotated

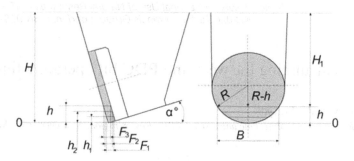

Figure 1. H - height of the blank with PDC; R - PDC radius; α - back angle; h - wear in height; F1≈F2≈F3 - the width of the diamond layer cutting part; B - the length of the diamond layer cutting part; 0-0 - cutting plane.

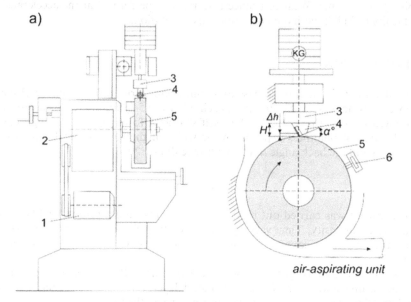

Figure 2. Installation for accelerated testing of materials AMT -2. a) hookup; b) unit scheme of PDC loading and abrasion.

by electric motor 1 through the reduction gear 2, the rotational speed of the wheel is 50 rpm. The sliding distance can be measured by means of distance counter 6 (Leusheva et al. 2019, Litvinenko & Dvoinikov 2019).

Hereby are specified the conditions, which were accomplished throughout, experiment process: the pressure on cutting edges of the wearing sample should not exceed 18 MPa – accordingly to the hardness conditions of used grinding wheel used (Sexton 2009; Kubasov 2013; Sergejchev 2015, Kamatov & Buslaev 2015). The abrasion speed should not exceed 2.24 m/s accordingly to the working conditions of the cutting edges of the bit with a diameter of 214 mm; and the sliding distance in case of wear should not be more than 942 m – accordingly to the conditions of AMT-2 facility operation within 20 minutes at a speed of 50 rpm (Bellin, 2010; Dverij, 1968; Tretyak, 2012, Vasilev et al. 2017).

Table 1

Experiment No.	1	2	3	4	5	6	7
Parameter							
Initial specific load, MPa	18	18	18	18	18	18	18
Initial axial load, kgf	22.5	51.0	65.5	77.2	87.3	96.1	104
Initial area of bluntness, mm^2	1.25	2.83	3.64	4.29	4.85	5.34	5.78
Experiment time, min	20	20	20	20	20	20	20
Sample size after the abrasion, mm	34.89	34.78	34.67	34.57	34.48	34.39	34.30
Wear of sample in height, mm	0.12	0.11	0,105	0.1	0.095	0,092	0.09
Final area of bluntness, mm^2	2.83	3.64	4.29	4.85	5.34	5.78	6.18
Final value of the specific load, MPa	8	14	15.3	15.9	16.35	16.63	16.83
Average value of the specific load, MPa	13	16	16.6	16.9	17.2	17.3	17.4
Intensity of wear in height and time, mm/min	0,006	0.0055	0.00525	0,005	0.00475	0.0045	0.0045

3 RESULTS

Authors has estimated initial area of bluntness and axial load, the sliding distance of PDC cutting element. The wear in height of cutting element was evaluated after the wear by using the grinding wheel, and a new value of area of bluntness was determined.

Final and average value of the specific load on the PDC during the wear process was determined. Then the intensity of wear in height and in time was measured. Then all the cycle was repeated. The results of the experiment are given in Table 1.

The average value of wear rate in height after seven measurements is $i_{h(t)PDC} = 0.00507$ mm/min with a coefficient of variation $K_V = 10,8\%$.

It seems certain that studying the wear resistance on rock formations in laboratory conditions is impossible, in view of the significant experiment data volume. Because of this reason we compare experimental results of wear research by means of AMT-2 at grinding wheels and at formations by an indirect method according to the results of tool wear with PDC during a well drilling with the depth of 85.2 m within 3 hours 50 minutes.

During research studies at AMT-2 with the duration of the experiment $t = 20$ min, and at the frequency of grinding wheel rotation $n = 50$ rpm, the sliding distance at the initial specific load $P = 18$ MPa is:

$$L_I = \pi \times D_{aw} \times n \times t = \pi \times 300 \times 50 \times 20 = 942 \text{ m} \tag{3}$$

Then, for the entire testing period (7 data points before wheel replacement) the overall sliding distance taking into account the reduce of the grinding wheel diameter was 6500 m and PDC wear in height was $\Delta h = 0,7$ mm. the similar wear of PDC would occur during well drilling with the depth of 85.2 m by a bit with diameter 112 mm at a speed of 250 rpm after 3.8 hours. That said, sliding distance at rock formation with cutting element is:

$$L_{II} = \pi \times d \times n \times t = \pi \times 112 \times 250 \times 230 \approx 20220 \text{ m} \tag{4}$$

The wear resistance of the cutting elements (I_{PDC}) should be defined as the ratio of sliding distance to PDC wear in height during the experiment. Then PDC wear resistance when tested in grinding wheels will be:

$$I_{PDC(I)} = \sum_{1}^{7} L_I \times (\Delta h_{PDC(I)})^{-1} = 6500 \times 10^3 \times 0,7^{-1} = 9,28 \times 10^6 \text{ mm/mm} \tag{5}$$

where – $\Sigma_1^7 L$ is sliding distance of PDC on the grinding wheel for all period of the experiment; $\Delta h = 0,7$ mm – is total wear of PDC in height.

PDC wear resistance at rock formation is the ratio of the sliding distance during the observation period to the PDC wear in height:

$$I_{PDC(II)} = \sum_1^7 L_{II} \times (\Delta h_{PDC(II)})^{-1} = 20220 \times 10^3 \times (0,7)^{-1} = 2,88 \times 10^7 \text{ mm/mm} \quad (6)$$

Clearly seen that the PDC wear resistance at rock formation is 3 times more than on the grinding wheel.

For calculations in practice, it is necessary to know the intensity amount of the PDC wear in time at rock formations:

$$i_{h(t)PDC(II)} = i_{h(t)PDC(I)} \times (a_2 \times a_1^{-1})^{0,3} \text{ mm/min} \quad (7)$$

where a_2 – is rock abrasiveness, mg; a_1 – is wheel abrasiveness, mg.

If $i_{h(t)PDC(II)}$ is known and allowable wear of the drilling bit diameter is known, we can define the time t of the tool according to the formula:

$$t = \Delta D \times 2 \times i_{h(t)PDC(II)} \text{ s} \quad (8)$$

where $\Delta D = 2h$, mm – is allowable wear of rock-cutting tool in diameter.

The maximum value of the area of bluntness of one cutting element area can be found according the formula:

$$F_Z^{max} = F \times B = F \times 2\sqrt{R^2 \times (R-h)^2} \text{ mm}^2 \quad (9)$$

where F – is PDC diamond layer thickness, mm; B – is calculated from the wear in height h, mm.

Then, it should be formed the drilling operational procedures a cyclic growth in the axial load P_1 at each interval: penetration; running in of a new tool; growth of the load as F_Z increases; loading P_2 growth to the limit value.

When determining the operating time of the rock-cutting tool (RCT), the main condition is the selection of the mode which provides that P_1 is always bigger than P_3 (bulk destruction mode) where:

$$P_3 = P_2 \times F_Z^{-1} \quad (10)$$

the average speed during the drilling time t shall can be found by calculating the drilling speed in each interval with help of the formula:

$$V_{ROP} = V_0 \times P_1 \times P_3^{-1} \text{ mm/s} \quad (11)$$

Then the operating time L of the bit should be calculated as per the expression:

$$L = V_{ROP} \times t \text{ m} \quad (12)$$

Summarizing all above-mentioned information, accelerated laboratory studies of PDC wear on grinding wheel AMT-2, make it possible to assign the intensity of the PDC wear in height over time, consequently giving an opportunity to calculate the mechanical frilling speed and the operating time of a tool.

REFERENCES

Arcimovich V.G., Lukash V.A. 1969. *Ob iznose tverdyh splavov pri burenii gornyh porod [About wear of hard alloys at drilling of rocks]*. Gornyj porodorazrushajushhij instrument: Kiev: Tehnika, Pp.35–42.

Bellin F., Dourfaye A., King W. 2010. *The Current State of PDC Bit Technology: Part 3*. World Oil. (November): Pp.67–71.

Dverij V.P. 1968. *Stendovye issledovanija razrushenija gornyh porod razlichnymi tipami lopastnyh dolot. Voprosy glubokogo burenija na neft' i gaz [Bench studies of rock destruction by different types of blade bits. Issues of deep drilling for oil and gas]*. Moscow: VIJeMS, Vyp.1. Pp.24–26.

Dvoynikov, M.V., Blinov, P.A. *Survey results of series-produced downhole drilling motors and technical solutions in motor design improvement //* International Journal of Applied Engineering Research 2016: 11(10), pp. 7034–7039.

Kamatov, K.A., Buslaev, G.V. *Solutions for drilling efficiency improvment in extreme geological conditions of Timano-Pechora region //* Society of Petroleum Engineers - SPE Russian Petroleum Technology Conference 2015.

Krapivin M.G., Rakov I.Ja., Sysoev N.I. 1990. *Gornye instrumenty 3-e izdanie, pererabotannoe i dopolnennoe [Mining tools 3rd edition, revised and expanded]*. Moscow: Nedra. 256p.

Kubasov V.V., Spirin V.I., Budyukov Yu.E. 2013. *Novy`e texnologii povy`sheniya rabotosposobnosti almaznogo porodorazrushayushhego instrumenta [New technologies of increase of efficiency of diamond rock cutting tool]*. Nauchno-prakticheskij zhurnal E`konomika XXI veka, innovacii, investicii, obrazovanie. Tula №2.

Leusheva E.L., Morenov V.A. & Martel A.S. *Combined cooling heat and power supplying scheme for oil and gas fields development //* Youth Technical Sessions Proceedings: VI Youth Forum of the World Petroleum Council - Future Leaders Forum (WPF 2019), June 23–28, 2019, Saint Petersburg, Russian Federation. – 2019, pp. 382–386.

Litvinenko V.S., Dvoinikov M. V. *Justification of the technological parameters choice for well drilling by rotary steerable systems //* Journal of Mining Institute 2019: 235, pp. 24–29.

Sergejchev K.F. 2015. *Almazny`e CVD-pokry`tiya rezhushhix instrumentov (obzor) [Diamond CVD coating of cutting tools (review)]*. Uspexi prikladnoj fiziki. Tom 3, №4. Pp.342–376.

Sexton T.N., Cooley C.H. 2009. *Polycrystalline diamond thrust bearings for down-hole oil and gas drilling tools*. 17th International Conference on Wear of Materials 19-23 April 2009. Las Vegas, USA. Pp.1041–1045.

Tretyak A.A., Litkevich Ju.F., Aseeva A.E. 2010. *Razrabotka metodiki rascheta narabotki porodorazrushajushhego instrumenta s almazno-tverdosplavnym vooruzheniem [Development of methodology for calculation of developments of rock cutting tool with a diamond-carbide armament]*. Stroitel'stvo neftjanyh i gazovyh skvazhin na sushe i na more. № 12. Pp.2–8.

Tretyak A.A., Litkevich Ju.F., Borisov K.A. 2017. *Opredelenie skorosti burenija i narabotki koronok novogo pokolenija, armirovannyh almazno-tverdosplavnymi plastinami [Determining a speed of drilling and the development of crowns of the new generation, reinforced by diamond-hard-alloy plates]*. Delovoj zhurnal Neftegaz.ru. № 3. Pp.46–49.

Tretyak A.Ya., Sysoev N.I., Burenkov N.N. 2012. *Raschet konstruktivnyh parametrov burovyh koronok, armirovannyh PDC [Calculation of design parameters of PDC-reinforced drill bits]*. Oil & Gas Journal Russia. № 5. Pp.66–69.

Vasilev, N.I., Bolshunov, A.V., Ignatiev, S.A. *Inertial mechanical reamer for borehole 5G-3 conditioning for penetration into subglacial lake Vostok //* International Journal of Applied Engineering Research 2017: 12(5), pp. 561–566.

Numerical analysis of stress-strain state of vertical steel tanks with defects

A.S. Dmitrieva, A.M. Schipachev & A.A. Lyagova
Saint Petersburg Mining University, St. Petersburg, Russia

ABSTRACT: Different defects may appear during operation of steel vertical tanks. Defects reduce the reliability of the tank and lead to accidents of varying severity. Oil tanks are structures of increased danger, their failures and accidents lead to serious economic, environmental and social consequences. The problem of ensuring operational reliability and durability of tank structures is relevant; its solution will lead to accident-free operation of tanks. The work is devoted to solving the problem of the safety operation of steel vertical tanks by the example of the analysis of residual life of oil tank RVSP-5000 with a crack and dent defects with allowance for their size, location in the tank structure and operating loads using the finite element modeling.

1 INTRODUCTION

Vertical steel tanks are structures that provide not only the storage of oil and petroleum products at oil tank farms, refineries, etc., but also ensure the safety and uninterrupted supply of the product through trunk pipeline systems. Oil steel vertical tanks are among the most dangerous objects in the system of main pipeline transport of oil. Steel vertical tanks for storage of oil and oil products are thin-walled metal structures working cyclically in a complex stress-strain state (SSS). Tanks are exposed to cyclically varying loads, temperatures and aggressive working fluids.

Constructive defects occur in the tanks while its operation. They can be caused by structural, technological and operational factors. Defects reduce the reliability of the tank and lead to accidents of varying severity. It should be noted that a significant part of the tank farms of Russia has practically exhausted its service life (25 - 30 years) or has defects, which are not permissible according to the standard documentation. On the one hand, overhaul or dismantling requires significant material costs. On the other hand, the occurrence of accidents at cylindrical steel tanks in most cases is accompanied by significant loss of oil, ground contamination and loss of life. Thus, the problem of prediction of operational reliability and durability of tank structures has contemporary importance; its solution will ensure accident-free operation of tanks. This problem leads to the need to develop methods for determining the actual danger of the defects discovered during the technical diagnosis of the tank, assessment of the stress-strain state of structural elements of tanks with defects, determination of terms and conditions for possible accident-free operation.

The conclusion on the industrial safety of the tank is issued in accordance with the current regulatory documents (RD 153-112-017-97 1997, RD 25.160.10-KTN-050-06 2005, RD 23.020.00.-KTN-296-07 2007, RD 08-95-95 1995). However, this approach means that more than half of the tanks should be immediately stopped and decommissioned due to the fact that they have deviations from regulatory requirements. Under this condition, the actual danger of detected defects or other deviations from the standards for a specific tank is not taken into account, taking into account the characteristics of its operation.

At present, there are many studies on the analysis of the stress-strain state of oil tanks (Gaisin et al. 2016, Mogil'ner 2009, Romanenko et al. 2004, Winterstetter et al. 2002). First of all, it should be noted that all the authors of these works point to the imperfection of analytical methods for calculating the stress-strain state of oil tanks that do not allow to estimate the actual residual life of the exploited tanks. This is due to the fact that calculations of the stress-strain state of the tank according to normative documents (RD 153-112-017-97 1997, RD 25.160.10-KTN-050-06 2005, RD 23.020.00.-KTN-296-07 2007, RD 08-95-95 1995, Safety manual of vertical cylindrical steel tanks for oil and oil products 2013) do not take into account the design features of the oil tank, operating conditions, external and technological factors, the presence and location of defects.

At present, numerous studies of the residual life of oil tanks with various types of defects are being carried out, new methods for calculating the stress-strain state of the tank using modern computer programs are proposed. The most common of these are programs that use the finite element method calculations. This method allows to obtain sufficiently correct results of strength calculations of the tank, setting of all boundary conditions, loads and the optimal number of finite elements of the mesh in tank simulation.

Analysis of accidents on steel vertical tanks (Accident Chronicle, Gaisin et al. 2016, Galeev et al. 2004, Kandakov et al. 1994, Shvyrkov et al. 1996) found that the cracks (22%) and dents (18%) are one of the most common defects in the geometry of the tank wall. Fragile destruction of oil and gas industry facilities with catastrophic consequences occurs during the formation and spreading of a crack in metal structures. The occurrence of defects during the operation of steel vertical tanks is an unavoidable process due to corrosion wear and low-cycle fatigue of the metal. Fatigue cracks appear as a result of cyclic loads in places of stress concentration.

Residual resource (life) of the tank is determined according to RD 153-112-017-97 on the parameters of low-cycle fatigue of the metal. The method is based on the Paris equation, which connects the crack growth rate and the stress intensity factor (SIF):

$$\frac{db}{dN} = C \cdot \Delta K^n. \tag{1}$$

where C & n = empirical coefficients; db/dN = crack growth rate; N = number of cycles; ΔK = Kmax - Kmin – the range of the stress intensity factor for one loading cycle, defined by formula:

$$\Delta K = Y_I \cdot \Delta\sigma \cdot \sqrt{\pi b}, \tag{2}$$

where $\Delta\sigma = \sigma_{max} - \sigma_{min}$ – the range of acting stresses; Y_I = K-calibration function, dimensionless coefficient, depending on the geometry of the body, crack parameters and loading conditions; b = crack depth.

The stress intensity factor (SIF) is a parameter that characterizes the SSS at the crack top and depends on the operating conditions, the geometry of the structure under and the size of the crack. To calculate the SIF of cracks in simple structures, for example, cylinders or plates, there are analytical dependencies presented in the book by Yu. Murakami "Stress Intensity Factors Directory". However, for cracks in the lower belts of a tank near the inlet/outlet nozzles and the corner weld such formulas are used incorrectly.

It should be noted that in the normative document (RD 23.020.00.-KTN-296-07 2007) rather strict criteria of allowable values of depth of the dent has been set. However, neither the geometric dimensions of dents (width, form) nor their location on the wall of the tank height are not taken into account. The experience of many years of accident-free operation of the tanks with the shape defects, which are not satisfying the requirements of the document (RD 23.020.00.-KTN-296-07 2007), shows a lack of sufficient validity of the imposed restrictions, which is also confirmed in (RD 23.020.00.-KTN-296-07 2007, RD 08-95-95 1995).

According to normative document (RD 23.020.00.-KTN-296-07 2007), when a dent or bulge is detected, the tank operation is possible only if the SSS calculations are made and it is proved that voltage limits in the wall of metal structure are absent and the defect does not lead to a loss of strength and stability of the tank wall.

The task of this research is to analyze the influence of the crack and dent defects on the stress-strain state of the tank using the finite element method realized in ANSYS and SIMULIA Abaqus software. In the simulation, the analysis of the SIF of vertical and horizontal cracks located on the inner and outer sides of the tank with variation in defect size has been performed.

The tank is a vertical cylindrical steel tank with pontoon, the capacity of the tank is 5000 m³. It is used for storage of gasoline. The parameters of the tank are presented in Table 1.

During the visual and measuring control of the tank the crack was found in the first tank belt. The sizes of the crack are: length - 10 mm and depth - 2 mm. This crack is taken into account to determine the residual service life of the tank. And in the tank wall at a height of 8483 mm from the bottom the dent was found with the following dimensions: the minor axis - 420 mm; major axis - 1100 mm; deflection (depth) - 39.1 mm.

2 SOLUTION PROCEDURES. CRACK EFFECT ANALYSIS

Stress-strain state calculations are performed by finite element method (FEM) in the ANSYS software. The geometry of the model was constructed as one quarter of a tank with two nozzles (Figure 1).

During the creation of the tank model the properties of steel were used: modulus of elasticity $E = 2.1 \cdot 10^{11}$ Pa, Poisson's ratio $\mu = 0.3$, density $\rho = 7850$ kg/m³. Actual thicknesses of the wall of the tank were assigned to Shell-elements. The following loads and boundary conditions were applied to the tank model: distributed weight of the roof and equipment of the tank, snow load, hydrostatic pressure corresponding to the filling level (10.26 m), rigid fixing of the tank bottom (encastre), symmetry condition, overpressure of oil vapors, the strength of the gravitational effect. Due to FEM nominal hoop, longitudinal and equivalent stresses, deformations were identified (Figure 2).

It should be noted that the maximum stresses fall on the first belts, and the stress concentrator is the nozzle.

Then the "crack" defect modeling is realized. For the simulation of a crack, it is expedient to use the method of submodeling. This allows to create mesh of sufficient density in a separate region of the model to obtain more accurate results. The application of the finite element method to describe crack growth usually involves the operative rearrangement of the finite element mesh during crack growth. The main advantage of submodeling is not only obtaining a more accurate result in the local area, but also the possibility of experimenting with various design options in the selected area. The basic principle of submodeling is to transfer the results (displacements) obtained from the calculation of the full model in the locations of the submodel boundaries to the nodes of the submodel with a more concentrated mesh. In the ANSYS software, for interpolating

Table 1. Vertical cylindrical steel tank RVSP-5000 specification.

Parameter	Value
Diameter of the tank	22.817 mm
Height of the tank	12 m
Height of the belt	1.5 m
Maximum level of gasoline	10.26 m
Thickness of the first belt	9.40 mm
Thickness of the second belt	7.30 mm
Thickness of the third belt	6.30 mm
Thickness of the fourth belt	5.80 mm
Thickness of the fifth - eighth belt	4.80 mm
Gasoline density	770 kg/m³
Steel	S235J2G3 ($\sigma_t = 245$ MPa, $\sigma_B = 370$ MPa)

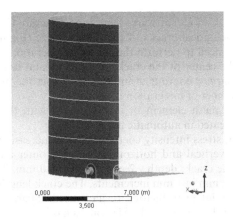

Figure 1. Geometry of the tank model.

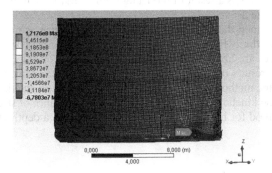

Figure 2. Distribution of hoop stress in the tank wall.

Figure 3. Creating a submodel in the first tank belt.

displacements from the full model to the submodel, the "Structural DOF" function is used, indi-cating the shell-solid type of submodeling.

A copy of the geometry of the tank construction was done in which a submodel was created, a plate 400 × 400 × 9.4 mm, "cut" from the first tank belt between the nozzles.

To simulate a crack in ANSYS software, there is a "Fracture-Crack" tool that automatically creates surface semi-elliptical cracks of the required size (Figure 4). To model a surface crack using this tool, you need to perform the following operations: to create a local coordinate system that will be the center of the crack; to set small and large semi-elliptic radii of the crack and radius of the largest contour of integration; to set the number of points along the front of the crack and number of integration contours.

The tetrahedron is the mesh element of the submodel with the defect. Mesh generation along the crack front is created in automatic mode.

The task was to obtain stress intensity coefficients with increasing crack depth, varying the location of the fracture: vertical and horizontal from the outer and inner sides of the tank wall. The initial size of the crack: depth - 2 mm, length - 10 mm. The growth of the crack is carried out to a depth of 7 mm in 1 mm increments. The crack length at each step is doubled.

To determine the stress intensity factors we have used "Sparse Solve". In the "Solution-Fracture Tool-SIF" menu the output of the SIF values has been set.

A vertical crack model with a depth L = 2 mm located on the outside of the tank wall with the results of the SIF distribution is shown in Figure 5.

The SIF values obtained as a result of modeling for all variations in the size and location of the crack are shown in Table 2.

Estimating the obtained data, we can conclude that with increasing depth of the crack, the SIF rises. In general, the SIF values of vertical cracks are greater than the horizontal ones. The highest SIF values are observed in vertical cracks located on the inside of the tank.

Table 3 shows the results of calculating the residual life of the tank before the formation of a galloping crack by the method (RD 153-112-017-97 1997). The calculation was carried out using SIF, obtained from the kinetic diagram of the cyclic crack resistance of steels (Pustovoi 1992) and the finite element method for horizontal and vertical cracks with a depth of 2 mm and a length of 10 mm.

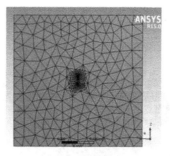

Figure 4. Vertical crack model.

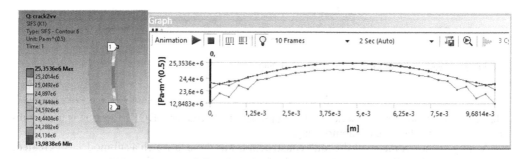

Figure 5. The SIF distribution in a vertical crack located on the outside of the tank.

Table 2. The SIF values of crack models.

Depth L, mm	$\Delta K_{I\max}$, $MPa\cdot$			
	Tank side			
	Outside		Inner	
	Location of the crack			
	Vertical	Horizontal	Vertical	Horizontal
2	25.354	22.243	26.800	21.584
3	26.907	22.551	28.192	21.872
4	28.132	22.900	29.342	22.148
5	29.441	23.189	30.258	22.346
6	31.134	23.467	31.255	22.481
7	31.909	23.614	31.983	22.619

Table 3. Residual life of tank RVSP-5000.

Method for determining SIF	Reference data, $\Delta K = 24.2$ MPa·m$^{1/2}$	FEM, vertical crack, $\Delta K = 26.8$ MPa·m$^{1/2}$	FEM, horizontal crack, $\Delta K = 21.6$ MPa·m$^{1/2}$
Residual life T, years	5.5	4.4	6.3

The results of the calculations justify the need to determine the exact SIF for a crack located in the wall of the tank. This coefficient has a significant effect on the value of the residual life of the tank. Knowing the specific location of the surface crack in the tank wall, it is possible to obtain the results of determining the residual service life, differing more or less from the value found without taking into account the location of the crack.

3 SOLUTION PROCEDURES. DENT EFFECT ANALYSIS

SSS calculation are performed by FEM in the SIMULIA Abaqus software. Creating a model of the tank is carried out similarly as described in paragraph 3. Due to FEM nominal hoop stress, longitudinal stress and equivalent stress, deformations were identified in the tank (Figure 6).

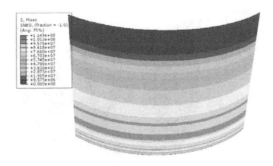

Figure 6. Distribution of equivalent stress in the tank wall.

Creating a dent in the tank model was reduced to a process of modeling of contact problem. Dent was simulated by subjecting the spherical surface on the tank wall. Contact problem was limited to the determination of the radius of the hemisphere, and the force, which with it will act on the surface of the tank wall. After defining the radius and strength steel hemisphere was created in PC Abaqus. It was positioned in the assembly model at the height 8483 mm from the bottom of the tank to create the deth with desired depth. Then the equivalent stress (von Mises) and the moving surface of the tank wall were determined (Figures 7, 8).

The maximum equivalent stresses occur in the center of the dent and amount to 213.3 MPa.

According to (RD 23.020.00.-KTN-296-07 2007), the restriction is set for the limit stresses in the wall of the RVS: the equivalent stresses are compared with the with the permissible stress, which are equal to the yield stress of steel:

$$\sigma_e \leq \sigma_\tau \tag{3}$$

It was found that the maximum equivalent stress arises at the center of the dent and is equal to 213.3 MPa. Consequently the condition (3) for a tank RVSP-5000 with a dent is satisfied. Thus, strength condition for the dent in the wall of the tank is executed and the area of the tank wall with a dent does not turn to inadmissible plastic deformation. In addition, we found the critical depth of the dent (58.05 mm), which leads to inadmissible plastic deformation, and it is necessary to put the tank out of action for the repair.

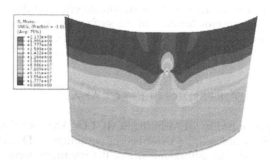

Figure 7.   Equivalent stress in the tank wall with the dent.

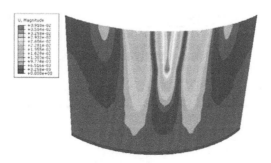

Figure 8.   Movement of the surface of the tank wall with a defect "dent".

4 CONCLUSION

Thus, calculation of the stress-strain state of the oil vertical cylindrical tank using the finite-element modeling allows to correctly estimate the residual life of the tank depending on the size and direction of the surface crack and predict a further accident-free operation life. It gives an opportunity to determine possible safe operation term of the tank before putting out of action to the scheduled repair. The analysis results show that the dent in the tank wall does not cause unallowable stresses.

Assessing the stress-strain state and determining the residual life of a specific tank, taking into account the conditions of its operation, structural features, the presence and location of defects, will significantly reduce repair cost, as well as reduce the risk of emergency situations in tank farms.

REFERENCES

Accident Chronicle. *Scientific and Industrial Journal "Safety in Industry"*. Available at: http://www.btpnad zor.ru/ru.

Gaisin, E. S. & Gaysin, M. Sh. 2016. Current state of the problems of ensuring the reliability of tanks for oil and oil products, *Transport and storage of oil products* 2. Available at: http://cyberleninka.ru/ article/n/sovremennoe-sostoyanie-problem-obespecheniya-nadezhnosti-rezervuarov-dlya-nefti-i-nefteproduktov.

Galeev, V. B. 2004. *Accidents of tanks and how to prevent them*. Ufa, Russia.

Kandakov, G. P., Kuznecov, V. V. & Lukijenko, M. I. 1994. Analysing the crash causes of the vertical cylindrical tanks, *Pipeline Transportation* 5: 15–16.

Mogil'ner, L. U. 2009. Calculation of allowable operating conditions of the tank wall with defects of geometry based on data of technical diagnostics, *Pipeline Transportation* 4: 64–72.

Pustovoi, V. N. 1992. *Metal structures of load-lifting machines. Destruction and forecasting of the residual life*. Moscow: Transport.

RD 153-112-017-97, 1997. Instruction for the diagnosis and assessment of residual life of vertical steel tanks. Ufa: Neftemontazhdiagnostika, 31 pp. (in Russian).

RD 25.160.10-KTN-050-06, 2005. Instruction on welding technology for construction and repair of steel and vertical tanks. Moscow: Transneft, pp. 259. (in Russian).

RD 23.020.00.-KTN-296-07, 2007. Evaluation manual of technical state of tanks of AK "Transneft". Moscow: Transneft, 135 pp. (in Russian).

RD 08-95-95. 1995. Regulation of system technical diagnostics of welded vertical cylindrical tanks for oil and oil products. Moscow: VNIImontazhspetsstroy, 19 pp. (in Russian).

Romanenko, K., Samofalov, M. & Šapalas, A. 2004. Linear and physical non-linear stress state analysis of local shape defects on steel cylindrical tank walls by the finite element method, *Mechanika* 46(2): 5–13.

Safety manual of vertical cylindrical steel tanks for oil and oil products. 2013. Moscow: CJSC «Nauchno-tehnicheskiy tsentr issledovaniy problem promyshlennoy bezopasnosti», 2013. 240 p. (in Russian).

Safina, I. S., Kauzova, P. A. & Gushchin, D. A. 2016. Evaluation of the technical condition of vertical steel tanks, *TechNadzor* 3. Available at: http://www.strategnk.ru/section/148/

Shvyrkov, S. A., Semikov, V. L. & Shvyrkov, A. N., 1996. Analysis of the statistic data of the tank destructions, *Theoretical and Engineering Developments* 5: 39–50.

Winterstetter, Th. A. & Schmidt, H. 2002. Stability of circular cylindrical steel shells under combined loading, *Thin-walled Structure* 40: 893–909.

Topical Issues of Rational Use of Natural Resources 2019 – Litvinenko (Ed)
© 2020 Taylor & Francis Group, London, ISBN 978-0-367-85720-2

Monitoring and control of the drilling string and bottomhole motor work dynamics

V.S. Litvinenko & M.V. Dvoynikov
Saint-Petersburg Mining university, Saint-Petersburg, Russian Federation

ABSTRACT: Based on the analysis of the research results of torsional vibrations of the drilling string and lateral beats of the bottomhole motor frame, the range of circular frequency of the top drive is determined depending on the geometric parameters of the gerotor mechanism.

As a result of performed computational experimental studies, the values of the critical circular frequency of the drilling string jointly with the bottomhole motor on an inclined direction of the well, taking into account the load on the bit, technical characteristics and composition of the drilling tool, were recommended. Recommendations on the choice of parameters of the drilling mode were set out.

1 INTRODUCTION

Hydrocarbon production by developing, for instance, offshore fields, as well as additional development of previously drilled areas, provides for the presence of complex profiles, the trajectories of which may contain curved sections that have a limited (minimum possible) radius or hold sections, of great length (more than 3000 m).

Drilling of such areas using only bottomhole motor (BHM) is difficult (Morenov & Leusheva 2017). The reasons for this are the large friction between the pipes and the walls of the well, which are usually accompanied by the impossibility of reducing the weight on the bit. The technological solution to this problem is the periodic or constant rotation of the drilling string (DS) by the rotor, or the top drive - a combined method of drilling.

In the process of well drilling, for example, in areas of stabilization of the zenith angle, in DS, jointly with BHM, various oscillations and vibrations occur that can damage its elements, as well as backs off and doglegs in the bottomhole assembly (BHA). Among the main forms of oscillations of DS and BHA can be identified the following (Morenov & Leusheva 2016):

- axial vibrations, caused by the work of the bit and the pulsation of drilling mud in the BHM bins;
- torsional vibrations, due to the drill string rotation and presented by the dynamics of the elastic–deformed rod;
- lateral beats of the BHM body caused by the design features of the gerotor mechanism.

The lateral oscillations of the BHA occur due to the eccentric arrangement of the rotor axis of the engine relative to the stator axis and made them around the latter - the planetary (translatory) movement (Baldenko et al. 2007).

For the purpose of a more detailed analysis of the DS and the BHM oscillations during their joint operation, preventing the coincidence of their frequencies, it is necessary to determine the optimal parameters of rotation of the top drive and the shaft of the BHM.

The frequency of DS rotation, which has certain geometrical parameters, elastic properties, as well as the ratio of the DS lengths and BHA, determines the frequencies of natural oscillations before the resonance.

These self-oscillations are determined by the stiffness of the DS and BHA system and the momentums of its inertia.

The frequency of forced torsional vibrations of the DS and BHA, taking into account the stiffness of the elements, can be determined by the equation (Bykov 2014):

$$f = \frac{1}{T}. \tag{1}$$

where $T = \frac{2\pi}{\omega}$- period of torsional oscillation, s.

The angular frequency of the DS is calculated based on the stiffness of the DS and BHA, taking into account the outer and inner diameters of the instrument:

$$\omega = \sqrt{\frac{C_{ds} \cdot (J_{ds} + J_{bha})}{J_{ds} \cdot J_{bha}}}. \tag{2}$$

where $J_{bha} = l_{bha} \cdot q_{bha} \frac{(d_{out.bha}^2 - d_{in.bha}^2)}{8}$, $J_{ds} = l_{ds} \cdot q_{ds} \frac{(d_{out.ds}^2 - d_{in.ds}^2)}{8}$ - the momentums of BHA and DS inertia, $C_{bha} = \frac{\pi \cdot G_{rigidity} \cdot (d_{out.bha}^4 - d_{in.bha}^4)}{32 \cdot l_{bha}}$, $C_{ds} = \frac{\pi \cdot G_{rigidity} \cdot (d_{out.ds}^4 - d_{in.ds}^4)}{32 \cdot l_{ds}}$- stiffness of BHA and DS respectively, $G_{rigidity}$ - rigidity modulus of instrument material, l_{ds}, l_{bha}- lengths of DS and BHA, q_{ds}, q_{bha}- weight of 1 meter DS and BHA(Bobrov 2000, Dvoynikov et al. 2017).

2 SIMULATION OF COMPUTATIONAL EXPERIMENT

Input parameters of the computational experiment:

– the length of the drilling string 500 m;
– diameter of the drilling string 127 mm;
– wall thickness (WT) of the drilling string from 9 to 16 mm;
– weight of 1 running meter is from 200 to 300 N.

BHA is represented by a BHM DGR 178.7/8.37 and drilling collar (DC) - 36 m and weighing 1 meter from 600 to 1200 N.

The results of the calculations (Figure 1) showed that the angular frequency depending on the wall thickness of the pipes and the weight of 1 meter of DS and BHA varies from 3.5 to 4.5 rad/s, and the frequency of torsional vibrations from 0.55 to 0.72 Hz

It was previously noted that when drilling with a BHM by jointly rotation of the drilling string in the stabilization sections, there are natural lateral and axial oscillations of the BHM, as well as forced lateral vibrations of the DS. The superposition of these oscillations leads to a resonance, which contributes to an abrupt increase in the amplitude of lateral oscillations of the BHA and DS, which leads to premature damage of the equipment and the back off of the tool elements.

To determine this effect, causing the coincidence of the frequencies of oscillations of DS and BHM, leading to an additional resonance of the system, calculations of the lateral beats of the case of the BHM gerotor mechanism were performed for various modes of operation.

The level of torsional vibrations of the engine, affecting the stability of its work, depends on the inertial F_{in} and hydraulic F_h forces acting on the rotor

$$F_{in} = mz_2^2\omega^2 e, \tag{3}$$

$$F_h = M_{ind}/ez_1 \tag{4}$$

M_{ind} – indicator momentum, $M_{ind} = M_{ind} - M_r$ (M_r – momentum of mechanical resistance); e - excentricity; z_1, z_2- number of stator teeth and rotor; m- rotor mass; ω- rotor angular frequency.

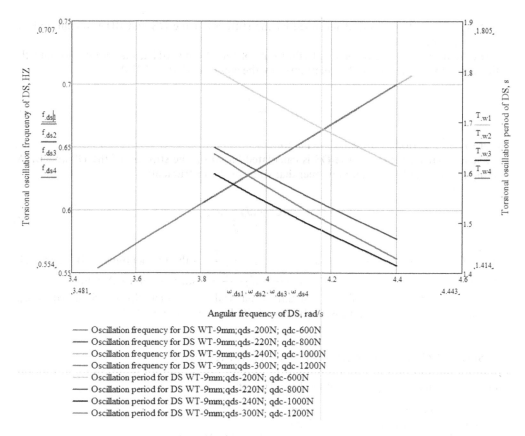

Figure 1. The functional connection of the frequency and period of DS torsional vibrations on the angular frequency of its rotation at drilling an inclined directional section of the well in the interval of 500 m.

The main regulatory technological parameters for optimizing axial and lateral vibrations are the weight on the bit and the flow-in. When drilling certain intervals of the well, as a rule, the flow-in is constant and is determined from three conditions: the necessary annular return velocity in the annular space for the removal of cuttings; high-quality bottom-hole cleaning; characteristics of the BHM.

It was noted above that the rotor axis rotates around its own axis, and also makes a portable movement around the axis of the stator counterclockwise. It should be noted that the frequency of the translatory (planetary) rotation of the axis of the rotor relative to the axis of the stator is higher than the frequency of rotation of the rotor around its own axis (Bobrov 2000).

The frequency of rotation of the translatory rotor rotation, which determines the frequency of vibration of the body, is defined as

$$\omega_p = z_2 \cdot \omega_r \qquad (5)$$

z_2 – number of rotor teeth; ω_r – rotor speed around its own axis (bit rotation frequency), s^{-1}.

When operating the BHM, namely its launch and subsequent loading, a torsion moment appears, leading to a disturbance in the spatial orientation (curving) of the rotor, its irregularity in speed rotation, the formation of additional standoffs and clearances in the working parts (WP) causing additional vibrations (Dvoynikov et al. 2017).

In general, the jamb momentum is equal to

$$M_c = \frac{P \cdot D \cdot t^2}{4\pi} \tag{6}$$

where D - stator diameter along teeth troughs (average rotor diameter), m; P- pressure in the motor, MPa; t- rotor pitch, m.

It should also be noted that an additional source of vibration in the engine is the pulsation of drawdown, leading to axial oscillations of the gerotor mechanism. The lateral and axial oscillations due to the causes noted above, are distributed along the entire length of the engine. Moreover, it was visually observed that in different engine operating modes the maximum and minimum amplitude of the beats are distributed unevenly along the hull.

Table 1 shows the calculated frequency of lateral oscillations of the BHA case, depending on the WP lobe configuration and the angular frequency of the motor shaft.

To exclude the possibility of self-oscillations of the DS and lateral beats of the BHM case, calculated taking into account the specific parameters of the tool length and dimensions (Figure 1) and comparing the critical angular frequency of the drilling string with which it is in a state of unstable equilibrium, it is recommended to adhere to the following angular frequencies of the top drive and the BHM shaft (Table 2).

The definition of the critical angular frequency of the drill string jointly BHM when drilling a hold section of the well in which it is in a state of unstable equilibrium is calculated by the formula

$$w_{\text{critical}} = \sqrt{\frac{\left(E.I - \frac{G.l^2}{\pi^2} + 0.013. \frac{q_H \cdot l^4}{f}\right) \cdot g \cdot \pi^4}{q.l^4}} \tag{7}$$

where w_{critical} - critical angular frequency, s^{-1};E - Young modulus, Pa; I - centroidal moment of inertia, m^4;G- weight on a bit, N; q - weight of 1 DS meter, N; q_H - weight depending on the angle of inclination of the well, N; 1- semi wave length (conditionally

Table 1. Frequency of lateral oscillations of BHM.

Motor shaft rotation frequency, rpm	BHM lobe configuration				
	4/3	5/4	6/5	8/7	10/9
	Case oscillation frequency, Hz				
20	1	1,33	1,67	2,33	3
40	2	2,67	3,33	4,67	6
60	3	4,00	5,00	7,00	9
80	4	5,33	6,67	9,33	12
100	5	6,67	8,33	11,67	15
120	6	8,00	10,00	14,00	18
140	7	9,33	11,67	16,33	21
160	8	10,67	13,33	18,67	24
180	9	12,00	15,00	21,00	27
200	10	13,33	16,67	23,33	30
220	11	14,67	18,33	25,67	33
240	12	16,00	20,00	28,00	36
260	13	17,33	21,67	30,33	39

Table 2. The recommended rotational frequency of the top drive, depending on the kinematic ratio of the WP of the BHM when harness the inclined section with a zenith angle of 50 degrees and set BHA parameters and drilling modes.

Hold section drilling interval,m	Rotation frequency of the top drive, rpm	Minimum permissible limit for shaft frequency of the BHM, rpm				
		WO lobe configuration of the engine				
		3/4	4/5	5/6	7/8	9/10
250-350	82-105	80	60	40	40	20
350-450	48-67	60	40	40	20	not limited
450-550	34-48	40	40	20	not limited	not limited
550-600	33-44	40	20	not limited	not limited	not limited

adopted taking into account the lower and upper shoulders and the angularity on the angle regulator), m; f – clearance, m.

Input parameters for the calculations: drilling interval from 250 to 500 m; BHM length - 12 m; the angle on the regulator angle of the BHM is 1.3 degrees; 35 kN weight on a bit; drill string diameter 127 mm; wall thickness of the drill string -10 mm; weight of 1 running meter - 300 N; DC - 12 m; weight of 1 meter 1000 N.

Reduced shaft frequency listed in Table 2 will lead to a superposition of the natural oscillations of the DS and the lateral oscillations of the engine.

3 CONCLUSIONS AND RECOMMENDATIONS

On the basis of the calculations carried out, the lateral oscillations of the frame of the BHM with different modes of its operation are determined. The results of the computational experiment allowed us to reveal the critical values of the angular frequency, the period and the natural frequency of torsional oscillations of the drill string when drilling a well in the hold sections. Recommendations are given on the choice of the optimal parameters of the joint work of DS and BHM, which do not allow the superposition of torsional and lateral vibrations during their joint operation.

However, for a more detailed study of the resonance caused by the superposition of the natural and forced frequencies of the torsional oscillations of the DS, at the hold sections, additional studies are needed in order to optimize the drilling parameters under the occurring impact loads at the lateral and axial vibration accelerations of the BHA taking into account its stiffness, as well as the physical and mechanical properties of the rock and the type of drill through tool (Litvinenko & Dvoinikov 2019).

REFERENCES

Baldenko D.F., Baldenko F.D., Gnoevoy A.N. Single screw hydraulic machine: In 2t. M .: "IRC Gazprom" LLC. - 2007 - V.2. Screw downhole motors - 470.

Bykov I.Y., Zaikin S.F., Perminov B.A., Perminov V.B. The dynamic properties of the drill string.. Construction of oil and gas wells on land and at sea. - 2014-№8. p.4-8.

Bobrov M.G. A study of transverse vibrations of screw downhole motors. Abstract of diss. the candidate tehn. Sciences. M .: OOO "Sigma". 2000. - p. 9 – 11.

Dvoynikov M.V., Syzrantsev V.N., Syzrantseva K.V. Designing a high resistant, high-torque downhole drilling motor. International Journal of Engineering, Volume 30, Issue 10, October 2017, Pages 1615-1621.

Litvinenko V.S., Dvoinikov M.V. . Justification of the Technological Parameters Choice for Well Drilling by Rotary Steerable Systems. Journal of Mining Institute. 2019. Vol. 235, p. 24-29. DOI: 10.31897/PMI.2019.1.24

Morenov, V., Leusheva, E. Development of drilling mud solution for drilling in hard rocks (2017) International Journal of Engineering, Transactions A: Basics 30(4), p. 620-626 DOI: 10.5829/idosi.ije.2017.30.04a.22

Morenov, V., Leusheva, E. Energy delivery at oil and gas wells construction in regions with harsh climate (2016) International Journal of Engineering, Transactions B: Applications 29(2), p. 274-279 DOI: 10.5829/idosi.ije.2016.29.02b.17

The calculation of endurance and gamma-percentage service life of wells

N.G. Fedorova, J.K. Dimitriadi, E.D. Voropaeva & N.A. Ryapolov
Institute of oil and gas, North-Caucasus federal university, Russia

ABSTRACT: In the operation practice of dangerous industrial objects, the peculiar significance has two temporal stages: service life and residual service life. A value of indicated periods has considerable economical component connected with the expenses on object maintenance.

The assigned service life is not an indicator of reliability and must be corrected in accordance with indicators of objects reliability achieved on practice.

It is expedient to receive the endurance indicators for wells by the calculation and experimentally due to the accumulation and the analysis of statistics.

It is shown that the definition of the standard service life of the casing pipes with connections "Premium class" demands the information about the value and the warrantable periods of preservation of the tightening force by the sealing unit "metal-metal" from pipes producers.

1   INTRODUCTION

In the operation practice of dangerous industrial objects, the peculiar significance has two temporal stages: service life and residual service life.

The industrial safety expert-examination is realized on completion of the service life; the technical condition, the possibility of the further operation and the residual service life of object are determined. Well service life is elongated within the limits of the residual service life; at the end of that well service life, the procedure of the industrial safety expert examination is repeated. A value of indicated periods has considerable economical component connected with the expenses on object maintenance.

The assigned service life of oil and gas wells (including underground gas storage wells) is practically considered 30 years (Control technology of technical condition of the production wells).

The assigned service life is not an indicator of reliability and must be corrected in accordance with actual indicators of well reliability.

Within this context it is expedient to receive the endurance indicators for wells by the calculation and experimentally due to the accumulation and analysis of statistics. We show the possibility of calculation method on the example of the calculation of the residual well service life.

2   METHODS

2.1   *The calculation of the residual well service life*

The classic approach states that service life indicators of the object are defined by service life indicators of its supporting structure (Bolotin, 1984).

For the wells of such design it is a support (that is combined structure, its elements are the casing pipes and the cement sheaths) (Merzlyakov et al. 2019, Merzliakov & Podoliak 2017).

The service life of the support is defined by its most loaded elements – the production casing and the cement sheath situated behind it (Manual to calculate the endurance and the residual service life of the wells).

Having considered the work of the production casing and the cement sheath (Kamenskih et al. 2018), we can point the next:

- the string and the cement sheath do not work together;
- they have different mechanisms of damage on the predictable period of service life.

In that case:

- service life indicators are separately calculated for the string and the cement sheath;
- service life of the support is determined as a minimum value from received values.

2.2 The residual service life of the casing strings

Design elements, which defined strength and endurance of the string, are the casing pipes.

Application of software in calculation practice based on the finite element method, in particular, allows one to receive the characteristics of the supporting power for damaged pipes.

The characteristics of the supporting power are individual for pipes diameter and for types of its damage. At bottom of fact, that are the expressions to calculate the coefficient of pipes supporting power decreasing (to external and internal overpressures) in the function of damage typical size and nominal side thickness (Leusheva & Morenov 2017).

Received characteristics can be used to predict service life indicators of pipes under condition that the type of damage which is considered by the characteristic is a dominating one on predictable period of service life.

Let's illustrate it on the next example.

2.3 The calculation of the residual service life for pipes with diameter 168.3 mm; the dominating mechanism of damage – common corrosive wear of the inner surface; the working load -inner excess pressure $P_{inner\ excess}$

The characteristic of the supporting power for pipes with diameter 168.3 mm at the common corrosion of the inner surface has the formula (see Equation 1 below) (Fedorova 2004):

$$K_2 = (0,0235 \cdot \delta_{actual} + 0,8038) \cdot \exp((0,0304 \cdot \delta_{actual} - 0,5455)u) \tag{1}$$

where K_2 = coefficient of the supporting power decreasing of damaged pipes to the inner excess pressure; δ_{actual} = actual thickness of pipes wall, mm; u = depth of corrosive damage, mm (δ_{actual} and u are defined at defectoscopy of the string in the moment of the calculation realization).

A criterion of the limiting condition of pipes is the decreasing of wall thickness up to maximum permissible value [u].Value [u] is calculated according to the transforming equitation (1) relatively to it (see Equation 2 below):

$$[u] = \frac{\ln\{[K_2]/(0,0235 \cdot \delta_{actual} + 0,8038)\}}{0,0304 \cdot \delta_{actual} - 0,5455} \tag{2}$$

where K_2 = permissible value of the coefficient of supporting power decreasing (see Equation 3 below):

$$[K_2] = n_2 \cdot P_{inner\ excess}/P_{yieldpoint}' \tag{3}$$

where $P_{yield\ point}'$ = inner excess pressure where maximum stresses in the wall of damaged pipes are equal the yield point of material, MPa (the value $P_{yield\ point}'$ is calculated, for example, in accordance with (Manual to calculate the damaged and situated in special operational conditions the casing pipes);$P_{inner\ excess}$ = working load on the predictable operational period, MPa (type of pressure – static (repeated-static); n_2 – safety factor coefficient to inner pressure (Instruction to design the casing pipes for oil and gas wells).

The residual service time T (years) of casing pipes is calculated according to the formula (4):

$$T = [u]/V \tag{4}$$

where V = corrosion velocity, mm/year.

$$V = \Delta\delta/\Delta t \tag{5}$$

where $\Delta\delta$ (mm) = the change of wall thickness to period of time Δt (year), during which it occurred.

Evidently, the prediction of the residual service time for pipes is realized taking into consideration its technical condition and individual reserve of the supporting power to the working load.

Everything connected with definition of service life indicators for the pipes of production casing is clear.

2.4 The calculation of the residual service time of the cement sheath

The calculation of the residual service time is realized by means of qualimetry methods and set of factors, its values are used to determine the technical condition of the cement sheath (in particular, the condition of its material – the cement stone) (Manual to calculate the endurance and the residual service life of the wells).

Among these factors are, for example, the condition of contacts "cement-string" and "cement-rock", the value of intercasing pressures. Evidently, the technical condition of the cement sheath is defined by indirect way according to the parameters values, in point of fact, which characterized its sealness.

It is well-known that the endurance of the cement sheath is linked with its strength (Bulatov 1990). That is why the application of parameters characterized its sealness for calculating the service life indicators of the cement sheath seems not entirely correct.

2.5 The analysis of statistics – data of the industrial safety expert-examination over a period from 2007 to 2011 years

The data analysis of the industrial safety expert-examination in a view of the endurance of wells, support and cement sheaths have illustrated the next:

- there are wells in the operation; wells ages are 40, 60 and more years;
- during indicated service life periods, the support as an object of capital construction has saved its consistency - according to the data of technical condition diagnostics for wells, mutual relocations of the casing pipes and the cement sheaths and the connected consequences are not observed.

Consequently, if the cement sheaths are considered as the constructions, which performed the power function – the unloading of the casing pipes from its own weight, so the field experience is the evidence that the cement sheaths endurance is not lower (but, theoretically, it is higher) than the endurance of the production casings.

We noticed that in the well the cement sheaths are situated between the casing pipes or between a casing pipe and a well side, i.e. between the supporting structures; at the same time, the necessary condition to destruct the cement sheaths will be the destruction of exactly supporting structures.

Table 1. The distribution of underground gas storage wells quantity according to the prolongation periods of the safe operation.

Elongating period of operation	Well quantity, %				
	Realization year of industrial safety expert-examination				
	2007	2008	2009	2010	2011
full	57.8 %	55.2 %	90.3 %	81.1 %	91.1 %
shortened	38.4 %	41.0 %	5.1 %	16.2 %	5.5 %
not prolonged	**3.8 %**	**3.8 %**	**4.6 %**	**2.7 %**	**3.4 %**

From above-mentioned information it follows that the support endurance must be determined by the production casing endurance as the most loaded and subjected to the effects (medium and mechanical) supporting structure.

The most important property of the support is the sealness. It is taken in consideration during the prolongation of the periods for the safe wells operation (Fedorova, 2017) when it is necessary not only the presence of the safety factors of supporting structures (to working loads) but also the realization of safety requirements, firstly connected with the support sealness (Shamshin et al. 2014).

2.6 The gamma-percentage service life of underground gas storage wells

Accumulated statistics allows one to determine the gamma-percentage service life of wells.

In the Table 1 the comparison of results of the industrial safety expert-examination 2007-2011 years according to the prolongation of the period for the safe operation of underground gas storage wells (Bulatov, 1990) is shown.

From data in Table 1, we see that wells quantity percent is practically stable, service life of which it is not prolonged. On average, it is about 4 % (3.7%).

This is the actual operational reliability indicator of underground gas storage wells in accordance of it the assigned service life of wells operation was corrected.

The gamma percentage service life of underground gas storage wells is 42 years if probable gamma 96%.

3 CONCLUSIONS

Considered above calculation of service life indicators for the casing pipes concerns its smooth (plain) part (Nutskova et al. 2019).

The increasing of wells service live periods makes to pay attention on the service life of threaded connections of the casing pipes from the standpoint of the preservation time of the tightening force.

Firstly, it concerns the connections "Premium class" and the tightening force of the elements of the sealing unit "metal-metal". The mechanism of tightening force decreasing and the creation of free distance between the elements of the sealing when running in hole of the pipes is cited in (Ryabokon et al. 1997).

Unsealing of the connection is linked with values correlation of the tightening force and the working loads. It should be pointed out that the value of the tightening force depends not only on the joint torque but also on design sizes of the sealing unit elements for the connection.

The necessity appears to receive the information not only about recommended joint torques and so on but also about the value and the warrantable periods of preservation of the tightening force by the sealing unit from producers of pipes with connections "Premium class".

Such information must be taken into consideration when the determination of the assigned service life of the casing pipes, in particular, with threaded connections "Premium class" is realized.

REFERENCES

Bolotin, V.V. 1984. *Prediction of the service life of machines and constructions*. Moscow: Mashinostroenie: 321.

Bulatov, A.I. 1990. *Generation and work of the cement stone in the well*. Moscow: Nedra: 409.

Control technology of the technical condition of the production wells. 2000. OAO "Gasprom", OOO "VNIIGAS". Moscow: 69.

Fedorova, N.G. 2004. Prediction of the service life for the casing pipes. *Gas industry* № 11. Moscow: 69-72.

Fedorova, N.G. 2017. Cement sheaths and well endurance. *Oil economy* № 8. Moscow: 86–88.

Instruction to design the casing pipes for oil and gas wells: instead Regulation document 39-7/1-0001-89. 1997. Moscow: VNIIGAS: 194.

Kamenskih, S., Ulyasheva, N., Buslaev, G., Voronik, A., Rudnitskiy, N. Research and development of the lightweight corrosion-resistant cement blend for well cementing in complex geological conditions, *Society of Petroleum Engineers - SPE Russian Petroleum Technology Conference 2018, RPTC 2018.*

Leusheva, E., Morenov, V. Research of clayless drilling fluid influence on the rocks destruction efficiency (2017) *International Journal of Applied Engineering Research* 12 (6), p. 945-949.

Manual to calculate the damaged and situated in special operational conditions the casing pipes. 2007. STO Gasprom 2-2.3-117-2007. Moscow: OOO "IRTS Gasprom": 55.

Manual to calculate the endurance and the residual service life of the wells. 2009. STO Gasprom 2-3.2-346-2009. Moscow: OOO "Gaspromekspo": 36.

Merzlyakov M.Y., Jennifer R.R. Hernandez & Zhapkhandayev Ch.A. Development of cement slurries for oil and gas wells lining in aggressive environment// *Youth Technical Sessions Proceedings: VI Youth Forum of the World Petroleum Council - Future Leaders Forum* (WPF 2019), June 23-28, 2019, Saint Petersburg, Russian Federation. – 2019, pp. 387-393.

Merzliakov, M.I., Podoliak, A.V. Terms of melting the permafrost when cementing of boreholes plugged with gas-liquid mixtures with hollow microspheres, *International Journal of Applied Engineering Research* 2017: 12(9), pp. 1874-1878.

Merzliakov, M.Y., Podoliak, A.V. Improving the efficiency of well cementing in permafrost regions by using gas-liquid cement mixtures, *International Journal of Applied Engineering Research*, Volume 12, Issue 9, 2017, pp. 1879-1882.

Nutskova M.V., Rudiaeva E.Y., Kuchin V.N., Yakovlev A.A. Investigating of compositions for lost circulation control, *Youth Technical Sessions Proceedings: VI Youth Forum of the World Petroleum Council - Future Leaders Forum* (WPF 2019), June 23-28, 2019, Saint Petersburg, Russian Federation. – 2019, pp. 394-398.

Ryabokon, A.A., Fedorova, N.G., Dubenko V.Ye. 1997. A mechanism of the threaded connections unsealing of the casing pipes with the face seal. *Collection of research papers VNIIGAS: The design of oil and gas wells*. Moscow: VNIIGAS: 93-100.

Shamshin, V.I., Fedorova, N.G., Dubenko, V.Ye. 2014. About periods definition of the safe operation of oil-and-gas wells. *Design of oil-and-gas wells onshore and offshore* № 3. Moscow: 30–32.

Topical Issues of Rational Use of Natural Resources 2019 – Litvinenko (Ed)
© 2020 Taylor & Francis Group, London, ISBN 978-0-367-85720-2

Improving the technology to restore well tightness in underground rock salt reservoirs

R.A. Gasumov & Yu.S. Minchenko
JSC «North Caucasus Scientific Research Project Institute of Natural Gases», Stavropol, Russia

ABSTRACT: Underground salt bed reservoirs are used for gas and liquid hydrocarbons storage in several regions of Russia and the near abroad. The article presents technological solutions for the elimination of leaks in the annular casing space of a salt well. Studies of the basic properties of a lightweight plugging solution are described. The basic requirements to the plugging solution for the installation of a cement bridge are defined. Studies of polymer dispersion are given based on the results of which a new component composition of the displacement fluid is proposed. The measures for intermediate bridge installation in the open hole of the technological well are proposed. The scheme for installation of a two-layer cement bridge is presented.

The application of recommendations based on the results of studies of displacement fluid and plugging solutions compositions as well as the developed technologies for their use in carrying out well repairs will help increase the service life of underground reservoirs in rock salt.

1 INTRODUCTION

In the world practice of creating underground hydrocarbon storages there has been a steady trend towards increase in the rate of underground reservoirs construction in impermeable rocks. The construction of underground reservoirs created in rock salt deposits using the geo-technological method (underground salt dissolution through boreholes) is especially intensive which is due to a number of economic and environmental benefits (Pyshkov & Salohin, 1993).

One of the most important problems in the field of gas storage is to ensure the tightness of underground gas storages (UGS) wells which ensures the reliability of their operation and production of additional recoverable gas. Need to maintain the operational status for operating stock of UGS wells is of paramount importance for solving these problems.

One of the main tasks of restoring the tightness of technological wells in underground rock salt reservoirs is the need to isolate a reservoir of a large geometric volume filled with a saturated sodium chloride solution (brine) from the technological well. Technically, this separation can be carried out with the help of an intermediate bridge installed below the production casing shoe in the open hole. To install the intermediate bridge various mechanical devices or displacement fluids (DF) can be used, the main purpose of which is to separate the brine from the plugging solution. Lightweight plugging solutions of various densities can be used as plugging solutions for cement bridge installation, providing the ability to transport and acquire the necessary structural and rheological properties for the formation of the required strength over the DF layer during cement stone hardening (Morenov & Leusheva 2017).

The aim of study was the development of compositions for DF and lightweight plugging solution to create an intermediate bridge in the open hole of the well to ensure the work on restoring the tightness of annular space of the production string.

It is obvious the installation of a cement bridge (Nutskova et al. 2017) in a brine-filled well without using any plugging solution-brine separator is impossible. During well drilling and repairing various dispersed systems with high structural and rheological parameters, including polymer based viscous-elastic compositions (VEC) are widely used to separate fluid flows.

2 METHODOLOGY

In accordance with the objectives the developed viscoelastic DF must meet the following requirements:

– be compatible with sodium chloride solutions (NaCl) of various concentrations, rock salt enclosing an underground reservoir as well as with other process liquids (for example, plugging solutions) used for the repair of technological wells;
– have the necessary density to ensure the possibility of the polymer VEC "pillow" formation and acquisition of structural and mechanical properties on the brine surface;
– have plastic strength and adhesion to rock salt providing the possibility of pumping the required volume of plugging solution to the VEC surface to install the intermediate cement bridge;
– have enough gelation time for transporting the polymer solution to the brine surface in the reservoir;
– be stable and steady when in contact with working environments (brine, cement solution) in specific conditions;
– maintain structural and mechanical properties for the time required to ensure thickening and setting of the plugging solution during the installation of the intermediate bridge;
– have an ingredient composition that ensures the environmental safety;
– be technological in preparation and use;
– not cause metal corrosion of casing, elements of technological equipment;
– be fire- and explosion-proof.

To select the ingredients for DF development the most satisfying these requirements studies have been carried out on the properties of aqueous dispersions based on acrylic polymers. Since the studied VEC by ingredient composition are polymer systems with time-varying rheological parameters (from a fluid state to the acquisition of rubber-like properties) depending on the specific composition for each polymer dispersion the most significant indicators were determined with a real possibility of their reliable assessment.

3 LABORATORY RESEARCH

Viscoelastic DF on the basis of high-molecular polymers: CM cellulose, polyvinyl alcohol, polyacrylamide, hypan and other polymers of acrylic series are of particular interest. Various compositions of these polymers are also used (Zimina & Kuznetsov 2019). Viscoelastic DF on a polymeric basis usually contain "crosslinking" reagents assuring particular structural and rheological characteristics, and differ from viscous DF (emulsions, clay solutions, etc.) in that they exhibit elastic properties during shear deformations (flow), the use of which is most effective for separating streams of various liquids (Bulatov et al. 1987).

However, taking into account the experience of developing the DF for various purposes of wells drilling and repairing it was determined that DF with the greatest plastic strength are formed on the basis of acrylic polymers - this is due to the structure of the macromolecules of the original polymer components.

The most promising polymer DF base was taken reagent «Aquapol 10».

Reagent «Aquapol 10» is an aqueous polyurethane dispersion produced by NPP Macromer Ltd. (Vladimir). In appearance, it is a milky-white liquid with a density of 1007 kg/m^3 with a mass fraction of 28–32% dry matter, pH 6.5–8.5 and viscosity at 25 °C about 100 mPa · c. Due to the additional stabilization «Aquapol 10» is a frost-resistant system: it can withstand at least 6 "freeze-thaw" cycles without changing its properties, while for other types of

Table 1. Change in the rheological parameters of the dispersion of DF polymers in time.

| Preparation time, min | Rheological parameters | | | Solid phase flotation |
	Plastic viscosity, mPa·c	Dynamic shear stress, dPa	Relative viscosity according to Field Viscometer-5,c	
60	95	20	–	observed
80	120	29	–	observed
100	140	48	180	negligible
120	144	57	200	no

polyurethane dispersions at temperatures below the freezing point of the water phase, irreversible coagulation and aggregation are possible, therefore such dispersions should be stored at temperatures above 0 °C.

As a result of the research DF composition on the basis of the «Aquapol 10» reagent was obtained best suited to the task and recommended for use in wells for the installation of intermediate bridges on the brine pad including vol.mas.:

Reagent «Aquapol 10»	100
Mud powder Benthokon «Super 200»	10
Polyacrylimide Praestol trademark 2510	0.14
Polyacrylimide AK 631 trademark A-930	0.04
Monasil	0.04

The interaction of ingredients in DF composition is described in detail in (Minchenko 2018, Perejma, Patent № 2475513 RF, MPK7 S 09K 8/40. 2012).

Change in the rheological parameters of the dispersion of DF polymers in time to the acquisition of a stable state (no flotation of the solid phase) is shown in Table 1.

When the dispersion of polymers reaches a plastic viscosity of 140 mPa c and a conditional viscosity within 180 c (Table 1), there is practically no trace of solid phase flotation in the DF (Nikolaev & Leusheva 2019). DF acquires the properties of a homogeneous system, which makes it possible in 100–120 minutes to consider it ready to be placed on the brine surface (saturated NaCl solution) in order to polymerize and form a displacement separating layer.

On the border of DF-brine contact a solidified NaCl part is formed almost instantly, above which the DF is in uncured gel state. In this case, the polymer DF gel is adjacent to the bounding surface of reservoir walls.

Polymerization of the upper DF layer at 40 °C occurs in 3 h of process sludge. As a result, a separating complex organic-mineral layer is formed on the surface of brine in the reservoir. The lower part of this separator is a solid, medium gel-like polymer system and the top is a thin rubber-like film adjacent to the lateral surface of the reservoir.

Studies have confirmed that DF provides for the formation of a separation barrier on the surface of brine allowing to with stand a column of lightweight plugging solution with a density of 1410 kg/m^3, 1.5–2 times the thickness of the DF layer on the brine. It should be recognized that this condition is feasible only when creating on the brine surface a separating DF layer with a thickness of at least 1–1.5 of the average diameter of reservoir at the brine surface level.

Figure 1 shows the formation of DF layer and above it a double-height column of plugging solution with a density of 1410 kg/m^3 in 1-liter polyethylene vessel (a) and 2-liter glass vessel (b).

4 RESULTS

The technical result which can be obtained using this DF is as follows: well sealing efficiency for underground reservoir filled with brine increases because of improved technological properties due to the high stability of its phase composition, increased bearing capacity and

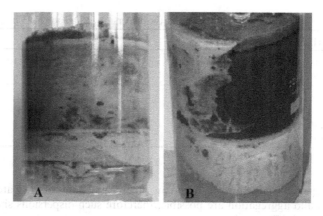

Figure 1.   Formation of the DF layer and plugging solution column.on the surface of saturated brine NaCl in polyethylene (A) and glass (B) vessels.

prevention contact of plugging solution used with brine as well as reducing material costs during the work.

Repair work technology (Gasumov et al. 2016.) with a cement bridge installation in the open hole of technological well in an underground rock salt reservoir includes: –lowering the fluid level in the well to the lower boundary of the intermediate bridge installation by pumping nitrogen or inert gases;

–DF pumping through the tubing string onto the brine surface in a volume that ensures the column height at least 1–1.5 times the reservoir diameter to prevent mixing of the lightened cement solution with the brine;

–installation of the first cement bridge of lightweight plugging solution in the given interval over the DF whose height 1.5–3 times the thickness of the DF layer, determined by its carrying capacity, as well as the ability to hold on its surface a second cement bridge of normal density plugging solution;

–installation of the second cement bridge of reinforced plugging solution with density of 1800 kg/m^{3}above the bridge from a lightweight cement solution in order to increase the total bearing capacity and ensure insulation works in the main casing with subsequent running and cementing of additional casing.

Installation of two-layer cement bridge according to the proposed technology (Perejma 2014) is performed in the order schematically presented in Figure 2.

Lightweight plugging solution for installation of the first (intermediate) cement bridge above the DF is prepared using as a mixing fluid a saturated NaCl solution (ρ = 1200 kg/m^3) of mixture of Portland cement, microspheres, metakaolin and gypsum followed by addition of a plasticizer(Kameskih & Ulyasheva & Buslaev & Voronik & Rudnitskiy 2018).

The density of the lightweight plugging solution allows to pump it into the tubing and transport it by self-spout onto the DF surface ensuring the height of the column 1.5–3 times the thickness of DF layer. The strength of the cement bridge of this plugging solution is sufficient (according to design values) to install the second cement bridge above it from reinforced plugging solution (RPS) with a density of 1800 kg/m^3.

In order to ensure the minimum time of contacting the plugging solution with the walls of rock salt reservoir, the changing rheological characteristics, in particular η и τ$_o$ are of no small importance after cement mixing as well as the acquisition time of the structural properties after which the intensity of interaction of plugging solution liquid phase with the walls of underground reservoir is significantly reduced and consequently their washing (Moradi & Nikolaev 2017).

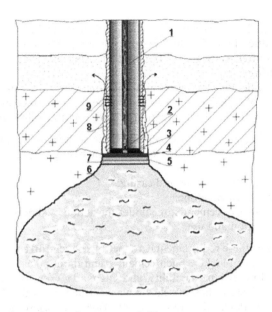

Figure 2.　Installation scheme for two-layer cement bridge.

Table 2.　Rheological parameters of plugging solutions.

Formulation	Time, h-min	Rheological parameters		
		η, mPa·c	τ_o, dPa	$CHC_{1/10}$, dPa
	Immediately after preparation at 20 °C			
	0–05	90	54	39/53
Composition № 2	after warming up to 40 °C			
	0-20	39	53	383/479
	1–30	47	421	immeasurable
	2–50	non-pumpable system was formed		

To determine the time of structure formation for lightweight plugging solution its rheological parameters were measured immediately after preparation at room temperature as well as at 40 °C corresponding to the temperature in the underground reservoir. To do this, the plugging solution was placed in a thermocouple of the rotational Fan 35 SA Baroid viscometer, and after it was heated to 40 °C, parameters were measured at certain intervals before the solution acquired coagulation structure (Perejma, 2014).

The results of determining the rheological parameters for plugging solutions at room temperature and 40 °C are shown in Table 2.

From the data of Table 2 it follows that the beginning of the structure formation of lightweight plugging solution at 40 °C occurs fairly quickly (in 20 minutes), and it fully acquires the properties of the coagulation structure in 1 hour and 30 minutes from mixing. This satisfies the condition of reducing time for liquid phase of plugging solution contact with the walls of rock salt reservoir to prevent their erosion and is sufficient for preparation and injection into the well.

1 – tubing string, 2 – inverted valve, 3 – drilled packer, 4 – plugging solution reinforced with basaltic fibre, 5 – lightweight plugging solution, 6 – brine (NaCl saturated solution), 7 – buffer system, 8 – behind-the-casing leakage, 9 – technological holes in the production string

After waiting for hardening the lightweight plugging solution a cement bridge of plugging solution with a density of 1800 kg/m^{3}is installed on the first (intermediate) bridge surface.

The plugging solution for bridge installation should meet the following requirements:

- increased compressive strength of the cement stone to hold the column of plugging solution while isolating the annular space of the main casing and cementing the additional casing;
- good adhesion to rock salt well walls and casing;
- low gas permeability of cement stone;
- increased crack resistance of cement stone under the action of shock loads;
- adaptability in preparation and use.

To the greatest extent these requirements are met by the reinforced plugging solution (RPS) with the addition of a gas blocker. Such cement stone differs from cement stone of "pure" portland cement by low gas permeability and high adhesion to metal. When tested for strength, the reinforced stone has an elastic-plastic fracture mode (Gasumov&Minchenko 2016,Scrivener&Wieker, 1992,TalabaniSovan,1999,Taylor, 1990), while the stone from "pure" cement is fragile with chipping-off parts of various sizes, which is confirmed by testing two-day cement stone for crack resistance.

5 CONCLUSIONS

Thus, the developed technology allows to install a three-layer intermediate bridge on a brine pad consisting of polyurethane based DF, a lightweight cement solution and a reinforced plugging solution, and to ensure running and cementing the additional casing string.

To ensure the tightness of the cement ring with alternating loads for UGS wells operation and prevent annular pressures the additional casing is to be cemented using reinforced plugging solution.

Complex of developed measures for organization of repair work and technologies will improve repair work efficiency in underground rock salt reservoirs.

REFERENCES

Bulatov, A.I. et al. 1987. *Displacement fluids used in well cementing*. Obz. inform. Ser. Burenie. Moscow: VNIIOEHNG, Iss. 8: 62.

Gasumov, R.A. & Minchenko, Yu.S. 2016. Improving the quality of well cementing using dispersion-reinforced cement materials. *Neftepromyslovoe delo*, № 8:53–57.

Gasumov, R.A. & Kashapov, M.A. & Minchenko, Yu.S. 2016. Technical and technological proposals for the mounting of wells of underground helio-storages in rock salts. *Stroitel'stvo neftyanyh i gazovyhskvazhin nasushe i na more*. № 7: 37–41.

Kamenskih, S., Ulyasheva, N., Buslaev, G., Voronik, A., Rudnitskiy, N. Research and development of the lightweight corrosion-resistant cement blend for well cementing in complex geological conditions, *Society of Petroleum Engineers - SPE Russian Petroleum Technology Conference* 2018, *RPTC 2018*.

Minchenko, Yu.S. 2018. Displacement fluid used in sealing the well of an underground reservoir filled with brine. *Bulatovskie chteniya. Vol. 3. Burenie neftyanyh i gazovyh skvazhin*: 191–193.

Moradi, S.S.T., Nikolaev, N.I. Free fluid control of oil well cements using factorial design, *Journal of Engineering Research* 2017: 5(1), pp. 220–229.

Morenov, V., Leusheva, E. Development of drilling mud solution for drilling in hard rocks (2017), *International Journal of Engineering, Transactions A: Basics* 30(4), p. 620–626.

Nikolaev N.I., Leusheva E.L. Low-density cement compositions for well cementing under abnormally low reservoir pressure conditions, *Journal of Mining Institute 2019*: 236, pp. 194–200.

Nutskova, M.V., Dvoynikov, M.V., Kuchin, V.N. Improving the quality of well completion in order to limit water inflows, *Journal of Engineering and Applied Sciences* 2017: 12(22), pp. 5985–5989.

Perejma, A.A &Trusov, S.G. & Minchenko, Yu.S. 2012. *Displacement fluid used for sealing well of underground reservoir filled with brine.*Patent № 2475513 RF, MPK7 S 09K 8/40.Patent holder PAO «Gazprom». № 2011138002/03.

Perejma, A.A. 2014. Technical and technological proposals for the mounting of wells in underground helium storages in rock salts. *Bulletinof the North Caucasian Federal University*, № 6: 61–68.

Pyshkov, N.N & Salohin, V.I. 1993. Methods of intensifying the creation of underground reservoirs in rock salt. *Transport i podzemnoe hranenie gaza*. Moscow: IRC Gazprom.

Scrivener, K.L. & Wieker, W. 1992. Advances in Hydration at low, ambient and elevated temperature: 449–482. *9th International Congress on the Chemistry of Cement*, Mew Dehli.

Talabani Sovan et al. 1999. *Energy Sources, V.21.№1-2:*163.

Taylor, H.P.W. 1990 *The Chemistry of Cement*. London: Academic Press.

Zimina D.A. & Kuznetsov R.Y. Development of cement composition with enhanced properties with the addition of microsilica, *Youth Technical Sessions Proceedings: VI Youth Forum of the World Petroleum Council - Future Leaders Forum* (WPF 2019), June 23-28, 2019, Saint Petersburg, Russian Federation. – 2019, pp. 399–404.

Topical Issues of Rational Use of Natural Resources 2019 – Litvinenko (Ed)
© *2020 Taylor & Francis Group, London, ISBN 978-0-367-85720-2*

The improving of the safety level of the equipment working under excessive pressure

K.Y. Kozhemyatov & Y.A. Bulauka
Polotsk State University, Novopolotsk, Belarus

ABSTRACT: The systematic data on the frequency and causes of crashes and refusals at one of the oil refineries in the Republic of Belarus for the period from 1969 to 2018 are presented in the article. Data are based on documents of official statistics. The ranking of the number of crashes and refusals at departments, types of processing equipment, causes and types of effects is given. Has been determined the dynamics of the time factor and the causes of crashes and refusals. The results of the analysis of the service life of equipment operating under excess pressure at the Belarusian oil refinery are presented, directions for improving industrial safety when working with this type of equipment are proposed in this study. It has been established that fittings with conditional passage up to DN100, the base metal and metal of the welds of the equipment body are the most wearing parts during the service life. Frequent replacement of fittings with a small conditional passage is connected with a small margin between the executive and rejection thicknesses of fittings' nozzles. The average service life of equipment operating under excessive pressure at the studied enterprise was established: types of equipment used the longest are columns (38.6 years), vessels (34.8 years), reactors (32.8 years) and the heat exchangers (31.2 years).

1   INTRODUCTION

The development of the oil refining industry is accompanied with an increase of the scale of production, the capacity of plants and equipment, the increasing complexity of technological processes and production management modes. Due to the complication and increase quantity of production volumes, accidents, in addition to human victims, are accompanied with huge negative environmental and economic consequences (Achaw, 2012; Anderson, 2005; Bulauka, 2011; Bulauka, 2013).

The enterprises of the oil refining industry are one of the most explosive objects. High fire and explosive level of these enterprises is explained by the presence of a large amount of hydrocarbons that circulates in the process. Emergency depressurization of technological equipment at refinery plant can cause a major accident with concomitant emissions of toxic substances, destruction and damage of expensive equipment, breakdown of technological processes, fires and explosions (Bulauka, 2016; Bulauka, 2017; Chang, 2006; Christou, 1999; Levesson, 2004; Pokrovskaya, 2016; Salmon, 2012; Zhang 2012).

Every year approx 20,000 major accidents occur in the oil and gas industry in the world, and in recent years there has been an increase of the accident rate in the oil refining industry (Dekker, 2002; Jang, 2012; Drogaris, 1993; Khan, 1999; Patterson, 2010; Tsunekawa, 2007).

Several examples of such accidents are:

– An accident in March 2005 at the isomerization unit of one of the largest US refineries owned by BP in Texas City. There was a powerful explosion, followed by a strong fire. 15 people were killed and over 70 were injured;
– On May 29, in 2008, a hydrogen-containing mixture exploded in a hydrogen compressor unit for the secondary processing of oil at the Kirishi Refinery, and then a fire occurred.

One person died on the spot, four died in hospital. The damage from the accident amounted to 107 million rubles;
- On August 7, 2011 there was a fire at the Khabarovsk refinery. Spilled fuel and pumping station burned on a total area of 50 m² 3 people were injured and 2 died;
- On June 15, 2014 at the gas fractionation unit of the Achinsk refinery there was a pass of hydrocarbon gas, which led to a major explosion and fire. 8 people died, 7 were hospitalized, the total number of injured was 24 people. The damage amounted to approximately $ 800 million.

In the CIS countries there is a targeted state policy in the field of industrial safety. However, the state of accidents at work place continues to be a complex socio-economic problem.

In this work, an a posteriori analysis of the state of accidents at the oil refinery enterprise of the Republic of Belarus, which is subject to official accounting, is carried out.

Table 1 provides an analysis of the causes of emergencies at the Belarusian refinery for the 55-year history of operation of the enterprise (according to archival materials and statistical reporting, 625 emergencies occurred at the enterprise).

The results of the analysis showed that unqualified and erroneous actions of staff, erroneous command transfer, inconsistent actions, low level of labor discipline, insufficient control over the work of increased danger, poor installation and repair work, violation of instructions contribute to the largest share in the overall structure of causes of emergency situations. or a project for carrying out work, carrying out fire work on an unprepared place, i.e. controlled causes associated with the "human factor", while the erroneous actions of refiners provoke significant social, environmental and economic risk.

On the second place are the reasons related to the failure of the equipment due to its physical deterioration:

- fatigue failure and wear (7,6%);
- destruction of bearings (friction pairs) (5,3%);
- damage of pump-compressor equipment seals (5,1%);
- pass of welded joints (2,2%).

Table 1. Dynamics of the causes of emergencies over ten-year periods at the Belarusian refinery.

| The reasons of accidents | % of total quantity during the period | | | | | |
	70/79	80/89	90/99	00/09	10/18	63/18
Violation of the rules of technological regulation, requirements of instructions, erroneous actions	46,2	27,9	24,2	17,6	4,6	24,5
The wearing out of the equipment	4,4	24,9	18,8	13,2	11,4	15,7
Poor technical condition of electrical equipment	4,4	15,7	10,2	25,2	4,6	13,7
Corrosion and erosion of the equipment	2,2	4,6	7,0	6,3	15,9	5,6
Malfunction of instrumentation and automation devices	6,6	2,0	7,8	6,9	6,8	5,2
Poor installation and repairing of the equipment	8,8	8,1	3,1	1,9	2,3	4,9
Natural phenomena	3,3	4,6	3,9	6,9	2,3	4,4
Seal damage	6,6	4,1	3,9	5,0	4,6	4,4
Burnout of pipes in furnaces due to local overheating and coke deposits	4,4	3,1	3,1	2,5	13,6	3,7
Design flaws	6,6	3,1	1,6	2,5	6,8	3,2
Welds' poor quality	-	4,1	2,3	1,3	9,1	2,6
Safety valves	2,2	2,0	2,3	1,3	6,8	2,1
Overflow of tanks, vessels and canalization	4,4	3,1	0,8	1,3	-	2,1
Solid particles ingress, accumulation of tar compounds	3,3	1,0	2,3	3,1	-	2,0
Water hammer, fluid entering the compressor cylinder	-	0,5	2,3	2,5	2,3	1,4
Design failures	1,1	1,0	2,3	0,6	-	1,1
Self-ignition of flows	1,1	-	4,7	-	-	1,1
Other	2,2	1,0	2,3	3,1	9,1	2,4

The third place is taken by the causes due to power failures and damage of the insulation of the power cables. The main cause of power outages is the aging and breakdown of insulation of electrical cables and technical devices.

The fourth rank place belongs to the reasons associated with corrosion and erosion wear of equipment, due to the presence of aggressive components in the oil and other flows. This fact is associated with the manifestation of corrosion typical for oil refineries: pitting, slotted, intergranular, stress corrosion cracking, stress cracking in a sulfide-containing medium, hydrogen embrittlement, corrosion under the influence of alkaline water containing hydrogen sulfide, erosion, etc.

Due to natural disasters, about 5% of incidents occurred (hurricane, lightning storms, long torrential rains, crystallization of the product at low temperatures with no heating, equipment defrosting, etc.), as well as the presence of animals near parts of electrical equipment.

Statistical analysis of accidents has shown that the main danger for the industrial territory of oil refining facilities is gas pollution, fires and explosions, and the formation of an explosive atmosphere in most cases occurs due to leaks of a flammable liquid or hydrocarbon gas.

The causes of fires were ignition or self-ignition of petroleum products during depressurization of equipment and communications, self-ignition of thermal insulation contaminated with petroleum products, poor preparation of fire work sites, and discharges of static and atmospheric electricity.

Most of the explosions occurred during the ignition of furnace burners in combustion chambers and gas ducts due to disruption of the ignition procedure. There have been cases of explosions in trays and collectors of a gravity circulating water supply system, as well as in tanks, vessels and compressor rooms during hot work.

Depressurization of equipment occurred due to corrosion of the metal, the formation of cracks in welds and heat affected areas, improper selection or installation of gaskets, insufficient tightening of fasteners for flange joints, as well as damage to the bearings of pumps and electric motors.

The tendency to a decrease in the number of fires and explosions during the selected observation period was revealed, which is probably related to the improvement of work to ensure the fire and explosion safety of equipment and the technological process at the enterprise. It is presented on the Figure 1.

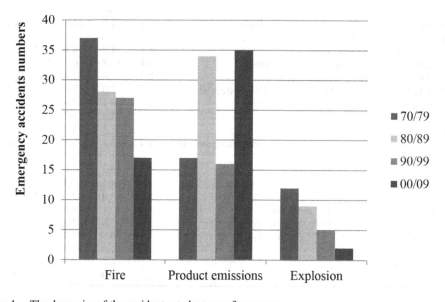

Figure 1.   The dynamics of the accident rate by type of consequences.

Table 2. Analysis of accidents and incidents depending on the type of technological equipment.

Types of equipment	Quantity of accidents and and incidents, % of total quantity Belarussian refinery	Japan refinery
Technological pipelines, connections, fittings	15,1	11,6
Pumps' and compressors' rooms	27,7	18,9
Vessels	13,1	15,0
Technological furnaces	7,8	11,4
Columns	3,1	11,2
Canalization	1,7	8,5
Tanks	2,9	3,8
Others	29,4	19,6

The ranking of accidents for a particular type of process equipment, shown in Table 2, showed that the most likely occurrence of accidents and incidents in the premises of pumping and compressor rooms is 13,6 and 14,1 % of the total number of recorded incidents, respectively. Comparison of the obtained data with statistical data on accidents in Japan showed that the distribution of the number of emergencies, depending on the type of process equipment in the refining of different countries. This is due to the difference in the applied technologies for the processing of crude oil and the instrumentation of the processes, the level of emergency protection systems, personnel training, etc.

The prevailing number of emergencies is recorded in the summer and winter periods. Probably, this pattern is associated with the withdrawal of plants in the summer period for scheduled maintenance, reconstruction of existing industries and the implementation of other construction and repair works. During the commissioning and repair work, about 15% of the incidents analyzed occurred. A significant number of emergency situations recorded in the winter. This fact is connected with the peculiarities of the operation of the refinery in cold climates, which is due to the high risk of equipment defrosting.

The maximum number of accidents and incidents falls on Friday, which is possibly related to the occurrence of a stage of developing fatigue of personnel servicing the technological processes.

Topographical analysis allowed to determine the places and areas with the highest probability of accidents and incidents. Analysis of the distribution of the number of accidents and incidents by units of the refinery showed that the first place (more than 58% of all cases) is occupied by the main production of petroleum fuels and aromatic hydrocarbons (as the most numerous production by the number of implemented technological processes), the second rank place (16% of all cases) occupies the main production of lubricating oils and bitumen, the third and fourth ranking places are occupied by auxiliary production of the oil refinery: ektrosnabzheniya (6% of all cases) and the commodity department (6%), respectively. danger in the operation of various technological installations is unequal. This is due to the technology, properties and quantity of processed raw materials, density of equipment placement.

Statistics show that the major accidents at oil refineries in most cases are due to leaks of flammable liquids and vapors or hydrocarbon gas. The reasons of their occurrence are follows:

– violation of the rules of operation, technological regulations (27% of cases);
– defects in construction and installation works, low quality of installation and repairing of equipment (21%);
– defects in the manufacture of equipment and materials (16%);
– deviation from the requirements of the design and technical documentation (12%);

– equipment deterioration, product leaks through gaskets, mechanical seals, oil seals, equipment corrosion, furnace burnout in furnaces (11%);
– constructive imperfection of equipment (5%);
– external natural and man-made impacts (5%);
– imperfection of design solutions, etc. (less than % 3).

Improving the resiliency and trouble-free operation of equipment under excessive pressure at oil refineries, are now particularly relevant directions to improve the level of industrial safety for the industry. This fact determined the purpose of this study.

2 METHODS OF RESEARCH

Equipment operating under excessive pressure at a Belarusian full-cycle oil refinery was taken as a research object; equipment such as columns, vessels and reactors has been studied; heat exchangers (including crystallizers), separators and filters.

The expert-statistical and analytical methods performed a comprehensive analysis of repairing documentation for the period 2008-2018 and terms of operating of equipment operating under excessive pressure at the Belarusian oil refinery.

3 RESULTS AND DISCUSSION

With the purpose of determination of the current state of equipment during shutdown repairs, as well as during technical diagnostics, and to determine the suitability for further operation and extend the life cycle of equipment that has worked out a standard period the following combination of non-destructive testing methods is usually used:

– visual inspection in accessible places;
– ultrasonic thickness measurement of housing elements and nozzles;
– ultrasonic defectoscopy of welds, as well as control of the continuity of the base metal;
– penetration test;
– hydrostatic test.

3.1 *Columns*

Columns at a research plant accounts for 6% of the total number of equipment. According to GOST 31838-2012, this type includes cylindrical vertical vessels of constant or variable cross-section, equipped with internal heat and mass transfer devices (plates or packing), as well as assistant units (liquid inlets, devices for set-up packed elements, etc.). providing a process (for example, rectification or direct heat exchange between steam (gas) and liquid, etc.). Analysis of repair documentation for the period 2008-2018 showed that the highest frequency of work on replacing nozzles with conditional passage up to DN100 (37% of the scope of work), replacing internal devices (23%), replacing nozzles with conditional passage DN100 and more (21%), while less frequently repairing base metal and metal of the body welds (about 19% of the work).

For the columns the replacement of internal devices usually means the replacement of the supporting elements of the plates welded to the vessel body and which have become unusable as a result of corrosive wear under the influence of the working medium. It is controlled visually and with the help of measuring devices. An effective way to increase safety is the wide use of the technology of welding support elements through shims welded directly to the body of the vessels. The shim plates for welding service platforms are shown in Figure 1.

If it is necessary to repair this unit, the support element is cut out and new one is welded to the shim plate without touching the body of the vessel. At the same time, there is no possibility of mechanical damage to the body when re-disassembling the support element. It also excludes its local overheating and changes in the metal structure as a result of overheating. It favorably affects to the period of further operation of this equipment. This method also can be applied to elements welded outside the vessels, such as service platforms, booms, supporting elements of level gauge columns, etc.

3.2 *Separators*

Separators make up 6% of the total number of studied equipment, belong to the settling equipment. Separators are similar in design to columns, but has smaller size and equipped with fewer plates, or an overflow plate for separating liquids by density. According to the maximum frequency of the types of repairs carried out by separators is work on replacing nozzles with conditional passage up to DN 100 (76% of the work); a small percentage falls on other types of repairs (repair of base metal and metal of body welds - 11% conditional passage of DN 100 or more - 9%, replacement of internal devices - 4%).

3.3 *Filters*

Filters make up 7% of the total number of the studied equipment, belong to the apparatus for the implementation of the filtering process. The most frequent repairs on filters are work for replacing nozzles with conditional passage up to DN100 (71% of works), a small percentage are for other types of repairs (repairing base metal and metal of body welds - 19% and replacing fittings with conditional passage DN100 and more - 10%). The reason for this is that this type of equipment has low average life.

3.4 *Reactors*

Reactors make up 2% of the total number of the studied equipment, belong to vessels in which chemical reactions take place including in the presence of a catalyst. As a rule, the presence of high pressure and temperature is usual. According to the maximum frequency of conducted repairs of reactors, work on the repair of the base metal and metal of the protective casing welds that were repaired (75% of the scope of work), 13% of the work comes from replacing the internal devices of the reactors.

The small number of repairs of this type of equipment is explained by the large margin of safety used in manufacturing, due to the high operating pressures and temperatures, and the difficulty of carrying out repair work in the field on thick buildings. When defects are detected, this type of equipment is changed entirely to ensure trouble-free operation during the overhaul period.

Almost a single type of repair that practiced on the equipment of this type is repairing of the protective case, which is designed to protect the sprayed concrete lining from the corrosive effects of the working medium in old type reactors. From the point of view of improving reliability it is advisable to switch to modern bimetallic reactors.

3.5 *Vessels*

Vessels make up 41% of the total number of studied equipment. According to TR CU 032/2013 vessels are hermetically closed containers (divided into fixed and mobile). This type of equipment is designed for conducting chemical, thermal and other technological processes, and also for the storage and transportation of gaseous, liquid and other substances. The replacement of nozzles with conditional passage up to DN100 (70% of the scope of work) has the highest frequency of repairs of vessels, 15% of the amount of work involved replacing nozzles with conditional passage of DN100 and more and repair of the base metal and metal of the body welds.

Figure 2 shows examples of corrosive wear out of the base metal and metal of the weld seams of vessels on a research plant.

3.6 Heat exchangers

Heat exchangers make up 37% of the total number of studied equipment. According to GOST 31842-2012 heat exchangers are the equipment designed to transfer heat under non-isothermal

Figure 2. Shim plates for welding service platforms.

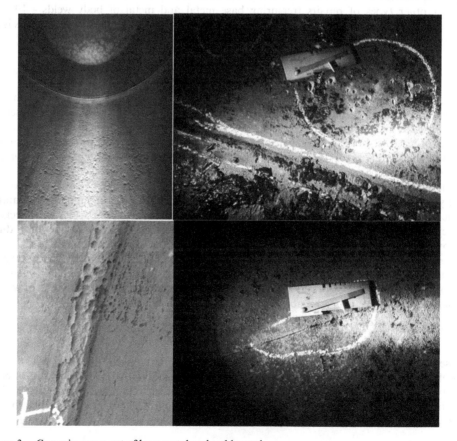

Figure 3. Corrosive wear out of base metal and weld metal.

operating conditions. According to the maximum frequency of the types of heat-exchange equipment repairs carried out, work on repairing the base metal and metal of the welds (31% of the work), work on replacing nozzles with conditional passage up to DN 100 (30% of the work) and replacing and repairing are allocated partition walls of distribution chambers (24% of the scope of work).

The main defects detected in tube bundles of heat exchangers include:

– corrosive wear out of pipe bundle pipes. It is controlled by hydraulic test in the annular space;
– corrosive wear out and thinning of the metal in the field of rolling tubes in tube plates. Controlled by hydraulic testing in the annular space.

Reducing the number of operational defects and extending the time for maintenance-free operation of heat exchange equipment can be achieved by introducing strict control over the operation of this equipment. Often, the gaps in the places of rolling of the pipes are caused by the deformation of the pipes by the excessive pressure in the annular space during the filling of the apparatus. Filling apparatus must be made starting with the tube bundle for the purpose of balance the pressure and reduce the probability of deformation of the tube.

Similar in construction are rigid tube-shell heat exchangers. Detected defects are the same as in shell-and-tube heat exchangers with removable tubular system. But at the same time there is no possibility of visual control of the inner surface of the case. To increase the level of control and detect defects at earlier stages, maps of wall thickness measurements with an increased number of control points are compiled for heat exchangers of these types compared to heat exchangers with a movable pipe system. In addition, rigid tube heat exchangers to compensate for temperature expansions are supplied with bellows expansion joints. Compensators are made of thin-walled stainless steel. At the same time, based on the practical experience of operating devices with a lens compensator, there was often a situation in which the wall thickness of the case had sufficient margin to the screening thickness, and the lens compensator became unusable due to corrosive erosion wear. Repair of heat exchangers of this type is impractical because of a non-dismountable design. So disassembly, replacement of worn out sections of the shell or tube bundle pipes, replacement of a compensator, and the following assembly are comparable in cost to new heat exchangers.

A specific type of heat exchange equipment is scraper chillers, which are additionally equipped with a scraper axle inside the internal tubes of heat exchangers. At the under study enterprise such equipment makes up only 1% of the total number of equipment, but the average lifetime of the chillers is significant - 49.2 years. The analysis of repair documentation for repairs carried out on chillers showed that the only type of repair is the replacement of internal or less often external pipes, as the main wear parts of this type of equipment. In addition to the corrosive effects of the environment, the wear of the inner tubes is influenced by the friction of the scrapers. At axial displacement of the axle during assembly, or uneven wear of the bearings, there is increased wear of the composite gaskets of the shaft scrapers, and then friction of the metal scraper on the inner surface of the pipes. At the same time, there is no possibility to control the condition of the scraper gaskets and the internal pipes of the chiller because the pipes are inaccessible for full visual inspection, it is impossible to conduct ultrasonic thickness gauging. Ultrasonic thickness gauging of external pipes, visual inspection of internal pipes in accessible places using a flashlight along the pipe axis, as well as a hydraulic test for tightness and density of internal pipes are used to assess the state of these technical devices. If a defect is suspected or traces of mechanical wear are detected, the inner tube is replaced. At the same time in such devices it is advisable to use pipes made of solid steel grades to reduce mechanical wear of the internal surface of the pipes.

Table 1 shows the average service life of equipment operating under excessive pressure at the Belarusian refinery.

So, the columns, vessels and reactors are operated the longest. At the same time, the service life of the vessels working under excessive pressure declared by the developer is usually 20 years.

Comprehensive analysis of the life cycle of equipment operating under excessive pressure at the Belarusian oil refinery showed that nozzles with conditional passage up to DN100, base

Table 3. Lifetime of equipment operating under excessive pressure at the refinery.

Type of equipment operating under overpressure	Average life at the refinery
Columnst	38,6
Vessels	34,8
Reactors	32,8
Heat exchangers	31,2
Separators	28,3
Filters	25,0

metal and metal of the body welds of various equipment are subject to a high risk of increased wear out during operation.

The reason for the frequent replacement of nozzles with a small conditional passage is associated with a small margin between the executive and rejecting thicknesses of nozzles. For example, a pipe with a nominal thickness of 4 mm is most often used for the nozzle of DN50 by strength calculation. In accordance with the instructions for revision, repair and rejection which adopted at the under study enterprise, the rejection thickness for the DN50 connection pipe is 2.0 mm, unless otherwise greater value indicated in the strength calculation. The executive wall thickness as a result of an error in the manufacture is often about 3.8-3.9 mm. At the same time, in practice such nozzles are rejected with a thickness of 2.5-2.7 mm, as approaching rejection to improve the reliability and infallibility of work during the overhaul period and to prevent the process installation from stopping due to the omission of the product. In this case, even with a corrosion rate of up to 0.1 mm/year the thickness of the nozzle is not enough even for the declared service life of the vessel (usually 20 years). From practical experience, the replacement of such nozzles during the repair with heavy walled ones of about 6-8 mm leads to trouble-free operation of these units throughout the life cycle of the equipment until the write-off.

Thus, as a way to increase the industrial safety of equipment operating under excessive pressure, it is proposed for the new equipment to coordinate with the developers an increase in the thickness of the fittings with conditional passage up to DN100. At the same time, despite the slight increase in the cost of such equipment, it is possible to achieve maintenance-free operation of the equipment even after the end of the designated service life. The reasons for repairing the base metal and weld metal are hidden metallurgical defects and weld defects that were not detected during the manufacture of the vessel (apparatus), as well as the aggressive influence of the working medium of the vessel (apparatus), the formation of stagnant zones, the accumulation of solid particles from the working medium (scale, contamination, etc.), which are monitored visually and using ultrasonic thickness gauging. Ultrasonic and penetrant test can be additionally performed in places suspicious of defects. In our opinion, in order to minimize the number of repairs of this type, it is necessary to strengthen the incoming control for newly installed equipment. Also it is necessary to strengthen control over the selection of material for a specific working medium and operating parameters such as temperature and pressure and to ensure strict adherence to process regulations. For newly designed equipment employ modern technical solutions to minimize the number of stagnant zones.

4 CONCLUSION

The results of research on the comprehensive life cycle analysis of equipment operating under excessive pressure can be effectively used to increase the level of industrial safety, reduce the risk of accidents at oil refineries and petrochemical plants.

REFERENCES

Achaw W, Boateng D. 2012. Safety practices in the oil and gas industries in Ghana. *Development and Sustainability* 1(2):456–465.

Anderson M. 2005 Behavioural safety and major accident hazards: Magic bullet or shot in the dark? *Process Safety and Environmental Protection* 83(2):109–116.

Bulauka Y. 2011. The analysis of industrial injuries at the oil refinery. *Bulletin of the Polotsk State University*. 3: 130–137.

Bulauka Y. 2013. The development of a comprehensive assessment of occupational risk by taking into account the total hazard of working conditions. *Hygiene & Sanitation*. 4: 47–54.

Bulauka Y. 2016. The problem of choice the most dangerous object for evaluation of explosion of process unit for refining and petrochemical industries. *Bulletin of the Polotsk State University*. 11: 125–129.

Bulauka Y. 2017. Factors of priori occupational risk for workers of oil refineries. *Bulletin of the Polotsk State University*. 3:135–140.

Bulauka, Y.A. et al. 2018. Emergency sorbents for oil and petroleum product spills based on vegetable raw materials *IOP Conference Series: Materials Science and Engineering*. 451 (1).- art. no. 012218.- DOI: 10.1088/1757-899X/451/1/012218.

Chang J., Lin C. 2006. A study of storage tank accidents.Loss Prev. Process Industries 19(1): 51–59.

Chettouh S., Hamzi R., Benaroua K. 2016. Examination of fire and related accidents in Skikda Oil Refinery for the period 2002–2013. *Loss Prev. Process Industries*. 41:186–193.

Christou M., 1999. Analysis and control of major accidents from the intermediate temporary storage of dangerous substances in marshalling yards and port areas. *Loss Prev. Process Industries* 12 (1):109–119.

Dekker S.W.A., 2002. Reconstructing human contributions to accidents: the new view on error and performance. *Saf. Res*. 33(3): 371–385.

Dien Y., Dechy N., Guillaume E. 2012. Accident investigation: from searching direct causes to finding in-depth causes-problem of analysis or/and analyst? *Saf. Sci*. 50(6):1398–1407.

Drogaris G. 1993. Learning from major accidents involving dangerous substances. *Saf. Sci*. 16(2):89–113.

Jang N., Koo J., Shin D. et al. 2012. Development of chemical accident database: considerations, accident trend analysis and suggestions Korean. *Chem. Eng*. 29(1): 36–41.

Katsakiori P., Sakellaropoulos G., Manatakis E. 2009. Towards an evaluation of accident investigation methods in terms of their alignment with accident causation models. *Saf. Sci*. 47(7): 1007–1016.

Khan F.I., Abbasi S.A. 1999. Major accidents in process industries and an analysis of causes and consequences. *Loss Prev. Process Ind*. 12(5): 361–378.

Kozhemyatov K., & Bulauka, Y. 2018. Analysis of the «Rules of procuring industrial safety of the equipment operating under excessive pressure» application at belarusian oil & gas refineries. *European and national dimension in research: materials of X Junior Researchers' Conference, Novopolotsk*, May 10-11, *2018* Polotsk State University: Novopolotsk: 137–139.

Leveson N., 2004. A new accident model for engineering safer systems. *Saf. Sci*. 42(4): 237–270.

Patterson M., Shappell S.A. 2010. Operator error and system deficiencies: analysis of 508 mining incidents and accidents from Queensland, Australia using HFACS. *Accid. Anal. Prev*. 42(4): 1379–1385.

Pokrovskaya S., Bulauka Y., Galkina D. 2016. Comparative analysis of the results of modeling consequences of accidents at dangerous production facilities refining industry using the software TOXI$^+$ ^{RISK}. *Bulletin of the Polotsk State University*. 3: 173–178.

Salmon P.M., Cornelissen M., Trotter M.J., 2012. Systems-based accident analysis methods: a comparison of Accimap, HFACS, and STAMP. *Saf. Sci*. 50(4): 1158–1170.

Tsunekawa K., Liu B.L., Gao J.M. et al. 2007. The initial researches on management of chemical work safety at home and abroad as the cooperation project between China and Japan. *Saf. Sci. Technol*. 3(5): 87–91.

Zhang D., Jiang K. 2012. Application of data mining techniques in the analysis of fire incidents, International Symposium on Safety Science and Engineering in China. *Procedia Eng*. 43: 250–256.

Analysis of well testing - well Pr2, Prušánky field

Z. Kristoň & A. Kunz
VSB – Technical University of Ostrava, Ostrava, Czech Republic

ABSTRACT: Well testing plays an important role at many stages of well's life by assessing both reservoir and well performance. Authors who have contributed to the topic include e.g. Matthews & Russel (1967) and Earlougher (1977).

This paper deals with a pressure build-up test, carried out in a natural gas reservoir, developed by a single vertical well (Pr2) that belongs to Prušánky field, Czech Republic. It aims to characterise the reservoir and well parameters and feasible flow and boundary models using the Weatherford's PanSystem sw. The reservoir was already depleted at the time of this study hence the comparison between the estimated OGIP and the cumulative production was made possible.

The study's main outcomes include estimation of important parameters such as reservoir permeability (125 mD) and skin factor (+6). Furthermore, the interpretation showed that selection of a two-cell compartmentalized reservoir model provided the most feasible estimation of the original gas in-place.

1 INTRODUCTION

Well testing provides the analyst with a wide variety of information about reservoir properties such as permeability, reservoir extent, heterogeneities etc. Also, it offers a useful tool to characterize the well itself by providing data such as wellbore storage, formation damage, evaluation of the efficiency of completion, and evaluation of workover and stimulation treatments, etc. In a nutshell well testing is a tool for assessing of both reservoirs and wells.

The studied reservoir is an uppermost baden sandstone layer developed by a single vertical well (Pr2) that belongs to Prušánky field. The field is part of the Vienna basin, lying at the border between Austria and the Czech Republic.

The goal of this study is to characterise the reservoir and well parameters and investigate selection of the flow and boundary models that correspond to this gas reservoir. Since the studied reservoir has been already depleted at the time of this study, it provided an opportunity to quality check the estimated OGIP of the reservoir (based on interpretation) by comparing it to the actual gas volume that has been recovered from the reservoir.

The outline of the study will be as follows: First of all, a brief insight to the underlying theory of pressure transient testing will be presented, followed by a summary of available information about geology, layer and fluid parameters. The study will also encompass distinct stages of analysing the data measured during the pressure build-up test, using the Weatherford's PanSystem sw. This includes preparing pressure and rate data for analysis and using different graphs for data visualisation and interpretation such as type curves, log-log plot, semi-log plot, and test overview plot. Finally, simulation of the test behaviour using the interpreted reservoir and well parameters will be performed in order to check the validity of the interpretation.

2 PRESSURE TRANSIENT WELL TESTING

This section contains a brief summary of the theoretical basis that enabled the development of methods used for the subjected analysis.

Pressure transient represents the case when flow velocity is not only a function of the radial distance to a well, but also of time. The radius of depression changes with time as well as the pressure distribution throughout the reservoir. (Bujok et al. 1985) Because we are dealing with the unsteady flow its description is more complex than for the steady flow situation. On the other hand, it offers a wider field of use than the steady flow case. In spite of this, according to Earlougher (1977), it would be unwise to oversell or undersell the value of pressure transient testing and analysis, furthermore he emphases the importance of selection of the appropriate methods for each problem.

2.1 Well testing – the inverse problem

A very practical viewpoint of the well testing in general is offered by Horne 1995. Well testing is refered to as an inverse problem (Figure 1), where known history of production changes is an input into the system that is unknown. The flow rate transient causes a corresponding output of pressure transient responses hence the parameters of the system can be studied.

2.2 The diffusivity equation

The fluid flow in porous media is a diffusive process. Its mathematical description is similar to that of other natural phenomenon, e.g. heat conduction or electrical potential distribution. (Matthiews and Russel 1967)

According to Matthiews and Russel (1967): The Darcy's law, The Mass Conservation Law and The Equation of State are the three principals from which the mathematical description of flow in porous media can be obtained. When combined we can derive a group of partial differential equations (PDE's) that describe various flow situations. See Equation 1 below.

The diffusivity equation:

$$\frac{\partial^2 p}{\partial r^2} + \frac{1}{r}\frac{\partial p}{\partial r} = \frac{\phi \mu c}{k}\frac{\partial p}{\partial t} \tag{1}$$

where r = radius; p = pressure; t = time; ϕ = porosity; μ = viscosity of the fluid; c = compressibility; k = permeability

Assumptions that are inherent to this equation are: single phase flow, applicability of the Darcy's Law, porosity, permeability and viscosity are constant, compressibility is constant and small, gravity and thermal effects are negligible.

Because the well Pr2 is a gas well the assumption of negligible compressibility of the fluid and constant viscosity are not valid. The version that takes into account the equation of state for non-ideal gas has to be used instead. See Equation 2 below.

$$\frac{1}{r}\frac{\partial}{\partial r}\left(\frac{p}{\mu z} \cdot r \frac{\partial p}{\partial r}\right) = \frac{\phi}{k}\frac{\partial}{\partial t}\left(\frac{p}{z}\right) \tag{2}$$

where: r = radius; p = pressure; t = time; ϕ = porosity; μ = viscosity of the fluid; k = permeability; z = gas compressibility factor

The three basic solutions for the steady flow are of major interest: Infinite reservoir, constant pressure outer boundary and bounded cylindrical reservoir. According to Matthiews

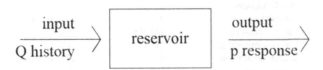

Figure 1. the inverse problem (Horne 1995, modified).

and Russel (1967) They can be combined, using superposition to yield solutions for any flow rate history.

All of the resulting equations for pressure change at the sandface are non-linear. From practical point of view, it's often practical to work with linear equations. To avoid dealing with p^2 we can make use of pseudopressures. This way the linear equations for liquids can be used for gas as well. (Dressler 1995)

The dimensionless variables of p_D, t_D, r_D simplify the reservoir models by embodying the reservoir parameters (such as k) thereby reducing the total number of unknowns, also they have the advantage of providing solutions independent of any unit system. (Horne 1995)

2.3 The importance of theoretical knowledge

For modern day analyst, though, it is not necessary to understand the process of solution of the pressure transmission equation. Solutions to this equation have been developed for a wide variety of specific cases, covering many reservoir configurations. These specific reservoir solutions are the models that we will use to match reservoir behavior, thereby inferring reservoir parameters we do not know in advance. (Horne 1995)

The onset of computing capabilities and development of specialized softwares featuring an interactive environment made the interpretation process much more efficient. Although most of the hand-writing is the past, knowledge of the underlying principals still enhances the capabilities of the analyst. Most importantly it enables an insight into advantages and shortcomings of the different interpretation techniques that have to be considered in the process.

3   RESERVOIR GEOLOGY AND MAIN PARAMETERS

The reservoir is situated in the western part of the Vienna basin, in the vicinity of an important Schrattenberg fault system. Both synthetic and antithetic faults associated with this system have had a major role in forming of the hydrocarbon reservoirs. According to the geological survey of this area, all reservoir traps investigated so far have been of structural-type origin. (Klímová et al. 2015) The Prušánky field comprises of 5 wells that develop neogene sedimentary formations and reach the maximum depth of around 1.6 km in the bottom of the baden sandstone. Above this horizon there are layers of age of sarmat, pannon and dak. The horizon investigated in this study is the top baden sandstone layer named 1. baden.

Apart from the flow rate and pressure data from gauges and flowmeters, other sources of information have to be introduced to make the well test analysis possible. They include particularly well logging and laboratory data. Some of the unknowns can be later computed via selected correlations. The layer parameters can be found in Table 1, the fluid parameters in the Table 2.

For gas viscosity correlation by Carr et al was used. Rock compressibility was computed based on the Hall's correlation.

Table 1.   Reservoir layer parameters.

Parameters	Values	Units
Net thickness	6	m
Porosity	0.26	dec. fraction
Layer pressure	138.75	bar
Layer temperature	51	°C
Water saturation	0.4	dec. fraction
Gas saturation	0.6	dec. fraction
C_r	4.7436e-5	Bar^{-1}

Table 2.   Fluid parameters.

Parameters	Values	Units
Check pressure	138.75	bar
Check temperature	51	degC
Gas specific gravity	0.567	sp. grav.
Gas density	98.7347	kg/m^3
Water salinity	20000	ppm
Rho w	1003.11	kg/m^3
U_g	0.0154723	mPa.s
B_g	0.0007448	m^3/m^{3}Vn
z	0.857508	
C_g	7.5507e-3	bar^{-1}
C_w	4.0786e-5	bar^{-1}
C_f	4.7436e-5	bar^{-1}
C_t	4.5942e-3	bar^{-1}

4   INTERPRETATION OF THE WELLTEST

The major goal of interpretation is to find feasible flow and boundary models and obtain qualitative and quantitative information about the reservoir parameters. There is not a single scenario that "fits all" to be followed to successfully finish this task. Instead various complementary methods which mostly provide different outcomes should be compared in order to find unbiased results.

4.1   Data preparation

The first step is the preparation of the flow rate and pressure data. After that we can visualize pressure (bar) and temperature (°C) on the Cartesian plot (Figure 2). The Cartesian plot should be also used to look for any irregularities. In Figure 2 we can see multiple peaks in both pressure and temperature curves. Because there is an anomaly in bottomhole

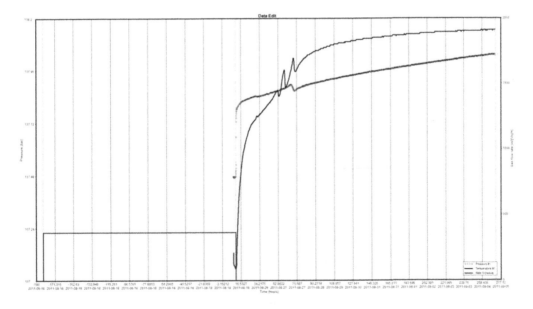

Figure 2.   Cartesian plot - data edit.

temperature as well, it infers a phenomenon inside the well rather than to be reservoir based. We can accept the probable scenario that the pressure dropped below the dew point during the production hence the condensate formed on the downhole gauge causing the anomaly. Although it will cause a major disruption in the interpreted data, it should not affect the interpretation process.

4.2 Model identification

The buildup can be divided into 3 stages: early-time, mid-time and late-time. Early time region is where wellbore storage (WBS) occurs, mid–time is the region where radial flow should occur and the late time is where boundaries manifestations can be found, if the test takes long enough time.

In this case the first investigated was the late time portion of the derivative in the type curve (TC) plot (Figure 3), since the boundaries seemed to be an easily recognizable feature. The best match was made with the curve that represented a model of 3 perpendicular boundaries. From the straight line part of the TC the first estimation of permeability (k = 97 mD) was done. Increase on the curve indicated the first boundary in the distance of roughly 240 m from the well. Since the mid-time region was not entirely clear a closer 1st boundary couldn't be ruled out. Although the tool (TC) wasn't able to find a perfect match, it indicated at least 3 boundaries were found during the test.

After the radial flow regime was identified, a different set of TC's was used to match the early time data (Figure 4). This tool didn't provide a good match to WBS, however the first estimation of skin factor was conducted. The value of skin factor was 2.68.

Next step, flow regimes were identified on log-log plot, (Figure 5). The unit slope provided a good match with the late-time portion of the derivative, meaning there was either water support or a compartmentalized reservoir system. For the early time the unit slope was used once again for the identification of the wellbore storage (WBS). The wellbore storage dominated area was distorted and the resulting values of WBS constant (Cs) and of well volume were clearly overestimated. Non-ideal WBS model was therefore accepted instead of the classic WBS model.

In the mid-time region, the radial flow was not clearly identified because the zero slope could be positioned anywhere along a wide strip within the derivative data points. The estimated values of permeability could vary between 90 and 130 mD. Hence the primary estimate of 97 mD from TC plot remained unchanged.

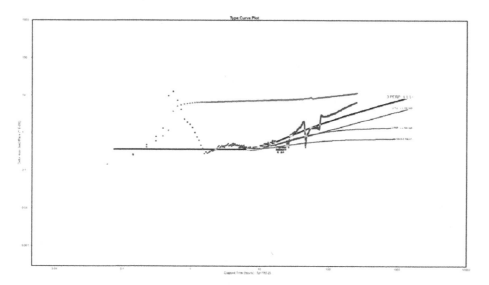

Figure 3.   TC matching – boundaries.

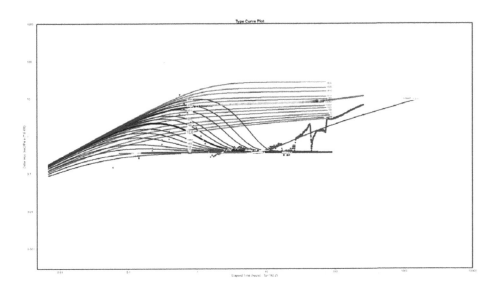

Figure 4.   TC matching – WBS.

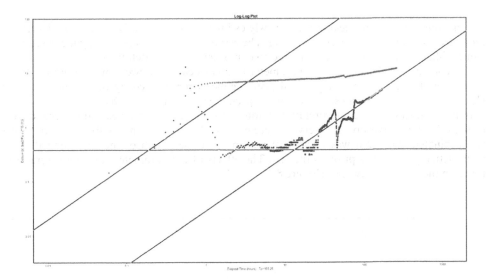

Figure 5.   Log-log plot - flow regimes identification.

Based on the information from the previous stages of the interpretation process the list of potential models was narrowed and several viable candidates, including e.g. closed system were simulation tested. During simulation-matching process the most feasible match was made with the radial homogenous flow model with non-ideal WBS and the 2-cell compartmentalized boundary model (Figure 6). The boundary model considered two compartments, divided by a semi-permeable barrier. The gas flowed from the neighboring cell to the main, which was penetrated by the well. The other three boundaries were sealing faults.

4.3   *Parameter estimation and verification of the test results*

As soon as the model was specified, the model parameters had to be refined to closely conform to the data. The boundaries were estimated in distances of 107 m, 107 m, 107 m and

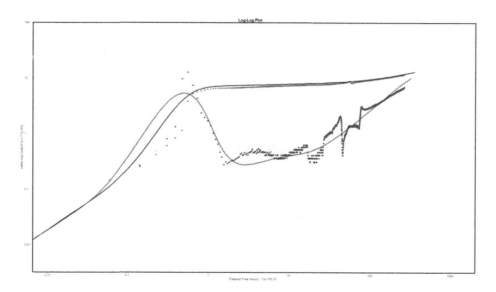

Figure 6.   Log-log plot – simulation.

275 m from the well. Permeability was newly estimated to be 125 mD which is an increase to the preceding estimate from TC matching. The estimation of skin also slightly increased to the value of 4.36. This value was later altered to 6 in order for the match the P_i.

For comparison the radial flow parameter estimates can be viewed on the semi-log (radial flow) plot (Figure 7) where the radial flow regime is represented by the straight line. The Horner method also offers estimation of its specific parameters (see Table 3).

In the final stage of the interpretation, parameters such as the volumes and respective OGIP of the cells that make up the reservoir were estimated and the transmissibility between them was evaluated. The goal was to achieve reasonably close match of the computed P_i to the value actually measured prior to the test. The effects of the parameter variations were monitored on the test overview plot (Figure 8).

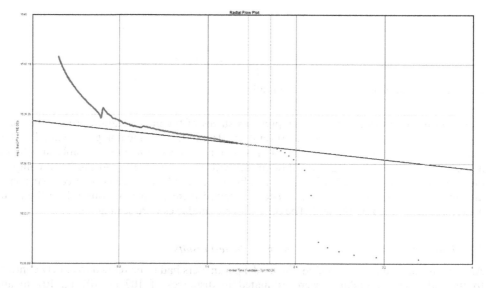

Figure 7.   Semi-log plot – Horner method.

Table 3. Horner method results.

Parameters	Values	Units
k	125.1219	mD
kh	750.7314	mD.m
p*	137.8589	bar
Rinv	1416.1135	m
FE	0.6773	
dps	0.1242	bar
S	4.36	

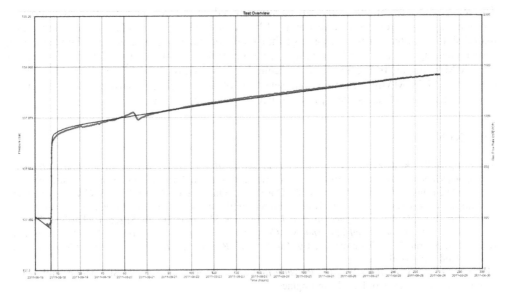

Figure 8. Test overview.

Table 4. Test results.

Parameters	Values	Units
Cs	1.05	m³/bar
V	139.0598	m³
Cphi	-1	bar
Tau	0.4	h
k	125	mD
kh	750	mD*m
S	6	
L1	107 m	m
L2	107 m	m
L3	107 m	m
L4	275 m	m
Area	81748.0135	m²
OGIP (main cell)	1.027e+007	m³ (Vn)
OGIP (neighbor cell)	6.9e+006	m³ (Vn)
Pi	138.7703	bar
Teff	15	m³/d/bar
dpS	0.171	bar

The OGIP of the main reservoir compartment was estimated at roughly 10 MMm^3 of gas. The neighboring cell's OGIP was estimated to contain 6.9 MMm^3 of gas. The final test results can be viewed in the Table 4.

The specific feature of this paper is in the fact, that the studied reservoir was already depleted at the time of this study. The estimated OGIP was 16.9 MMm^3 while the amount of gas actually extracted from the reservoir during its production period was 14.53 MMm^3. This gives us the recovery factor of 86 %. We can therefore check the validity of the estimate and compare it to different analyzed reservoir models such as closed system.

5 CONCLUSION

The main outcomes of the study can be summarized as follows: 1) estimation of important parameters such as reservoir permeability (125 mD) and skin factor (+6). 2) The positive estimated skin indicated formation damage that might be decreased by acidizing the well and in turn increasing the well's deliverability. 3) Comparing the estimated OGIP of the reservoir (16.9 MMm^3 of gas) to the actual gas volume that has been recovered from the reservoir until depletion (14.53 MMm^3) showed that selection of a two-cell compartmentalized reservoir model instead of a closed system provided better estimation of gas in-place.

REFERENCES

Bujok, P, Muller K & Zeman V. 1985. *Výzkum vlastností kolektorských vrstev a sond na podzemních zásobnících plynu.* Ostrava,. Skripta pro postgraduální studium. Vysoká škola báňská v Ostravě.

Dressler M., Kaňa J.; Krčál T., Novotný M., Obruča M., Onderka V., Staněk I. 1995. *Uskladňování plynů a kapalin v kolektorských vrstvách.* Sylabus pro rekvalifikační a inovační kurz. GeoGas s.r.o. Brno.

Earlougher, R.C., Jr. 1977. *Advances in Well Test Analysis.* SPE monograph, SPE 5, Richardson, TX.

Horne R.N. 1995. *Modern Well Test Analysis, Computer-Aided Approach.* Petroway, Inc. Palo Alto, CA.

Klímová L. Horáček J., Piškulová Z., Buchta, Š. 2015 *Revize výpočtu zásob ložiska Prušánky k 31. 3. 2015* Hodonín, MND a.s.

Mattews, G. S. & D.G. Russell. 1967. *Pressure Buildup and flow Tests in Wells.* SPE Monograph Series. Vol. 1. Dallas: Society of Petroleum Engineers.

NOMENCLATURE

sw = software
MMm^3 = million cubic meters (of gas)
WBS = wellbore storage
TC = type curve/s
r = radius (m)
p = pressure (bar)
t = time
ϕ = porosity (decimal fraction)
μ = viscosity of the fluid (mPa.s)
c = compressibility
z = gas compressibility factor
k = permeability (mD)
kh = permeability-thickness (mD.m)
p* = false pressure (bar)
Rinv = radius of investigation (m)
FE = flow efficiency
dps = additional pressure drop due to skin

S (s) = skin factor
V = well volume (m^3)
Cphi = Phase redistribution pressure parameter
Tau = time duration of the WBS (h)
L1 = distance to boundary 1 (m)
L2 = distance to boundary 2 (m)
L3 = distance to boundary 3 (m)
L4 = distance to boundary 4 (m)
OGIP = Original gas in place (MMm^3)
Pi = Initial reservoir pressure (bar)
Teff = transmission efficiency (m^3/d/bar)
Ug =gas viscosity (mPa.s)
Bg = gas formation volume factor
Cg = gas compressibility (bar^{-1})
Cf = fluid compressibility (bar^{-1})
Ct = total compressibility (bar^{-1})
Cr = rock compressibility (bar^{-1})

Topical Issues of Rational Use of Natural Resources 2019 – Litvinenko (Ed)
© 2020 Taylor & Francis Group, London, ISBN 978-0-367-85720-2

Engineering design of the ejector system for liquefied natural gas (LNG) vapor discharge

V.A. Voronov & Y.V. Martynenko
Saint-Petersburg Mining University, St. Petersburg, Russia

ABSTRACT: The article offers a set of technological solutions to ensure safety during storage and low-tonnage production of liquefied natural gas (LNG). Tanks are equipped with gas discharge pipelines and safety valves to prevent planned and emergency pressure, through which the excess steam phase is discharged to the system torch or into the atmosphere. The efficiency of gas-equalizing tank strapping has been proven theoretically, including the ejector system, in order to reduce the loss of liquefied natural gas and energy consumption during its safe storage. The most productive aerodynamic scheme of the flow part of the liquid-gas ejector was chosen to significantly reduce the investment in construction and operation of cryogenic tanks. The efficiency and the principal possibility of implementing the method of a vapor discharge with an embedded ejector is confirmed by computational simulation in the ANSYS software.

1 INTRODUCTION

Much attention is given to operations of vapor phase dumping not only due to ecological and fire safety, but also because of the significant losses of the product. However, failure to observe safety regulations can lead to heavy losses, including cases of damages and leaks. Overpressure in the storage tank caused by damaged tools, pressure safety valves or operators' errors is one of the reasons of liquefied gases leakage.

LNG vapor dumping from the tanks through the gas-discharge pipeline and pressure safety valves takes place in the following emergency situations (SP 240.1311500.2015, 2016):

- in case of total loss of vacuum in the insulation cavity or destruction of thermal insulation;
- when the pressure regulator jams in the open position;
- when the flowing pressure is exceeded in emergency situations.

Filling tanks which already contain some remains of fuel with a new portion of LNG with density and temperature different from these parameters of the remaining fuel can lead to the formation of stratified macro-layers in the reservoir and to intensive vapor generation. This situation also requires an accident-prevention dumping system.

Cryogenic tanks provide not only emergency, but also periodic gas discharges, such as the discharge of cold vapors during tanks refueling or during technological operations.

For this reason, the pneumatic-hydraulic circuit of cryogenic reservoir should provide an emergency shutdown system (ESD system), which controls the pressure and level of LNG in the tank.

According to safety rules, dumping of the excess of LNG vapor phase, which becomes burnt in the flare, happens in case of the excess of surplus pressure relatively to nominal by the given number. LNG is discharged into the gas bleeder into the atmosphere directly, if the required pressure reduction does not occur. This leads to significant losses of expensive fuel. In case the reservoir is equipped with treating iron, the separation of vapor phase excess is made by compressors, and that leads to additional energy costs.

2 EJECTOR SYSTEMS FOR VAPOR DUMPING

Pump-ejector systems are suggested as an alternative to existing control and dumping systems. They are included into the treating iron of the reservoir in purpose of operational pressure maintaining by dumping the fuel to separator using the energy of high vapor phase pressure. Such systems are distinguished primarily by simplicity, small capital investments and the absence of the need for additional expenditure of energy from outside.

The choice of the pump jet system depends on the parameters of the exhausted fluid such as pressure, compression ratio, flow rate. The liquid-gas jet is of the greatest interest for cryogenic LNG tanks. The operating principle of this device is based on the following: low-pressure liquid phase comes to mixing chamber because of its vacuum area caused by high-pressure flow of the fluid. As pressure safety valves in LNG reservoir trip when the vapor exceeds the pressure level, there is no need in high-pressure flow.

There is a suggestion to include liquid-gas ejector 3 to the treating iron of cryogenic LNG reservoirs so that the liquid phase would come to the mixing chamber because of the high-pressure flow of vapor phase, which occurs during the emergency or periodical dumping. The two-phase flow after passing liquid-gas ejector is going to the separator 4. After that, gas phase can be used for technological supply of the reservoir park, while LNG (if it fits the required pressure and temperature parameters of the system) can be transported back to reservoir 1. Thus, pumping with the economically efficient fluid jet device can decrease the volume in the reservoir and lessen the excess pressure (Figure 1) (Voronov & Martynenko, 2018).

3 METHOD FOR CALCULATING THE DETERMINING PARAMETERS FOR THE EFFICIENCY OF THE APPLICATION OF LIQUID-GAS EJECTOR

The main indicators which show the efficiency of liquid-gas ejector usage are the coefficient of ejection, capacity and efficiency factor depending on the chosen aerodynamic configuration of the device. Scientific works show different effective aerodynamic configurations for the calculation of geometrical design factors of the flow part of ejector, however the calculation of fluid jet devices with cryogenic fuel as a work environment has never been done before (Donec, 1990).

The main indicators of liquid-gas ejector:

• inner displacement rate of liquid, transported by liquid-gas ejector (equation 1):

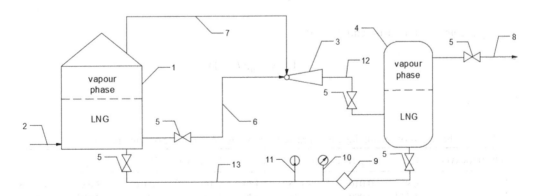

Figure 1.   Schematic diagram of the ejector system for LNG vapor dumping.
1 - LNG tank, 2 - LNG pipeline, 3 - liquid and gas ejector, 4 - separation tank, 5 - gate valve, 6 - LNG supply line to the ejector, 7 - steam line to the ejector, 8 - pipeline for technological needs, 9 - a filter, 10 - a manometer, 11 - a thermometer, 12 - a line for supplying a two-phase flow, 13 - LNG piping.

$$Q_f^{optm} = \frac{Z.P_0.T_f}{(P_g - P_s).T_0} \cdot Q_{f0},$$ (1)

where z-LNG compressibility factor; P_2-saturated vapor pressure of working fluid, MPa, P_g-gas pressure (LNG vapor), MPa; T_f-liquid temperature (LNG), K; Q_{f0}-flow rate of pumped liquid, м³/u; T_0, P_0 -absolute indictors of temperature and pressure, K, MPa.

• reduced working fluid pressure (equation 2):

$$\bar{P} = \frac{P_g - P_s}{P_{f-}P_s},$$

Where P_f — the absolute pressure of the LNG at the $\in$ of the ejector, MPa. (2)

• coefficient of ejection (inner coefficient of displacement rate (equation 3)):

$$u_{optm} = u_{max} \cdot \left[1 - \exp(-B_{optm} \cdot \sqrt{\bar{P} - 1})\right]$$ (3)

where u_{max}-maximal coefficient of ejection, depending on the type of aerodynamic system (Table 1), B_{optm}-empirical constant, depending on the type of aerodynamic system (Table 1). coefficient of pressure recovery (equation 4):

$$\psi_{optm} = \psi_{max} \cdot \left[1 - \exp(-a_{optm} \cdot \sqrt{\bar{P} - 1})\right]$$ (4)

where ψ_{max} -maximal coefficient of pressure recovery, depending on the type of aerodynamic system (Table 1), a_{optm}-empirical constant, depending on the type of aerodynamic system (Table 1).
• flow rate of working gas flow (equation 5):

$$Q_g = \frac{Q_f^{optm}}{u_{optm}}$$ (5)

• gas compression ratio (equation 6):

$$\varepsilon = 1 + \psi_{optm} \cdot (\bar{P} - 1)$$ (6)

Table 1. The results of calculating the characteristics of the ejector at the optimal regim.

Schematic type	u_{optm}	Q_g, м³/u	φ_{optm}	ε	P_{mix},MPa	N, MW	N_{reduce}	η, %
1	0,955	0,010	0,146	1,140	1,166	0,016	3,489	15,938
2	1,114	0,009	0,082	1,079	1,138	0,014	2,992	9,871
3	1,401	0,007	0,068	1,06	1,131	0,011	2,380	10,132
4	1,812	0,005	0,045	1,044	1,121	0,008	1,840	8,631

• mixture pressure on the way out of liquid-gas ejector (equation 7):

$$P_{mix}^{optm} = P_f + \psi_{optm} \cdot (P_g - P_f)$$
(7)

efficiency coefficient (equation 8):

$$\eta = \frac{Q_f \cdot P_f}{Q_g \cdot (P_g - P_{mix})} \cdot \ln \frac{P_{mix}}{P_g}$$
(8)

power, spent on compressing while the ejector is working, with accuracy up to the efficiency coefficient of the pump (equation 9) (SP 240.1311500.2015, 2016):

$$N = P_g \cdot Q_g$$
(9)

reduced power spent on compressing (equation 10):

$$\bar{N} = \frac{N}{Q_f \cdot (P_f - P_s)}$$
(10)

The difference of the suggested technological scheme is that the working fluid for ejector is gas, because due to its energy LNG is passing to jet device and then makes a two-phase flow. That's why the described calculation method pays attention to pressure and flow rate parameters' changes relatively to gas and liquid.

4 THEORETICAL CALCULATION OF THE CHARACTERISTICS OF LIQUID-GAS EJECTOR AND SELECTION OF THE AERODYNAMIC SCHEME.

The following parameters were determinate for the calculations:
• LNG mixture composition $(CH_4 = 95,5\%., C_2H_6 = 2,3\%; N_2 = 1,7\%$ and others.) (GOST R 56851-2016, 2016);
• LNG compression ratio in the enter to the ejector $z = 0,03671$;
• LNG temperature in the enter to the ejector $T_f = 140k$;
• absolute LNG pressure in the enter to the ejector $P_f = 1,102 MP_a$;
• vapor pressure in the enter to the ejector (after pressure increase) $P_g = 1,547 MP_a$;
• fuel flow rate $Q_f 0 = 4,9 M^3/u$;
• heavy vapor pressure $P_s = 0,6375 MPa$.

Starting parameters correspond to national standard (GOST R 56851-2016), where thermodynamic LNG properties are listed for different mixture types. There was considered a case which caused the discharge of the gas phase due to temperature increase from 120 K to 140 K and, correspondingly, to the increase of saturated vapor pressure.

The choice of the aerodynamic scheme of the ejector depends on the most effective efficiency factors, the ejection factor and the reduced power. It is recommended to choose the aerodynamic scheme, which provides the highest efficiency at the design value of the compression ratio.

The results of calculating the main parameters, taking into account the coefficients for the optimal regime, are presented in Table 1.

For a clear analysis of the results, the dependences of the mixture pressure at the outlet from the ejector, the reduced power and the efficiency of the ejection coefficient on the four aerodynamic schemes under consideration are plotted (Figure 2).

Thus, the aerodynamic scheme №1 is the most effective, since its geometric characteristics make it possible to obtain the highest values of efficiency, power and pressure of the mixture.

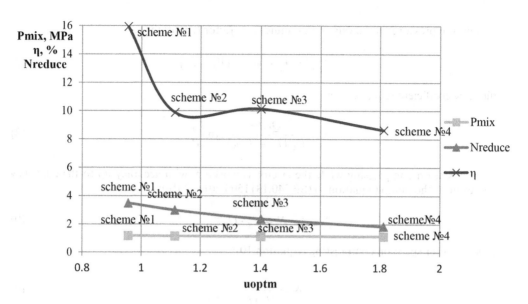

Figure 2. The results of calculating the characteristics of the ejector at the optimal mode.

According to the chosen scheme, the parameters of the ejector were calculated at the limiting and disruptive operating conditions of the liquid-gas ejector and a pressure characteristic was constructed (Figure 3).

5 CALCULATION OF GEOMETRICAL PARAMETERS

To determine the geometric parameters of the selected ejector circuit, it is necessary to calculate the nozzle diameter (equation 11) and the diameter of the working chamber (equation 12). The remaining dimensions of the flow section are selected according to Table 6 (Donec, 1990).

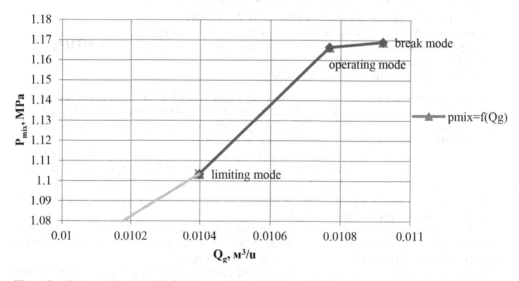

Figure 3. Pressure characteristic for the aerodynamic scheme No.1.

$$d_0 = \sqrt{\frac{4Q_g}{\pi \cdot \mu_{noz} \cdot i}} \cdot \sqrt{\frac{\rho_g}{2(g-f)}}, \tag{11}$$

where μ_{noz}-coefficient of nozzle flow rate ($\mu_{noz} = 0,79$ for scheme №1), i-number of nozzles ($i=7$).

$$d_{wc} = d_0 \cdot \sqrt{i} \cdot m \tag{12}$$

where m- coefficient for scheme 1(m=3,2).

The initial parameters were used to calculate the geometric dimensions of the flowing part of the liquid-gas ejector. The results are shown in Table 2.

The calculated geometric values correspond to the constructive scheme of the flow part of the ejector, shown in Figure 4.

6   COMPUTATIONAL SIMULATION OF PROCESSES IN ANSYS SOFTWARE

The next stage of the research included the computational simulation of process, namely the interaction of liquefied natural gas flows and excess vapors in the liquid-gas ejector. The

Table 2.   Geometrical dimensions of the flowing part of liquid-gas ejector.

d_0,m	a,m	D_0,m	D_1,m	d_2,m	d_{wc},m	l_{noz},m	l_1,m	l_2,m*	l_{diff},m*
0,0075	0,0023	0,1012	0,0237	0,0316	0,0355	0,142	0,112	0,379	0,056

* $l_1 = \frac{d_{wc}-d_2}{2tg\frac{\gamma_1}{2}}$, $l_{diff} = \frac{2.d_{wc}-d_{2y}}{2tg.\frac{\gamma_{diff}}{2}}$.

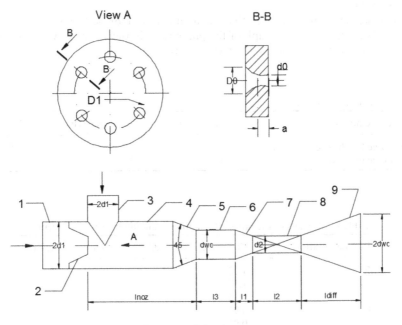

Figure 4.   Constructive diagram of the flow part of the ejector.
1 – nozzle inlet working gas; 2 – nozzle; 3 – receiving pipe; 4 – the pre-chamber; 5 – confuser; 6 – mixing chamber; 7 – a convergent phase; 8 – cylindrical portion; 9 – diffuser area.

ANSYS Fluent software was used for this purpose. It is one of the best tools for gas and hydrodynamics computer simulation and includes a large set of carefully verified models that provide relatively fast accurate results. Start conditions and dimensions taken for the process simulation were calculated in the previous stages of the research.

3D model that reproduces the calculated parameters liquid-gas ejector was constructed using SolidWorks simulation automation systems. After that, the resulting geometry was imported into the ANSYS Mechanical software to construct the computational gridwhich was built as follows: part of the pre-chamber was determined as accurately as possible, the size of the element did not exceed five millimeters in general and ten millimeters in the compressing pipe. These dimensions of computational part elements were chosen on the basis of preliminary calculations on the study of the convergence of the results of calculations of the gas flow with turbulence parameters close to 10% to the parameters obtained in the task. The total number of cells did not exceed eight million, which allowed to obtain a solution in a short time. The resulting mesh work was imported into the ANSYS Fluent software.

The stationary process was modeled in ANSYS Fluent, which was based on the calculation of pressures and velocities in absolute form. Multiphase flows were described in the Volume of fluid (VOF) model with an explicit solution scheme. Energy conservation and calculation equations were also used. The common model k-omega sst was used as a viscosity model. This model provides good results in the simulation of internal and jet flows.

Natural gas in gaseous and liquid phases models were loaded as materials. On the entrance sections nozzle excess vapour and LNG conditions has been specified, which are given in Table 3. On the outlet nozzle of the liquid-gas ejector the project pressure was set to 1,16 MPa.

As a result of excess vapors and liquefied natural gas mixing process modeling, parameters of the absolute pressure were obtained in different zones of the structural part of the liquid-gas ejector. The calculated parameters are presented in Table 4, where P1 is the pressure in the ejector from the gas nozzle to the jet, MPa, P2 - the pressure in the ejector in the pre-chamber, MPa, P3 - the pressure in the confuser, MPa, P4 - the pressure in the mixing chamber, MPa, P5 - the pressure in the confuser section, MPa, P6 - the pressure in the cylindrical section, MPa, P7 - the pressure in the diffuser section, MPa.

The pressure distribution in the liquid-gas ejector zones is shown in the graphs for visual analysis (Figure 5, 6).

The numerical experiment proved that the liquid-gas ejector performs its core function. In the graph of velocities and the graph of the phase ratio in the steady flow, the dissolution of LNG in the flow of excess vapor is observed (Figure 7, 8).

Table 3.   Start conditions.

The name of the nozzle	Speed, m/s	Pressure, MPa	Temperature, K
The inlet nozzle for the excess vapour	18	1,547	140
The inlet nozzle for LNG	1,5	1,102	120

Table 4.   Results of the flows interaction simulation in ANSYS Fluent.

Part designation	The average value of the absolute pressure, MPa
P_1	1,22
P_2	1,17
P_3	1,15
P_4	0,97
P_5	0,92
P_6	0,89
P_7	1,15

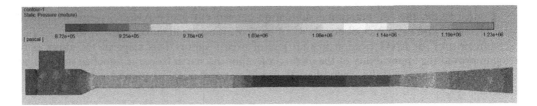

Figure 5.   The graph of the pressure distribution in the liquid-gas ejector.

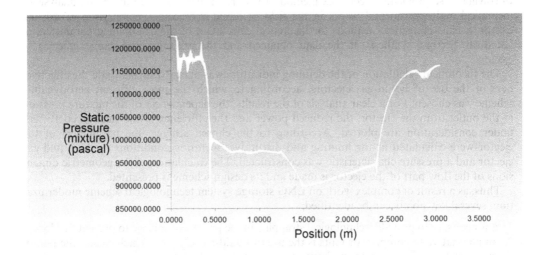

Figure 6.   The diagram of pressure distribution to the length of the liquid-gas ejector.

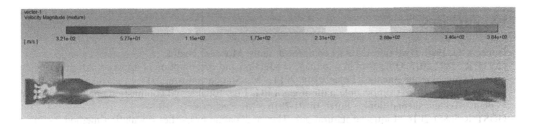

Figure 7.   The graph of speeds in the liquid-gas ejector.

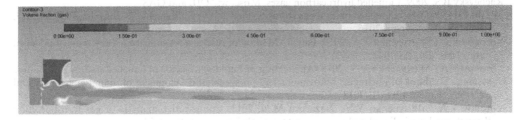

Figure 8.   The graph of phase relations in steady-state flow in the liquid-gas ejector.

Consequently, the computational simulation carried out in the ANSYS software showed the efficiency and the principal possibility of implementing the liquid-gas ejector with specified characteristics for LNG ejection in the flow of excess vapor. This stage of the study once again proved the applicability of the method of vapor discharge from the liquefied natural gas tank using the energy of the steam phase flow and an economical jet-pump.

7 CONCLUSIONS

In solving these problems, a complex method of research is used: generalization and analysis of theoretical and experimental works in the field of liquid-gas ejectors, comparison of existing ejector systems, numerical methods for theoretical calculation of the determining parameters of liquid-gas ejectors, synthesis of the data obtained and the existing aerodynamic schemes of ejectors.

The theoretical calculation of the defining indicators was made to substantiate the effectiveness of the use of liquid-gas ejectors, according to which the most efficient aerodynamic scheme was chosen. For a clear analysis of the results, the dependences of the mixture pressure at the outlet from the ejector, the reduced power and the efficiency of the ejection coefficient under consideration are plotted. According to the chosen scheme, the parameters of the ejector were calculated at the limiting and disruptive operating conditions of the liquid-gas ejector and a pressure characteristic was constructed. The calculation of the geometric dimensions of the flow part of the ejector is made and its design scheme is presented.

Thus, as a result of complex work on LNG storage system technological scheme modernization, several key points can be identified:

1. it is necessary to provide technological strapping in the process of spillage to prevent fuel loss;
2. an alternative to compressor units is the use of liquid-gas ejectors, which ensure the pumping of fuel into the separator without energy costs from the outside;
3. the use of a gas equalization system in LNG tank farms with an implemented ejector will ensure safety in fuel storage, while reducing capital and operating costs.

REFERENCES

Donec K.G. 1990. Hydraulic drive jet compressor units. Moscow: Nedra.

GOST R 55892-2013. Objects of low-tonnage production and consumption of liquefied natural gas. General technical requirements. - Enter. 01-06-2014. - M.: Standartinform, 2014.

GOST R 56021-2014. Combustible natural gas LPG. Fuel for internal combustion engines and power plants. Technical conditions. - Enter. 01-01-2016. - M.: Techexpert, 2016.

GOST R 56851-2016 "Liquefied natural gas. Thermodynamic properties calculation method." - Introduced 18.01.2016. - Moscow: Standartinform, 2016.

Morozova N. In. Korshak A. A. 2013. Universal characteristics of liquid-gas ejectors. Oil and Gas business (electronic scientific journal), №6:368–383.

Rachevsky B.S. 2009. Liquefied hydrocarbon gases. Moscow: "OIL and GAS".

SP 240.1311500.2015 "Liquefied natural gas storages. Fire safety requirements." - Introduced 31.08.2015. - Moscow: Tekhexpert, 2016.

Voronov V.A., Martynenko Y.V. 2018. Patent "Method of resetting a vapor from a reservoir of liquefied natural gas" RU 2677022.

Voronov V.A., Martynenko Y.V. 2018. Application of liquid-gas ejector in liquefied natural gas storage systems. The book of abstracts "Innovations and prospects of development of mining machinery and electromechanics: IPDME-2018": 120.

Voronov V.A., Martynenko Y.V., Nazarova M.N. 2017. Comparative analysis of gas transportation through pipelines in liquid and gas-air conditions, Neftegaz.ru, №10:20–22.

Topical Issues of Rational Use of Natural Resources 2019 – Litvinenko (Ed)
© 2020 Taylor & Francis Group, London, ISBN 978-0-367-85720-2

Study of water-containing ability of gas-liquid cement mixtures

M.Y. Merzlyakov, I.A. Straupnik & D.V. Serbin
Saint-Petersburg Mining University, Saint-Petersburg, Russia

ABSTRACT: The aim of this paper is to determine the methods of cement mixtures water yield decreasing and sedimentation resistance improving. For that purpose we have run experimental studies to learn the degree of aeration and hydroxyethyl cellulose influence on water-containing ability of the cement mixtures. We have found that the addition of solid fillers in the gas-liquid cement mixtures composition prevents the liquid flow out from their cellular structure. There is information on the effect of cement mixture mechanical activation on the filtrate loss value. We have solved the problem of obtaining a stable cement composition by using various methods to reduce water yield.

1   INTRODUCTION

The quality of well construction depends on wide specter of geological and technical factors. Proper accounting of these factors allows increasing well life time. It is very important to make the right cement materials selection which provides to sustain reliable sealing of the space between casing pipes and borehole wall.

The cement mixtures must meet the following requirements, first of all, geological conditions, technology and economic specifics of ongoing drilling project.

One of the main factor which affects on the quality of well lining is the cement mixtures water-containing ability. The intensification of sedimentation and water yield processes does not allow the cement to reach the design elevation. That is why the risks of reservoir fluids annular flows occurrence and clogging of the productive strata are increasing. In the wells having high level of the drift angels the process of cement water loosing provokes the creation of extended canals which break the casing sealing. In the permafrost conditions the water-containing cement volumes can be the reason of casing pipes collapse, so the well lining strength and life time decrease.

The practice of gas-liquid cement mixtures (GLCM) using show that they are reliable and effective materials for the qualitative annulus well space sealing in low reservoir pressure conditions, during fluid losses and permafrost presence (Detkov, 2012).

2   RESEARCHES METHODS

Base liquid and the knitting substance are prepared for the purposes of density, water yield, multiplicity factor and water-containing ability. After that they mechanically mix together and with the air forced by means of the propeller mixer. Mixtures hashing continues within 5 minutes before obtaining uniform structure with sizes of bubbles less than 1 mm. GLCM compositions are filled in to a filter-press cell (water yield) or in measured cylinders (water separation) after establishment of values of their multiplicity factor and density. Definition of water yield of cement mixtures is carried out with a pressure of equal 0.7 MPa.

GLCM have small mobility because of special cellular structure. Nevertheless, hashing of mixes at a low gradient of pressure or when forcing by the pump happens decrease in their viscosity. Thus, cement mixtures become more fluid (Detkov, 2005). Proceeding from the aforesaid, it is possible to claim that GLCM spreadability does not affect success of their pumping. It is recommended to define mixes mobility on a AZNII cone without gas phase inclusion in their structure (Detkov, 2005).

Researches of cement mixtures properties are made by standard techniques with use of methods of experiments planning. On the basis of the obtained experimental data correlation dependences of varied factors influence on the studied parameters are established.

3 RESEARCH OF AERATION AND WATER-CONTAINING ADDITIVES PRESENCE INFLUENCE TO CEMENT MIXTURE FILTRATE LOSS

The concentration and the type of foaming agent can affect on the integrity and the quality of the GLCM and the cement stone made of it as well (Gorbach, 2014). A gas-phase involving becomes possible due to foaming agent specific properties. The foaming agents are surface active agents (SAA). The SAA adding allows increasing GLCM sedimentation resistance and prevents the coalescence of gas bubbles (Detkov, 2005). It is necessary to admit that SAA concentration reduction is limited by the cement mixture ability to hold the gas-phase within its own structure. On the other hand, high foaming agent concentration leads to increase in setting and curing time and degrade the concrete quality. The optimal concentration of SAA in cement mixtures is 0.5 to 2% (Yakovlev, 2000). However, the increase in the solid-phase specific surface in GLCM leads due to adsorption to reduction of the SAA molecules proportion needed for cellular structure creation. In our research we have used a term 'multiplicity factor' for the evaluation of gas content. It can be counted the following way:

$$k = \frac{\sum V_{GLCM}}{\sum V_{CM}} \qquad (1)$$

where $\sum V_{GLCM}$ = the total volume of the GLCM after aeration, m^3; $\sum V_{CM}$ = the total volume of the GLCM before aeration, m^3. The optimal GLCM multiplicity factor is in the range from 1.5 to 5 (Yakovlev, 2000).

Water yield of cement mixtures is called: "low" if the volume of released filtrate does not exceed 50 sm^3 per 30 min; "medium" – from 50 to 500 sm^3 per 30 min; "high" – more than 500 sm^3 per 30 min (Apaev, 2018). The laboratory studies show that the involving of gas-phase into the cement mixture allows to improve its water-containing ability and to reduce the volume of released filtrate (Figure 1). Tested GLCM had the multiplicity factor equal to 3.6. Gas-phase injection leads to increasing of mixture total volume that let us to reduce the water

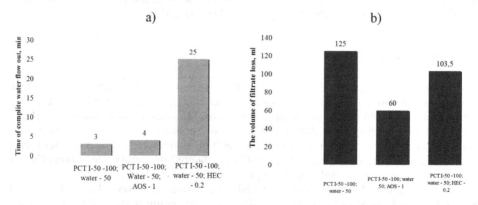

Figure 1. The dependence of the full water drainage time (a) and the amount of released filtrate (b) on the cement mixture composition (components content, mass parts, m.p.).

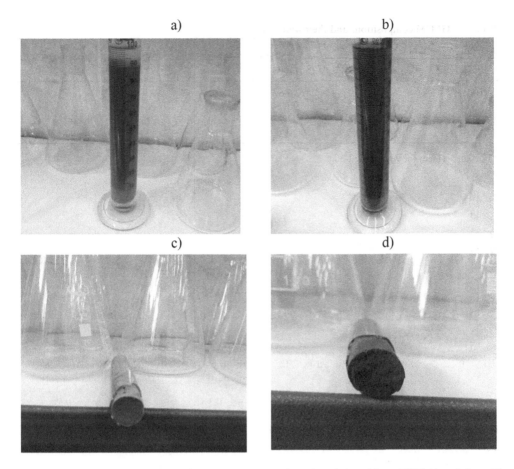

Figure 2. Investigation of cement mixtures water separation: a) and c) – without HEC; b) and d) – with adding of 0.2% HEC.

quantity for its preparation. The cellular structure formed by alpha olefin sodium sulfonate (AOS) increases GLCM water-holding ability.

Besides, reduction of the filtrate volume can be reached using water-holding additives, for example, polymers. Based on the research we decided to use hydroxyethyl cellulose (HEC).

According to Figure 1 aeration and HEC combined use allow to improve stability of liquid-phase in the cement mixture and to reduce the amount of free water by gas-phase.

The HEC using allows getting the mixture with improved sedimentation resistance properties and without water yield. The Figure 2 shows the cement mixture without structure forming additives. Three hours after GLCM mixing water yield was about 10%. The same mixture was put in a horizontally placed container. After 48 hours we analyzed the formed cement stone (Figure. 2c). Its cross section in the upper part was slightly flattened that indicates about water separation during the process of solidification. The reverse result was achieved by adding HEC (Figures 2b, 2d). There are some compositions of the GLCM in the Table 1. The graph on the Figure 3 shows a dependence of the water yield from the HEC concentration. Due to HEC concentration increasing GLCM air entrainment, mobility and water yield decrease. In the first mixture during first 12 minutes free water completely filtered out but in the others mixtures it was possible to significantly increase the water-containing ability and maintain a low density through the use of AOS and HEC.

The HET adding allows increasing the stability of cellular structures having reduced syneresis.

Table 1. GLCM compositions and their water yield.

#	Composition, m.p.				Multiplicity factor	Spreadability, sm	Water yield, ml/30'	Specific water yield, ml/30'
	PCT I-50*	Water	HEC	AOS				
1	100	50	0.2	1	2.36	18.0	-	-
2	100	50	0.3	2	2.67	17.0	26	113.52
3	100	50	0.4	2	2.42	16.5	21	84.00
4	100	50	0.5	2	2.12	16.0	13	46.90

* Portland cement for well-lining I-50.

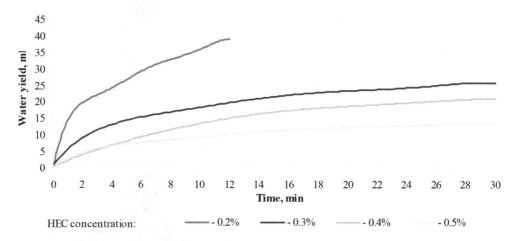

Figure 3.   The influence of the HEC concentration in the GLCM on the value of water yield.

We have determined the value of specific water yield for greater objectivity:

$$B' = \frac{B}{n} \tag{2}$$

where B – water yield measured by the means of a filter-press at a pressure difference of 0.7 MPa, sm^3 per 30 min; n – the ratio of the mixing liquid volume to the cement mixture volume, $n = V_L/V_{CM}$.

We have established that low HEC concentrations (less than 0.3 m.p.) led to GLCM complete water yield during first 25 min after the start of the experiments. Using of the polymers and aeration allowed reducing the filtrate loss (down to 13 sm^3 per 30 min).

4   STUDY OF THE INJECTED FILLERS EFFECTING ON GLCM WATER-CONTAINING ABILITY

A research (Yakovlev, 2000) shows that GLCM have high sedimentation stability. An air-entraining additives and solid-phase content effect on mixture stability. The presence of solid particles with certain properties, structures, shapes and sizes helps to reduce water losses from the GLCM cellular structure (Merzlyakov, 2015). As such additives it is possible to use highly dispersed silica, alumina or other materials with the crystal and chemical similarity with the GLCM components.

In our researches we have used aluminosilicate hollow microspheres (ASHM), polypropylene fiber, microsilica and expanded perlite or circulate (Figure 4).

854

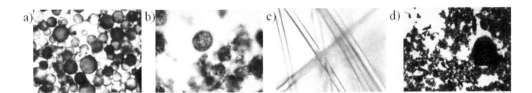

Figure 4. Microphotographs of the GLCM additives: a – ASHM (×400); b – circulate (×400); c – polypropylene fiber (×400); d – microsilica (×100).

ASHM are hollow silicon-containing balls which are produced by coal combustion. This additive is characterized by low cost, low density, high melting point, high strength and chemical resistance. ASHM are widely used as a facilitating additive in cement mixtures.

Perlite is an igneous rock produced by volcanic eruptions. In our country there are perlite deposits in the North Caucasus and Kamchatka, Buryatia and Magadan region. The main feature of volcanic glasses is their porosity caused by the presence of dissolved water. As a result of the volcanic glass fast heating, it softens and swells due to the water conversion into steam. Circulite like the ASHM is used as facilitating additives to the cement mixtures.

Polypropylene fiber is a fibrous material made of polypropylene granules. It is used as a reinforcing additive to cement-based building materials; it increases their water resistance, crack resistance, adhesion and resistance to mechanical stress and influences of aggressive environment. Increasing of the fiber concentration and its length leads to a preparation deterioration of cement mixtures due to the formation of clumps. The optimal fiber content in the mixture should not exceed 1 % (Bekbaev, 2017). We have used the fiber 2 mm length.

Microsilica is fine silica of dark gray color. The analysis of the used microsilica (Figure 4-d) has showed the presence of isometric (spherical), lamellar and needle-shaped particles. Adding microsilica can improve the cement materials properties (strength, frost resistance, water resistance, etc.), as well as reduce their cost.

The adding of these additives in the GLCM helps to reduce water yield (Table 2, Figure 5). According to this criterion, mixtures containing circulate and microsilica have showed the best results, mixtures with ASHM and polypropylene fiber – the worst. Circulate and microsilica represented by fine particles of different shapes have a high specific surface area that binds free liquids in the mixture structure. However, its density increases, gas involving ability and spreadability reduce. The addition of polypropylene fiber, on the contrary, allows reducing the mixtures density and slightly affecting their mobility. Even a small concentration of this fibrous material contributes to a positive effect.

Table 2. Properties of the GLCM with solid additives.

#	Composition, m.p.								k			
	PCT I-50	Water	HEC	AOS	ASHM	C.*	P.f.*	M.*		ρ, kg/m^3	Spreadability, sm	B/B', ml/30'
1	100	50	0.5	1	-	-	-	-	2.12	870	16.0	13.0/ 46.9
2	90	50	0.5	2	10	-	-	-	1.79	765	15.0	11.0/ 41.3
3	95	50	0.5	2	-	5	-	-	1.38	935	14.0	9.5/ 30.9
4	90	50	0.5	2	-	10	-	-	1.18	935	<10.0	3.5/ 11.4
5	90	50	0.5	2	-	-	-	10	1.50	965	14.5	8.5/ 26.8
6	99.5	50	0.5	2	-	-	0.5	-	2.00	765	15.5	8.5/ 34.0

* C. = circulate; P.f. = polypropylene fiber; M. = microsilica.

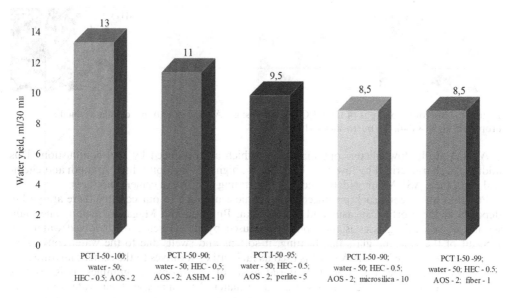

Figure 5.   The influence of the GLCM composition on water yield.

5   STUDY OF THE INFLUENCE OF MECHANICAL ACTIVATION ON GLCM WATER YIELD

The structure formation of cement stone occurs under the influence of various physical and chemical processes: dissolution of the binder particles, hydration, coagulation etc. Depending on the degree of hydration, there are four stages of structure formation (Kruglickij, 1974), the analysis of each other allows to understand which processes prevail over others (Figure 6):

The *first stage* is associated with intensive hydration, at which the maximum speed of the process is achieved. Coagulation of colloidal particles begins, and by the end of this stage, a spatial framework of the coagulation structure formed cement particles covered by hydrated formations. The electrical conductivity, heat dissipation, pH increase and the system most compacted state is reached. The value of the contraction reaches its maximum value by the end of the first stage due to the rearrangement of the structural elements and the liquid in the cement mixture.

In the *second stage*, the spatial coagulation structure is formed. At the beginning of this stage, the hydration and processes associated with the construction of the cement stone frame slow down, hydration has small values, and destructive phenomena occur, which are expressed by a decrease in the value of the module of rapid elastic deformation on the structure curve. Presumably, this destruction is due to the transformation of hydrate structures into thermo-dynamically more stable forms.

The *third stage* proceeds with the formation of a spatial crystallization framework. There is an intensification of the structure formation process, which is indicated by a high increase in the degree of hydration, modulus of elasticity and heat dissipation.

During the *fourth stage*, there is the greatest increase in the strength of the formed structure and a slight increase in elasticity and hydration. Presumably, the increase in strength is caused by the development of the structure of hydrosilicate materials. There are a crustification of the cement stone main crystallization skeleton and formation of new minerals in it. Internal stresses grow causing destruction and the decline of elasticity and strength.

In the research (Detkov, 2003) and earlier ones V.P. Detkov presents studies of the mechanical activation effect on the cement mixture properties. According to this method, the cement mixtures were allowed to stand for 60...80 minutes, after that they were stirring for 5...10

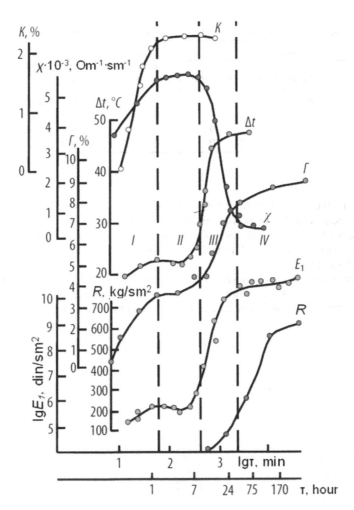

Figure 6. The complete curve of the cement stone structure formation process (E1) during hardening under normal conditions for 28 days and the curves of heat release (Δt, °C), changes in the degree of hydration (Γ), tensile strength (R), specific electric conductivity (χ) and contraction (K) (Kruglickij, 1974).

minutes. The formed thickened structure was destroyed until a homogeneous consistency was obtained. Activation occurred at the end of the first stage of structure formation, when the spatial dispersion structure formed by astringent materials is most susceptible to activating effects. According to the results of the experiments, it was found that there were minor changes in the spreadability before and after the activation, the setting time and water yield decreased, sedimentation stability and density increased. In the formed cement stone gas permeability decreased, strength and adhesion increased.

Figure 7 shows the dependence of the total liquid flow out of the cement mixture on the time after activation. There is a decrease in the amount of filtrate released due to the binding in the process of hydration of the free liquids in the mixture and the formation of a spatial framework of the coagulation structure that prevents the water filtration from the mixture.

Different plasticizers can be used to preserve the mobility of cement mixtures and reduce the amount of liquid for their preparation. We have used lignosulfonate (LS) as a plasticizing additive in the studied GLCM (Table 3). Over time, there is a decrease in the multiplicity of GLCM, which entails an increase in the mixtures density. At the same time, there is an increase in the mobility of the mixture and a decrease in water-containing ability. This

Table 3. The influence of LS and mechanical activation on the GLCM properties.

# PCT I-50	Composition, m.p.					Time after acti-vation	k	ρ, kg/m3	Spreadability, sm	B/B', ml/30'
	Water	HEC	AOS	Fiber	LS					
1 100	40	0.5	2	-	0.5	-	1.70	1000	14.0	5.5/20.3
2 100	40	0.5	2	-	0.5	60	1.38	1230	15.0	7.5/21.8
3 100	45	0.5	2	0.5	0.5	-	1.74	880	15.5	9.5/35.3
4 100	45	0.5	2	0.5	0.5	60	1.62	990	16.0	11.0/36.7

phenomenon is due to the fact that during the period of 60 minutes of exposure there is an increase in dispersion of lignosulfonate particles, which leads to an increase in the interaction of the plasticizer with the liquid and solid phases of the mixture and a decrease in the efficiency of the foaming additive. The addition of polypropylene fiber to the GLCM has not given us the desired results.

It is not advisable to use mechanical activation and lignosulfonate together to reduce the level of GLCM water yield. In the future, we want to use other plasticizing and water-containing additives to increase the mobility of cement mixtures and minimize their filtrate loss.

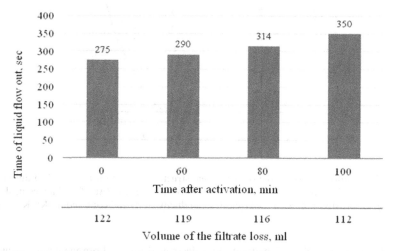

Figure 7. Dependence of the total liquid flow out of the cement mixture (PCT I-50 and water, water/cement ratio is 0.5) on the time after activation.

6 CONCLUSION

Based on the results of the research we can conclude the following:

- GLCM, unlike conventional cement mixtures, have a lower concentration of free fluid in its volume;
- The cellular structure prevents the liquid flow out from the cement mixture;
- The addition of solid fillers in the GLCM composition contributes to better retention of the liquid in the structure of the mixture;
- The water-containing ability of the mixtures is affected by the amount, shape and size of the solid phase;

- The use of cement mixtures mechanical activation can reduce their water yield;
- GLCM containing fine particles of lignosulfonates have better mobility, but over time their ability to involve the gas-phase and bind the liquid-phase contained in the cellular structure decreases.

We plan to continue the research, which will be aimed at the development of new light cement mixtures with a low water yield to ensure high-quality well lining in difficult mining and geological conditions.

REFERENCES

Apaev A.A. Issledovanie fil'tracionnyh svojstv tamponazhnyh rastvorov / A.A. Apaev, A.A. Kabdushev // Molodoj uchenyj. – 2018. – №18. – S. 39-42.

Bekbaev A.A. Issledovanie armirovannyh oblegchennyh tamponazhnyh materialov / A.A. Bekbaev [i dr.] // Nanotekhnologii v stroitel'stve: nauchnyj internet-zhurnal. – 2017. – № 4. – t. 9. – S. 131-148.

Detkov V.P. Fiziko-himicheskaya mekhanika – osnova dlya razrabotki tekhnologii cementirovaniya skvazhin v usloviyah Krajnego Severa / V.P. Detkov, A.R. Hismatulin // Stroitel'stvo neftyanyh i gazovyh skvazhin na sushe i na more. – 2003 – № 7. – S. 31-37.

Detkov V.P. Izolyacionnye raboty v skvazhinah razlichnogo naznacheniya: Monografiya / V.P. Detkov. – Krasnodar : EHkoinvest, 2012. – 484 s.

Detkov V.P. Ocenka davleniya sil poverhnostnogo natyazheniya v aehrirovannom tamponazhnom rastvore/V.P. Detkov, A.R. Hismatulin // Stroitel'stvo neftyanyh i gazovyh skvazhin na sushe i na more. – 2005. – № 5. – S. 28-32.

Gorbach, P.S. Vliyanie penoobrazovatelya na svojstva peny i penobetona / P.S. Gorbach, S.A. Shcherbin // Vestnik TGASU. – 2014. – № 5. – S. 126-126132.

Kruglickij N.N. Fiziko-himicheskaya mekhanika tamponazhnyh rastvorov / N.N. Kruglickij [i dr.]. – Kiev : Naukova Dumka, 1974. – 288 s.

Merzlyakov M.Y. Issledovanie tekhnologicheskih svojstv aehrirovannyh tamponazhnyh sostavov s vklyucheniem v nih polyh alyumosilikatnyh mikrosfer / M.Y. Merzlyakov, A.A. Yakovlev // Vestnik PNIPU. Geologiya. Neftegazovoe i gornoe delo. – 2015. – № 14. – S. 13-17.

Yakovlev A.A. Gazozhidkostnye promyvochnye i tamponazhnye smesi (kompleksnaya tekhnologiya bureniya i krepleniya skvazhin) / A.A. Yakovlev. – SPb : SPGGI (TU), 2000. – 143 s.

Topical Issues of Rational Use of Natural Resources 2019 – Litvinenko (Ed)
© 2020 Taylor & Francis Group, London, ISBN 978-0-367-85720-2

Innovative approaches to light-alloy drill pipes modification for drilling in abnormal operating conditions

M.N. Nazarova
Saint Petersburg Mining University, Saint-Petersburg, Russian Federation

A.I. Shakirova, R.A. Ismakov, A.Kh. Agliullin & N.K. Tsenev
Ufa State Petroleum Technological University, Ufa, Russian Federation

ABSTRACT: Abnormal drilling conditions are often due to near ultimate loads such as high temperatures, vibrations, and harsh environment. In this context an issue of a need for equipment modification, which not only extends the service life, but ensures safe drilling, is thrown into sharp relief. Development and use of a drill string made of aluminum pipes are of great practical importance to improve drilling efficiency and failure-free operation in different mining and geological conditions. Light-alloy drill pipes have a number of advantages, including the possibility for transportation to remote drilling areas, and development of long vertical wells. During well construction the drill string interacts with aggressive environment, which affects pipe operational performance and drilling process as a whole. Superalkalinity drill fluids with pH more than 10 impact on materials the tubular billets are made of. In order to increase the efficiency of pipe serviceability, not only composition of drilling fluids, but also the pipe material processing methods shall be modified and changed. This article presents the existing methods of light-alloy drill pipe material processing and the ways to increase the corrosion resistance of D16T, the most common alloy, to aggressive environment

1 INTRODUCTION

Special attention in the drilling technology is recently given to application of usual commercial light-alloy drill pipes (LADP) and high-strength drill pipes made of D16T, 1953T1, AK4-1T1 aluminum alloys, which advantages are not only almost 3 times lower specific gravity as compared to steel pipes, but demonstration of increased resistance in aggressive environment. LADP doesn't need for additional surface protection that reduces drilling costs significantly. A need for failure-free operation under adverse climatic and geological conditions has appeared along with well development in the Arctic zones and High North regions. The existing alloys currently don't solve these problems in full. In order to solve the above problems modification aluminum alloys structure is needed that is capable of improving corrosive resistance (Jian et al., 2014; Braham, 1997; Nadzhafobodi, 2013), one of the service life values.

The purpose of the article is to increase LADP corrosive resistance in aggressive environment by improving its material. The study of corrosion rate and weight loss of D16T alloy sample in aggressive environment depending on interaction time is the problem of the article.

1.1 Developing SPD technics

Strengthened submicrocrystalline structure of aluminum alloy, which the LADP are made of, can be obtained on the basis of severe plastic deformation methods (SPD), an equal-channel-angle pressing (ECAP) with small diameter samples for experiment and laboratory tests and on a commercial scale by local shear pressing (LSP) method. Both

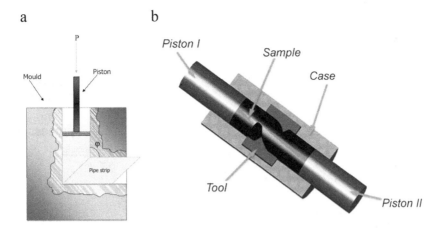

a b

Figure 1. Severe plastic deformation methods, a – equal-channel-angle pressing, b – local shear pressing.

methods represent shear strain, differing against each other in that equal-channel-angle pressing requires repeated pressing, and LSP allows for forming deformations in entire molded piece (pipe length is 9-12 meters) (Figure 1 a, b) (Islamgaliev et al., 2003; Bakhtizin et al., 2015; Furukawa et al., 2015).

The samples of bearing alloy D16T were obtained after using SPD method and after being processed by ECAP method. In order to determine corrosion rate of the samples before and after processing the procedure consisted of examination of 6 samples of aluminum alloy in aggressive environment with pH=10, which is close to real conditions. The test samples with precomputed surface area and measured weight on an analytical balance, are put into measuring flask with clayless biopolymer solution with pH=10 (Water + Na2CO3 (0.1%) + NaOH (0.2%) + bactericide (1.5%) + PAC -LV (0.4%) + Xanthan gum (0.8%).

In order to determine the loos in weight of the samples the rust is well removed by putty knife and hair brush. In case of a film or hard rust removal a special anti-corrosion agent is used. After that the samples are rinsed with tap and distilled water and thoroughly dried and weighed by analytical balance. Further processing of the results lies in calculation of corrosion rate of the samples using the formula 1 below (Latypov et al., 2017; Landau & Lifshits, 2005).

$$V_{cor} = (m_1 - m_2)/(S \cdot t), \; mg/m^2 \cdot hr \tag{1}$$

where m_1 is a sample weight before test, mg; m_2 is a sample weight after test, mg; S is a sample surface area, m_2; t is for time of interaction with environment, hr.

Sample surface area is calculated by formula 2:

$$S = \pi \cdot h(D_{outer} + d_{inner}) + \pi/2 \cdot (D_{outer} - d_{inner}), 10^{-6} \cdot m^2 \tag{2}$$

where h is sample height, mm; D_{outer} – outer diameter of the sample, mm; d_{inner} – inner diameter of the sample, mm.

The obtained results of weight loss and corrosion rate dependence on aggressive environment contact time are shown on the figures below 2-3, accordingly

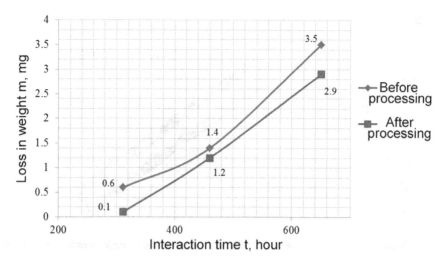

Figure 2. Loss in weight of D16T samples before and after processing depending on time of interaction with environment pH=10.

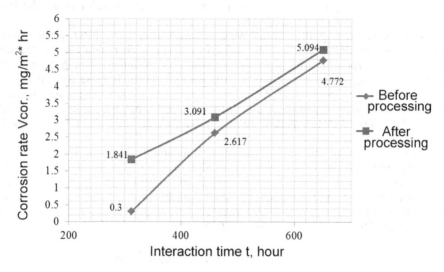

Figure 3. Corrosion rate of D16T samples before and after processing depending on time of interaction with environment pH=10.

2 CONCLUSIONS

During laboratory tests the results of loss in weight and corrosion rate of d16T aluminum alloy samples before and after ECAP were obtained. It was found that the best results were obtained after use of severe plastic deformation methods. Based on the results the corrosion rate decreased almost in 1.5 times as compared to reference material with alkaline medium for 650 hours of interaction.

The severe plastic deformation is recommended for possible improvement of corrosion resistance of aluminum alloys on a commercial scale.

REFERENCES

Bakhtizin R.N., Ismakov R.A., Nazarova M.N., Tsenev N.K., Yakhin A.R. 2015. Causes and features of the destruction of the elements of the layout of the drill string in modern conditions of geological and technical operation. Services in oil production. Ufa, USPTU: 25-33.

Braham V.J. 1997. Corrosion of Aluminium in Contact with Cutting Fluids/Electrochemistry of Corrosion. Newcastle University Library.

Furukawa M., Berbon P., Horita Z., Nemoto M., Tsenev N.K., Valiev R.Z., Langdon T.G. 2015. Production of Ultrafine-Grained Metallic Materials Using an Intense Plastic Straining Technique. Materials Science Forum. Volume: Towards Innovation in Superplasticity. 171-177.

Islamgaliev R.K., Yunusova N.F., Valiev R.Z., Tsenev N.K., Perevezentsev V.N., Langdon T.G. 2003. Characterisics of superplasticity in an ultrafine-grained aluminum alloy processed by ECA pressing. Scripta materialia, № 49, p. 467-472.

Jian L., SJianhua S., Xinmiao L., Yongqin Z., Li P. 2014. Development and Application of Aluminium Alloy Drill Rod in Geological Drilling. IGEDTC, *New International Convention Exposition Center; 23-25 May 2014.* Chengdu Century City: China.

Landau L.D., Lifshits E.M. 2005. Electrodynamics of continuous media. Science. Moscow: Russian Federation.

Latypov O.R., Tyusenkov A.S., Cherepashkin S.E., Bull D.E. 2017. Polarization studies of metals and alloys. Tutorial. Ufa, USPTU.

Nadzhafobodi R.A., Maceami A., Sharifi H. 2013. Corrosion of aluminum alloys., Lambert Academic Publishing. 320.

Information and kinetic approach to the estimation of the strength of the vessels working under pressure

V.V. Nosov
Peter the Great St. Petersburg Polytechnic University, St. Petersburg, Russia
Saint-Petersburg Mining University, St. Petersburg, Russia

E.V. Grigorev & E.R. Gilyazetdinov
Saint-Petersburg Mining University, St. Petersburg, Russia

ABSTRACT: Nowadays, the relevant task is to increase the useful life, reliability and efficiency of vessels working under pressure. The condition of pressure vessels that have worked for long periods is determined by the degree of development of defects formed during operation in welded joints and steel sheet elements. Due to the long useful life of vessels and devices, and the increasing accident rate of failures connected with the formation of operational defects in metal, the routine technical inspection of the structure becomes insufficient due to the large intervals between surveys and the use of outdated local control methods (Homera 2015-Chernyak 1975).

1 INTRODUCTION

The development time of the lamination is commensurable with the periodicity of the technical examination which necessitates the timely detection of damage to the metal of the equipment and the need to develop and improve methods for assessing the residual life based on a detailed analysis of the primary diagnostic information received from the test object during a routine technical examination. Obviously, different areas of the vessels undergo damage differently and by the volume of the same object defects of the type of separation can occur at different stages of their development.

Estimation of the resource of long-term working technical objects is currently carried out mainly by calculations for endurance or by methods of mechanics of crack development, including empirical coefficients that are unspecified characterizing the individuality of the state of the object. In an attempt to improve the accuracy of estimation based on non-destructive testing methods, due to metrological and strength heterogeneity, the uncertainty of the statistical approach and the insufficiency of the reference analogy are also encountered. A perspective approach here is based on the micromechanical model of AE parameters, combining statistical and physical approaches to diagnostics, as well as additionally "drawing" information on macro-, micro-, and nano-levels of strength studies to substantiate the choice of valuable diagnostic AE indicators.

According to the results of preliminary studies of AE, described in (Nosov 2016a, 2016b), the following conclusions were made:

1. Strength characteristics, parameters of the process of destruction and AE materials depend on the result of competition simultaneously occurring in the material processes of destruction and plastic deformation of structural elements.

2. The resource of the majority of long-loaded materials, structures and structures is determined by the process of micro-cracking, occurring in conditions of elastic deformation.

3. Destruction consists of two stages - finely dispersed (scattered throughout the object or locally grouped in the defect area) accumulation of the concentration of micro-cracks, and

enlarged localized discontinuity (formation or growth of the crack), flowing elastically or plastically.

4. The first stage consists of two stages with different kinetic and dissipative properties: kinetically inhomogeneous, associated with the destruction of equal-strength structural elements with increased dissipative properties, and kinetically homogeneous micro-crack formation with moderate dissipative properties sufficient for stress relaxation.

5. Acoustic emission of elastically deformed materials is mainly associated with the process of micro-cracking. The number of signals from plastic deformation of overstressed structural elements is relatively small. To reduce their destabilizing effect on the results of resource prediction, frequency and amplitude filtering should be used.

6. The heterogeneity of the strength state of complexly loaded objects is associated with the non-uniformity of the field of mechanical stresses and the structural heterogeneity of the weld material.

7. For samples made without pronounced concentrators, the stage of non-uniform destruction is the longest (up to 60% of the duration of the first stage). For samples with defects, the stage of kinetically inhomogeneous fracture does not exceed 30%. After the formation of cracks, the samples are destroyed homogeneously, as with a stress concentrator.

8. The destruction of samples with a stress concentrator is kinetically homogeneous. After the formation of a crack, its development occurs with the unloading of the stressed zones of the weld and plastic restructuring of the material structure.

9. The effect of a non-developing crack on the fracture process occurs elastically (without prolonged AE signals) and contributes to reducing the degree of heterogeneity of fracture.

10. The consequence of the strength inhomogeneity of the states of the structural elements is a decrease in the amplitude of the signals during loading and the irreproducibility of the AE activity upon repeated loading (the Kaiser effect).

11. The latency of the AE with repeated loading of structures is informative and indicates the remoteness of the moment of accumulation of the critical concentration of micro-cracks near the concentrator and the absence of a connection of dangerous defects in the material.

12. Valuable diagnostic indicators X_{AE}, Y_{AE}, W_{AE} are proposed and diagnostic indicators of the state of difficultly loaded control objects are formulated, which can become the basis for creating a methodology for assessing the performance of objects operating under conditions of uncertain stress-strain and structural state.

2 DESCRIPTION OF INFORMATION AND KINETIC APPROACH

From the position of the micromechanical model (Nosov 2012), predicting the resource for safe operation according to the cyclic strength criterion will be reduced to predicting the moment t' of accumulation of the critical concentration of micro-cracks C'. At the stage of homogeneous destruction, the time dependence of the concentration C(t) of micro-cracks is described by the law

$$\frac{dC}{C_0 - C} = \frac{dt}{\theta_{cp}(t)} \tag{1}$$

where C_o = the initial concentration of structural elements in the material before destruction, is the average waiting time for the destruction of one structural element, given by the Zhurkov's formula (Buylo 2017), with the initial condition C(0) = 0.

$$C(t) = \frac{C_0 t}{\tau_0} \exp\left(\frac{\gamma\sigma - U_0}{KT}\right) \tag{2}$$

where τ_0 = the period of atomic oscillations (a relatively stable value approximately equal to 10^{-13} s); U_0 = the energy of sublimation (the energy of detachment of an atom when a body transitions from a solid to a gaseous state); γ = a structurally sensitive parameter; σ = the effective stress; t = the current time; K = the Boltzmann constant; T = the absolute temperature.

Similar to (2), with uniform loading in the regime of constant speed $\dot{\sigma}$ stress growth, when $\sigma = (\dot{\sigma}t)$,

$$C(t) \approx \frac{C_0 KT}{\tau_0 \gamma \dot{\sigma}} exp\left(\frac{\gamma \dot{\sigma}t - U_0}{KT}\right). \tag{3}$$

The parameters τ_0 and U_0 are the most conservative and weakly depend on the state of the structure; they are determined by the characteristics of the interatomic interaction of the structural element; the values of the parameter γ are a characteristic of the nanostructure of the material, which is weakly sensitive to its chemical nature, but together with stresses σ, the parameter γ reflects the strength individuality of the structural element, and the parameter $\gamma\sigma$ / (KT) included in Zhurkov's formula is actually a parameter of its strength state.

According to the micromechanical model with correct AE testing, the primary parameter of AE (for example, the total number of registered pulses N_Σ) will be proportional to the concentration of microcracks and, at the stage of uniform destruction, be described by time dependence:

$$N(t) = k_{AE}C(t) = k_{AE}\frac{C_0 KT}{\tau_0 \gamma \dot{\sigma}} exp\left(\frac{\gamma \dot{\sigma}t - U_0}{KT}\right) \tag{4}$$

where k_{AE} = the acoustic emission coefficient (the proportion of signals that have passed amplitude, frequency and time filtering).

The micromechanical model allows the parameter γ to be determined directly from the AE experiment, since it associates the primary AE information with the damage parameters. From (4), it follows that at a constant and known stress growth rate, when $\sigma = (\dot{\sigma}t)$, the stage of uniform destruction will be characterized by a straight line plot on the graph of the number of pulses versus stresses $\ln N_\Sigma (\sigma)$, represented in semi-logarithmic coordinates.

In assessing the endurance of vessels and pipelines, calculations are based on approximated fatigue curves and limit stress diagrams based on the results of destructive fatigue tests of standard samples and described by equations with empirical coefficients (Doc. 26.260.004-91. 1992).

The parameters of the homogeneous destruction of the temporal dependence of the logarithm of the number of AE pulses $\ln N_\Sigma (t)$ determine other diagnostic parameters

$$X_{AE} = \frac{d}{dt}\ln(N_\Sigma(t)) = \gamma\dot{\sigma}/KT \tag{5}$$

$$W_{AE} = \frac{d\ln N_\Sigma}{dK_N} = \gamma\sigma/KT \tag{6}$$

where K_N = the load factor (the ratio of the diagnostic load to the working).

Prediction of the resource N of the object being diagnosed by the criterion of cyclic strength according to the approach is proposed to be conducted on the basis of the Zhurkov fatigue curve equation (Nosov 2012)

$$N = \frac{N_B}{exp\left(\frac{\gamma\sigma}{KT}\right)} \tag{7}$$

$$N_B = \omega_N \frac{C^*\tau_0}{C_0} exp\left(\frac{U_0}{KT}\right) \tag{8}$$

Where N_B = constant of the material, temperature and frequency ω_N of its loading, is determined from the results of cyclic testing of samples and the construction of a fatigue curve. The parameters τ_0 and U_0 included in the formula are the most conservative and do not depend on the state of the structure.

The formula (7) for calculating the resource of a technical object in an expeditious, non-destructive way and the parameters determined on a specific object is one of the results of the approach that has received theoretical and experimental substantiation.

The article describes an information-kinetic approach to the assessment of the strength state of structural materials by acoustic emission parameters, combining macro-, micro-, and nano-levels of strength research, outlines the sequence for evaluating the strength state of test objects. The described approach allows the assessment of the resource of the object through the parameters evaluated directly on the diagnosed object during AE testing and eliminates the need to be guided by empirical coefficients of fatigue curves obtained during destructive testing of samples with properties different from the properties of a real object.

3 EXPERIMENTS AND RESULTS

To demonstrate the above-described method for predicting the lifetime of technical objects with structural heterogeneity, we give an example of AE tests of the hydrogen sulfide purifier absorber with a solution of monoethanolamine, described in detail in (Homera 2015). Figure 1 shows a fragment of the apparatus and the areas on which the ultrasonic inspection was additionally carried out.

The primary AE information obtained from the hydrostatic tests of the absorber contained data on the time, amplitude and number of emissions of each recorded pulse, as well as the numbers of sensors that sequentially recorded each AE event. During hydrotesting, 4,126 events were recorded, each of which was accompanied by signals on 4–7 sensors (a total of 23,245 signals).

The average number of signals, average and maximum values of X_{AE}, Y_{AE}, the minimum static safety factor of (9) and the minimum and maximum residual resources were determined by the absorber zones, after which the correlation coefficients of parameters with belt number were found (Table 1).

$$S_{st} = \frac{\sigma_{usm}}{\sigma_w} \approx \frac{\sigma_{uss}}{\sigma_w} \frac{Y_R}{Y_{AE}} \tag{9}$$

where σ_{usm} = the ultimate strength of the material of the product, σ_{uss} = the ultimate strength of the sample of material of the product, σ_w = the nominal value of the operating voltage in the belt, Y_R = parameter of curve fatigue.

The largest number of signals (maximum activity) was recorded by channels №№TAE 1 (N_Σ = 2116 signals), 3 (N_Σ = 2116), 4 (N_Σ = 1809), 2 (N_Σ = 1510), 5 (N_Σ = 1480), 30 (N_Σ = 1461), 6 (N_Σ = 1377) - all on belt № 8, 20 (N_Σ = 1241)-belt №. 7, the most "dangerous" was the zone in the area of TAE № 30 with the minimum ordinate of the installation, equal to 12 cm (Figure 1a) and X_{AE} = 0.097 s^{-1}, which demonstrates the effectiveness of the information and kinetic approach in relation to assessing the state of the absorber considered. The information content of the described parameters X_{AE} and Y_{AE} in relation to the strength assessment was also repeatedly confirmed when analysing the results of destructive testing of samples of welded joints of various shapes with various defects (Nosov 2016b).

Further analysis of the results of AE tests was conducted with respect to specific sources of AE signals using their AE localization and UT identification in the following sequence:

1. Localization of AE sources;
2. Isolation of areas of accumulation of AE sources at various stages of loading;
3. Determination of the time dependence of the AE parameters on each of the selected areas and the allocation of temporary areas of uniform destruction;

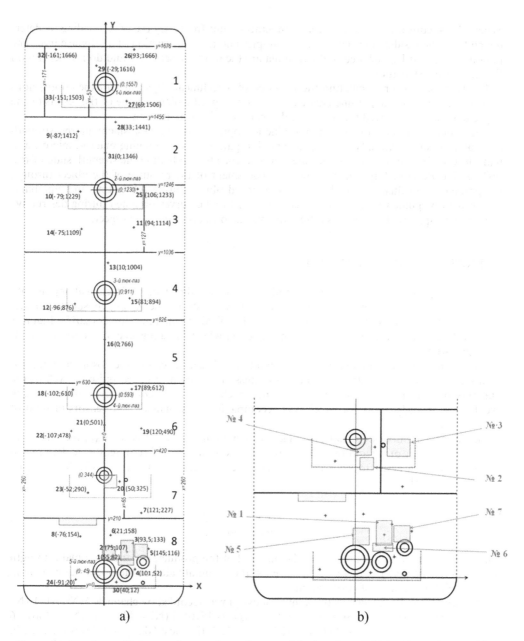

Figure 1. Numbers of belts, numbers and coordinates of the location of TAE, (a) numbers and locations of sections of additional ultrasonic testing (b) of hydrogen sulfide cleaning absorber.

4. Processing data on the parameters of the micromechanical model;

5. Discussion and comparison of the obtained results with experimental data obtained during ultrasonic testing and previous surveys;

6. Prediction of the duration of damage accumulation to critical values.

Post-processing of AE information was solved in the program MathCad. For all 4126 AE events, the coordinates of the sources were determined, which were then plotted on the body scan in the AutoCad Mechanical program (Figure 2). The figure shows the zones of accumulations of AE sources located on the 7th and 8th belts. In higher zones, signal sources are

Table 1. Correlation of the main control parameters with the number of the absorber belt.

Parameter which determines X_{AE}	$N_{\Sigma avg}$*	X_{AEavg}, c^{-1}	X_{AEmax}, c^{-1}	Minimum static safety factor	Maximum residual resource, years	Minimum residual resource, years
Number of signals	0,896	0,984	0,925	-0,8455	-0,965	-0,965
Total amplitude of signals	-	0,705	0,956	-0,8071	-0,959	-0,959
Amount of emissions	-	0,657	0,693	-0,5924	-0,828	-0,828

* $N_{\Sigma avg}$= average number of signals registered by one channel of the belt.

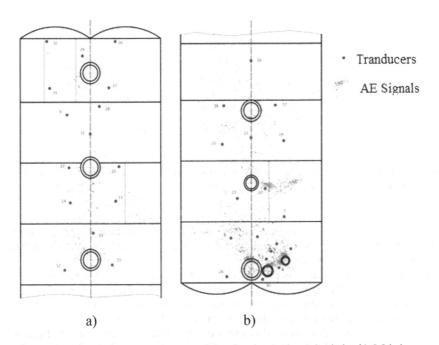

a)   b)

· Tranducers

AE Signals

Figure 2. Location of AE sources on the scan of the absorber body: a) 1-4 belts; b) 5-8 belts.

dispersed, which, from the point of view of the micromechanical model, indicates the occurrence of non-localized destruction in these areas, which does not pose a danger to the integrity of the structure.

Y_{AE} values were determined for individual zones with known defect characteristics. For this purpose, a picture of the location and a scheme for conducting a UT were combined, and parts of the AE event clusters at sites №№ 1–7 were considered (Figure 3).

The same time range for the stage of homogeneous destruction and close values of the Y_{AE} strength state indicator (0.0155 MPa^{-1} and 0.0135 MPa^{-1} for belt 7 and 0.0701-0.0772 MPa^{-1} for belt 8) are signs of the same damage structures within one belt, while higher values of the Y_{AE} parameter indicate a more unfavourable strength state of belt 8. If for sections 3 and 4 this is confirmed by the USI results that revealed irregular discontinuity in both areas, then for the lower belt full control USI house conducted only for the most active site of AE №1. From our point of view, the most dangerous zone in the area of TAE№30 (by activity of the channel of this TAE was in sixth place) by the survey was missed. Localization showed that this source was between the 30th and 24th

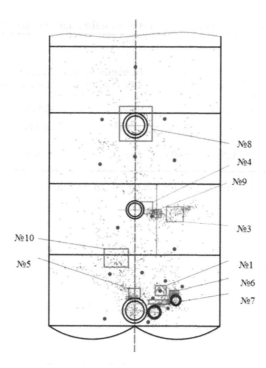

Figure 3. Areas of AE-activity in zones №№ 1-7 of the control of ultrasound and areas of additional analysis № № 8,9.

TAE. However, practice has shown (Nosov 2016b) that the damage to the metal of the vessel always increases with decreasing level, which confirmed the representativeness of the parameters X_{AE} and Y_{AE}.

Moreover, other location clusters of AE events, visually observed on the scan of the absorber body with the applied location pattern of AE sources, were also considered. Figure 3 shows three additional analysed plots. Cluster number 8, as can be seen from the figure, is located around the manhole. Presumably, the acoustic emission signals at the site under consideration were also initiated by bundles, since the manhole is located near the mass transfer device, where, as a rule, the processes of development of the bundles take place more actively than in the areas more distant. The analysis of the time dependence of the signals according to the above method indicates the strength state of section №. 8, similar to sections №№. 3,4. On the plot № 10, allocated on the body of the apparatus, the parameter Y_{AE} is equal to 0.0091 MPa^{-1}. In comparison with all considered clusters, section №10 is characterized by the most favourable strength state, however, its value close to critical $Y_R = 0.011$ MPa^{-1} for a defect-free metal indicates the presence of a developing defect in or near the weld. In this case, the type of defect may differ from delamination and may be a crack or a weld defect of a technological nature. As numerous examples of AE analysis of the behaviour of defects in welds have shown, the time dependences of damage accumulation in them have the same character, and the degree of danger of such defects can be estimated by the parameters of the micromechanical model (Nosov 2015a, 2015b).

The described examples demonstrate the promising application of the proposed approach for the detailed analysis of the primary AE information from the test object.

4 CONCLUSION

The article describes an information-kinetic approach to the assessment of the strength state of structural materials by acoustic emission parameters, combining macro-, micro-, and nano-levels of strength research, outlines the sequence for evaluating the strength state of test objects. The described approach allows the assessment of the resource of the object through the parameters evaluated directly on the diagnosed object during AE testing and eliminates the need to be guided by empirical coefficients of fatigue curves obtained during destructive testing of samples with properties different from the properties of a real object. The given example of the use of the micromechanical model in the diagnosis of vessels working under pressure demonstrates its effectiveness in terms of assessing the strength state and residual life with defects of both fatigue and chemical origin under conditions of hydrogen absorption.

REFERENCES

Buylo, C.I. 2017. Physico-mechanical, statistical and chemical aspects of acoustic emission diagnostics: monograph. Rostov-on-Don:PublishingHouse of the Southern Federal University.

Chernyak, Ya.S., Durov, V.S. 1975. Repair work at oil refining and petrochemical enterprises.Moscow: Chemistry.

Document 26.260.004-91. 1992. Methodical instructions.Prediction of the residual life of the equipment to change the parameters of its technical condition during operation.

Homera, V.P., Rastegaev, I.A. 2015. On the issue of early diagnosis of dissemination in the walls of pressure vessels by ultrasonic and acoustic emission methods, *Control. Diagnostics* 1:82-90.

Nosov, V.V. 2012. *Diagnostics of machines and equipment.Tutorial.* St. Petersburg: "Lan" publishing house.

Nosov, V.V. 2016a. Principles of Optimization of Acoustic Emission Monitoring Technologies for Industrial Object Strength. *Defectoscopy* 7: 52-67.

Nosov, V.V., Samigullin, G.Kh.,Yamilova, A.R., Zelensky N.A. 2016b. Micromechanical model of acoustic emission as a methodological basis for predicting the destruction of welded joints *Oil and Gas Business* 14(1): 244-253.

Nosov, V.V., Nominas, S.V., Zelensky, N.A. 2015a´Evaluation of the strength of pressure vessels based on the use of the phenomenon of acoustic emission *Scientific and Technical Bulletin of St. Petersburg State Polytechnic University* 2(219): 182-190.

Nosov, V.V., Potapov, A.I. 2015b Acoustic Emission Testing of the Metal Structures under Complex Loading *Russian Journal of Nondestructive Testing* 51(1): 50–58.

Topical Issues of Rational Use of Natural Resources 2019 – Litvinenko (Ed)
© 2020 Taylor & Francis Group, London, ISBN 978-0-367-85720-2

Research of pressure gradient in a gas well working with a foaming agent

V.A. Ogai, A.I.U. IUshkov, N.Y. Portniagin, V.O. Dovbysh, J.M. Azamesu & A.A. Voropaev
Industrial University of Tyumen, Tyumen, Russia

ABSTRACT: In the process of development of gas fields as depletion of deposits there are a number of related problems. One of these problems, which eventually occur in almost all wells of any field, is the accumulation of fluid at bottom hole, which leads to a decrease in their flow rates or a complete stop. Currently, a large number of Cenomanian gas deposits in Russia are at the final stage of development, gas production from such deposits is accompanied by "self-kill" wells a column of liquid. One of the methods of solving this problem is the introduction of foaming agent into the well. For the study and subsequent modeling of gas-liquid flows with surfactant, taking into account the influence of pressure, temperature, water-gas ratio and other key parameters, an "Experimental facility to simulate gas-liquid flow and dynamic processes in the tubing of a gas well" was developed and implemented.

1 INTRODUCTION

In the process of development of gas and gas condensate fields as depletion of deposits there are a number of related problems. One such problem is the accumulation of fluid at the bottom hole, which leads to a decrease of their production rates, or well shutdown. It should be noted that the problem occurs in almost all wells completed as high-permeability and low permeability reservoirs (Lea et al. 2003, Lee et al. 2004).

As accumulating on the bottom hole of the well fluid can be condensation water is passed into the liquid phase in the upper parts of the wellbore, flowing down the tubing string coming from the reservoir brine of natural origin or technical water. These fluids can enter the well bottom hole at the same time in different quantities.

Fluid accumulation occurs due to insufficient flow rate of gas-liquid mixture in the production string and tubing strings, including the perforated interval or filter. This problem may occur if the increase in the share of liquid extracted from the reservoir fluid. Large amounts entering the bottom hole fluid are not in time be brought to the surface at the same flow rates of gas in the wellbore. Emerging hydraulic resistance leads to a decrease of well production on the gas and the accumulation of fluid on the bottom hole with a gradual complete killing of well hydrostatic pressure. In addition, due to the flooding of the reservoir gas and gas flow rate fall as a result of decreasing gassy bulk of reservoir rock and reduce permeability of the reservoir gas during the growth of the share of the liquid in the reservoir formation. Another common reason for "self-kill" wells is a gradual decline in production rate (velocity) of gas due to the inability to further reduce bottom hole pressures after the natural decrease in pressure in the reservoir. To continue to further reduce the mouth of well and bottom hole pressure, mainly, does not allow the compressor equipment. As a result, even at a low concentration of liquid in gas (e.g., condensation water), it is not made from the well and gradually builds up, leading to a drop in the rate of gas production and well shutdown.

Turning to the description of the problem of liquid loading at Russian fields, it should be noted that in recent years about 80% of natural gas is extracted from Cenomanian gas deposits in Russia (Sarancha et al. 2015). Considerable part of the Cenomanian gas deposits are in the later phase of development, which leads to the manifestation of the described problems (Izyumchenko et al. 2018).

The choice of wellbore intervention (WI) to combat the accumulation of liquid is due to both the technological features of the production process at a particular field, and the economic efficiency of their application. In addition to the previously mentioned WI, the technology of introducing foaming surfactants into the well is widespread in the world, which is characterized by a relatively low level of capital investments and a high level of efficiency, including economic efficiency (Kalwar et al. 2017, Rauf. 2015, Peyton et al. 2013, Schinagl et al. 2007). Surfactants can be introduced into the well as solid rods or pumped as liquid solutions. With proper selection of the composition (Huang et al. 2012) surfactants can be used to remove water (condensation and formation), gas condensate from both vertical and horizontal wells, at different pressures and temperatures (including high) (Alzhanov et al. 2018, Gcali et al. 2018, Omrani et al. 2016). In the Russian Federation considerable experience of application surfactant in various regions is accumulated: on fields of the North Caucasus, Krasnodar Krai, the Orenburg region, Far North (Yamburg, Urengoy, Bear, etc.) (Koryakin. 2016). It should be noted that in the production of Cenomanian gas at the bottom of wells, only water accumulates mainly (more often condensation), which creates favorable conditions for the use of surfactants in such wells (Panikarovskij et al. 2017).

The process of the well deliquification when entering the foaming agent is due to its interaction with the fluid and the upward flow of gas, which leads to the formation of foam, reducing the density of the gas-liquid mixture and the surface tension between the liquid and gas. As a result, the critical gas velocity required for fluid removal and well deliquification is reduced. In the Russian gas industry, the technology of injection of liquid foaming agents into the well has recently become widespread, which, in comparison with the introduction of solid surfactant rods, allows to ensure a stable concentration of the reagent in the accumulating fluid and automation of the technological process. The technology is also recommended for use if the value of the minimum gas flow rate of the well for the removal of liquid exceeds the maximum allowable flow rates, excluding abrasive wear of equipment and destruction of the bottom-hole formation zone (Izyumchenko et al. 2018). In this case, the injection of the foaming agent will remove the accumulation of liquid and will further facilitate the cleaning of the well face from the sand with foam, eliminate the formation of sand and clay plugs.

As production technique of Cenomanian gas wells with introduction of a foaming agent is becoming more common it's necessary to take into account a differential pressure in well. Today, in hydrodynamic simulation of gas field exploitation, the problem is the lack of multiparameter models (functional relations) describing the polyphase foamed flow in gas well with surfactants. These models represent multidimensional arrays (VFP-tables) and characterize differential pressures between the bottom and the mouth of well.

2 STUDY SUBJECT AND RESEARCH METHODS

Colossal experience in modeling two-phase gas-liquid flows in the conditions of Cenomanian gas wells with relatively low formation pressure has been accumulated by LLC "Gazprom VNIIGAZ". Especially for this was build experimental facility with a height of production string 30m and a working pressure up to 4 MPa (Nikolaev et al. 2013). A feature of the facility is the ability to maintain a constant level of gas and liquid consumption over a long period of time, which allows investigating gas-liquid flows in both unstable and steady-state conditions. The results obtained at the facility affected the idea of the hydrodynamics of vertical gas-liquid flows and showed significant differences from the results obtained previously by international research teams in the course of bench studies and theoretical justification (Izyumchenko et al. 2013). Contradictions are due to the fact that the physical parameters characteristic of Cenomanian gas wells at the final stage of development have specific features such as relatively low liquid content in the flow (WGR<10^{-5}), low formation pressure (up to 4 MPa), large diameters of production string (114 mm и 168mm) and etc. Previously, experimental studies with regard

to these physical features were almost not carried out. On the basis of theoretical justification, supported by a large number of experimental studies it was revealed that for the
conditions of the Cenomanian gas wells dimensionless pressure drops i in a steady vertical gas-liquid flow can be represented as a sum of two independent summands, one of
which is determined by the gas rate and is independent of the liquid flow rate and other
is determined by the liquid flow rate and is independent of the gas rate (Nikolaev et al.
2013):

$$i = \frac{\lambda}{2} Fr^* + kBu \qquad (1)$$

where λ – hydraulic resistance coefficient of the pipe for single-phase gas, k – dimensionless
empirical coefficient, and modified Frud's argument Fr^* and S.N.Buzinov's argument Bu
determined by the formulas:

$$Fr^* = \frac{\rho_g}{\rho_0} \bullet \frac{u^2}{gd} \qquad (2)$$

$$Bu = \left(\frac{\sigma}{\rho_l gd^2}\right)^{\frac{1}{2}} \bullet \left(\frac{w^2}{gd}\right)^{\frac{1}{3}} \qquad (3)$$

where ρ_a– gas density at operating conditions; ρ_0, ρ_l – pressure independent liquid density;
u – gas rate at operating conditions reduced to a unit cross-sectional area of the pipe; g – gravity acceleration; d – pipe diameter, w – average velocity of the liquid phase, σ – gas-liquid
interfacial tension.

For an example confirming the self-similarity of the arguments, Figure 1 demonstrates
the experimental dependence of the dimensionless pressure drop on the argument Bu at
$Fr^*=1.5$for different pipe diameters, liquid flow ratefrom 2 to 500l/h and pressure from
0.3 to 3 MPa; from the results obtained, k = 9.6.

It should be noted that the experiments conducted in LLC "Gazprom VNIIGAZ"
with surfactant in the vertical water-gas flow showed that after adding the surfactant,
the elevator characteristics shift to the left (Figure 2). At relatively low gas velocities,
pressure drop decrease and at relatively high gas velocities increase. According to the

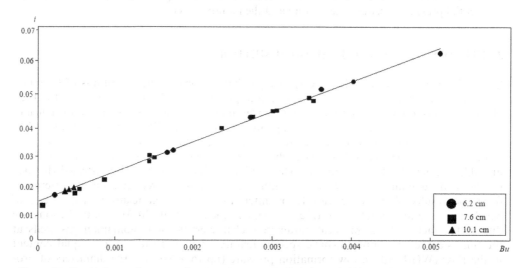

Figure 1.   Dependence of pressure drop on argument Bu for pipes of different diameter.

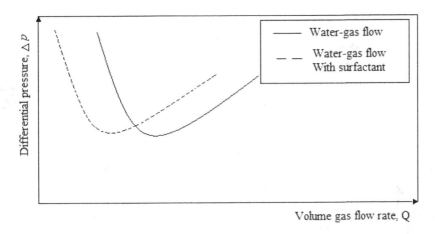

Figure 2.   The effect of surfactants on the characteristics of the elevator.

authors of the article (Izyumchenko et al. 2013), there is no acceptable theory that allows carrying out relevant technological calculations, taking into account the effect of surfactants on the characteristics of the gas-liquid lift, this issue has a high level of relevance and is for further study.

Among the current international experience of research of gas-liquid flows, including foaming agents, experimental work carried out at Delft Technical University (Delft, the Netherlands) with the support of large companies such as Royal Dutch Shell and ExxonMobil (van Nimwegen et al. 2016). The result of this work was a qualitative explanation of the effect of surfactants on the gas well deliquification from the liquid and the development of a mechanistic model of the gas-liquid flow with surfactants, which allows to calculate the pressure gradient. It should be noted that the error of the developed model in comparison with the experimental results was less than 25% (van Nimwegen et al. 2017).

Similar experimental work on small and large scale installations was carried out at the University of Tulsa (Tulsa, USA) (Kelkar et al. 2015).

Within the framework of the work, the authors managed to theoretically justify the model for calculating the differential pressures of a gas-liquid lift from surfactants, according to them this is the first model that was presented in open literature sources. The results of the calculations on the derived model were compared with the results of experiments on a large scale installation, the error of the comparison results was 30% with a reliability of 90%. It is also noted that the obtained correlation of the differential pressures has a significant error and can be improved by properly taking into account the various properties of the foamer and the resulting foam.

Comparison of the obtained models of pressure gradient prediction during the well is working with surfactants (including comparison with the model developed by TNO) is presented in (van't Westende et al. 2017). It should be noted that the results of calculations on the models have significant differences, and one of the main conclusions of the article is the need to conduct experiments in conditions close to the well to improve the model.

3   RESULTS

The analysis of the results of best practices in the study of gas-liquid lifts, including foamer in the flow, showed that to be able to more accurately simulate pressure drop in the wellbore of a gas well working with surfactants, it is necessary to conduct experimental bench studies taking into account the characteristics of the physical parameters of specific conditions (pressure, temperature, gas-liquid ratio, etc.).

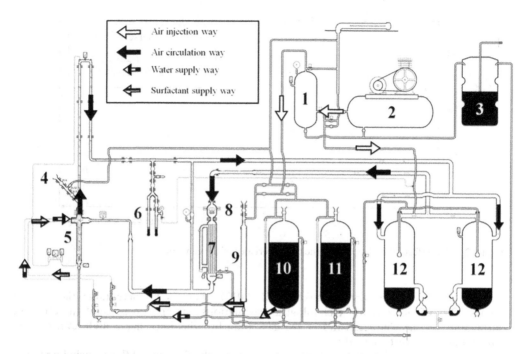

Figure 3. Schematic diagram of the Experimental stand. 1 - Surge receiver; 2 - Compressor; 3 - Containers for recycling; 4 - Node of input the solid surfactant; 5 - Lift column; 6 - Node of sampling foam; 7 - Node of heat-exchanging; 8 - Blower; 9 - Tank of input the liquid surfactant; 10 - Tank for storage and supply of water (liquid); 11 - Tank for storing and supplying liquid defoaming; 12 - Separators for separation of spent gas-liquid, foamed mixture.

Based on the analysis, taking into account the specific features of the development of Cenomanian gas wells in the Russian Federation, as well as the relevance of the problem of their "self-kill", the authors of the article designed and implemented an "Experimental stand to simulate a gas-liquid mixture and dynamic processes in the gas well".The schematic diagram of the experimental stand is shown in Figure 3.

Table 1. Technical specifications of the facility.

Specification	Unit	Value
Tubing length (base)*	m	6
Outer/inner diameter of tubing (base)*	mm	50/42
Working pressure in the system (not more than)	MPa	1.5
Working range of temperature	°C	15...50
Max superficial gas velocity at pressure 1.5 MPa **	m/s	10
Volume flow of water	l/h	3...1200
Volume flow of surfactant solution	l/h	3...1200
Absolute error of pressure sensors	kPa	0.8
Absolute error of water and surfactant solution flowmeter	l/h	0.01
Absolute error of gas flowmeter	m³/h	0.1
Working fluid	Water, air (distilled water and brine)	

*It is possible to change the diameter of the tubing and the tubing length.
**For the base diameter of tubing

With the help of the facility it is possible to conduct experiments related to the simulation of dynamic processes occurring in a gas well working with liquids, including those in interaction with surfactants and other non-aggressive chemicals, and to obtain digital data. Types of possible experimental studies: a) study of the processes of "self-kill" of the well with liquid; b) study of the processes of deliquification the well from the liquid gas flow; C) study of the processes of removal of liquid using liquid surfactants of different types and concentrations.

It should be noted that with the help of the stand it is possible to reproduce physical conditions similar to the conditions of Cenomanian gas wells of fields with falling production: automatic keeping of pressure up to 1.5 MPa, temperature in the range of 15-50 °C, water-gas ratio (including low WGR$<10^{-5}$). The basic technical specifications of the facility are presented in Table 1.

The process of conducting experiments is fully automated, when modeling processes; it is possible to maintain fixed values of WGR, pressure, temperature, surfactant flow, etc. Control is carried out with the help of PLC and specially developed software.

"Operator panel" is the workplace of the person conducting the experiment. With full functionality for stand management, the software allows:

a) to set of experiment settings;

b) to record telemetry data about of gas, liquid and surfactant solution flows;

c) to capture the telemetry data with the readings of pressures, temperatures at different points of the flow;

d) to output the results of observation of the amount of separated liquid in time;

4 CONCLUSIONS

Summing up, it should be noted that currently there is a final testing of algorithms for stand control and preparation for a long series of experiments in automatic mode. Research on the experimental setup will differ from the international experience of experimental work presented in the article in that it will take into account the pressures effect (up to 1.5 MPa) and temperature effect (up to 50°C) on the characteristics of the gas-liquid flow, including foaming agents. It is planned to further analyze the results and create a computational model of pressure losses in the gas well segment with reformatting in VFP-tables, reflecting the nature of stationary and non-stationary processes occurring in the wellbore. The implemented system of video recording of the flow will allow to determine the flow regimes and the direction of flow of liquid, foam.

REFERENCES

Alzhanov, Y., Karami, H., Pereyra, E., & Gamboa, J. 2018. Efficacy of Surfactants in Rich Gas Shale Wells. SPE Artificial Lift Conference and Exhibition - Americas.

Gcali, C., Karami, H., Pereyra, E., & Sarica, C. 2018. Surfactant Batch Treatment Efficiency as an Artificial Lift Method for Horizontal Gas Wells. SPE Artificial Lift Conference and Exhibition - Americas.

Huang, F., & Nguyen, D. 2012. Optimized foamers for natural gas well deliquification: A statistical design approach. Fuel 97: 523–530.

Izyumchenko, D.V., Mandrik, E.V., Melnikov, S.A., Ploskov, A.A., Moiseev V.V., Haritonov A.N., & Pamuzhak S.G. 2018. Operation of gas wells in conditions of active water and sand production. Vesti gazovoy nauki № 1 (33): 235–241.

Izyumchenko, D.V., Nikolaev, O.V., & SHulepin S.A. 2013. Gas-liquid flows in vertical pipes: paradoxes of hydrodynamics. Vesti gazovoy nauki № 4 (15): 36–45.

Kalwar, S. A., Awan, A. Q., Rehman, A. U., & Abbasi, H. S. 2017. Production Optimization of High Temperature Liquid Hold Up Gas Well Using Capillary Surfactant Injection. SPE Middle East Oil & Gas Show and Conference.

Kelkar, M., & Sarica C. 2015. Gas Well Pressure Drop Prediction under Foam Flow Conditions. RPSEA 09122-01 Final Report: 192.

Koryakin, A.Y. 2016. Complex solutions of problems of development and operation of wells of the Urengoy producing complex. M.:272.

Lea, J. F., & Nickens, H. V. 2004. Solving Gas-Well Liquid-Loading Problems. Gulf Professional Publishing: 314.

Lea, J. F., Nickens, H. V., & Wells M. 2003. Gas well deliquification solutions to gas well liquid loading problems. Journal of Petroleum Technology № 04 (56): 30–36.

Nikolaev, O.V., Borodin, S.A., & SHulepin S.A. 2013. Experimental study of the similarity of vertical gas-liquid flows in the operation of flooded gas wells. Vesti gazovoy nauki № 4 (15): 76–83.

Omrani, P. S., Shukla, R. K., Vercauteren, F., & Nennie, E. 2016. Towards a Better Selection of Foamers for the Deliquification of Mature Gas Wells. International Petroleum Technology Conference.

Panikarovskij V.V., & Panikarovskij E.V. 2017. Operation of gas wells at the late stage of field development. The journal "Oil and Gas Studies" № 5: 85–89.

Peyton, S.H., Neve, S.L., & Krevor, C. 2013. Investigation of Batch Foamer Efficacy and Optimisation in North Sea Gas Condensate Wells. SPE Candidate Paper.

Rauf, O. 2015. Gas Well Deliquification–A Brief Comparison between Foam Squeeze and Foam Batch Approach. Journal of Industrial and Intelligent Information, Vol. 3, № 1: 45–47.

Sarancha A.V., Sarancha I.S., Mitrofanov D.A., & Ovezova S.M. 2015. Technology of production of low-pressure Cenomanian gas. Modern problems of science and education № 1 (1): 211.

Schinagl, W., Caskie, M., Green, S. R., Docherty, M., & Hodds, A. C. 2007. Most Successful Batch Application of Surfactant in North Sea Gas Wells. Offshore Europe.

van Nimwegen, A.T., Portela, L.M, & Henkes, R.A.W.M. 2016. The effect of surfactants on upward air-water pipe flow at various inclinations. International Journal of Multiphase Flow № 78:132–147.

van Nimwegen, A.T., Portela, L.M, & Henkes, R.A.W.M., 2017. Modelling of upwards gas-liquid annular and churn flow with surfactants in vertical pipes. International Journal of Multiphase Flow.

van't Westende, J. M. C., Henkes, R. A. W. M., Ajani, A., & Kelkar, M. 2017. The use of surfactants for gas well deliquification: a comparison of research projects and developed models. BHR Group.

Topical Issues of Rational Use of Natural Resources 2019 – Litvinenko (Ed)
© 2020 Taylor & Francis Group, London, ISBN 978-0-367-85720-2

Research materials for cooking heat-resistant cement mixtures on a slag base

A.S. Oganov, S.L. Simonyants & S.F. Vyaznikovtsev
Gubkin University, Moscow, Russia

ABSTRACT: Considered the creation of cement materials for the attachment of high-temperature wells. Traditional Portland cement cannot be used at temperatures above 100 ° C due to thermal corrosion of hardening products due to interphase and intra-phase recrystallization, while the permeability of the cement stone significantly increases. The results of laboratory studies of heat-resistant cements obtained using cement slag-based materials are given. The physic-mechanical and strength properties of slag sand mixtures were investigated. The phase composition of the hardening products of cement materials was investigated by X-ray analysis. High thermal resistance of cement stone was obtained.

1 INTRODUCTION

The increase in drilling depths leads to an increase in bottom hole temperatures in wells, which, along with abnormally high reservoir pressures, the aggressiveness of reservoir fluids and other geological and technological factors, significantly reduces the reliability and durability of lining of constructed wells due to thermal destruction and corrosive destruction of cement stone(Kozhevnikov et al. 2015, Moradi & Nikolaev 2017, Kamenskih et al. 2018). The scope of traditional cements is mainly limited to temperatures below 100 ˚C. At higher temperatures, the strength decreases over time and the permeability of the cement stone increases. Therefore, at high temperatures, it is necessary to use heat-resistant grouting materials that ensure the reliability and durability of the constructed well.

Aggressive reservoir fluids (Merzlyakov et al. 2019), primarily hydrogen sulfide, quickly destroy the lining of the well, and the cement block is the most vulnerable element of the lining. Analysis of the mechanism and kinetics of corrosion processes shows that an increase in the corrosion resistance of cement stone can be provided by controlling the phase composition of the hardening products, controlling the structure of the pore space, as well as inhibiting the inner surface of the pores (Morenov et al. 2018). Taking into account the influence of various factors on the properties of the solution and the resulting stone, the component composition of the cement material and reagents for controlling the technological properties of the solution and the stone were substantiated. Studies have allowed us to develop formulations of base cement slurries based on corrosion-resistant (Nikolaev & Leusheva 2019), heat-resistant and weighted cement mixtures, taking into account geological and technical conditions and casing cementing technologies in some gas wells (Nikolaev & Lyu Khaoya 2017).

Traditional Portland cement cannot be used at temperatures above 100°C due to thermal corrosion of hardening products due to interphase and intra-phase recrystallization, while the permeability of the cement stone significantly increases (Zimina & Kuznetsov 2019). The basis of these phenomena is thermal corrosion of cement, the essence of which consists in the recrystallization of hardening products.

2 RESEARCH METHODOLOGY

The formation of thermodynamically stable compounds in the process of hardening of the cement material in hydrothermal conditions is the most important condition for the heat resistance of cement. In addition, it is necessary that these compounds have good structure-forming properties, because without this it is impossible to obtain high strength and low permeability of the resulting porous body. High structure-forming properties have crystals with a high degree of dispersion and with a pronounced ability to form intergrowth contacts. Stable compounds should be formed in the early stages, since subsequent recrystallization processes will weaken the already formed structure of the cement stone (Agzamov & Izmukhambetov 2005, Bulatov 1990, Daniushevsky et al. 1987).

Cement-sand mixtures at low temperatures are characterized by slowing compared with conventional cement thickening and setting of the cement slurry. At high temperatures, only a slight retardation of thickening is observed as compared with cement without an additive, which does not preclude the use of retarders. The properties of a cement-sand mixture under identical hydration conditions are most influenced by its mineralogical composition, grain size and shape, sand pretreatment, etc. Quartz sands are the best, having a high percentage of silica sufficient to bind calcium hydroxide when hydrating cop.The use of slags in the composition of cement cements is based on a significant increase in their chemical activity with increasing temperature. At temperatures of 120–250°C, suspensions of even such inactive minerals as γ-2CaO-SiO2 and anorthite harden. Depending on the content of CaO and SiO2 in the slag, hydronates and low-base or highly basic hydrosilicates are formed as hydration products. It was the most common type of high temperature cement (Moradi & Nikolaev 2016, 2018).

At the same time, the tightening of requirements for the quality of lining wells and the protection of subsoil led to the fact that many cement materials (Nutskova et al. 2019), meeting the criterion of heat resistance, did not meet other requirements for cement slurries and the requirements of cementing technology. Against the background of the obvious advantages of cement slag materials (thermal stability, corrosion resistance, etc.) there are also serious disadvantages. The properties of the slag are determined by a number of factors: the type of raw material and pig iron smelted, its chemical composition, the technology of slag production and granulation. All this determines the stability of the properties of slags, and, consequently, of grouting cements based on them, obtained by various factories.

The disadvantages of cement slurry based on slag cements include low water holding capacity and high permeability of stone compared to other types of binders (silica-siliceous, portland cement, etc.). The water loss (Nutskova et al. 2017) of pure slag solutions at temperatures of 120-200°C after 1 minute is more than 80%. Reduction of the water-cement ratio leads to intensive thickening of the solution, deterioration of its pumpability, whereas in areas with an increased water-cement ratio a loose stone with high permeability is formed. These drawbacks limit the use of slag cements when attaching oil and gas wells.

3 EXPERIMENTAL

Experimental work on the study of the properties of slag sand mixtures (SHPTS) to obtain heat-resistant cement, used at temperatures up to 250 ° C, was carried out in the laboratory of technological liquids of the Department of drilling at Gubkin University. A self-dissolving slag obtained in the production of low-carbon ferrochrome was used. The results of laboratory tests are shown in Table 1.

According to the presented data, it can be concluded that the composition of the slag – sand mixture in the mass ratio of slag: sand is 50:50 is the most acceptable. On the basis of the selected composition of the slag – sand mixture, research was carried out on its use at high temperatures from 150 to 250° C. Accordingly, the slag-sand mixture was called "ACTIVE" - the slag-sand grouting mixture activated.

Table 1. Physic-mechanical properties of slag sand mixtures.

Composition	Water-cement	Cone spreading, mm	Density, kg/m3	Water separation, мл	Sieve residue, %	The time of thickening the cement slurry, min	Cement strength, MPa	
							bending	under compression
60:40	0,50	225	1820	1,5	15,8	114	5,6	10,6
50:50	0,47	230	1830	2,0	17,0	122	6,6	13,5
40:60	0,45	240	1840	3,5	19,1	102	6,0	11,2

Table 2. Strength properties of cement stone based on slag-sand mixture and process water at high temperatures.

Cement	Water-cement	Cone spreading, mm	Density, kg/m^3	Temperature,°C	Pressure, MPa	Cement stone strength after 8 hours, MPa	
						bending	under compression
ACTIVE-150	0,45	210	1840	150	60,0	4,2	10,6
ACTIVE-200	0,45	220	1850	200	70,0	4,8	12,9
ACTIVE-250	0,45	210	1850	250	80,0	5,1	14,2

Activation obtained by disintegrator preparation. According to the temperature of use, cements are designated as follows:
- for temperatures from 100 to 150°C - ACTIVE-150;
- for temperatures from 150 to 200°C - ACTIVE-200;
- for temperatures from 200 to 250°C - ACTIVE-250.

These cements were investigated at the time of thickening under dynamic conditions. To ensure the beginning of thickening for at least 90 minutes, a time delay setting agent was selected for the cement. The strength properties of slag sand mixtures mixed with process water were tested under static conditions with holding samples for 8 hours. The results are shown in Table 2. Research results show that slag-sand cement can be used at high temperatures from 150 to 250°C.

To ensure suffusion-sedimentation stability of cement slurries based on self-disintegrating slag, it is necessary to introduce special additives into it: stabilizers, plasticizers, defoamers, etc., which are stable in a calcium environment at high temperatures, and shut on a solution of sodium chloride with a density of 1070 kg/m3. As an effective stabilizer, it was proposed to use Diasel FL from Chevron Fillips, which is a neutral powder loss factor for cement mortars and can be used at bottomhole temperatures up to 232°C. The density of the powder is 1680 kg/m^3. The reagent is completely soluble in water. Compatible with cement mortars, shut in fresh and saline water. The effectiveness of the reagent increases when used with plasticizers.

4 RESULTS AND DISCUSSION

The phase composition of the hardening products of cement materials was investigated by X-ray analysis (X-ray diffraction), which is based on the study of X-ray diffraction scattering of individual atoms that are members of structurally-ordered formations (molecule, crystal). For the studies, samples of cementing stone of cement ACTIVE-200 were obtained under the following conditions:

- temperature - 200 °C;
- pressure - 70.0 MPa;
- curing time - 1 and 7 days;
- cooking technology - disintegrator.

X-ray structural analysis of the samples was carried out on a DRON 407 X-ray diffractometer with the use of the program for controlling the shooting process and data processing program. The results of X-ray phase analysis showed that neither free calcium hydroxide nor highly basic calcium silicate hydroxide were detected in the test samples. This indicates that even for a short period of hardening, the processes of formation of low-base calcium silicate silicates have been completed and the principle of their one-step synthesis has been implemented. Analysis of the obtained results indicates the impossibility of the occurrence of interphase recrystallization processes in these samples and indicates a high thermal resistance of the cement stone.

REFERENCES

Agzamov, F.A. & Izmukhambetov, B.S. 2005. Durability of cement stone in corrosive media. *St. Petersburg: Nedra.*

Bulatov, A.I. 1990. Formation and operation of cement stone in the well. *Moscow: Nedra.*

Daniushevsky, V.S. & Aliev, R.M. & Tolstoy, I.F. 1987. Reference guide to grouting materials. *Moscow: Nedra.*

Kamenskih, S., Ulyasheva, N., Buslaev, G., Voronik, A., Rudnitskiy, N. Research and development of the lightweight corrosion-resistant cement blend for well cementing in complex geological conditions, *Society of Petroleum Engineers - SPE Russian Petroleum Technology Conference 2018, RPTC 2018.*

Kozhevnikov, E.V., Nikolaev, N.I., Melekhin, A.A., Turbakov, M.S. Studying the properties of cement slurries for cementing oil wells with long horizontal section drilled with rotary steerable system, *Neftyanoe Khozyaystvo - Oil Industry* 2015: (9), pp. 58–60.

Merzlyakov M.Y., Jennifer R.R. Hernandez & Zhapkhandayev Ch.A. Development of cement slurries for oil and gas wells lining in aggressive environment, *Youth Technical Sessions Proceedings: VI Youth Forum of the World Petroleum Council - Future Leaders Forum (WPF 2019), June 23–28, 2019, Saint Petersburg, Russian Federation.* – 2019, pp. 387–393.

Morenov, V., Leusheva, E., Martel, A. Investigation of the fractional composition effect of the carbonate weighting agents on the rheology of the clayless drilling mud (2018) *International Journal of Engineering, Transactions A: Basics* 31(7), p. 1152–1158.

Moradi, S.S.T., Nikolaev, N.I. Free fluid control of oil well cements using factorial design, *Journal of Engineering Research* 2017: 5(1), pp. 220–229.

Moradi, S.S.T., Nikolaev, N.I., Leusheva, E.L. Improvement of cement properties using a single multi-functional polymer, *International Journal of Engineering, Transactions A: Basics* 2018: 31(1), pp. 181–187.

Nikolaev N. I., Lyu Khaoya, Results of cement-to-rock contact study, *Journal of Mining Institute 2017:* 226, pp. 428–434.

Nikolaev N.I., Leusheva E.L. Low-density cement compositions for well cementing under abnormally low reservoir pressure conditions, *Journal of Mining Institute 2019:* 236, pp. 194–200.

Nutskova M.V., Rudiaeva E.Y., Kuchin V.N., Yakovlev A.A. Investigating of compositions for lost circulation control, *Youth Technical Sessions Proceedings: VI Youth Forum of the World Petroleum Council - Future Leaders Forum (WPF 2019), June 23–28, 2019, Saint Petersburg, Russian Federation.* – 2019, pp. 394–398.

Nutskova, M.V., Dvoynikov, M.V., Kuchin, V.N. Improving the quality of well completion in order to limit water inflows, *Journal of Engineering and Applied Sciences* 2017: 12(22), pp. 5985–5989.

Tabatabaee Moradi, S., Nikolaev, N.I. Optimization of cement spacer rheology model using genetic algorithm, *International Journal of Engineering, Transactions A: Basics* 2016: 29(1), pp. 127–131.

Tabatabaee Moradi, S.S., Nikolaev, N.I. Assessment of the cement failure probability using statistical characterization of the strength data, *Journal of Petroleum Science and Engineering 2018:* 164, pp. 182–188.

Zimina D.A. & Kuznetsov R.Y. Development of cement composition with enhanced properties with the addition of microsilica, *Youth Technical Sessions Proceedings: VI Youth Forum of the World Petroleum Council - Future Leaders Forum (WPF 2019), June 23–28, 2019, Saint Petersburg, Russian Federation.* – 2019, pp. 399–404.

Topical Issues of Rational Use of Natural Resources 2019 – Litvinenko (Ed)
© 2020 Taylor & Francis Group, London, ISBN 978-0-367-85720-2

Rheological and microrheological study of microsuspension with nanodiamonds

M.I. Pryazhnikov, E.I. Mikhienkova & A.V. Minakov
Siberian Federal University, Krasnoyarsk, Russia

ABSTRACT: In this paper, an experimental study of the viscosity and rheology of microsuspensions with nanodiamonds on a rotational viscometer was carried out. Microrheology of suspensions was studied by Rheolaser MASTER™. A water-based clay solution was used as a microsuspension. The nanodiamonds mass concentration was ranged from 0.125 to 2%. The effect of nanodiamond concentration on the rheological and viscoelastic properties of microsuspensions was studied.

1 INTRODUCTION

Currently there is an active interest in nanofluids (Devendiran & Amirtham 2016). Nanofluid is a suspension based on the base liquid (water, ethylene glycol, etc.) and nanoparticles smaller than 100 nm. Nanofluids are used in many fields of science and industry (Raja et al. 2016).

Such properties of nanosuspensions as viscosity and rheology (Mahbubul et al. 2012, Murshed & Estellé 2017) are of great interest for researchers. Most nanofluids are non-Newtonian fluids for which rheological parameters need to be investigated. However, in most works only viscosity of nanofluids is studied.

The effect of nanoparticles on the viscosity and rheology of microsuspensions has not been well studied as distinct to nanosuspensions. A typical example of microsuspensions is drilling mud, which is often used in practice (Morenov et al. 2018). Recently, the influence of nanoparticles on drilling fluids, which are also microsuspensions, has been actively investigated (Minakov et al. 2018a,b). Drilling mud is a water-based clay microsuspension (Blinov et al. 2017).

In this paper, the study of nanodiamonds concentration effect on clay microsuspension was carried out. Rheology of microsuspension was investigated by rotational viscometer OFITE 900, and microrheology was investigated using Rheolaser MASTER™ (Melekhin et al. 2016).

2 MATERIALS AND METHODS

2.1 Sample preparation

Clay suspension was prepared as follows. Clay powder (bentonite of Taganskiy deposit) was added to the distilled water and stirred intensively for 30 minutes with use of high-speed 20000 rpm stirrer (OFITE 152-18–Prince Castle). The clay suspension was kept for two days after preparation for the final clay swelling (Leusheva & Morenov 2017). Then, the necessary amount of nanosuspension with nanodiamonds was added to clay suspension. A standard two-step method was used for the preparation of nanosuspension (Rudyak et al. 2016). The clay mass concentration was 5%.

The powder of nanodiamonds was produced by Joint Stock Company Federal Research & Production Center ALTAI. The size of primary particles is 4-6 nm, it was defined by X-ray phase analysis. The aggregate size is 20-2500 nm and it is obtained by using a scanning electron microscope. The nanoparticles mass concentrations ranged from 0.125 to 2%.

2.2 Rotational viscometer OFITE 900

Rotational viscometers are the most commonly used instrument for studying viscosity and rheology of suspensions. The rheological properties of suspensions were studied with an OFITE 900 viscometer. The range of shear rates was $0.01–1022$ s^{-1}. The error in the measurement of viscosity was not lower 2%. All measurements presented below are performed at 298 K.

2.3 Rheolaser MASTER™

Microrheology samples investigated with Rheolaser MASTER™. The technology used in Rheolaser MASTER™ (http://www.formulaction.com/) is based on one of the dynamic light scattering methods called Diffuse Wave Spectroscopy (DWS). DWS is a passive microrheology method (Weitz & Pine 1993, Mason 1997).

The samples were contained in 20 ml glass vials and were illuminated with laser (at wavelength 650 nm). The radiation scattered by the sample particles forms the so-called speckle field due to the interference of photons scattered on individual particles. The CCD camera registers a constantly changing speckle field due to the Brownian motion of the particles. The relative decorrelation (RDC) and the mean quadratic displacement (MSD) are calculated as functions from time using a proprietary algorithm, when the speckle field is changed.

If suspension is a viscous Newtonian fluid, the dependence of mean square displacement of particles on the relaxation time MSD is linear in accordance with the Einstein-Smoluchowski diffusion equation (Mason 2000). For the movement of particles in an ideally elastic medium, MSD output to the plateau observes, it corresponds to the equilibrium of the thermal energy kT and the energy of the elastic interaction. The MSD curve is S-shaped for the general case of a viscoelastic sample. Figure 1 shows dependences of relative decorrelation and MSD of clay suspension with 1 wt % nanodiamonds on time.

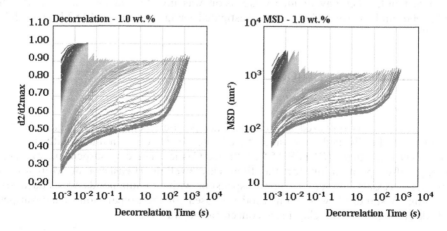

Figure 1.   RDC and MSD of clay suspension with 1 wt %nanodiamonds.

3 RESULTS

3.1 *Rheological studies*

The dependences of the viscosity coefficient and shear stress on the shear rate were obtained (Figure 2) at a temperature of 298 K using a rotational viscometer. Figure 2 shows that the addition of nanodiamonds to clay suspension significantly increases viscosity. The viscosity coefficient increases with nanoparticles concentration. A significant change in viscosity coefficient and rheology of microsuspension is observed even at low nanodiamonds concentrations. The density of suspension does not change, which is very important for practical application.

The analysis of viscosity coefficient μ dependence on shear rate showed that rheology of microsuspensions with nanoparticle addition is well described by the Power-law model (Nutskova & Kupavyh 2016):

$$\mu = K\dot{\gamma}^{n-1} \tag{1}$$

where K = the consistency factor; $\dot{\gamma}$ = the shear rate; and n = the flow index.

The study of rheological parameters of suspensions with nanoparticles showed that flow index decreases with the increase of nanodiamonds concentration, and consistency index, on the contrary, increases. Thus, nanodiamond additive enhances the rheological properties of microsuspension.

3.2 *Microrheological studies*

Rheolaser calculates several parameters characterizing viscoelastic properties of sample. The slope of the MSD curve at elastic plateau is characteristic of the ratio between solid-like behaviour and liquid-like behaviour of sample. SLB is calculated as the slope of MSD curve in logarithmic coordinates. Indeed, the lower is this slope, the slower is the motion of particles, meaning sample is more solid than liquid.

SLB was obtained from time for the studied suspensions (Figure 3a). At the initial moment microsuspensions are mostly viscous (liquids). However, in a short time suspensions acquire elastic properties. Microsuspension without nanoparticles acquires elastic properties after about 320 s. A significant reduction in the time at which sample behaves mainly as an elastic/solid body is observed already at low concentrations of nanoparticles. The addition of 0.25 wt % nanodiamonds reduce the transition time to 210 s.

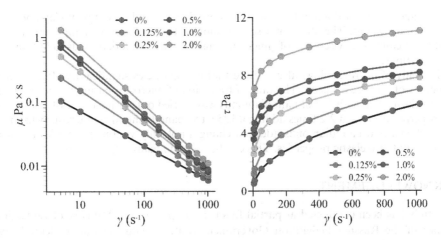

Figure 2. Dependence of viscosity coefficient (left) and shear stress (right) of microsuspension on shear rate at different concentrations of diamond nanoparticles.

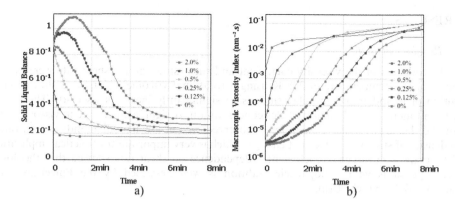

Figure 3.   SLB (a) and MVI (b) of microsuspensions with nanoparticles.

It is also convenient to use a parameter such as the macroscopic viscosity Index (MVI) for the analysis of viscous and elastic properties, which is calculated as the inverse of slope of MSD curve in linear coordinates. The slope of MSD curve after the plateau is a characteristic of sample macroscopic viscosity. At long correlation times, particles can move quickly or slowly. The slower the particles move, the higher is macroscopic viscosity of sample. Conversely, the faster the particles move, the lower is viscosity, as particles can move a longer distance in the same time period.

The macroscopic viscosity index MVI is proportional to macroscopic viscosity μ of sample and gives an indication of the evolution of sample viscosity as a function of time.

From the analysis of the data of MVI (Figure 3b) it can be concluded that the rate of transition from viscous to elastic properties for sample with nanodiamonds addition increases. Rate increases by 1.2 times and by 3.2 times adding 0.125 wt % and 0.5 wt % nanodiamond respectively (Figure 3b). It should also be said about the values of MVI microsuspensions at the initial time. The dependence of MVI on nanodiamonds concentration is clearly seen on Figure 3b. The nanodiamonds addition increases MVI with increasing concentration. This corresponds to the result obtained by rotational viscometer.

4   CONCLUSION

In this paper, effect of diamond nanoparticles on viscosity and rheology of clay microsuspension was carried out. The study of viscosity showed that addition of nanodiamonds significantly increases the viscosity of microsuspension with nanodiamonds addition at low concentrations.

The study of microrheology also showed that nanoparticles significantly affect viscoelastic properties of microsuspension. The time of transition of microsuspension properties from viscous to elastic with addition of nanodiamonds was carried out. Transit time of viscous to elastic properties decreased 2.2 times adding 0.125 wt % nanodiamond to microsuspension. Thus, it was shown that nanodiamonds addition changes not only viscous properties of microsuspensions, but also elastic properties. This can be useful in many applications.

ACKNOWLEDGEMENT

The study has been conducted at partial financial support of the Ministry of Education and Science of the Russian Federation Government contract with Siberian Federal University (No. 16.8368.2017).

REFERENCES

Blinov, P.A., Dvoynikov, M.V., Sergeevich, K.M., Rustamovna, A.E. Influence of mud filtrate on the stress distribution in the row zone of the well, *International Journal of Applied Engineering Research* 2017: 12(15), pp. 5214–5217.

Devendiran, D.K. & Amirtham, V.A. 2016. A review on preparation, characterization, properties and applications of nanofluids. *Renewable and Sustainable Energy Reviews* 60: 21–40.

Leusheva, E., Morenov, V. Research of clayless drilling fluid influence on the rocks destruction efficiency, *International Journal of Applied Engineering Research* 2017: 12(6), pp. 945–949.

Mahbubul, I.M. et al. 2012. Latest developments on the viscosity of nanofluids. *International Journal of Heat and Mass Transfer* 55: 874–885.

Mason, T.G. et al. 1997. Diffusing-wave-spectroscopy measurements of viscoelasticity of complex fluids. *Journal of the Optical Society of America A* 14: 139–149.

Mason, T.G. 2000. Estimating the viscoelastic moduli of complex fluids using the generalized Stokes-Einstein equation. *Rheologica Acta* 39: 371–378.

Melekhin, A.A., Chernyshov, S.E., Blinov, P.A., Nutskova, M.V. Study of lubricant additives to the drilling fluid for reducing the friction coefficient during well construction with rotary steerable system, *Neftyanoe Khozyaystvo - Oil Industry* 2016: (10), pp. 52–55.

Minakov, A.V. et al. 2018a. The effect of nanoparticle additives on the rheological properties of the suspension. *Technical Physics Letters* 44(9): 3–11.

Minakov, A.V. et al. 2018b. A study of the influence of nanoparticles on the properties of drilling fluids. *Colloid Journal* 80(4): 418–426.

Morenov, V., Leusheva, E., Martel, A. Investigation of the fractional composition effect of the carbonate weighting agents on the rheology of the clayless drilling mud, *International Journal of Engineering, Transactions A: Basics* 2018: 31(7), pp. 1152–1158.

Murshed, S.M.S. & Estellé, P. 2017. A state of the art review on viscosity of nanofluids. *Renewable & Sustainable Energy Reviews* 76: 1134–1152.

Nutskova, M.V., Kupavyh, K.S. Improving the quality of well completion in deposits with abnormally low formation pressure, *International Journal of Applied Engineering Research* 2016: 11(11), pp. 7298–7300.

Raja, M. et al. 2016. Review on nanofluids characterization, heat transfer characteristics and applications *Renewable and Sustainable Energy Reviews* 64: 163–173.

Rudyak, V.Ya. et al. 2016. *Technical Physics Letters* 43(1): 23–26.

Weitz, D.A. & Pine, D.J. 1993. Diffusing-wave spectroscopy. In W. Brown. *Dynamic Light scattering* 652–720. New York: Clarendon Press.

Topical Issues of Rational Use of Natural Resources 2019 – Litvinenko (Ed)
© 2020 Taylor & Francis Group, London, ISBN 978-0-367-85720-2

Rational gas inflow restriction technologies during the development of oil rims

N.A. Rovnik, V.A. Lushpeev & D.S. Tananykhin
Saint-Petersburg Mining University, Saint-Petersburg, Russia

I.V. Shpurov
Saint-Petersburg State University, Saint-Petersburg, Russia

ABSTRACT: Rationale development of the reservoir with a massive gas cap is relevant for a lot of oil and gas companies. The main problems of such challenge are the formation of gas cones, which lead to gas breakthroughs to production wells. The article describes the methods of the development of reservoir with a gas cap. The article considers three options for oil recovery from the gas cap reservoir using the example of the tNavigator training model: option 1 - operation of vertical production wells, option 2 - operation of horizontal exploitation wells, option 3 - formation of an insulating screen at the gas-oil contact.

1 INTRODUCTION

A significant part of the hydrocarbon reserves in Russia are oil/gas/condensate fields. About 6 billion tons of oil is concentrated in oil rims. Oil reserves within oil/gas/condensate fields are not developed efficiently, due to the negative impact of gas breakthrough from the gas cap. Due to the complexity of the development, oil reserves of oil/gas deposits they are referred to hard-to-recover.

The efficiency of the development oil and gas condensate fields depends on many factors: exploration level of oil and gas reserves; type of deposits; the quality level of technical equipment of the oil and gas industry; methodology of technology section.

An analysis of the articles on the research topic showed that most of the works are devoted to the technologies for limiting the cone formation of gas by increasing the filtration resistance or controlling the operating mode of the well with the help of inflow control devices. Some companies have their own invention sat gas restriction technologies, but the general underlying principles of action are similar.

Some authors describe the whole cycle of implementation of gas inflow restriction technology: from laboratory experiments to pilot-industrial tests (Dorofeev, Taldykin, Kalugin, Bochkarev, 2008). In the works of Presnyakov A.Yu., Lomakina I.Yu., Nigmatullina T.E., Al-Dhafeeri, A.M., Nasr-El-Din, Ali E., Bergren, F.E., DeMestre the features of gas supply insulation in conditions of carbonate reservoirs are considered L.A. Tomskaya, I.I. Krasnov, D.A. Marakov, I.S. Tomskii, cite in their article the classification of technologies for isolating gas breakthroughs by the type of insulating agent. The authors sing out crystalline hydrates, foams, plugging compositions, polymer solutions and etc.

A large number of articles are devoted to mathematical simulating of the gas cone-forming processes (Ali, Bergren, DeMestre, Biezen, Van Eijden, 2007). Dorofeev N.V., Taldykin S.A., Kalugin A.A., Bochkarev A.V. consider the theory of the cone formation of Musket - Czarny. In the work of Gian Luigi Chierici, Giuseppe M. Ciucci, Giuseppe Pizzi a method of potentiometric simulating was proposed.

Analyzing these articles, we can conclude that mathematical simulating of cone formation is aimed at the derivation of formulas of two main parameters: critical (maximum gas-free) production rate and gas breakthrough time. Some articles describe the testing of mathematical regularities in hydrodynamic models in the conditions of the feilds of Prudhoe Bay, Yu. Korchagin, Troll et al. (Lushpeev, Vasyanovich, 2014).

In the course of the literary analysis, the authors found out that the topic of justifying the effectiveness of technologies for limiting gas inflow during the development of oil rims in the software tNavigator was poorly investigated. Authors singled out only one work where the simulation of the insulating screen during developing the oil rim with horizontal wells is presented, however, comparison of the technology with other methods is not presented in the article. For this reason, authors aimed to demonstrate the process of simulating the gas cap of the reservoir, and suggest ways to reduce the restriction of gas inflow to oil wells in the tNavigator software and presented their analysis and comparison of efficiency.

Within the framework of this goal, the following tasks were set:

1. Consider the main methods of limiting gas inflow
2. Describe the simulating of the gas cap in the tNavigator software using the example of the training model;
3. Consider several options for developing a gas-cap reservoir by simulating them in the tNavigator software, identifying the most relevant and suggesting options for improving it.

The process of simulation the technologies in hydrodynamic simulators reduces the risk of incorrect methods choice for an object. The work done by the authors can be used in the development of oil rims and the design of technologies for limiting gas inflow.

2 METHODS

Methods for developing an oil reservoir with a massive gas cap can be divided into two categories: first one is developing a field with a system for gathering, preparing and transporting gas, and second one is development without a system for gathering, preparing and transporting gas (Lomakina, Nigmatullin Presnyakov, Razyapov, Sorokin, 2014) (Figure 1).

In this article, authors focused on the development of oil rims in the absence of a system for gathering, preparing and transporting gas. Under such conditions, usually a cycling process is used or techniques for limiting cone formation are considered. Techniques for preventing gas breakthroughs in production wells are usually based on increased filtration or on controlling drawdown pressure in the oil production process.

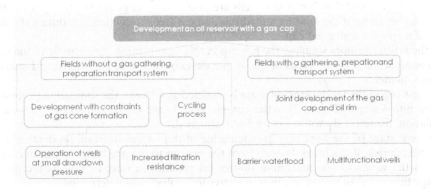

Figure 1. Scheme of methods for developing an oil reservoir with a gas cap (Lomakina, Nigmatullin Presnyakov, Razyapov, Sorokin, 2014).

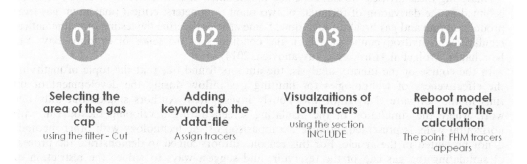

01	02	03	04
Selecting the area of the gas cap	Adding keywords to the data-file	Visualzaitions of four tracers	Reboot model and run for the calculation
using the filter - Cut	Assign tracers	using the section INCLUDE	The point FHM tracers appears

Figure 2.   Stages of simulating the gas cap in the tNavigator software.

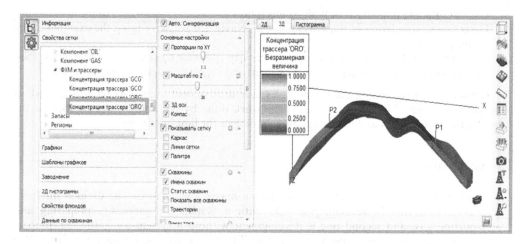

Figure 3.   View of the reservoir after the calculation with a gas cap.

The formation of different kinds of insulating screens at the interface of GOCs with the help of foams, polymers, crystalline hydrates, silicon organic compounds, etc., belongs to the first (Aasheim, Groenning, Mjaavatten, Saelid, 1984). The second category includes inflow control devices, which are usually divided into active and passive, in recent years, adaptive inflow control systems have also been identified as a separate group.

It should be noted that horizontal wells are also a technology for limiting gas inflow, they have a higher value of the limiting anhydrous and gas-free production rate during the development of reservoirs with a gas cap.

In the article, authors simulate the gas cap in tNavigator using the example of the reservoir training model, and then 3 types of the oil rim development were considered and analyzed on the same model.

The gas cap in the tNavigator software is modeled as a tracer, the simulation process can be divided into 4 stages (Figure 2). After completing all four stages and starting the model for calculation, the result shown in Figure 3 was obtained.

The next stage of the work was the simulating of three versions of oil production in conditions of high risk of gas breakthrough from the gas cap to the productive wells, in each of the variants the values of the maximum oil-free oil production rate, as well as the gas breakthrough time were compared, the most effective technology was determined.

Option 1 - without additional measures to limit the gas inflow - one production well in the oil part of the deposit. *Option 2* - horizontal production well. *Option 3* - creating a water screen at the gas-oil contact boundary to limit the flow of gas by adding an injection well.

3 RESULTS AND DISCUSSIONS

The simulation of the model was set by default and was carried out for three years in all types. *Option 1 is a single production vertical well* (Figure 4)

To analyze the calculation, authors unlead from the model: the daily gas production schedules from the gas cap, the accumulated production table and the oil and gas flow rates (Figure 5-7)

According to Figures 7-9, the accumulated oil production in December 2015 is 76 thousand m³, the accumulated water production is 32.7 thousand m³. Gas from the gas cap begins to break through 7 months after the start of the well, and by the end of 2015 the daily gas production from the cap increases from 2.27 * 10-15m3 / day to 1.6 * 10-6 m3 / day, the critical value of the gas-free oil production rate is 72.5 m3/day. This dynamic suggests the need to implement measures to limit gas inflows from the gas cap, in order to reduce the risk of drilling wells out of operation due to high gas contamination.

Type 2 - horizontal production well

Horizontal wells are also a technology for limiting gas inflow.They have a higher value of the limiting anhydrous and gas-free production rate for the development of deposits with a gas cap. Therefore, the article considered a variant, where instead of the usual vertical well D_1, a horizontal well (Figure 8, 9) was drilled.

After calculating the model, the following well performance indicators were analyzed: cumulative oil and water production at the end of 2015, water cut, recovery gas factor from the gas cap, duration of the gas-free oil production period, critical value of the gas-free oil production rate and comparison with the vertical production well.

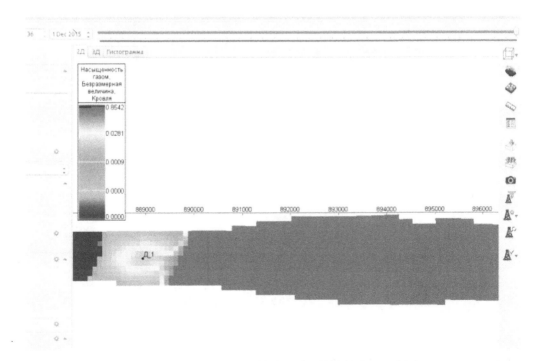

Figure 4.   Location of production well D_1 on the training model. Gas saturation after presented the calculation of the model (2D view).

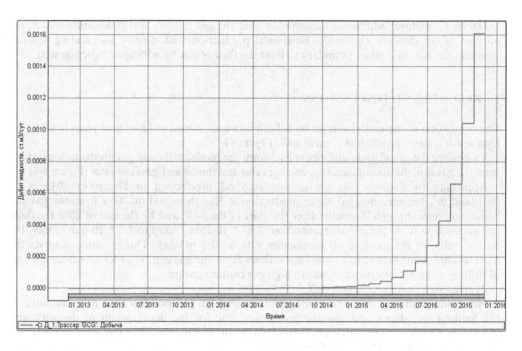

Figure 5. Gas recovery from the gas cap (Tracer GGG).

36	1 Dec 2015		
Графики	Сводные графики		
	Д_1 Накопл. нефть, тыс. ст.м3	Д_1 Накопл. вода, тыс. ст.м3	Д_1 Трассер, тыс. ст.м3/сут	Д_1 Трассер, тыс. ст.м3/сут
1 Dec 2012	0	0	0	0
1 Jan 2013	3.24807	1.2273	0	15.297
1 Feb 2013	5.93576	2.29841	0	16.3639
1 Mar 2013	8.12863	3.20144	0	15.3775
1 Apr 2013	10.4634	4.17517	0	14.1851
1 May 2013	12.6713	5.10302	0	13.245
1 Jun 2013	14.9172	6.05185	0	12.532
1 Jul 2013	17.0658	6.96328	2.27165e-15	12.0304
1 Aug 2013	19.2647	7.8993	9.28736e-15	11.6531
1 Sep 2013	21.4453	8.83042	3.30343e-14	11.366
1 Oct 2013	23.5447	9.72857	2.06747e-13	11.1305
1 Nov 2013	25.699	10.6527	1.10832e-12	10.9233
1 Dec 2013	27.7681	11.5428	2.50143e-12	10.7493
1 Jan 2014	29.8952	12.4598	5.54546e-12	10.5879
1 Feb 2014	32.0124	13.3744	1.17141e-11	10.4407
1 Mar 2014	33.9176	14.1986	2.18542e-11	10.319
1 Apr 2014	36.0211	15.1098	7.36314e-11	10.1948
1 May 2014	38.0493	15.9897	2.15166e-10	10.0864
1 Jun 2014	40.136	16.8969	3.81957e-10	9.98848
1 Jul 2014	42.1493	17.7734	6.48609e-10	9.90311
1 Aug 2014	44.2236	18.6778	1.10545e-09	9.82524
1 Sep 2014	46.2921	19.581	1.84696e-09	9.75527
1 Oct 2014	48.2886	20.4539	2.97856e-09	9.69448
1 Nov 2014	50.3463	21.3547	4.84593e-09	9.63943
1 Dec 2014	52.3328	22.2255	7.64454e-09	9.59202
1 Jan 2015	54.3805	23.1243	1.2191e-08	9.54928
1 Feb 2015	56.4232	24.022	1.93187e-08	9.51041
1 Mar 2015	58.2643	24.8321	2.89069e-08	9.47755
1 Apr 2015	60.298	25.728	4.55324e-08	9.44381
1 May 2015	62.262	26.5942	7.03987e-08	9.41339
1 Jun 2015	64.2871	27.4883	1.10917e-07	9.38429
1 Jul 2015	66.243	28.3528	1.71788e-07	9.35819
1 Aug 2015	68.26	29.2454	2.71047e-07	9.33326
1 Sep 2015	70.2732	30.1371	4.27542e-07	9.31067
1 Oct 2015	72.2178	30.9993	6.61852e-07	9.29019
1 Nov 2015	74.2236	31.8895	1.04174e-06	9.27056
1 Dec 2015	76.1614	32.7504	1.60666e-06	9.25286

Figure 6. Results of well D_1 data after 3 years of operation.

	Дебит нефти, ст.м3/сут	Массовый ... т/сут	Дебит воды, ст.м3/сут	Массовый ... т/сут
1 Dec 2012	0	0	0	0
1 Jan 2013	96.4894	82.0159	37.2022	37.3882
1 Feb 2013	84.6769	71.9754	33.9768	34.1467
1 Mar 2013	79.0808	67.2187	32.448	32.6102
1 Apr 2013	75.7021	64.3468	31.5025	31.66
1 May 2013	73.7797	62.7127	30.9599	31.1147
1 Jun 2013	72.5259	61.647	30.6069	30.7599
1 Jul 2013	71.6504	60.9029	30.364	30.5158
1 Aug 2013	70.9314	60.2917	30.1682	30.319
1 Sep 2013	70.3274	59.7783	30.006	30.1561
1 Oct 2013	69.9034	59.4179	29.9293	30.0789
1 Nov 2013	69.4148	59.0026	29.7974	29.9464
1 Dec 2013	68.9096	58.5732	29.6262	29.7744
1 Jan 2014	68.5598	58.2758	29.5383	29.686
1 Feb 2014	68.2445	58.0078	29.4596	29.6069
1 Mar 2014	67.9852	57.7874	29.3957	29.5427
1 Apr 2014	67.8244	57.6508	29.3968	29.5438
1 May 2014	67.4684	57.3482	29.2689	29.4153
1 Jun 2014	67.2491	57.1618	29.2191	29.3652
1 Jul 2014	67.0487	56.9914	29.174	29.3199
1 Aug 2014	66.8538	56.8258	29.1321	29.2778
1 Sep 2014	66.667	56.667	29.0925	29.2379
1 Oct 2014	66.4937	56.5196	29.0562	29.2015
1 Nov 2014	66.3206	56.3725	29.0204	29.1655
1 Dec 2014	66.1594	56.2355	28.9873	29.1323
1 Jan 2015	65.997	56.0974	28.9542	29.099
1 Feb 2015	65.8379	55.9622	28.9218	29.0664
1 Mar 2015	65.6988	55.844	28.8936	29.0381
1 Apr 2015	65.5502	55.7177	28.8637	29.008
1 May 2015	65.4112	55.5996	28.8358	28.9799
1 Jun 2015	65.2728	55.4819	28.808	28.9521
1 Jul 2015	65.1432	55.3717	28.782	28.926
1 Aug 2015	65.0136	55.2616	28.756	28.8998
1 Sep 2015	64.8876	55.1544	28.7306	28.8742
1 Oct 2015	64.7696	55.0542	28.7068	28.8503
1 Nov 2015	64.6517	54.9539	28.683	28.8264
1 Dec 2015	64.5413	54.8601	28.6608	28.8041

Figure 7. Oil and water flow rates of well D1 in the period 2012-2015 years.

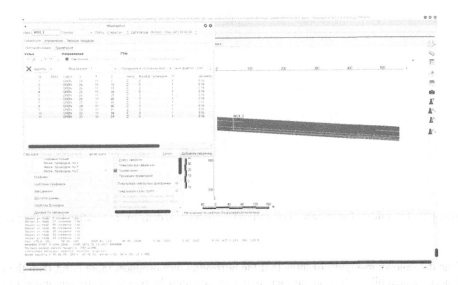

Figure 8. Horizontal Well Trajectory.

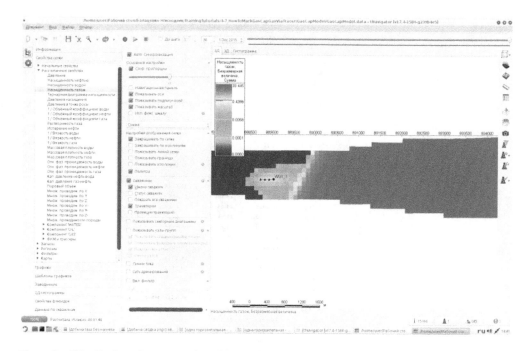

Figure 9.   Distribution of gas saturation after starting the model calculation.

	Накопл. на... тыс. ст.м3	Накопл. не тыс. ст.м3	Накопл. вода, тыс. ст.м3	WU1_1 Трассер 'OCG' ст.м3/сут	WU1_1 Трассео 'OCO' ст.м3/сут
1 Dec 2012	0	0	0	0	0
1 Jan 2013	5.00668	0	2.47088	0	0
1 Feb 2013	9.14361	0	4.6137	0	0
1 Mar 2013	12.611	0	6.46346	0	0
1 Apr 2013	16.2516	0	8.45389	0	0
1 May 2013	19.658	0	10.3529	0	0
1 Jun 2013	23.0997	0	12.3038	0	0
1 Jul 2013	26.39	0	14.2027	0	0
1 Aug 2013	29.7534	0	16.1694	0	0
1 Sep 2013	33.0812	0	18.1295	0	0
1 Oct 2013	36.2833	0	20.0334	0	0
1 Nov 2013	39.5768	0	22.0078	0	0
1 Dec 2013	42.7522	0	23.9236	0	0
1 Jan 2014	46.0226	0	25.9077	0	0
1 Feb 2014	49.288	0	27.8951	0	0
1 Mar 2014	52.2294	0	29.692	0	0
1 Apr 2014	55.4733	0	31.682	0	0
1 May 2014	58.6052	0	33.608	0	0
1 Jun 2014	61.8341	0	35.5977	0	0
1 Jul 2014	64.9518	0	37.5224	0	0
1 Aug 2014	68.166	0	39.5099	0	0
1 Sep 2014	71.373	0	41.496	0	0
1 Oct 2014	74.4698	0	43.4164	0	0
1 Nov 2014	77.6629	0	45.3991	0	0
1 Dec 2014	80.7464	0	47.316	0	0
1 Jan 2015	83.9259	0	49.2948	1.25167e-12	0
1 Feb 2015	87.0985	0	51.2716	1.78706e-12	0
1 Mar 2015	89.9585	0	53.0554	2.43452e-12	0
1 Apr 2015	93.1182	0	55.0282	3.41495e-12	0
1 May 2015	96.1698	0	56.9354	4.68899e-12	0
1 Jun 2015	99.3166	0	58.9041	6.4624e-12	0
1 Jul 2015	102.356	0	60.8072	8.73219e-12	0
1 Aug 2015	105.49	0	62.7716	1.1845e-11	0
1 Sep 2015	108.618	0	64.7339	1.59475e-11	0
1 Oct 2015	111.639	0	66.6309	2.10157e-11	0
1 Nov 2015	114.755	0	68.589	2.90192e-11	0
1 Dec 2015	117.765	0	70.482	3.66178e-11	0

Figure 10.   The values of accumulated oil and water production, daily gas production from the gas cap.

The data shown in Figure 10 show that the horizontal well produces more effect than the verti-cal, accumulated oil production has increased approximately by two times and amounted to 117.7 thousand m^3, while the accumulated water production amounted to 70.5 thousand m^3. The

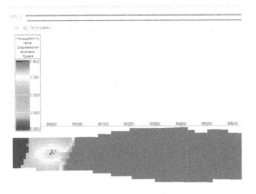 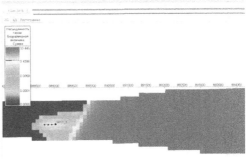

(a) - Distribution of gas saturation during the production of oil with a vertical well

b) - Distribution of gas saturation during the production of oil with a horizontal well

Figure 11. (a) - Distribution of gas saturation during the production of oil with a vertical well. b) Distribution of gas saturation during the production of oil with a horizontal well.

period of gas-free oil production increased from 7 months to 2 years, and the value of gas-free production rate was 103.2 m^3/day.

By comparing Figures 11a and 11b, it is possible to trace the decrease in the amount of gas that breaks out from the gas cap when the oil is extracted by a horizontal well.

Type 3 - formation of a water barrier at the GOC

The technology of forming a water screen at the gas-oil contact by injecting water with the injection well has been successfully applied at the Vankor oil and gas condensate field. The formation of a water barrier at the boundary of the gas-oil contact reduces the relative phase permeability through the gas, which leads to a decrease in the amount of gas that breaks out from the gas cap.

To create a barrier, an injection well was added, the injectivity of the well was set at 300 m^3/day, the operation period 2012-2015 (Figure 12 -14).

As a result, with the addition of an injection well and the formation of a waterproof screen, the cumulative oil production amounted to almost 80 thousand m^3, while the accumulated water production amounted to 33, 4 thousand m^3. The period of gas-free oil production increased from 7 months to two and a half years compared to a vertical well, and the value of gas-free production rate was 69.5 m3/day. Figure 15a, b shows a comparison of gas saturation without formation a waterproof screen (15 a) and with it formation (15 b)

Analyzing Figure 15 a and b, it can be concluded that the waterproof screen allows to increase the breakout time for gas for a while, but at some point, the gas begins to curve around the barrier and break into the oil-saturated part of the deposit, so it is necessary to select ways of improving this technology.

4 CONCLUSION

The results obtained during the calculation of the oil reservoir model with a gas cap, are described in Table 1

The table shows that the greatest gas breakthrough time is reached when the waterproof screen is formed and is equal to 2.5 years, when oil is extracted without a screen with the help of

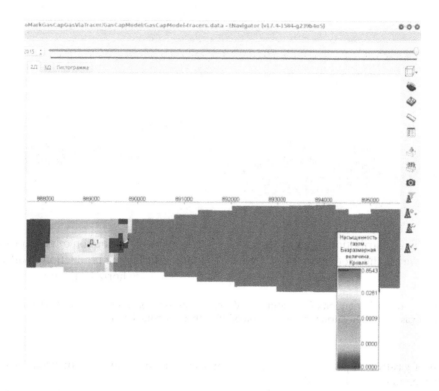

Figure 12. Distribution of gas saturation after the addition of the injection well and starting the calculation of the model.

a vertical production well, gas begins to break out of the cap after 7 months. The critical value of the gas-free production rate is 69.5 m^3/day when adding the injection well, and without a barrier 72.5 m^3/day. As for the type with a horizontal production well, the gas breakthrough start time is 2 years, but the value of the gas-free production rate is 103.5 m^3/day, and the accumulated oil production is 156 thousand m^3, which is much larger than in the first and third types.

It is possible to improve the efficiency of the third option and increase the time of the gas breakthrough by selecting the optimal length of the horizontal part of well, and also by using adaptive inflow control systems (AICS) that enable the flow profile to be aligned and the gas breakthrough in the well to be laid. In the Russian Federation there already exists a similar device. This device has already been successfully tested at the horizontal wells of the Yu. Korchagin field (Bochkarev, Dorofeev, Kalugin, Taldykin, 2008).

The effectiveness of the first and third variants considered in the work can also be improved by forming special insulating screens of complex composition at the gas-oil contact boundary. The formation of waterproof screen does not always justify itself, because even if the fixed screen (barrier) is not permeable to gas, then after a certain time the gas envelops this screen and breaks to the bottom of the producing well. A more effective way is to simultaneously form an insulating gel screen with the formation of a water barrier (Figure 16).

In the case of a vertical well in the intertubular space, a packer is installed at the gas-oil contact level. On the intertubular space, a gel is injected into the gas-saturated zone, which propagates along the gas-oil contact, and forms an impermeable zone. Then, water is pumped into the gas-saturated part also through the annulus and a liquid barrier is formed. Then oil recovery from the oil rim begins. In this case, water injection continues, the liquid barrier separates the gas further from the bottom of the well, and reduces the risk of its breakthrough to the production well (Mostafa, Ismail, 1993).

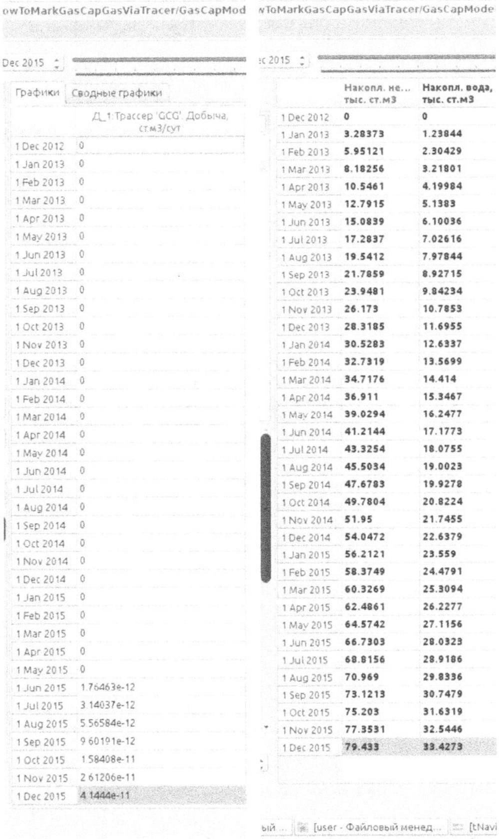
owToMarkGasCapGasViaTracer/GasCapMod

Dec 2015

Графики | Сводные графики

	Д_1:Трассер 'GCG'. Добыча, ст.м3/сут
1 Dec 2012	0
1 Jan 2013	0
1 Feb 2013	0
1 Mar 2013	0
1 Apr 2013	0
1 May 2013	0
1 Jun 2013	0
1 Jul 2013	0
1 Aug 2013	0
1 Sep 2013	0
1 Oct 2013	0
1 Nov 2013	0
1 Dec 2013	0
1 Jan 2014	0
1 Feb 2014	0
1 Mar 2014	0
1 Apr 2014	0
1 May 2014	0
1 Jun 2014	0
1 Jul 2014	0
1 Aug 2014	0
1 Sep 2014	0
1 Oct 2014	0
1 Nov 2014	0
1 Dec 2014	0
1 Jan 2015	0
1 Feb 2015	0
1 Mar 2015	0
1 Apr 2015	0
1 May 2015	0
1 Jun 2015	1.76463e-12
1 Jul 2015	3.14037e-12
1 Aug 2015	5.56584e-12
1 Sep 2015	9.60191e-12
1 Oct 2015	1.58408e-11
1 Nov 2015	2.61206e-11
1 Dec 2015	4.1444e-11

wToMarkGasCapGasViaTracer/GasCapMode

:c 2015

	Накопл. не... тыс. ст.м3	Накопл. вода, тыс. ст.м3
1 Dec 2012	0	0
1 Jan 2013	3.28373	1.23844
1 Feb 2013	5.95121	2.30429
1 Mar 2013	8.18256	3.21801
1 Apr 2013	10.5461	4.19984
1 May 2013	12.7915	5.1383
1 Jun 2013	15.0839	6.10036
1 Jul 2013	17.2837	7.02616
1 Aug 2013	19.5412	7.97844
1 Sep 2013	21.7859	8.92715
1 Oct 2013	23.9481	9.84234
1 Nov 2013	26.173	10.7853
1 Dec 2013	28.3185	11.6955
1 Jan 2014	30.5283	12.6337
1 Feb 2014	32.7319	13.5699
1 Mar 2014	34.7176	14.414
1 Apr 2014	36.911	15.3467
1 May 2014	39.0294	16.2477
1 Jun 2014	41.2144	17.1773
1 Jul 2014	43.3254	18.0755
1 Aug 2014	45.5034	19.0023
1 Sep 2014	47.6783	19.9278
1 Oct 2014	49.7804	20.8224
1 Nov 2014	51.95	21.7455
1 Dec 2014	54.0472	22.6379
1 Jan 2015	56.2121	23.559
1 Feb 2015	58.3749	24.4791
1 Mar 2015	60.3269	25.3094
1 Apr 2015	62.4861	26.2277
1 May 2015	64.5742	27.1156
1 Jun 2015	66.7303	28.0323
1 Jul 2015	68.8156	28.9186
1 Aug 2015	70.969	29.8336
1 Sep 2015	73.1213	30.7479
1 Oct 2015	75.203	31.6319
1 Nov 2015	77.3531	32.5446
1 Dec 2015	79.433	33.4273

ый ... [user - Файловый менед... [tNavi

Figure 13. Daily gas production from the gas cap and accumulated oil and water production for the period 2012 – 2015.

Such a technology can be a method of improvement for type 2 (horizontal well).Two horizontal wellbores are drilling (Figure 17). One horizontal wellbore is located in the gas-saturated zone above the gas-oil contact, and another one - in the oil-saturated zone (Presnyakov et al., 2014).

An insulating agent is injected through the upper horizontal wellbore, which forms a screen at the gas-oil contact. Then water is pumped through the same horizontal wellbore to form a barrier. The lower horizontal wellbore is used for an oil production, which begins some time after the formation of the liquid barrier (Mostafa, Ismail, 1993).

In conclusion, it should be noted that in spite of the fact that the calculation results showed the greatest technological efficiency when using horizontal wells, it is impossible to exclude the economic aspect from attention. The cost of drilling horizontal wells is much higher than the cost of drilling vertical, which subsequently can offset the profit from the extra oil produced. Therefore, before considering such types for development, it is necessary to conduct a detailed technical and economic analysis.

When using different methods of developing oil/gas/condensate fields, it becomes very important to control the state of reservoir fluids and the movement of the boundaries of the rims, which will prevent gas breakthroughs and identify and eliminate the shortcomings of the adopted system. This is what ultimately determines the efficiency of the development of the field.

	Дебит нефти, ст.м3/сут	Массовый ... т/сут	Дебит воды, ст.м3/сут	Массовый ... т/сут
1 Dec 2012	0	0	0	0
1 Jan 2013	95.6652	81.3154	36.9712	37.156
1 Feb 2013	83.8686	71.2883	33.7969	33.9659
1 Mar 2013	78.7992	66.9793	32.389	32.5509
1 Apr 2013	76.1668	64.7418	31.6191	31.7772
1 May 2013	74.7739	63.5578	31.2322	31.3884
1 Jun 2013	73.8811	62.7989	30.985	31.1399
1 Jul 2013	73.2553	62.267	30.8105	30.9646
1 Aug 2013	72.7525	61.8396	30.6691	30.8225
1 Sep 2013	72.3417	61.4904	30.5535	30.7062
1 Oct 2013	72.0017	61.2015	30.456	30.6082
1 Nov 2013	71.7003	60.9452	30.3677	30.5196
1 Dec 2013	71.4466	60.7296	30.2917	30.4432
1 Jan 2014	71.2186	60.5358	30.2204	30.3715
1 Feb 2014	71.017	60.3644	30.1548	30.3056
1 Mar 2014	70.8535	60.2254	30.0999	30.2504
1 Apr 2014	70.6921	60.0883	30.0436	30.1938
1 May 2014	70.5519	59.9691	29.9929	30.1429
1 Jun 2014	70.4215	59.8583	29.9441	30.0938
1 Jul 2014	70.3059	59.76	29.8993	30.0488
1 Aug 2014	70.1976	59.668	29.8559	30.0052
1 Sep 2014	70.0982	59.5835	29.8148	29.9639
1 Oct 2014	70.0102	59.5087	29.7772	29.9261
1 Nov 2014	69.9259	59.437	29.7408	29.8895
1 Dec 2014	69.8499	59.3724	29.7072	29.8557
1 Jan 2015	69.7775	59.3108	29.6742	29.8225
1 Feb 2015	69.7112	59.2545	29.6428	29.791
1 Mar 2015	69.6557	59.2073	29.6156	29.7637
1 Apr 2015	69.5975	59.1579	29.5865	29.7345
1 May 2015	69.5459	59.114	29.5595	29.7073
1 Jun 2015	69.4971	59.0726	29.5325	29.6802
1 Jul 2015	69.4535	59.0355	29.5072	29.6547
1 Aug 2015	69.4109	58.9993	29.4816	29.629
1 Sep 2015	69.372	58.9662	29.4566	29.6039
1 Oct 2015	69.3376	58.937	29.4328	29.58
1 Nov 2015	69.3053	58.9095	29.4088	29.5558
1 Dec 2015	69.2766	58.8851	29.3857	29.5326

Figure 14.   Oil and water flow ratesof the well D_1 in the period 2012-2015.

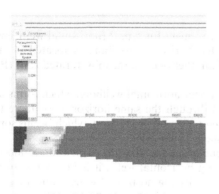 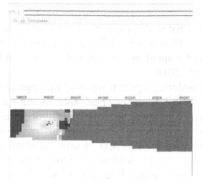

(a) - Distribution of gas saturation during the production with a vertical well

b) - distribution of gas saturation during the production with a vertical well with the formation of a water barrier

Figure 15.   (a) Distribution of gas saturation during the production with a vertical well. b) - distribution of gas saturation during the production with a vertical well with the formation of a water barrier.

Table 1. Results of calculations of three types of oil recovery from the oil rim.

Type	Time of beginning gas breakthrough, months	Critical value of gas-free production rate, m³/day	Accumulated oil production (2012-2015), thousand m³
1 (vertical production well)	7	72,5	76
2 (horizontal production well)	24	103,5	155,5
3 (water barrier)	30	69,5	80

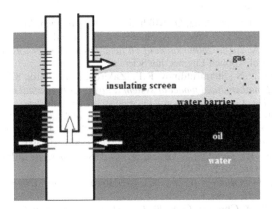

Figure 16. The process of simultaneous formation of a gel screen and a water barrier at the gas-oil contact (Mostafa, Ismail, 1993).

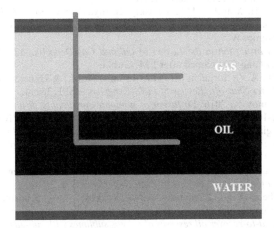

Figure 17. Location of the injection and production wellbores (Mostafa, Ismail, 1993).

REFERENCES

Al-Dhafeeri, A.M., Nasr-El-Din, H.A., Al-Mubarak, H.K., Al-Ghamdi, J. 2008. *Gas Shutoff Treatment in Carbonate Reservoir for Oil Wells in Saudi Arabia*. SPE 114323.
Ali, E., Bergren, F. E., DeMestre, P., Biezen, E., & Van Eijden, J. 2007. *Effective Gas-Shutoff Treatments in a Fractured Carbonate Field in Oman*. SPE 102244.

Denney, D. 2011. *Simulating, Simulation, and Optimal Control of Oil Production Under Gas-Coning Conditions.*SPE-143520.

Dorofeev NV, Taldykin SA, Kalugin, AA. 2008. *Causes And Ways of Minimizing Gas Breakthrough Into Mining Wells At Mine's Deposit. Yu. Korchagin.*: 21-35 Neftepromyslovoe delo, № 7.

Eoff, L., Vasquez, J., Recio, A., Reddy, B. R., & Pascarella, N. 2015. *Customized Sealants for Water/Gas Shutoff Operations in Horizontal and Highly Deviated Wellbore Completions.* SPE 174263.

Gian Luigi Chierici, Giuseppe M. Ciucci, Giuseppe Pizzi. 1964. *A Systematic Study of Gas and Water Coning By Potentiometric Models.* SPE 871 PA.

Hatzignatiou, D.G. Mohamed, F. 2001. *Water and Gas Coning in Horizontal and Vertical Wells.* PETSOC-94-26.

Herring, G.D., Milloway, J.T., & Wilson, W.N. 1984. *Selective Gas Shut-Off Using Sodium Silicate in the Prudhoe Bay Field, AK.* SPE 12473.

Lushpeev, V.A., Valerievna Tyukavkina, O., Mikhailovich Vasyanovich, M. 2014. *Method of determining the cause of water cut wells.*: 71-77 World Applied Sciences Journal.

Lushpeev, V., Margarit, A. 2018. *Optimization of oil field development process based on existing forecast model.*: 89-94. Journal of Applied Engineering Science.

Lushpeev, V.A, Fedorov, V.N., Sharafutdinov, R.F., Zakirov, M.F., Zatik, S.I. 2006. *Estimation of parametrical sensitivity of mathematical model of system a layer - A horizontal borehole at the direct hydrodynamic problem solving.:120-124.* Neftyanoe khozyaystvo - Oil Industry.

MacDonald, R.C. 1970. Methods for Numerical Simulation of Water and Gas Coning.SPE 2796 PA

Mjaavatten, A., Aasheim, R., Saelid, S.,& Gronning, O. 2006. *A Model for Gas Coning and Rate-Dependent Gas/Oil Ratio in an Oil-Rim* Reservoir *(Russian).* SPE 102390.

Mjaavatten, A., Aasheim, R., Saelid, S., & Groenning, O. 1984. *Supercritical Production From Horizontal Wells in Oil-Rim Reservoirs.* SPE-25048-PA.

Mostafa, Ismail. 1993. *Evaluation of Water and Gas Pattern Flooding Using Horizontal Wells in Tight Carbonate Reservoirs.* SPE 25599.

Onwukwe, S.I., Obah, B., & Chukwu, G.A. 2012. *A Model Approach of Controlling Coning in Oil Rim Reservoirs.* 163039-MSSPE.

Presnyakov A.Yu., Lomakina I.Yu., Nigmatullin T.E., Razyapov R.K., Sorokin A.S. 2014. *Integrated approach to the choice of technology for water and gas inflow restriction in the conditions of the Yurubcheno-Tokhomskoye field.*: 77-83. Oil industry, No. 6.

Tomskaya LA, Krasnov II, Marakov DA, Tomskiy IS, Inyakin VV. 2016. *Isolation technologies of gas supply restriction in oil wells of Western Siberia fields.*: 69–73. Bullet in of the North-Eastern Federal University . M.K. Ammosov.

Suslova AA. 2015. *Gas Insulation in the Layers of Oil and Gas Deposits.*: 123–127. Thesis, the Russian State University of Oil and Gas named after I.M. Gubkin.

Zhdanov, S.A., Amiyan, A.V., Surguchev, L.M., Castanier, L.M., & Hanssen, J.E. 1996. *Application of Foam for Gas and Water Shut-off: Review of Field Experience.* SPE 36914.

Tananykhin D., Khusainov R. 2016. *Diffusion of nonionic surfactants diffusion from aqueous solutions into viscous oil.*:1984-1988. Petroleum Science and Technology. doi:10.1080/10916466.2016.1233245.

Tananykhin D.S., Shagiakhmetov A.M.2016. *Justification of technology and fluids for treatment of the unconsolidated carbonate reservoirs.*: 744-748. International Journal of Applied Engineering Research (ISSN: 09734562), 11(1).

Zhdanov, S.A., Amiyan, A.V., Surguchev, L.M., Castanier, L.M., & Hanssen, J.E. 1996 *Application of Foam for Gas and Water Shut-off: Review of Field Experience.*SPE 36914.

Yazkov, A.V., Gorobets, V.E., Zagaynov, A.N., Yazkov, A.V., Kudrin, P.A., Poushev, A.V., Milushkin, A.M. 2006. *The use of horizontal wells with complex completion as one of the ways of efficient processing of hard-to-recover oil reserves of subgas sublimes with sub-gas water.* SPE 181907.

Topical Issues of Rational Use of Natural Resources 2019 – Litvinenko (Ed)
© 2020 Taylor & Francis Group, London, ISBN 978-0-367-85720-2

Energy-efficient small-scale liquefied natural gas production technology for gas distribution stations

V.A. Voronov & A.Yu. Ruzmanov
Saint Petersburg Mining University, Saint Petersburg, Russia

ABSTRACT: Gas distribution stations are designed for the main gas flow throttling. The gas pressure is reduced by the reduction unit, which contains throttle valves. Most of the transmitted by compressors at the compressor station the main gas flow energy lost during this process. However, there is energy extraction and useful application possibility. The simplest way to use the main gas flow power is to generate electricity. However, one of the most perspective direction is to produce liquefied natural gas (LNG). During work on the solution of a main gas flow excess energy usage for LNG production problem, the highly efficient method for LNG producing was developed. The method efficiency is formed due to the optimal use of energy obtained by expanding the main gas flow.

1 INTRODUCTION

Gas distribution stations are designed for the main gas flow throttling. The gas pressure is reduced by the reduction unit, which contains throttle valves. In this way, the pressure is reduced from 3-6 MPa either to 1.2 MPa (for industrial customers) or to 0.6 MPa (for the domestic customers gas distribution lines). Most of the transmitted by compressors at the compressor station main gas flow energy lost during this process. However, there is energy extraction and useful application possibility.

The most suitable energy recovery method for gas distribution stations is an expansion turbine. Turboexpanders (Figure 1) is a facility are designed for the excess pressure energy conversion. Such equipment is simple in operation and has high-reliability level (Davydov 1987).

2 MAIN GAS FLOW ENERGY EXTRACTING

Energy is transferred to the turboexpander runner due to flow and runner blades (Figure 2) interaction. Modern turboexpander can be either single(two)-stage with reactive gas expansion nature type or centripetal radial type or radial-axial type. From the energy saving point of view, the most promising technology is a radial-axial type turboexpander. Turboexpanders with one-sided half-opened radial-axial runners are the best option for gas distribution stations with high gas flow rates (Davydov 1987). These turboexpanders are the easiest to operate and have the highest reliability due to high runners strength characteristics.

At the moment there is a large number of patented gas distribution stations with turboexpander construction, but all of them have the same problem is the parallel to the station turboexpander connecting (STC "MTT" 2011). It leads to incomplete the main gas flow excess energy extraction. A much more effective way to connect the turboexpander to the gas distribution station lines is to connect the unit in parallel to the station reduction unit. This solution allows to reduce the turboexpander installation time, provides the ability to ensure the minimum additional pipelines length and, most importantly, let to use the entire main gas flow energy (Voronov 2018).

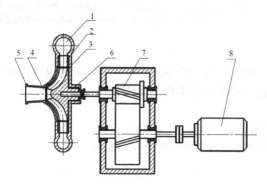

Figure 1. Turboexpander: 1 – inlet nozzle; 2 - distributor; 3 - runner; 4 - diffuser; 5 - body; 6 - pressure breakdown labyrinth; 7 – reduction gear; 8 – electric generator.

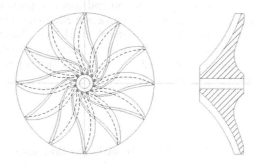

Figure 2. Runner.

3 THE NET POWER USAGE

The turboexpander unit net power usage when it is put into gas distribution station operation is one of the key issues from both economic and technological points of view. The simplest way to use turboexpander unit net power is to generate electricity. However, one of the most perspective net power realization direction is to produce liquefied natural gas (LNG). A large amount of experience in the turboexpanders application in such tasks framework and low temperature (about 200K) at the outlet of the turboexpander contribute to the implementation of such project. During work on the solution of the main gas flow excess energy usage for LNG production problem, the highly efficient method for LNG producing was developed (Figure 3). The method efficiency is formed due to the optimal usage of energy obtained by expanding the main gas flow (Ruzmanov 2018).

The turboexpander unit net power is used to generate electrical energy. After expansion in the turboexpander the gas flow is divided into the technological flow intended for producing LNG and the production flow intended for supply to the customers. The technological flow is directed into LNG integrated expander and flash cycle. The production flow is odored after additional heating and directed to the gas distribution station output. The discharge of the boil-off gas from the LNG storage tanks is carried out in the condensate collector following the flash exchanger.

4 THE METHOD EFFECTIVENESS

The turboexpander thermodynamic (Pavlenko, 2007) and structural calculations (Bumagin 2007) was performed to confirm the method efficiency, as well as LNG performance

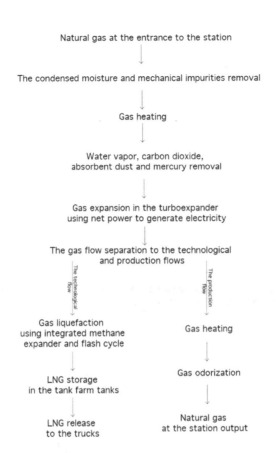

Figure 3.   LNG producing method.

calculation for the following gas distribution station options (R Gazprom 2014 & STO Gazprom 2014):

– inlet pressure: 5.4 MPa;
– outlet pressure: 0.6 MPa;
– average gas consumption: 21760 m³ per hour.

The method effectiveness is confirmed by the graphs shown in Figure 4. In addition to high energy efficiency this solution allows providing increased station reliability due to the availability of an alternative reduction method by using the gas distribution station reduction unit.

The gas flow separation into the technological flow intended for LNG production and the production flow intended for supply in the gas distribution pipeline net after passing the turboexpander allows providing significant gas flow cooling from 273-281K (at 5.2-5.4 MPa) to 195-200K (at 0.6 MPa). This makes it possible to reduce LNG production energy consumption, which is confirmed in the graphs shown in Figure 5 (Tseytlin 2007).

The graphs shown in Figure 5 confirm that for the gas temperature of 290 K and pressure of 5 MPa the specific energy required to liquefy natural gas more than 4 times exceeds the same value for the turboexpander output gas conditions of 0.6 MPa at a temperature of 200 K. Thus, natural gas liquefaction energy consumption after gas passing through the turboexpander unit is reduced more than 4 times.

The main gas flow reduction is performed by using the turboexpander unit and compression within the main gas liquefied cycle is performed by using the compressor as a separate device. It allows increasing the method reliability in comparison with the methods providing for the expander-compressor unit usage.

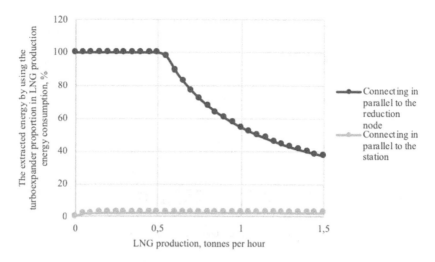

Figure 4. The dependence of the extracted energy by using the turboexpander proportion in LNG production energy consumption for the LNG production.

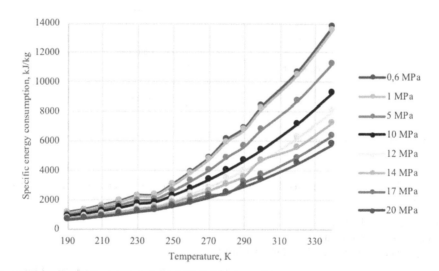

Figure 5. The dependence of LNG production specific energy consumption on temperature for different gas pressure values.

The turboexpander and the integrated methane expander and flash cycle expander net power is used for electric energy generation. This electrical energy is then used to cover the need for liquefied gas equipment in electricity.

5 SMALL-SCALE LNG PLANT BASED ON THE METHOD

Based on the developed small-scale LNG production method the plant has been designed. The scheme of the plant is shown in Figure 6 (Ruzmanov 2018).

The gas from the gas pipeline enters to the gas distribution station inlet and then goes to the gas distribution station treatment unit (1), where it passes the initial cleaning from dust and drip moisture, then gas is heated in the heating unit (2), if necessary.

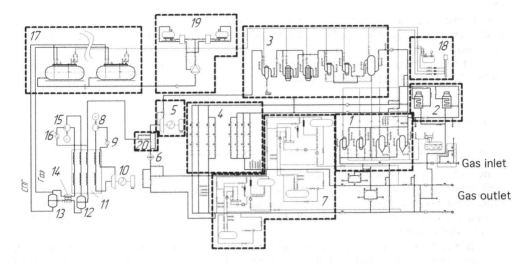

Figure 6. Small-scale LNG plant on the gas distribution station. 1 – treatment unit; 2 – heating unit; 3 – additional treatment unit; 4 – reduction unit; 5 – turboexpander and electrical generator; 6 – additional heating unit; 7 – odorization unit; 8 – main refrigeration cycle compressor with electric motor; 9 – main heat exchanger-refrigerator; 10 – main refrigeration cycle expander and electrical generator; 11 – main heat exchanger; 12 – condensate tank; 13 – condensate tank; 14 – flash exchanger; 15 – LNG vapor liquefaction cycle compressor with electric motor; 16 – LNG vapor liquefaction cycle refrigerator; 17 – LNG tank farm; 18 – flare unit; 19 – LNG trucks loading unit; 20 – gas separation unit.

If turboexpander unit (5) is in the operation, then the gas passes through the additional gas treatment unit (3), which provides for the drop liquid removal to avoid the hydrates formation in the cryogenic section. Further purification from water vapor and carbon dioxide (CO_2) is carried out in desiccant devices filled with adsorbent. Molecular sieves are used as an adsorbent. The dried and purified natural gas passes through dust collector type filters for adsorbent dust removal. Mercury vapor removal provides by using specialized membrane adsorber. If the turboexpander isn't being used, then gas distribution station reduction unit (4) is used.

The gas flow is separated in the gas separating unit (20) at the technological and the production flows after gas expansion in the turboexpander and electrical generator unit (5).

The production flow is directed to the gas odorization unit (7) through the additional heating device (6) designed to increase the gas temperature to the value required by normative documentation. After this, the gas is recorded and the flow is directed to the gas distribution networks to consumers.

The technological gas flow at a low temperature is directed to the main heat exchanger (11), then it appears in the condensate tank (12). The technological gas flow cooling is provided by using the main heat exchanger (11) with refrigerator (9), compressor (8) and expander (10). The main heat exchanger cooling efficiency is increased by heat exchange with the LNG vapor from the condensate tanks (12 and 13). Boil-off LNG vapor is re-liquefied in the LNG vapor liquefaction cycle with the compressor (15) and the refrigerator (16).

When LNG flows from the condensate tank (12) to the condensate tank (13) the LNG vapor is subjected to heat exchange with LNG from the condensate collector (12) within the flash exchanger (14) before entering the LNG vapor liquefaction cycle. It allows to significantly reduce energy costs within the LNG vapor liquefaction cycle.

From the condensate tank (13) LNG flows into the tank farm (17), which consists of cryogenic tanks equipped with the necessary set of shut-off and safety valves. Storage in the tank farm is carried out at constant pressure and temperature.

The constant pressure is maintained by the regular LNG vapor removal from the tanks into the LNG vapor discharge line, then it is mixed with the LNG vapor in the condensate tank

905

(13). However, in case of exceeding the regulated pressure in the tanks, it is provided that the LNG vapor stream is discharged into the flare unit (18), which is also used during the purging equipment process.

The LNG distribution to the road tankers is carried out by using special cryogenic pumps within the LNG trucks loading unit (19).

6 CONCLUSIONS

Thus, as a result of the work on the LNG production method was established:

– the turboexpander connection in parallel to the reduction unit has an advantage in terms of efficiency over the method of connection in parallel to the entire station;
– the main gas flow separation on the technological and production flows after the turboexpander allows to reduce the LNG production specific energy consumption due to the flow additional cooling;
– net power usage for power generation and its subsequent consumption due to refrigeration cycles compressors actuation allows to ensure the mutual independence of the compressors and expander work. It positively affects plant reliability.

REFERENCES

Bumagin, G.I. 2007. Calculating machines. The calculation of the turbo expander. *Cryogenic machinery.* Omsk: OmSTU Publisher.
Davydov, A.B., Kobulashvili, A.S. & Sherstyuk A.N. 1987. Calculation and design of turbines. Moscow: Machinery.
Pavlenko G.V., Volkov, A.G. 2007. The gas dynamic calculation of the centerline of the gas turbine. Kharkiv: HAI.
R Gazprom 2-6.2-600-2011. 2012. "Application of turbo-expander energy mustache tings and calculation of the main parameters.". Moscow: JSC "Gazprom".
Ruzmanov A. Yu., Voronov V.A. 2018. e.a. Sposob poluchenija szhizhennogo prirodnogo gaza v uslovijah gazoraspredelitel'noj stancii [The way of liquefied natural gas producing at the gas distribution station conditions]. Patent RF, no 2665088.
Ruzmanov A.Yu., Voronov V.A. 2018. e.a. Ustanovka szhizhenija prirodnogo gaza (SPG) v uslovijah gazoraspredelitel'noj stancii (GRS) [Natural gas liquefaction (LNG) utility for gas distribution station (GDS)]. Patent RF, no 2673642.
STC "MTT". 2011. Review of modern turbo-expander designs [Text]. St. Petersburg: STC "MTT".
STO Gazprom 2-3.5-748-2013. 2014. "Turbo-expander units. Typical technical requirements.". Moscow: JSC "Gazprom".
Tseytlin, A.M. 2007. Minimal energy expenses for liquefied natural gas production by using one-stage expanding refrigeration cycles. Astrakhan: AGTU.
Voronov, V.A., Ruzmanov, A. Yu. & Samigullin, G. H. 2017. Turboexpander application at gas distribution stations in order to produce liquefied natural gas. *NEFTEGAZ.RU journal, #10, 2017.* Moscow: NEFTEGAZ.RU.

Topical Issues of Rational Use of Natural Resources 2019 – Litvinenko (Ed)
© 2020 Taylor & Francis Group, London, ISBN 978-0-367-85720-2

Physical and mathematical modelling of waterflooding low-permeability reservoirs using surfactants

A.S. Sukhikh, M.K. Rogachev & A.N. Kuznetsova
Saint Petersburg mining university, Russia

A.A. Maltcev
Saint Petersburg mining university, Russia
Peter the Great St. Petersburg Polytechnic University, Russia

ABSTRACT: This paper has reviewed the current state of development of oil fields with a low-permeable, clay-rich reservoir. Considered criteria for the applicability of the use of surfactants for enhanced oil recovery. On the basis of laboratory results, the effectiveness of using surfactants in the waterflooding system has been proved. According to the results of hydrodynamic modeling, additional oil production from the use of surfactants in the waterflooding system on a specific case was predicted. The results of the multivariate calculations of the injection of the surfactant solution are presented.

1 INTRODUCTION

Currently, in our country, most fields with traditional reservoirs are already developed or are in the late stages of development. There is a decrease in reservoir permeability, a rapid approach to the shale threshold of oil reservoirs introduced into the development. Such layers are characterized by low values of oil recovery factor, their development may be unprofitable. One of the ways to increase the profitability of such reserves may be the use of such a method of enhanced oil recovery as waterflooding with the use of surfactants. The reasons are features of the geological structure of deposits (degraded properties of the reservoir and reservoir fluid, the manifestation of the rheological properties of the fluid, a relatively low concentration of reserves in the area and in the section, small reservoir sizes, low permeability, etc.) as well as remoteness from the main fields and areas of oil production. These factors increase the risks associated with the development of deposits with hard-to-recover reserves, and reduce the economic attractiveness of projects for their development.

Traditionally, an effective method of developing oil reservoirs is flooding, implemented through a system to maintain reservoir pressure. However, in the case of flooding of a low-permeable reservoir, the economic efficiency of this method is significantly reduced. One of the ways to increase the profitability of such reserves may be the use of such a method of enhanced oil recovery as waterflooding using surfactants.

From a technical point of view, based more on laboratory than field tests, chemical flooding of oil reservoirs is one of the most successful methods of enhanced oil recovery (EOR) in depleted reservoirs with low reservoir pressure. However, as is well known from technical literature, chemical flooding hardly pays for itself, and in most cases turns out to be unprofitable (Surguchev, 1985; Schramm, 2000; Shturn, 2004). Initially, the goal of chemical flooding was to extract additional amounts of oil left over from traditional flooding. Therefore, chemical flooding is considered as a tertiary method of oil production (Lake 1989; Khavkin, 1993). However, in the late 1980s, oil companies came to the conclusion that this method was unprofitable. Subsequently, the number of studies in this area was significantly reduced. Nevertheless, a number of research groups continued to look for ways to improve this technology by simplifying the waterflooding process, increasing the effectiveness of surfactants, and developing new reagents.

One of such research groups from the St. Petersburg Mining University was able to obtain a new surfactant composition (patent RU 2655685), which showed the synergistic properties of its constituent substances in reducing the interfacial tension at the phase boundary and inhibiting clay swelling (Kuznetsova, Sukhikh, 2018; Rogachev, 2016). The authors cite laboratory studies of the resulting composition, according to the results of which the composition is characterized as highly effective. This composition effectively reduces the interfacial tension, effectively inhibits clay swelling. Also, according to the results of filtration tests, the composition is characterized by low adsorption.

2 METHODS

After successfully conducting laboratory studies of the surfactant composition and obtaining positive results on its supposed effectiveness, it is necessary to evaluate the profitability of using this composition in flooding low-permeable, clay-filled reservoirs. For such an assessment, it was decided to perform a hydrodynamic simulation, which was carried out in a commercial hydrodynamic simulator tNavigator (Rock Flow Dynamics).

To simulate the injection of a surfactant solution into the oil reservoir, a small reservoir of the N deposit (Western Siberia) was selected, which is represented by Upper Jurassic deposits. The calculations were carried out using the geological and hydrodynamic model, which was created using all the information on the deposit. Based on the geological model of the reservoir, a hydrodynamic model was created taking into account the data obtained from the results of core studies. Core surveys were carried out on samples obtained from the well that uncovers this reservoir. Also, based on the results of in-depth tests, a PVT model was created. Table 1 shows some parameters characterizing the selected reservoir. Also shown below is the permeability distribution of the modeled formation in space (Figure 1).

After receiving the reservoir model, the model was adapted. Adaptation of the model was made based on actual data. Adaptation results are presented in Figure 2.

To simulate chemical flooding, some laboratory studies are needed to correctly account for the effect of surfactants on the properties of rock and fluids. It is necessary to establis (Petrov, 2004; Moreno, 2018):

- Surfactant adsorption vs surfactant concentration;
- Oil-water IFT vs surfactant concentration;
- Capillary de-saturation curve (indirect measurements, but can be calculated from dependence capillary number vs residual saturation);
- Modification of relative permeability.

Table 1. The main parameters of the layer UV1-1.

Parameter	Values
The average depth of the roof, m	2905-2983
Type of deposit	layer-uplifted
Collector type	sandstone
Mean total thickness, m	13.1
Effective permeability, mD	2.7
RF current, u. s.	0.099
RF project, u. s	0.37
Start of development	01.2013

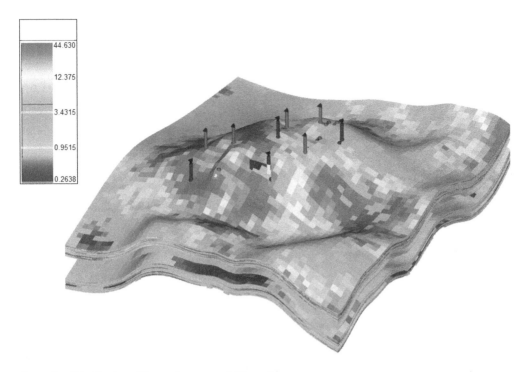

Figure 1. Distribution of formation permeability, mD.

All data, except for the Capillary de-saturation curve (CDC) can be obtained from the results of laboratory tests. To get a CDC, you must perform a series of calculations (Sheng, 2011). Capillary number is calculated by the Equation 1:

$$N_c = \frac{v \cdot \mu}{\sigma} \tag{1}$$

where N_c – capillary number; v - velocity of filtration; μ - viscosity of fluid; σ – interfacial tension.

However, if we consider that the filtering rate:

$$v = \frac{k \cdot \Delta P}{\mu \cdot L} \tag{2}$$

then Equation 1 takes the following form:

$$N_c = \frac{k \cdot \Delta P}{\sigma \cdot L} \tag{3}$$

According to the results of laboratory tests, the dependence of the residual saturation on the capillary number is constructed. The obtained data are normalized and reduced to the following form (Figure 3).

After obtaining the required data set, it is necessary to determine the technology of surfactant injection into the oil reservoir, namely, to choose a well through which the surfactant injection modeling and reacting wells will be carried out. In the model of this reservoir for injection of surfactant, well 1 was selected, reacting wells - 2, 3, 4 (Figure 4)

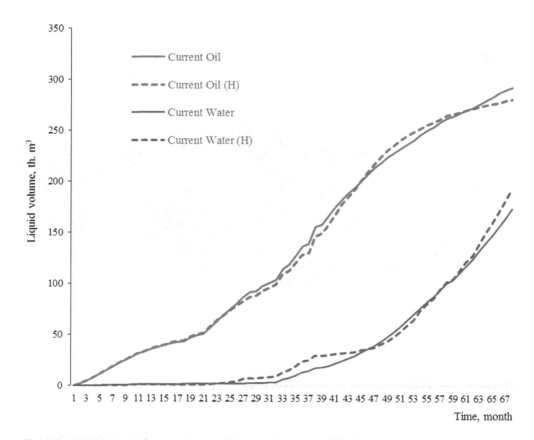

Figure 2. Model adaptation results.

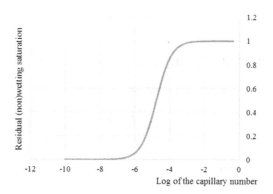

Figure 3. Capillary de-saturation curve.

3 RESERCH

After the adaptation of the model and the choice of surfactant injection technology, a series of multivariate hydrodynamic calculations was performed. In these calculations, the parameters of the surfactant concentration in the injected solution, the volume of the injected surfactant rim (pore volume) and the residual oil saturation at the time of the

Figure 4. Technology implementation of waterflooding with surfactants.

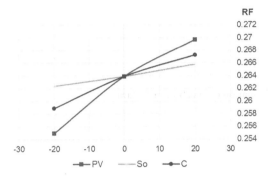

Figure 5. Sensitivity analysis of recovery factor.

start of surfactant injection were varied. The last parameter was changed due to the different start of surfactant injection.

According to the obtained data, the sensitivity factor recovery factor (RF) was analyzed, which showed that RF is most dependent on the pore volume of the surfactant composition (Figure 5).

All the results of the calculations were compared by technical and economic indicators - economic recovery factor (ERF), profitability index (PI), and RF. The results are presented in Figure 6.

According to the results of calculations, it is clear that the best values of the parameters correspond to different variants of calculation. The best ERF corresponds to the variant with the latest onset of surfactant injection. The highest PI was obtained as a result of the calculation with the earliest onset of surfactant injection. The option with the lowest surfactant concentration corresponds to the highest value of the chemical efficiency coefficient. And the highest RF value is obtained in the version with the largest rim of the pumped surfactant solution. Therefore, the final choice of the best result was difficult. However, if we compare the best results for each indicator among themselves, then the set of parameters makes the best option obvious (Figure 7).

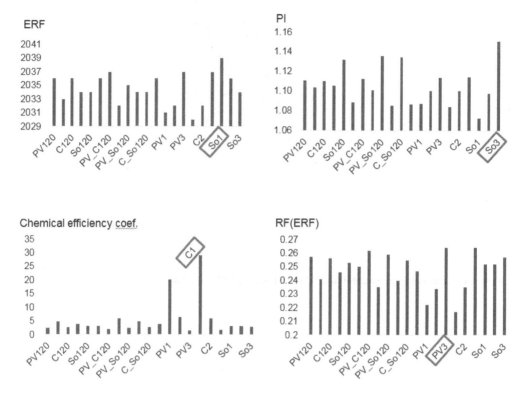

Figure 6. Results of simulation of injection surfactant.

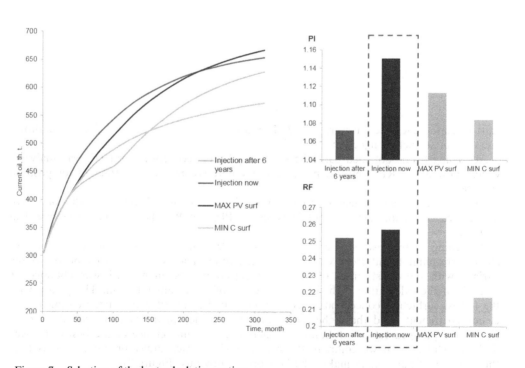

Figure 7. Selection of the best calculation option.

The option with surfactant injection at the moment with the greatest residual oil saturation (early start of injection) is characterized by the best PI value and high RF value. This is the best technical and economic option.

4 CONCLUSIONS

The above algorithm is a good example of the design of chemical flooding. It includes laboratory tests of chemical composition on a real core and calculations of forecast technical and economic indicators of a deposit. Investigated in this work, the composition according to the results of laboratory tests is characterized by high efficiency. Projected reservoir indicators confirm the possible technological and economic efficiency of this method of enhanced oil recovery.

According to the results obtained in the course of this work, we can say that the use of technology to injection surfactant into the reservoir under the conditions of modern technological challenges will make it possible to develop low-permeability clay reservoirs cost-effectively and increase the final oil recovery.

REFERENCES

Khavkin A.Ya. 1993. New directions and technologies for developing low-permeability layers, *Oil Industry*, - №3.

Kuznetsova A.N., Rogachev M.K., Sukhikh A.S. 2018. Surfactant solutions for low-permeable polimictic reservoir flooding. *IOP Conf. Series: Earth and Environmental Science 194 (2018) 042011.*

Lake L.W. 1989. Enhanced Oil Recovery. Prentice Hall, Inc., New Jersey.

Moreno J. E., Flew S., Gurpinar O., Liu Y., Gossuin J. 2018. Effective use of laboratory measurements on EOR planning. *Offshore Technology Conference.* - Houston, Texas, USA.

Petrov N.A. et al. 2004. Cationic surfactants - effective inhibitors in the technological processes of the oil and gas industry. - St. Petersburg: *Nedra*, - 408.

Rogachev M.K., Kuznetsova A.N. 2016. Research and development of solutions for surfactants for the flooding of low-permeability polymictic reservoirs. *Scientific and technical journal "Engineer-oilman".* Vol. 1: 49–53.

Sheng J.J. 2011. Modern chemical enhanced oil recovery. – Oxford, U.K. *Elsevier.*

Shturn L.V. 2008. Features of the development of oil fields in Western Siberia with low-permeability reservoirs. *Scientific and Technical Journal "Territory of Neftegaz"*, No. 2, from 64–69.

Schramm L. 2010. Surfactants: Fundamentals and applications in the petroleum industry. – Cambridge: *Cambridge University Press.* – 589.

Surguchev M. L. 1985. Secondary and tertiary methods of enhanced oil recovery. – Moscow: *Nedra.*

Topical Issues of Rational Use of Natural Resources 2019 – Litvinenko (Ed)
© 2020 Taylor & Francis Group, London, ISBN 978-0-367-85720-2

Preserving the quality of petroleum products when mixed in tanks

R.R. Sultanbekov & M.N. Nazarova

Saint-Petersburg Mining University, Saint-Petersburg, Russian Federation

ABSTRACT: The article analyzes the problem of the formation of a total sediment in the mixture due to the incompatibility of petroleum products, which occurs when petroleum products are mixed in tanks. The problem of determining compatibility before mixing petroleum products in tanks is considered. Also reviewed are existing methods for determining compatibility. The proposed solution for accurate determination of the compatibility of the mixed petroleum products. Studies of the effect of mixing on sedimentation.

1 INTRODUCTION

In tank farms and fuel terminals often mix different types of petroleum products. Objects for the storage of petroleum products must operate efficiently in order to provide consumers with quality energy resources. The technological process is continuous, therefore often in tanks with residues, a new product is drained due to the lack of free tanks. In some cases, the mixture occurs intentionally to obtain another brand of petroleum product with a lower viscosity, which can be used to heat industrial and residential facilities and as marine fuels. However, during technological operations on tanks, there is always a residue of petroleum products, which can affect the quality when mixing different types of petroleum products.

In recent years, the world has established a tendency for deeper refining, namely the use of residual products of cracking processes. However, along with the deepening of refining, the quality of petroleum products is reduced due to the increase in the proportion of asphalt-resinous products in the composition of heavy fuels. Therefore, the stability and compatibility of petroleum products is an important parameter. Since when mixing different types of dark oil there is a risk of precipitation (Mitusova et al. 2005).

A high total sediment in petroleum products adversely affects the operation of engines and fuel systems, contributes to their wear and disruption, and also leads to clogging of filters and separators (Karimov & Mastobaev 2012). Also, precipitation of the total sediment contributes to the active accumulation of the "dead" residue in the tanks, which may lead to a deterioration in the quality of the oil being drained and reduce the useful volume of the tanks. The permissible total sediment content is governed by standards, in Russia it is GOST R 33360-2015, which complies with international standards ASTM D4870-IP 375, IP 390 and ISO 10307. According to ISO 8217, in fuels the total sediment content should not exceed 0.1%.

There are several actual problems that are caused by the incompatibility of petroleum products during technological operations on tanks:

- Deterioration of the quality of petroleum products;
- Promotes wear and clogging of process systems;
- An increase in the intensity of accumulation of the "dead" residue on the tanks, which leads to a decrease in the useful volume of the tanks;
- Economic losses due to lower prices for petroleum products.

2 MATERIALS AND METHODS

Sediment formation occurs due to the imbalance and stability of fuels, because the proportion of aromatic hydrocarbons decreases and the proportion of paraffins and asphaltenes in the mixture increases products (Mitusova et al. 2018; Kondrasheva et al. 2018). However, when mixed, there is a risk of static electricity and precipitation due to fuel incompatibility (Sultanbekov & Nazarova 2019). Research is underway to determine and improve the stability of marine fuels (Kondrasheva et al. 2019; Sultanbekov & Nazarova 2019).

Manifestations of "incompatibility" when mixing petroleum products are associated with the emergence of strong intermolecular interactions caused by changes in the structural group composition and the relative ratio of the concentrations of high molecular compounds of petroleum products, which leads to the formation of associates of molecules, bulk colloidal particles of various shapes and structures .

There are several methods for assessing the stability and compatibility of the fuel. The simplest is the method of determining the stability and compatibility of the spot according to GOST R 50837.7. The essence of the method is the visual assessment of the core and stain stains on the paper filter, formed by a drop of the sample under test conditions, and comparing the stain with standard stains. However, this method does not have high accuracy and has a large error.

There are known methods for determining toluene and xylene equivalent according to GOST R 50837.3 and GOST R 50837.4, respectively, which make it possible to judge the stability of petroleum products.

Toluene and xylene equivalents are criteria for the stability of the dispersed structure, showing the degree of aromaticity of the fuel, which is necessary to preserve asphaltenes in a dispersed state.

The value of the xylene equivalent, less than or equal to 25/30, is a criterion for straightness.

A common drawback of both methods is the low accuracy due to the subjectivity of the visual assessment of the presence of a darker spot inside the entire spot from a drop of a dilute portion of fuel with a mixture of toluene or xylene with n-heptane.

The method for determining straightness in mixed fuels by the indicator "spot number" (GOST R. 50837.7-95. Method for determining stability and compatibility by spot).

This method involves taking a sample of fuel in an amount of at least 500 cm3, preparing it, mixing the sample with the distillate component in a 1: 1 ratio, heating the mixture (up to 100 °C, estimating the precipitate released for 1 hour. By the look of the stain on the chromatographic paper Number from №1 to №5 is assigned.

The disadvantage of this method is the relatively low accuracy and accuracy of determination of secondary residual products of oil refining in composite fuels, associated with the subjectivity of visual assessment of the type of spot and the need to use foreign materials.

Existing methods for assessing the stability and compatibility of petroleum products do not have high accuracy and have a large error.

To increase the stability of petroleum products, it is possible to use dispersing additives or, thanks to an accurate method for determining compatibility before mixing fuels, it is possible to select the necessary components of petroleum products (Mitusova et al. 2017).

To determine the compatibility of petroleum products, an algorithm has been developed that allows obtaining results accurately and quickly.

The principal difference of the method used is that to determine the compatibility of several types of petroleum products, first of all, it is necessary to carry out tests to determine the total sediment with preliminary chemical aging (Total Sediments Accelerated - TSA) or thermal aging (Total Sediments Potential – TSP) according to the method of GOST R 33360-2015 for each component of the mixture, which are planned for mixing. Further, after determining the total sediment for each of the components, these petroleum products should be thoroughly mixed in the desired proportion and then perform tests to determine the total sediment of the prepared mixture of petroleum products.

The experiments were carried out in a laboratory installation of Total Sediment Tester of Seta Clean (Figure 1), to which a vacuum pump is connected, an oil bath was also used to

Figure 1.   - Laboratory Equipment Seta Clean Total Sediment Tester.

Table 1.   – Determination of the total sediment of the components at mixture ratios of 1:1.

Component №1. Fuel oil 100, low-ash (STO 05747181-2013 with changes 1-2).
Quality indicators: $\rho_{15} = 989,0 kg/m^3$; $v_{50} = 590,5 \frac{mm^2}{s}$ (kinematic vis cos ity at 50 Deg C); $TSA = 0,03\%$;

№	Component 2	TSA, %
1	Marine Fuel Brand IFO - 30 (STO 72468105-002-2010). $\rho_{15} = 1010,0 kg/m^3$; $v_{50} = 12,3 \frac{mm^2}{s}$; $TSA = 0,01\%$	**0,13**
2	Distillate medium gas condensate, type I (STO 05766575-129-2016). $\rho_{15} = 874,0 kg/m^3$; $v_{40} = 3,4 \frac{mm^2}{s}$; $TSA = 0,00\%$	0,02
3	Gas Condensate Fuel, Brand 100 GKT (TU 0252-060-05780913-98) $\rho_{15} = 933,4 kg/m^3$; $v_{50} = 21,4 \frac{mm^2}{s}$; $TSA = 0,05\%$	**0,33**
4	Burner Oil (Dark) (TU 0252-038-48418772-2003) $\rho_{15} = 1008,4 kg/m^3$; $v_{50} = 14,45 \frac{mm^2}{s}$; $TSA = 0,01\%$	0,02
5	Heavy Vacuum Gas Oil (TU 0258-001-76453499-2015). $\rho_{15} = 935,0 kg/m^3$; $v_{50} = 14,1 \frac{mm^2}{s}$; $TSA = 0,01\%$	0,04
6	Heavy Pyrolysis Oil, Brand A (TU 2451-183-72042240-2013). $\rho_{15} = 1034,0 kg/m^3$; $v_{50} = 17,2 \frac{mm^2}{s}$; $TSA = 0,03\%$	**0,18**
7	Marine Ecological Fuel, Brand A (STO 00044434-033-2014) $\rho_{15} = 892,8 kg/m^3$; $v_{50} = 13,4 \frac{mm^2}{s}$; $TSA = 0,01\%$	0,02
8	Marine Ecological Fuel, Brand A (STO 00148599-034-2017). $\rho_{15} = 852,0 kg/m^3$; $v_{50} = 19,5 \frac{mm^2}{s}$; $TSA = 0,01\%$	**0,24**
9	Fuel Oil 100 IFO – 180 (TU 0252-002-10416871-2013) $\rho_{15} = 951,3 kg/m^3$; $v_{50} = 165,7 \frac{mm^2}{s}$; $TSA = 0,03\%$	0,04
10	Fuel Oil 100 IFO - 380 (GOST 10585-2013). $\rho_{15} = 959,0 kg/m^3$; $v_{50} = 269,2 \frac{mm^2}{s}$; $TSA = 0,02\%$	0,03
11	Residual product processing (tar, oil) (STO 73171028-003-2015) $\rho_{15} = 967,4 kg/m^3$; $v_{50} = 9210,8 \frac{mm^2}{s}$; $TSA = 0,05\%$	0,07
12	Fuel Oil 100 (GOST 10585-2013). $\rho_{15} = 978,5 kg/m^3$; $v_{50} = 561,7 \frac{mm^2}{s}$; $TSA = 0,02\%$	0,03
13	Fuel Oil 100, Ash (GOST 10585-2013). $\rho_{15} = 999,0 kg/m^3$; $v_{50} = 679,9 \frac{mm^2}{s}$; $TSA = 0,03\%$	**0,12**
14	Fuel oil 100 low-ash (GOST 10585-2013) $\rho_{15} = 984,0 kg/m^3$; $v_{50} = 695,1 \frac{mm^2}{s}$; $TSA = 0,03\%$	0,04
15	Fuel oil 100 type I (GOST 10585-2013) $\rho_{15} = 833,1 kg/m^3$; $v_{50} = 4,66 \frac{mm^2}{s}$; $TSA = 0,02\%$	**0,17**

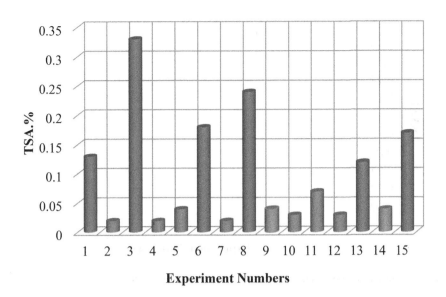

Figure 2.   - The content of total sediment components at a mixture ratio of 1:1.

Figure 3.   – Filters after laboratory tests №6.

Table 2.   - Determination of total sediment at different ratios of the mixture of components.

Component №1. Fuel oil 100, low-ash (STO 05747181-2013 with changes 1-2).
Quality indicators: $\rho_{15} = 989,0 kg/m^3$; $\upsilon_{50} = 590,5 \frac{mm^2}{s}$ (kinematic vis cos ity at 50 Deg C); $TSA = 0,03\%$;

Component №2. Heavy pyrolysis resin, grade A (TU2451-183-72042240-2013).
Quality indicators: $\rho_{15} = 1034,0 kg/m^3$; $\upsilon_{50} = 17,2 \frac{mm^2}{s}$ (kinematic vis cos ity at 50 Deg C); $TSA = 0,03\%$,

№	Component mixing ratio, %	TSA, %
1	№1 – 90, №2 – 10	0,08
2	№1 – 80, №2 – 20	0,10
3	№1 – 70, №2 – 30	0,13
4	№1 – 60, №2 – 40	0,15
5	№1 – 50, №2 – 50	0,18
6	№1 – 40, №2 – 60	0,15
7	№1 – 30, №2 – 70	0,12
8	№1 – 20, №2 – 80	0,09
9	№1 – 10, №2 – 90	0,06

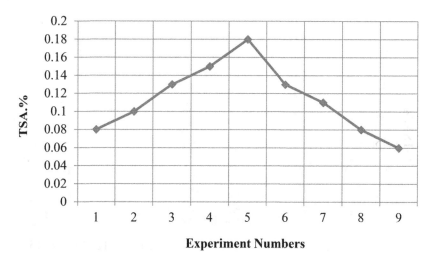

Figure 4. - Analysis of petroleum products for compatibility with different ratios of components.

aging the sample and scales with an accuracy of 0.01 and 0.0001 g. All laboratory vessels and reagents meet the required requirements.

Experiments were conducted to determine the total sediment with various types of petroleum products using the new method presented. On the basis of the conducted experimental tests (Table 1), the compatibility of various types of petroleum products was obtained. The compatibility of fuel oil 100 (Component 1) with other types of petroleum products (Component 2) has been considered.

According to the results of experiments No. 2, 4, 5, 7, 9, 10, 11, 12 and 14, petroleum products are compatible and stable when mixed (Figure 2), in these fuels there is a low content of paraffins and asphaltenes. "Incompatibility" was determined in experiments No. 1, 3, 6, 8, 13 and 15 when mixing the presented types of petroleum products, an active precipitation of the total precipitate occurs. It follows that ash (experiments 6, 13), with a high content of paraffins (experiments 1, 3, 8, 15) and asphaltenes (experiments 3, 6) residual oil products behave unstably and are prone to sedimentation.

In experiment No. 6, when mixed, precipitation is higher than the required norms (Figure 3). Studies of the effect of mixing ratio of incompatible petroleum products. Component №1. Fuel oil 100, component number 2. Heavy pyrolysis resin, grade A. The results are presented below.

In the experiments (Table 2), compatibility analyzes were performed with different ratios of the mixture of petroleum products (Figure 4). As can be seen from the results, the fuels are incompatible, and the maximum precipitation occurred at a mixing ratio of 50% to 50%. It is also worth noting that in experiments 1, 8, 9, the total sediment value does not exceed the permissible norms. Thus, if necessary, it becomes possible to select the ratio of the mixed oil products, which will allow maintaining the acceptable quality of the mixture.

3 CONCLUSIONS

The problem of "incompatibility" of petroleum products is extremely relevant, because due to precipitation of the total sediment on the tanks during operation, the useful volume decreases and there is a need to clean the tanks, also when mixing fuels, because of the "incompatibility", the quality of the whole product deteriorates in the first place increases the risk of equipment wear. Therefore, to solve this problem, an algorithm has been developed to test the compatibility of petroleum products based on GOST R 33360-2015, the introduction of which will quickly determine the compatibility of petroleum products prior to their actual mixing in tanks and will help preserve the quality of the stored product.

REFERENCES

GOST 33365–32015. Topliva ostatochnye. Opredelenie pryamogonnosti. Metod opredeleniya stabil'nosti i sovmestimosti po pyatnu. - M.: Standartinform, 2016.

GOST R 50837. 3–95. Topliva ostatochnye. Opredelenie pryamogonnosti. Metod opredeleniya toluol'nogo ehkvivalenta. – IPK Izdatel'stvo standartov, 1996.

GOST R 50837. 4–95. Topliva ostatochnye. Opredelenie pryamogonnosti. Metod opredeleniya ksilol'nogo ehkvivalenta. – IPK Izdatel'stvo standartov, 1996.

ISO 8217Fuel Standar. – 2017. – URL https://https://www.wfscorp.com/sites/default/files/ISO-8217–2017-Tables-1-and-2-1-1.pdf (date of the application10.01.2019).

Karimov, R. M., Mastobaev, B. N. Vliyanie soderzhaniya parafinov, smol i asfal'tenov na tovarnye kachestva neftej//Bashkirskij himicheskij zhurnal.-2012.-Tom 19.

Kondrasheva N.K., V.A. Rudko, D.O. Kondrashev, R.R. Konoplin, K.I. Smyshlyaeva, and V. S. Shakleina. Functional influence of depressor and depressor-dispersant additives on marine fuels and their distillates components. Petroleum Science and Technology 2018 36 (24), 2099–2105. DOI: 10.1080/10916466.2018.1533858.

Kondrasheva N.K., V.A. Rudko, D.O. Kondrashev, V.S`. Shakleina, K.I. Smyshlyaeva, R.R. Konoplin, A.A. Shaidulina, A.S. Ivkin, I.O. Derkunskii, O.A. Dubovikov. Application of a Ternary Phase Diagram To Describe the Stability of Residual Marine Fuel / // Energy & Fuels. 2019. Vol. 33, № 5. P.4671–464675.

Mitusova T.N., Averina N.P., Pugach, I.A.. Ocenka stabil'nosti i sovmestimosti ostatochnyh topliv. Mir nefteproduktov №1, 2005, 33–35.

Mitusova T.N., N.K. Kondrasheva, M.M. Lobashova, M.A. Ershov, and V.A. Rudko. Influence of dispersing additives and blend composition on stability of marine high-viscosity fuels. Journal of Mining Institute 2017 228, 722–725. DOI: 10.25515/PMI.2017.6.722.

Mitusova T.N., N.K. Kondrasheva, M.M. Lobashova, M.A. Ershov, V.A. Rudko, and M.A. Titarenko. Determination and Improvement of Stability of High-Viscosity Marine Fuels. Chemistry and Technology of Fuels and Oils 2018 53 (6), 842–845. DOI: 10.1007/s10553-018-0870-6.

Sultanbekov R.R., Nazarova M.N. Research effect of humidity in vapor space of the vertical steel tank for storing oil and oil products on the generation of static electricity. Gornyy informatsionno-analiticheskiy byulleten'. (2019);4/7:498–506. [In Russ] DOI: 10.25018/0236-1493-2019-4-7-498-506.

Sultanbekov R. R., Nazarova M. N. Determination of compatibility of petroleum products when mixed in tanks. EAGE., Tyumen, (2019), DOI: 10.3997/2214-4609.201900614. Available at: http://earthdoc.eage.org/publication/publicationdetails/?publication=96369 (Accessed 30 March 2019).

Topical Issues of Rational Use of Natural Resources 2019 – Litvinenko (Ed)
© 2020 Taylor & Francis Group, London, ISBN 978-0-367-85720-2

Mud cake removal efficiency of spacer fluids

S.Sh. Tabatabaee Moradi
Sahand University of Technology, Tabriz, Iran

N.I. Nikolaev, Y.V. Lykov & A.A. Petrov
Saint-Petersburg Mining University, Saint-Petersburg, Russia

ABSTRACT: Spacer fluids were initially used in the well construction process to separate drilling mud and cement slurry, which are considered as two chemically incompatible fluids. Beside separating these two fluids, spacers should completely remove the drilling mud remaining from the casing pipe and formation rock surfaces to guarantee a high quality bond between the cement and adjacent surfaces. There are several practical methods in the oil and gas industry to evaluate the mud removal efficiency of the spacers. However, these methods are not capable of assessing the spacer efficiency in removing mud cake, which usually forms on the formation rock surface. Presence of even a thin mud layer on the rock surface can significantly decrease cement bond quality. In this work, a method is proposed to evaluate the mud cake removal efficiency of a high density spacer fluid.

1 INTRODUCTION

An effective cementing operation guarantees a long producing life of the well and less difficulties in future workover operations. To achieve an effective cementing operation, the whole drilling mud, which presents in the annular space between the casing and surrounding rock formations, should be removed before pumping the cement into the wellbore (Tabatabaee moradi & Nikolaev 2016; McClure et al., 2014). At the same time it is important to take into account temperature fluctuations in the wellbore (Belsky at el. 2018).

Over the years, it has been proved that using spacers is the most efficient way to remove the entire drilling fluid from annular space (Pernites et al.,2018). Generally, the spacers are made up from the following main components (Gordon et al., 2008):

- Water as the base of the spacer.
- Weighting agents to adjust the spacer density in the required interval (Tehrani et al., 2014).
- Viscosifying agents to provide the system stability (Li et al., 2016).
- Surfactants to increase the mud removal efficiency of the spacer (Deshpande et al., 2016).

Researchers have suggested several methods to evaluate the mud removal efficiency of the spacer fluids. Kefi et al. (2014) used the Reverse Emulsion Test (RET) also known as Spacer Surfactant Screening Test (SSST) to optimize the cement spacer formulation for enhanced removal of non-aqueous drilling fluids. In SSST, the amount of spacer, which is needed to invert an oil-external emulsion (oil based and synthetic based drilling muds) into a water-external emulsion is evaluated by measuring the electrical conductivity of the spacer and mud mixture. However, this method can be only used in the case of oil based or synthetic based drilling muds.

Some researchers (Haut & Crook, 1982) use Rotor Test (RT) to assess the mud removal efficiency of the spacers during wellbore cleaning process. In RT, the sleeve of a viscometer is placed in drilling mud container for a specific time. The sleeve coated with the mud layer, will be dried, installed on the viscometer and rotated inside the cup containing the designed spacer. The weight differences of the sleeve before and after placing in the mud container and after placing in the spacer container are used to calculate the removal efficiency of the spacer.

In the work of Pernites et al. (2015), the water wetting capacity of the spacer fluids is evaluated using Casing Coupon Test (CCT). In CCT a casing coupon or sometimes silica glass is coated with drilling mud film and then will be placed in a spacer container for a specific duration of time. The contact angle comparison on the mud coated coupon and spacer-washed coupon makes it possible to conclude about the water wetting capacity of the spacer and therefore its mud removal efficiency.

The RT and CCT can only be used to evaluate the mud removal efficiency of the spacer with respect to a metallic surface, i.e. these tests are only capable to simulate removal of the drilling mud remaining from the casing surface by the spacer. However, for a good cementing bond between the cement and the surrounding rock formations, the spacer should effectively remove the mud cake, which usually forms on the rock surfaces. None of the above-mentioned tests are capable of testing spacer effectiveness in removing mud cake.

In this paper, a simple laboratory test based on the RT is suggested to evaluate the mud cake removal efficiency of an aqueous spacer fluid.

2 MATERIALS AND METHODS

A simple water based drilling mud composition is considered in this work as presented in Table 1. The spacer compositions is presented in Table 2. Three different surfactants, i.e. OP-10 (non-ionic), Sodium dodecyl sulfate (anionic) and benzalkonium chloride (cationic) are selected and used in 0.5-1 % concentration to enhance the removal efficiency of the spacers.

To assess the mud removal efficiency of spacer, a simple RT is designed as follows

- The rotor of the viscometer (Figure 1) is placed in the drilling mud container for 10 minutes.
- The coated rotor is left to dry for 2 minutes.

Table 1. Drilling mud composition.

Reagents	Mass fraction, %
Water	60.9
Bentonite	8
Barite	31
Viscosifying agent	0.1

Table 2. Spacer composition.

Spacer composition (mass fraction, %)	Density, kg/m3
Water (59), Polymer (8), Weighting agent (33)	1600

Figure 1. Viscometer rotor after being in drilling mud container for 10 minutes.

- The coated rotor is then installed on the viscometer and is rotated in the spacer container for 2 minutes at the speed of 300 rpm.
- The weight differences of the rotor before (W1) and after (W2) placing in mud container and after rotating in the spacer container (W3) are used in the following formula to calculate the mud removal efficiency:

$$MRE = \left(\frac{W_2 - W_3}{W_2 - W_1} \right) \times 100 \qquad (1)$$

where the MRE is the spacer mud removal efficiency, %.

To assess the mud cake removal efficiency of spacers, it has been proposed in this work to use the same principles as the rotor test with the exception of using a cylindrical rock sample instead of rotor. Synthetic cylindrical rock sample is prepared by mixing the Portland cement, silica sand and water in specific proportions (10%, 85% and 5% respectively) and letting the mixture to be consolidated inside a cylindrical mold under pressure. The rock samples will be placed in the drilling mud container for 10 minutes. Then a special tool is used to mount the rock sample, with the mud cake formation on its surface, on the viscometer. The rock sample will be rotated in the spacer container for 2 minutes at the speed of 300 rpm. The weight differences are used in equation 1 to evaluate the mud cake removal efficiency of the spacer fluids.

3 RESULTS AND DISCUSSION

Figure 2 shows the results of the rotor test on spacer compositions with different non-ionic, anionic and cationic surfactants. As it is evident from the presented results, it can be concluded that the spacer containing 0.5% non-ionic surfactant (OP-10) has the most efficient removal capacity.

Spacer fluid with 0.5% OP-10 surfactant is then used to remove the mud cake from the surface of the rock samples. However, due to the roughness of the rock surfaces, the result is not satisfactory (MRE = 50.35%).

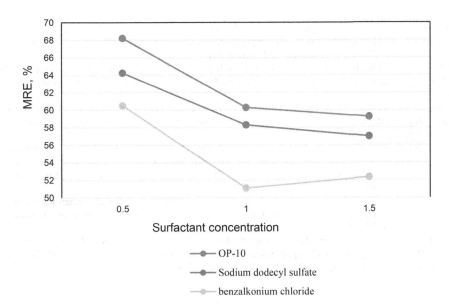

Figure 2. Mud removal efficiency of the spacer fluids.

4 CONCLUSIONS

Following conclusions can be made:

- There are three main methods (SSST, RT, CCT) in the oil and gas industry to evaluate effectiveness of the spacer fluids. However, none of these methods are capable of assessing the mud cake removal efficiency of the spacers. Therefore, a new method based on the rotor test is proposed to evaluate mud cake removal efficiency of the spacers.
- Using the RT and the new proposed method, spacer composition with OP-10 surfactant and 5% silica sand is the most optimum spacer between the developed compositions.
- The main focus of this work was on the development of a method to evaluate the mud cake removal efficiency of the spacer fluids. However, it is also necessary to conduct other necessary experiments like compatibility tests of the spacer with the drilling mud and cements and spacer rheological characteristics.

REFERENCES

Belsky, A.A., Dobush, V.S., Morenov, V.A. & Sandyga, M.S. 2018. The use of a wind-driven power unit for supplying the heating cable assembly of an oil well, complicated by the formation of asphalt-resin-paraffin deposits. Journal of Physics: Conference Series, Volume 1111, 1, 21 December 2018, p. 1–8.

Deshpande, A., Dumbre, M., Palla, V.G.R. & Rayaprolu, A. 2016. Selection and Optimization of Spacer Surfactants for Enhanced Shear-Bond Strength. *in International Petroleum Technology Conference*, Bangkok, Thailand, 14–16 November 2016.

Gordon, C., Lewis, S. & Tonmukayakul, P. 2008. Rheological Properties of Cement Spacer: Mixture Effects. *in AADE Fluids Conference and Exhibition*, Houston, Texas, USA, 8–9 April 2008.

Haut, R.C. & Crook, R.J. 1982. Laboratory Investigation of Lightweight, Low-Viscosity Cementing Spacer Fluids. *Journal of Petroleum Technology*, 34 (08).

Kefi, S., Pershikova, E., Docherty, K., Barral, Q., Droger, N., De La Mothe L.R. & Khaifallah, I. 2014. Successful cementing based on new design methodology for displacement of non-aqueous drilling fluid. In *IADC/SPE Drilling Conference and Exhibition, Society of Petroleum Engineers*, Fort Worth, Texas, USA, 4–6 March 2014.

Li, L., Alegria, A., Doan, A.A., Kellum, M.G. & Castanedo, R. 2016. Application of a Novel Cement Spacer with Biodegradable Polymer to Improve Zonal Isolation in HTHP Wells. *in Offshore Technology Conference*, Houston, Texas, USA, 2–5 May 2016.

McClure, J., Khalfallah, I., Taoutaou, S., Vargas Bermea, J.A. & Kefi, S. 2014. New Cement Spacer Chemistry Enhances Removal of Nonaqueous Drilling Fluid. *Journal of Petroleum Technology*, 66(10).

Pernites, R., Khammar, M. & Santra, A. 2015. Robust spacer system for water and oil based mud. In *SPE Western Regional Meeting, Society of Petroleum Engineers*, Garden Grove, California, USA, 27–30 April 2015.

Pernites, R., Padilla, F., Clark, J., Gonzalez, A. & Fu, D. 2018. Novel and High Performing Wellbore Cleaning Fluids with Surprisingly Flat Viscosity over Time and Different Temperatures. *in SPE Western Regional Meeting, Society of Petroleum Engineers*, Garden Grove, California, USA, 22–26 April 2018.

Tabatabaee Moradi, S.Sh. & Nikolaev, N.I. 2016. Considerations of well cementing materials in high-pressure, high-temperature conditions. *International Journal of Engineering, Transaction C: Aspects* 29(9): 1214–1218.

Tehrani, A., Cliffe, A., Hodder, M.H., Young, S., Lee, J., Stark, J. & Seale, S. 2014. Alternative Drilling Fluid Weighting Agents: A Comprehensive Study on Ilmenite and Hematite. *in IADC/SPE Drilling Conference and Exhibition*, Fort Worth, Texas, USA, 4–6 March 2014.

The new type of strainer construction

A.A. Tretyak & V.V. Shvets
Platov South-Russian state polytechnic university (NPI), Novocherkassk, Russia

ABSTRACT: This article describes the colmatation process that occurs after strainer exploitation. The design of the self-cleaning strainer is shown. The technology of self-cleaning strainer operation for hydrogeological wells is described. Completed experimental laboratory work on the optimization of water magnetizing technology.

1 INTRODUCTION

At the present time there are hundreds of different types of strainer construction which succeeds at their main function – fluid filtering, including petroleum, but after certain amount of time they lose their through capacity, in other words they colmatation. Strainers of hydrogeological wells colmatation cations of calcium carbonate hardness salts. Strainers of oil and gas wells usually colmatation paraffine and tar products (Kunshin et al. 2019, Litvinenko & Dvoinikov 2019).

Water is filled with various dissolved mineral salts that enriches water with ions, which come from the natural sources of limestone, gypsum and dolomite. Water hardness is caused by calcium and magnesium salts. The hardness of natural water does no harm to our health, the calcium salts help to get out cadmium that harms heart-vascular system. But high hardness makes water unavailable for every-day use. One of the traditional methods of waters cleansing involves the usage of ion exchange resin. Ions of calcium and magnesium from water replace the ions of sodium, but this process of regeneration has a side effect: softened water contains increased amount of sodium (Morenov & Leusheva 2019). Another method is based on the use of reverse-osmosis membranes, but they lower the contents of all salts evenly. Alternative method of preventing calcium and magnesium bicarbonates scaling is water magnetization.

Proto-crystals are formed under the influence of magnetic force field in the following way. Magnetic field orientates and polarizes dipoles of water (Syzrantseva et al. 2016). That leads to alteration of the water structure, which manifests in changing the type of dipole bonds; double bonds are formed instead of single bonds.

As a result hydrated ions of Ca^{2+} and CO_3^{2-} form ions, and later, molecules. Other ions of Ca^{2+} and CO_3^{2-} from the solution join the newly formed molecules and form congestion points, that would later become the centers of crystallization.

The main advantage of the magnetization method is that after crossing lines of magnetic force cations of hardiness salt will form in the body of water and not in strainer itself.

2 MATERIALS AND METHODS

The formation of colmatant in the hydrogeological well strainers is caused by the disruption of the chemical balance of the shelf and occurs at the water abstraction stage. The disruption of the chemical balance is connected to desorption of the free carbon dioxide after changes to its partial pressure. Usually colmatation is multicomponent, containing calcite $Ca(CO_3)$, siderite $Fe(CO_3)$, magnesite $Mg(CO_3)$, pyrolusite MnO_2 and other hardly soluble compounds which clog filtering mesh and causes wells to malfunction (Bernal 1933; Tretyak 2006; Gavrilko 2017).

The new type of strainer construction makes it possible to eliminate colmatant formation, increase well capacity and intensify filtering process. We offer a design for the new type of strainer construction (pic. 1-3). The effectiveness of water magnetization relies on the Lorentz force, which affects ions, polarized molecules and colloidal particles and changes solution's structure.

As was discovered by the authors, Lorentz force density is affected by many factors, such as directional distribution of flux density, geometrical size, concentration and mobility of ions and charged particles, velocity profile in a non-magnetic tube, magnetic properties of the solution and environment (Vasilev et al 2016, Urshulyak et al. 2016).

The well strainer works in the following way. Water goes through perforation holes 5 on non-magnetic perforated carcass 7 and turns magnetic by permanent magnetic rings 4. After crossing, the lines of magnetic force cations of hardiness salt will be formed in the body of water, outside of the strainer. Pic.2 shows the formation of the force F by the upper water flow.

1 - bearing rod; 2 - spacing sleeve (made from nonmagnetic materials – brass or bronze); 3 - sleeve bearing (made from materials such as "Maslyanite"; 4 - webbed permanent magnet

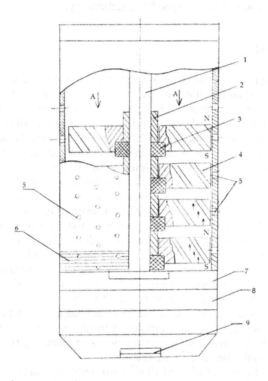

Pic 1.   A new type of strainer construction.

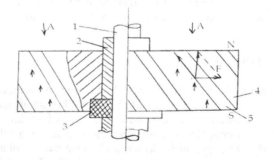

Pic 2.   The formation of the force F by the upper water flow.

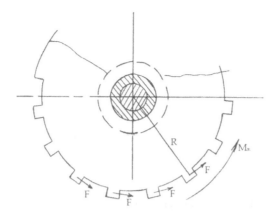

Pic. 3.   The moment of rotation $M_\kappa = \sum_1^r F \cdot R$, that spins the webbed magnet.

(with tilted ribs); 5 - perforation holes; 6 - capron rope; 7 - strainer's carcass; 8 - bottom sub; 9 - circulating valve.

1 - bearing rod; 2 - spacing sleeve (made from nonmagnetic materials – brass or bronze); 3 - sleeve bearing (made from materials such as "Maslyanite"); 4 - webbed permanent magnet (with tilted ribs); 5 – magnet's rib stiffeners.

In such a manner, magnetic field polarizes molecules of hardiness salt and splits them into atoms. Positively charged ions form more porous aragonite instead of calcite, which is destroyed more easily and washed away by water currents in the strainer (Arnol'd, 1969; Coey, 2000; Tretyak, 2011).

Proto-crystals are formed under influence of magnetic force field in the following way. Magnetic field orientates and polarizes dipoles of water. That leads to alteration of the water structure, which manifests in changing the type of dipole bonds; double bonds are formed instead of single bonds, which prevents hardiness salt formation on the internal surface of the strainer.

As a result hydrated ions of Ca^{2+} and CO_3^{2-} form ions, and later, molecules. Other ions of Ca^{2+} and CO_3^{2-} from the solution join the newly formed molecules and form congestion points that would later become the centers of crystallization. The formation of colmatant in the hydrogeological well strainers is caused by the disruption of the chemical balance of the shelf and occurs at the water abstraction stage. The disruption of the chemical balance is connected to desorption of the free carbon dioxide after changes to its partial pressure. Usually colmatant is multicomponent, containing calcite $Ca(CO_3)$, siderite $Fe(CO_3)$, magnesite $Mg(CO_3)$, pyrolusite MnO_2 and other hardly soluble compounds which clog filtering mesh and causes wells to malfunction. The new type of strainer construction makes it possible to eliminate colmatant formation, increase well capacity and intensify filtering process, by the usage of the permanent magnetic field.

Slowly spinning permanent sign-variable magnetic field cause sign-variable effect on mud fluid (water, petroleum) which increasing the effect of magnetization.

After analyzing the existing magnet types and preliminary modeling the four-poled magnet construction was chosen as the base of magnetization system.

The choice of the two magnet poles placement, placed above each other, was based according to the possibility of their synergy to create magnetic field in mud not only in the inside diameter of the permanent magnets, but in the parts of the tube where magnets are not present as well. The dotted line on pic.4 shows flux path that isolates between the poles of the permanent magnet.

According to the analysis of the literature sources (Bunyakin 2013; Tretyak 2000; Kovalenko 2003; Patent №247877 RU) and preliminary modeling the conclusion was made that the bigger cross-section of the permanent magnet rings, the higher magnetic field intensity will be in its surroundings.

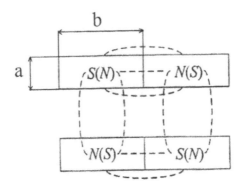

Pic. 4.   Relative position of the S and N magnet poles.

Two variants of the materials were considered in the modeling process:

1. NdFeB magnets (Neodymium-iron-born) – rare-earth magnets, that boast high mag-netic characteristics and low cost. The downsides of the NdFeB magnets are: rela-tively high temperature instability, high corrosion level. To reduce the corrosion level permanent NdFeB magnets are covered with zink, nickel, copper or combination of those materials.
2. SmCo magnets (Samarium-cobalt) have combination of extremely high magnetic proper-ties: high residual flux density, coercitive force, high temperature stability and resistance to corrosion. The downsides of SmCo magnets are high price and fragility. The signature property off this alloy is magnetic efficiency, which is the reason why permanent alloy SmCo magnets were used for strainer construction.

The effectiveness of the water magnetizing relies on the Lorentz force, which affects ions, polarized molecules and colloidal particles and changes solution's structure.

3   RESULTS

As was discovered by the authors, Lorentz force density is affected by many factors, such as directional distribution of flux density, geometrical size, concentration and mobility of ions and charged particles, velocity profile in a non-magnetic tube, magnetic properties of the solution and environment. Effectiveness of the magnetization depends on the water's nature, pH value and on magnetization mode – magnetic field intensity, amount of magnetic poles, current speed.
Changes in one of the parameters lead to changes of optimal characteristics of other parameters.
After the experimental research, following results were acquired:

1. The invention of self-cleaning well strainer with installed permanent SmCo alloy magnets.
2. For the maximal total magnetic intensity, the size of the permanent magnets should max-imal: a=10mm, B – 70mm, and the distance between magnets must be minimal.
3. In laboratory environment, it was estimated that the strainer does not susceptible to colmation.
4. This research about strainer construction and magnetization of water from hydrogeological wells does not claim to be fully completed despite solving a considerable amount of ques-tions. Scientific works should continue working in this direction and their promising out-look causes no question, especially in the oil and gas wells industry. At this moment scientific knowledge allows to claim that changes of water and petroleum properties under the influence of magnetic field is one of the biggest problems of our century and deserves the most thorough attention and research.

REFERENCES

Arnol'd P.P. 1969. *Raschet i proektirovanie magnitnykh sistem s postoyannymi magnitami [Projecting magnet systems with permanent magnets]*. Moscow: Energiya. p.184.

Bernal J.D., Fowler R.H. 1933. *A theory of water and ionic solution with particular reference to hydrogen hydroxyl ions*. Journal of Chemical Physics V. 1 (8), p.515.

Bunyakin A.V. 2013. *Tree-level discrete quantum model of ideal water chain in and constant magnetic field*. International Journal of Quantum Mechanics Research V.1, N 1 October. Pp.1-18.

Coey J.M.D., Cass S. 2000. *Magnetic water treatment*. Journal of Magnetism and Magnetic Materials. V. 209. Pp.71-74.

Gavrilko V.M., Alekseev B.C. 2017. *Fil'try burovykh skvazhin [Well strainers]*. Feniks. p.367.

Kovalenko A.S. 2003. *Kompleksnaya obrabotka burovykh rastvorov fizicheskimi polyami [Complex processing of drilling mud with physics fields]*. Izvestiya VUZov Severo-Kavkazskii region. Tekhnicheskie nauki. №4. Pp.103-103104.

Kunshin A.A., Dvoynikov M.V. & Blinov P.A. *Topology and dynamic characteristics advancements of liner casing attachments for horizontal wells completion* // Youth Technical Sessions Proceedings: VI Youth Forum of the World Petroleum Council - Future Leaders Forum (WPF 2019), June 23-28, 2019, Saint Petersburg, Russian Federation. – 2019, pp. 376-381.

Morenov, V., Leusheva, E. *Influence of the solid phase's fractional composition on the filtration characteristics of the drilling mud* (2019) International Journal of Engineering, Transactions B: Applications 32(5), p. 794-798.

Litvinenko V.S., Dvoinikov M. V. *Justification of the technological parameters choice for well drilling by rotary steerable systems* // Journal of Mining Institute 2019: 235, pp. 24-29.

Patent №247877 RU. Tretyak A.Ya., Burda M.L., Litkevich Yu.F. *Skvazhinnyi fil'tr [Well strainer]*. Opubl. 10.04.2013. Byul. №10.

Syzrantseva, K., Arishin, V., Dvoynikov, M. *Optimization of the damping element of axial vibrations of the drilling string by computer simulation* // Journal of Engineering and Applied Sciences 2016: 11(10), pp. 2312-2315.

Tretyak A.Ya., Burda M.L., Shaikhutrinov D.V., Onofrienko S.A. 2011. *Vybor optimal'nogo magnitnogo pol-ya s tsel'yu regeneratsii fil'trov gidrogeologicheskikh skvazhin [The choice of the optimal magnetic field for regeneration of hydrogeological strainers]*. Izvestiya VUZov Severo-Kavkazskii region. №4. Pp.121-124.

Tretyak A.Ya., Chikhotkin V.F., Pavlunishin A.P. 2006. *Tekhnika i tekhnologiya sooruzheniya gidrogeolog-icheskikh skvazhin [Hydrogeological well construction technology]*. YuNTs RAN. p.408.

Tretyak A.Ya., Sidorenko P.F., Kovalenko A.S. 2000. *Rastvor dlya vskrytiya vodonosnogo plasta [Solution for water bearing bed drilling]*. Izvestiya VUZov Severo-Kavkazskii region. №3. Pp.94-96.

Vasilev, N.I., Bolshunov, A.V., Dmitriev, A.N., Podoliak, A.V. *Round-trip assembly for investigations of subglacial lake Vostok* // International Journal of Applied Engineering Research 2016: 11(9), pp. 6376-6380.

Urshulyak, R.V., Buslaev, G.V., Bortey, P.D. *Laboratory tests and analysis of core samples from ruptured carbonate deposits in the timan-pechora oil and gas province exposed to acid fracturing liquids* // Society of Petroleum Engineers - SPE Russian Petroleum Technology Conference and Exhibition 2016.

Topical Issues of Rational Use of Natural Resources 2019 – Litvinenko (Ed)
© 2020 Taylor & Francis Group, London, ISBN 978-0-367-85720-2

Drilling mud for drilling in abnormal conditions

A.Ya. Tretyak, Yu.M. Ribalchenko & S.A. Onofrienko
Platov South-Russian state polytechnic university (NPI), Novocherkassk, Russia

ABSTRACT: This article presents the state of the art, research tasks and the formulation of the drilling fluid for abnormal operation conditions. We have proposed a principally new high-inhibited polymer-mud fluid that has the improving filtration and rheological properties. In this article we have given the mechanism of quality control, preparations technology and technical and economic indexes of the practice of the tailored composition of the multiple-functional drilling fluid, that is treated gradually in the ultrasonic field and in the magnetic field.

1 INTRODUCTION

One of the main factors that defines drilling efficiency is the drilling mud quality. One of the main drilling goals – support of the walls of the borehole is achieved thanks to drilling mud (Mikheev, 1979; Tretyak A.Ya., 2011; Patent № 2303047 RU). Inhibited fluids are used to lower the intensity of the transition of mud to the drilling fluid and increase walls stability. Their structure contains inorganic electrolytes or polyelectrolytes. Soaking capacity and dispersion are lowered thanks to: adding plurivalent metals and salts to solution, turning it into hydroxide; processing with alkalis, increasing mud capacity of the fluid; usage of modified lignosulphates; processing with polymer materials.

According to detailed borehole drilling studies, geological abnormalities and drilling incidents drilling mud does not satisfy to high standards of inclined and horizontal drilling (Chudinova et al. 2019). As a result, this leads to creation of low quality mud crust on the walls of the well, which has low antifilter and inhibitor qualities, which further leads to sticking.

Stuff members of the 'oil and gas technologies and engineering' faculty of the South Russian State Polytechnic University (NPI) offer a multi component, highly molecular inhibition drilling mud with complex qualities, having high greasing, reinforcing and anti-sticking qualities with good rheological parameters. Drilling mud is designed to drilling in ground and water surface and on the ledge of inclined and horizontal wells comprised of highly viscous mud shales prone to soaking and softening, with a chance of damaging well's hole (Tretyak A.A., 2011; Tretyak, 2014; Tretyak, 2016).

2 MATERIALS AND METHODS

The objective is achieved thanks to drilling mud (Blinov & Dvoynikov 2016). It consists of polyanionic cellulose, potassium chloride, barium sulfate, bishofite, ferrochrome lignosulfonate, potassium methylsiliconate, potassium acetate, foam suppressant, water, also containing marble aggregate, vegetable fat wastes, hydrophobisator and sulfanolum. The percentage of the components is: marble aggregate 5-10%, polyanionic cellulose 2-10%, sulfanolum 2-5%, potassium chloride 2-5%, potassium methylsiliconate 1-4%, potassium acetate 1.5-2%, bishofite 2-5%, ferrochrome lignosulfonate 1-5%, hydrophobisator 2-5%, barium sulfate 0.5-5%, foam suppressant 0.5-1%, with liquid body being the rest; liquid state contains vegetable fat wastes and water with the next mass correlation in percents 55/45-80/20.

The technical result is an increase of fortifying, greasing and anti-sticking qualities of the drilling mud with inhibition properties while also increasing recovery efficiency by increasing inhibition and water-repellent qualities of the mud, and as a result – no gutter and sticking formation, fortification of the inclined, horizontal and sub-horizontal well holes.

At the moment patent for inventing mentioned drilling mud is acquired (Morenov & Leusheva 2019). Increase of inhibition properties is obtained by upgrading fortifying parameters. Synergic effect mechanics scientifically proved needed percentage of every element. The result is acquired by adding soaking reagents – inhibitors: potassium chloride, bishofite, potassium acetate, hydrophobisator, ferrochrome lignosulfonate, potassium methylsiliconate. The combination of these inhibitors provides the most synergically efficient combination for drilling in abnormal conditions. Synergism is the well-functioning mechanism of the mud components.

Plastic viscosity, yield point and mud filtration were identified to be dependent on the concentration of the following inhibitors: potassium chloride, bishofite, potassium acetate, hydrophobisator, potassium methylsiliconate, ferrochrome lignosulfonate (Morenov & Leusheva 2017).

Vegetable fat may be represented by the wastes of soy, sunflower, cotton, corn, and other oils. Sulfanolum does the function of emulsifier. Marble aggregate does the structure-forming function (temporally colmatant). Polyanionic cellulose works as a filter regulator. Foam suppressor is usually represented by "Penta-465". Barite works as weighting compound and added in the 0.5 – 5 percentage.

With decreased water loss and chemical consumption rate being the goal, ultrasonic and magnetic field processing were advised (Figure 1) (Patent № 2582197 RU).

Supersonic vibrations influence not only dispersion phase, but water itself, causing destruction of molecules' bonds, because they are much weaker then inner bonds and make for a total of 21,5 J/mol. It is known that interaction of water with active centers of mud minerals may result due to formation of hydrogenium and molecular bonds. Under influence of ultrasonic water obtain additional hydrogenium bonds, which helps to form tighter and more intensive interactions with mud particles, which leads to decreasing water loss and increase of fluid viscosity.

The cause of the speed boost of the particles' coagulation is magnet field influence on double electron fields that surrounds mud particles – adsorptive and diffusional. Coagulation of mud particles happens due to penetration of electron barriers that are made by electron fields of coagulative particles. Penetration of this barrier happens due to kinetic energy of particles of different configuration and masses in places with minimal barrier thickness (Vasilev et al. 2016).

Magnetic field changes adsorption potential with its energy, which leads to changes in the fluid structure. It takes more defined form with areas where fluid is fully immobilized. Magnetic field processing also results in elongating and straitening of macro molecules. Even more water is adsorbed by their surface, which leads to decreasing water loss and increase of fluid viscosity.

Complex (magnetoacoustic) processing firstly leads to destruction of structure by ultrasonic, and then with the magnetic influence to formation of cluster structure, which leads to

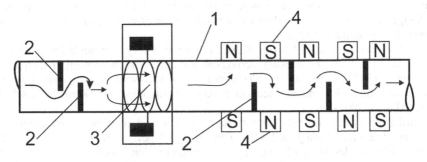

Figure 1.   The instrument for drilling mud processing
1 –polyvinylchloride tube, 2 – deflectors, 3 –supersonic emitter, 4 – permanent magnets.

decreasing water loss and increase of fluid viscosity. Preliminary ultrasonic processing makes this process more effective.

Complex (magnetoacoustic) processing leads to optimal combination of water loss and viscosity that cannot be achieved by using fields separately. This also makes processing twice as faster. This means that complex field processing is the most perspective way.

The offered multi component, highly molecular, inhibition drilling mud with complex properties has very high greasing and fortifying characteristics, good rheological property, viscosity – 40s, water loss – o sm^3/30 min, friction coefficient - 0, 06 (according to KTK – 2 device), has high inhibition qualities, zero filtration, upgraded structurely-rheological characteristics, anti-sticking and ecological properties. Experimentally proven in a lab conditions the event of synergic effect of the complex processing of drilling mud by multiple inhibitor reagents.

The offered drilling mud is created in field conditions with already presented equipment. The first step is making a mixture of marble aggregate and water that is then processed by polyanionic cellulose. All other components are added in the stirrer with constant stirring. The order of adding chemicals is next: wastes of vegetable fat, potassium chloride, potassium methylsilico-nate, potassium acetate, potassium methylsilicate, bishofite, sulfanolum, hydrophobisator, foam suppressant, barium sulphate. Rheological characteristic research is standard. Drilling mud must be processed with following chemicals after 4-staged cleaning. High pressure disperser is used.

The mechanics of inhibition are as following: upon adding inhibition materials to the fluid physical and chemical reaction of mud and cation occur, which replace free, negatively charged ions in the crystalline grid of the mud particles (Nutskova & Kupavyh 2016). Adsorption on the mud surface of the inhibition reagent leads to soaking resistance reduces soaking and softening of the active mud.

Further experiments have established correlation of plastic viscosity, yield point and filtration of the drilling mud, and concentration of inhibition materials: potassium chloride, potassium acetate, potassium methylsiliconate, bishofite. The research covered not only structural and mechanical properties, but also moisturizing properties. Anti-sticking characteristics were evaluated. Nonlinear mathematical models were made with the "Brandon method" program in MathCAD.

In our case we have two lines that shows the number of surfaces that are described by equations with identical parameters, and crossing point describes the borders of the surfaces.

Completed scientific research helped to identify that the usage of the six inhibitors in one fluid helped to achieve synergy, increasing the inhibition properties of the fluid, with every reagent complementing each other and increasing fortifying properties of the mud. Also, the mentioned combination of chemical reagents helps to replace sodium cations from the mud formations, transforming sodium mud to calcium mud, which helps to reduce hydration, soaking and fluctuation, decreasing the number of caving's and sloughs.

The advantage of the formula lies in the increasing the number of ions K^+ from 800 to 1200 mg/l. This show the direction of osmosis from shelf to well with considerably low isotonic coefficient K_{kp} = 1.31. The presence of ions of potassium and magnesium ups it to 4.7. In such a manner, dissociation of electrolyte increase the number of active osmosis particles in drilling mud. Usage of this drilling mud allows successful drilling of wells up to 3000m in the surfaces represented by unstable mud shales.

The optimal drilling mud is №8 (see Table 1), which has very high inhibition properties, zero filtration, has upgraded anti-sticking, ecological, structure and rheological characteristics that he saves up to 130 °C. In the lab and filed conditions of Krasnodar Region coast synergic effect of the drilling mud with complex reagent processing was identified (Koshelev 2004; Tretyak 2006; Chikhotkin 2007).

Invented hydrocarbon drilling mud with high filtration and greasing properties has following parameters: filtration of drilling mud is 0/30 min, crust stickiness – 0, friction coefficient – less than 0.1, crust thickness – less than 0.5 mm, water and oil correlation in percentages % - from 55/45 to 80/20, mud density from 1.1 to 1.2 g/sm^3, apparent viscosity –35/40 sec, plastic viscosity – 20-40 mPa·s, yield point – 15-20/20-30 dPa, sand substance – less than 0.5%, Ca^{2+} more than 1600 mg/l, Cl – more than 30000 mg/l.

Table 1. Drilling mud formulas.

Mud	№ 1	№ 2	№ 3	№ 4	№ 5	№ 6	№ 7	№ 8
Chemical reagents (%)								
Marble aggregate	5.0	6.0	6.0	7.0	8.0	9.0	9.0	10.0
Polyanionic cellulose	5.0	6.0	6.5	7.0	7.0	8.0	9.0	10.0
Potassium chloride	1.5	2.0	2.0	3.0	3.5	4.0	4.5	5.0
Ferrochrome lignosulfonate	1.0	1.5	2.0	2.5	3.0	4.0	4.5	5.0
Potassium methylsilicate	1.2	2.0	2.0	2.5	3.0	3.5	3.5	4.0
Potassium acetate	1.5	1.5	1.5	2.0	2.5	3.0	3.0	4.0
Bishofite	2.0	2.0	2.5	2.0	3.5	4.0	4.0	4.5
Sulfanolum	2.0	2.0	2.5	3.0	3.0	4.5	4.5	5.0
Hydrophobisator	2.0	2.5	3.0	3.5	3.5	4.0	4.0	5.0
Foam suppressant,	0.5	0.5	0.5	1.0	1.0	1.0	1.0	1.0
Barium sulfate	0.5	1.0	1.0	2.0	2.0	3.0	4.0	5.0
Oil/water proportions and other	55/45	60/40	65/35	65/35	70/30	75/25	75/25	80/20
Drilling mud characteristics								
Density, g/sm^3	1.16	1.18	1.19	1.20	1.20	1.21	1.21	1.22
Viscosity, sec	30	32	33	34	35	36	37	40
Water loss, sm^3/30 min.	3.5	2.5	1.5	1.5	1.5	1.0	0.5	0
Friction coefficient	0.15	0.14	0.17	0.12	0.12	0.08	0.09	0.06

3 RESULTS

The completed studies led to following conclusions:

1. Usage of this drilling mud allows successful construction of prospecting boreholes for oil and gas with more than 3000 m depth with horizontal endings in the surfaces, represented by unstable mud shales and self-dispersing mudstone.

2. It was experimentally proven that synergic effect of the mud's components – reagents work better together than separately.

3. The offered formula has very high inhibition efficiency, reduces hydration and swelling processes.

4. The offered mix of components helps the drilling mud to successfully prevent, slow and stop deformation processes of the hole, decrease cavernosity.

5. The offered drilling mud has upgraded rheological, greasing and anti-sticking properties while saving energy and having considerably ecologically safe compounds. It also decrease the risk of differential sticking, upgrade rheological profile of the flush system in the ring space and increase stability of the system. All this helps to successful function of the hydraulic flushing of the well.

REFERENCES

Blinov, P.A., Dvoynikov, M.V. The process of hardening loose rock by Mud Filtrat. *International Journal of Applied Engineering Research* 2016: 11(9), pp. 6630–6632.

Chikhotkin V.F., Tretyak A.Ya., Rybalchenko Yu.M., Burda M.L. 2007. *Burovoi rastvor i upravlenie ego reologicheskimi svoistvami pri burenii skvazhin v oslozhnennykh usloviyakh [Drilling mud and rheological control in abnormal drilling conditions]*. Burenie i neft'. № 7–8. Pp.58–58160.

Chudinova I.V., Nikolaev N.I. & Petrov A.A. Design of domestic compositions of drilling fluids for drilling wells in shales, *Youth Technical Sessions Proceedings: VI Youth Forum of the World Petroleum Council - Future Leaders Forum (WPF 2019), June 23–28, 2019, Saint Petersburg, Russian Federation:* 2019, pp. 371–375.

Koshelev V.N. 2004. Obshchie printsipy ingibirovaniya glinistykh porod i zaglinizirovannykh plastov Common principles of inhibition of drilling mud and muddy shelves. Stroitel'stvo neftyanykh i gazovykh skvazhin na sushe i na more. №1. Pp.13–15.

Mikheev V.L. 1979. Tekhnologicheskie svoistva burovykh rastvorov Technollogical properties of drilling mud. *Moscow: Nedra*. p.239.

Morenov, V., Leusheva, E., Influence of the solid phase's fractional composition on the filtration characteristics of the drilling mud, *International Journal of Engineering, Transactions B: Applications* 2019: 32(5), pp. 794–798.

Morenov, V., Leusheva, E. Development of drilling mud solution for drilling in hard rocks (2017), *International Journal of Engineering, Transactions A: Basics* 30(4), p. 620–626.

Nutskova, M.V., Kupavyh, K.S. Improving the quality of well completion in deposits with abnormally low formation pressure, *International Journal of Applied Engineering Research* 2016: 11(11), pp. 7298–7300.

Patent № 2303047 RU. A.Ya. Tretyak, V.A. Mnatsakanov, B.C. Zaretskii, S.A. Shamanov, P.A. Frolov, V.F. Chikhotkin, Yu.M. Rybalchenko. Vysokoingibirovannyi burovoi rastvor [Highly inhibited drilling mud]. *Opubl.20.07.2007 Byul. № 20*.

Patent № 2582197 RU. Tretyak A.Ya., Rybalchenko Yu.M., Shvets V.V., Lubyanova S.I., Turuntaev Yu.Yu., Borisov K.A. Burovoi rastvor (Drilling mud). *Opubl.20.04.2016 Byul. № 11*.

Tretyak A.A., Rybalchenko Yu.M. 2011. Biopolimernyi rastvor dlya oslozhnennykh uslovii bureniya Biopolymer drilling mud for drilling in abnormal conditions. *Oil and Gas journal Russia. № 11*. Pp.52–57.

Tretyak A.A., Rybalchenko Yu.M., Lubyanova S.I., Turuntaev Yu.Yu., Borisov K.A. 2016. Burovoi rastvor dlya stroitel'stva skvazhin v slozhnykh usloviyakh Drilling mud for well drilling in abnormal conditions. *Neftyanoe khozyaistvo. № 2*. Pp.28–31.

Tretyak A.Ya., Rybalchenko Yu.M. 2006. Teoreticheskie issledovaniya po upravleniyu burovym rastvorom v oslozhnennykh usloviyakh Theoretical research about drilling mud control in abnormal conditions. *Izvestiya VUZov Sev.-Kav. region, tekhnich. Nauki. № 7*. Pp.56–61.

Tretyak A.Ya., Rybalchenko Yu.M., Burda M.L., Onofrienko S.A. 2011. Biopolimernyi vysokoingibiruyushchii burovoi rastvor dlya sooruzheniya naklonno-napravlennykh i gorizontal'nykh skvazhin Biopolymer highly inhibited drilling mud for construction of tilted and horizontal wells. *Vremya koltyubinga. № 2-3*. Pp.66–74.

Tretyak A.Ya., Savenok O.V., Rybalchenko Yu.M. 2014. Burovye promyvochnye zhidkosti: uchebnoe posobie Drilling mud: teaching guide. *Novocherkassk: Lik*. p. 374.

Vasilev, N.I., Dmitriev, A.N., Podoliak, A.V., Lukin, V.V., Turkeev, A.V. Maintaining differential pressure in boreholes drilled in ice and the effect of ice hydrofracturing, *International Journal of Applied Engineering Research* 2016: *11(19), pp.* 9740–9747.

Topical Issues of Rational Use of Natural Resources 2019 – Litvinenko (Ed)
© 2020 Taylor & Francis Group, London, ISBN 978-0-367-85720-2

Hydrodynamic analysis of the PDC drill bits

A.Ya. Tretyak, K.A. Borisov & S.A. Getmanchenko
Platov South-Russian state polytechnic university (NPI), Rostov region, Novocherkassk, Russia

ABSTRACT: Authors have estimated geometrical characteristics of stabilizing construc-tions of the cutting – shearing type drill bits. Multiple experiments regarding optimal hydraulic construction of the drill bits' shafts for most effective water filtering ability were made, with calculations and equations being present. The usage of the offered construction will help effective whole drilling, increase drill bit's efficiency with upgraded purification system and bit's cooling, increase stability, durability, drilling efficiency; it will help to achieve high cutting speed in medium and upper medium hardness rock and in abrasive rock.

1 INTRODUCTION

At the current stage of drilling and prospecting technology development, where it has achieved high levels of complexity, the problem of finding reserves for increasing its efficiency has become especially important. In a complex of goals for prevention of this problem, the high importance of questions regarding the connection of hydraulic process and increasing of rock destruction efficiency is undebatable (Bugaev 1978; Khappel 1976; Shlikhting 1969; Chikhot-kin 1996; Grossu 2015).

High efficiency of diamond drilling bit as well as rational settings (axial weight of the bit and rotary speed) is achieved thanks to according amount of flush fluid, which help to cool down, clean the bit, and help to regulate rock destruction efficiency (Morenov & Leusheva 2017).

2 METHODS

At present time the efficiency of flush fluid affecting bottom hole can be achieved in multiple ways: increasing the amount of flush fluid without changing size and numbers of flushing ports; increasing the numbers and decreasing the size of hydraulic nozzles; changing the shape of hydraulic nozzles (Dvoynikov et al. 2017, Gorelikov et al. 2019).

We have made a research about optimal flushing port construction that will provide the most effective usage of water to affect bottom hole and flush mud from the head of the drilling bit.

The process of the destruction of the rock is accompanied by the Rehbinder effect.

It is considered that the effect is equal to the highest hydrodynamic pressure in the area of rock destruction.

Figure 1 shows stabilizing two-tier cutting-shearing type, for the proper work of the offered drill bit it essential to use known mud motor and planetary redactor that transfer received head-on rotation to lower drill-in tier with rotational speed ω_1 and to upper drill-out tier with rotational speed ω_2.

The drill bit works in a following way: in a process of rotation of the lower drilling bit 1 with the rotational speed ω_1 and upper drill-out tier 2 with rotational speed ω_2 upon axial movement of the drilling bit cutting wings 3 of the lower drill-in tier 1 dig in the bottom hole, widening the well to the diameter d, which is connected to the placement of the outer cutters 4 of the lower drill-in tier 1, that are placed in the upper parts of the cutting wings 3 of the lower drill-in tier 1;

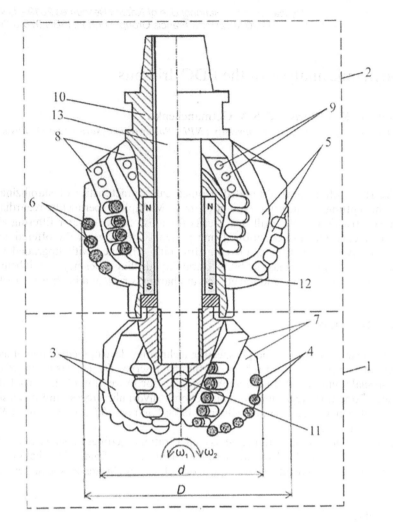

Figure 1.   Stabilizing two tier drilling bit of cutting-shearing type.

And cutting wings 5 of the upper drill-out tier 2 with cutters 6 of the upper drill-out tier 2 wider the well upon reaching the diameter D, that is connected to the placement of the outer cutters 6 of the upper drill-out tier 2, while stabilizers 7 of the lower drill-in tier 1 and stabilizers 8 of the upper drill-out tier 2 enlarge the are of contact with the wells walls. This helps to prevent rotation problems that are the main reason of PDC breakdown.

Stabilizers of the upper drill-out tier are implanted with highly resistant poles that have high calibrating effect, the number of cutters on the wings of the lower drill-in tier is equal to number of cutters of the upper drill-out tier, that is coaxially implanted with permanent magnet ring.

Drilling mud is flowing from the internal channel of the drilling bit 13 in the upper drill-out tier 2, which purpose is the drilling mud translocation, that goes from the shank adapter 10 in the upper drill-out tier 2 to the flushing ports 11 in the lower drill-in tier 1 of the drilling bit, lowing thru the inner channel of the drill bit 13 further thru the implanted in of the upper drill-out tier 2 permanent magnet ring 12, going thru the magnetizing effect, that leads to upgrade of the drilling mud qualities (viscosity, density, water loss, yield point).

For the prevention of the torsion motion drilling fluid must be freed from transferring rotative moment from the drill bit. This becomes possible if the rotative moments of the lower drill-in tier and upper drill-out tier are equal by modulus but opposite by direction; for that result the number of cutters on the lower drill-in tier must be equal to the number of cutters on the upper drill-out tier. By equating by modulus the values of rotative moments on drill-in and drill-out tiers, it is possible to prevent creation of torsion motion in the drilling assembly and decrease the amount of damage to PDC done by torsion motion (Vlasyuk 2013; Budyukov 2007; Tretyak 2017).

Outer magnetic field relocates electron clouds of ions and polarizes electron clouds of water molecules. This changes the energy of interaction of ions with closest water molecules and polarization of water by ions, which leads to changes in the fluid structure. Macromolecules of polymers are highly molecular materials with linear structure. After processing fluid with magnetic field macromolecule length is increased. As a result, freer water is being adsorbed, which increases viscosity and decreases water loss, increases density and yield point, overall boosting drilling mud characteristics.

The test of drill bits armored with comb shaped PDC helped to achieve higher mechanical cutting speed as against drill armored with the standard PDC, which helps to decrease drilling time and lower the cost of well drilling.

The drilling bit was designed and constructed specially for drilling of hard and medium hard rock. Thanks to upgraded stabilization, effectively placed cutters, improved hydraulics and high quality PDC cutter s the drill bit has exceptional reliability and provides high drilling speed in medium, hard and abrasive rocks. The claim for invention was made for the construction of the drill.

PDC armored drill bits are known for their hydraulic efficiency. To a great extent this efficiency is achieved thanks to their unique shape, profile and construction of hydraulic nozzles. Hydraulic flows help to boost mechanical cutting speed, if the mud is flushed from the bottom hole in time. If the flushing efficiency is low, mud will continuously go thru the PDC cutters and only the elevate to the surface. Hydraulic flows that go from the nozzles cool the bit and protect it from heat damage. The main hydraulic equation is:

$$H = \frac{P_g}{1.714} \tag{1}$$

where: H = hydraulic horsepower; P = pressure of the water current, KPa; g = flow rate, L/min; 1,714 = conversion of pressure to hydraulic horsepower.

For determining parameters of the water monitor affecting bottom hole made of laminated rock we will look upon the known parameters of speed and water monitor flow pressure.

Bottom hole made of laminated rock and limestone is most effectively destroyed by the currents with speed of $V_S \geq 70$ m/s. Upon flowing of water thru the flooded opening (drill bit in a well) speed of the current is:

$$V_S = \varphi\sqrt{2g\Delta P} \tag{2}$$

where: φ = 0,7÷0,8 –speed coefficient; g = 9,81 m/s^2-gravity acceleration; $\Delta P = P_P - P_W$; P_W = well pressure, MPa; P_P = pump pressure, MPa.

Knowing the expense Q, pump pressure P_p, highness H of the water column in the well and starting area of section S we can find the diameter of the exit hole of the water monitor currents using the expense equation:

$$Q = \mu S\sqrt{2g\Delta P} \tag{3}$$

where: μ = 0,7 –expense coefficient; S = area of section of two currents with the starting diameter d, sm^2,

$$S = \frac{10^2 \cdot Q}{\mu\sqrt{2g\Delta P}} \tag{4}$$

Diameter of the exit hole of the water monitor nozzles is found by using the equation:

$$V_s = 10\sqrt{\frac{2S}{\pi}}$$

(5)

Rotative moment of drill bit is found by using the equation:

$$M_r = F \cdot l \cdot \cos\gamma$$

(6)

where: F = horsepower of the water monitor current,

$$F = \frac{S}{2}10^2\Delta P$$

(7)

where: S = area of section of two currents, m²; ΔP = exit pressure of the current, MPa; 1 –Arm of force F of the bit axis, m; γ = inclination angle of water current to the area of bottom hole, degree.

3 CONCLUSIONS

For the drill bit with the diameter D = 215,9 mm and with distance between currents L = 0,2 m with the expense of the pump work Q = 50 L/s and working pressure P = 10 MPa in a well with the deepness H = 500 m rotative moment is M_S = 357 N·m. This provides stable bit work in the rocks with the contact hardness of P_C = 450 MPa with axial movement h = 1 mm/spin.

The important part of designing of the PDC armored drill bits is analysis of hydraulic configuration of the drill bit (Solovev, 1997; Tretyak, 2011; Karakus, 2014; Jiren, 2014).

The important role in the process of the hydrodynamic analysis of the drill bit's work is the quantity and construction of the contiguously allocated hydraulic nozzles. The most effective are changeable hydraulic nozzles for drilling in different hardness rocks (Figure 2).

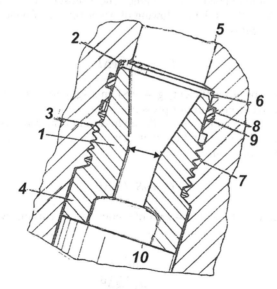

Figure 2.   Design of the changeable hydraulic nozzle.

The middle section of the hydraulic cylinder 1 of the body 2 is provided with external thread 3, the lower section 4 is without it and has a bigger diameter then middle section 1. The slush nozzle of the drilling mud discharge port 5 has a form of a well with widened bottom 6, with internal thread 7 according to external thread 3 of the slush nozzle. The upper section of the well 6 is implanted with the ring cavity 8, with installed elastic sealing ring 9.

The cross section of the lower section 10 of internal channel in a body 2 of the slush nozzle has a form of a hexagon compatible with socket wrench. This form of a lower section allows fast and easy change of the slush nozzle. The lower sealing is achieved thanks to tight fit of the tilted parts of the body and well that happens with the help of rotative moment received from the socket wrench. Upper sealing is achieved with help of elastic ring 9.

According to the calculations soft rocks are advised to be drilled with the minimal pressure and maximal pump capacity, with the starting speed V_m of the stream being 90 m/s for drilling soft rocks according to drillability scale, $V_m \geq 70$ m/s for medium rocks, $V_m \geq 50$ m/s for hard rocks.

REFERENCES

Budyukov Yu.E., Vlasyuk V.M., Spirin V.I. 2007. *Almaznyi instrument dlya bureniya napravlennykh i mnogostvol'nykh* skvazhin *[Diamond drilling tool for drilling designed and multi hole wells]*. Tula: GriFiK. P.176.

Bugaev A.A., Livshits V.N., Ivanov V.V. 1978. *Sinteticheskie almazy v geologorazvedochnom burenii [Synthetic diamonds in geologic exploration drilling]*. Kiev: Naukova dumka. P.231.

Chikhotkin V.F., Bogdanov R.K., Zakora A.P. 1996. *Vliyanie konstruktivnykh osobennostei promyvoch-nogo kanala impregnirovannoi koronki na razrushenie gornykh porod [The effect of constructional character-istics of the flushing port of the impregnated crown to destruction of the rock]*. Institut sverkht-verdykh materialov NAN Ukrainy. Kiev. № 19. Pp.68–78.

Dvoynikov, M., Syzrantsev, V., Syzrantseva, K. *Designing a high resistant, high-torque downhole drilling motor*//International Journal of Engineering, Transactions A: Basics 2017: 30(10), pp. 1615–1621.

Gorelikov, V. G., Lykov, Y. V., Gorshkov, L. K., Uspechov, A. M. Investigation of Thermal Operational Regimes for Diamond Bit Drilling Operations//International Journal of Engineering, Transactions B: Applications 2019: 32(5), pp. 790–793.

Grossu A.N. 2015. Burovoe doloto rezhushchego tipa s gidromonitornym privodom dlya skvazhinnoi gidro-dobychi zheleznykh rud. [Cutting type drill bit with hydraulic motor drive for downhole hydraulic produc-tion of iron ores]. Izvestiya VUZov. Severno-Kavkazskii region. Tekhnicheskie nauki. №1. Pp. 107–110.

Jiren T., Yiyu L., Zhaolong G., Binwei X., Huijuan S. et al. 2014. *A new method of combined rock drilling*. International Journal of Mining Science and Technology. Vol. 24. Iss. 1. Pp. 1–6.

Karakus M., Perez S. 2014. *Acoustic emission analysis for rock-bit interactions in impregnated diamond core drilling*. International Journal of Rock Mechanics and Mining Sciences. Vol. 68. Pp.36–43.

Khappel D., Brenner G. 1976. *Gidrodinamika pri malykh chislakh Reinol'dsa [Low Reynolds number hydrodynamics]*. M.: Mir. P.630.

Morenov V., Leusheva E. *Development of drilling mud solution for drilling in hard rocks* (2017), International Journal of Engineering, Transactions A: Basics 30(4), p. 620–626.

Shlikhting G. 1969. *Teoriya pogranichnogo sloya [Boundary layer theory]*. Moscow: Nauka. P.742.

Solovev N.V., Chikhotkin V.F., Bogdanov R.K., Zakora A.P. 1997. *Resursosberegayushchaya tekhnolo-giya almaznogo bureniya v slozhnykh geologicheskikh usloviyakh [Resource-saving technology of PDC drill-ing in abnormal geological conditions]*. Moscow: Izd-vo VNIIOENGP. P.329.

Tretyak A.A. 2011. *Tekhnologiya bureniya skvazhin koronkami, armirovannymi almazno- tverdosplavnymi plastinami [Drilling technology of PDC armored drills. Prospect and protection of mineral resources]*. Razvedka i okhrana nedr. № 12. Pp. 63–65.

Tretyak A.Ya., Popov V.V., Grossu A.N., Borisov K.A. 2017. *Innovatsionnye podkhody k konstruirovaniyu vysokoeffektivnogo porodorazrushayushchego instrumenta [Innovational ways of drilling instruments]*. GIAB. № 8. Pp. 225–2230.

Vlasyuk V.I., Budyukov Yu.E., Spirin V.I. 2013. *Tekhnicheskie sredstva i tekhnologii dlya povysheniya kachest-va bureniya skvazhin [Technical materials and technologies for increasing drilling of wells qual-ity]*. Tula: Grif i K. P.176.

Author Index